Intensity-Modulated Radiation Therapy

Yasumasa Nishimura • Ritsuko Komaki
Editors

Intensity-Modulated Radiation Therapy

Clinical Evidence and Techniques

Springer

Editors
Yasumasa Nishimura
Department of Radiation Oncology
Kinki University Faculty of Medicine
Osaka-Sayama, Osaka, Japan

Ritsuko Komaki
Department of Radiation Oncology
The University of Texas MD Anderson
Cancer Center
Houston, TX, USA

ISBN 978-4-431-55485-1 ISBN 978-4-431-55486-8 (eBook)
DOI 10.1007/978-4-431-55486-8

Library of Congress Control Number: 2015936030

Springer Tokyo Heidelberg New York Dordrecht London

Printed on acid-free paper

Springer Japan KK is part of Springer Science+Business Media (www.springer.com)

Preface

Our goal with radiation therapy (RT) is to improve the therapeutic ratio, which means to kill cancer cells without increasing normal cell-kill adjacent to the cancer cells. A significant advance in RT to improve the therapeutic ratio has been partly achieved by application of intensity-modulated radiation therapy (IMRT). IMRT has shown improvement of conformality to give a higher dose to the target volume and a lower dose to the surrounding normal tissue, especially in organs at high risk for toxicity. IMRT has provided us the opportunity to give a simultaneous boost to intensify the dose to the gross tumor volume (GTV) and a decreased dose to the clinical target volume (CTV).

Since the start of IMRT in the mid-1990s, much clinical evidence of the advantage of IMRT has been collected. Now, IMRT is a standard RT technique and is widely used for many tumor sites. The number of patients treated by IMRT is increasing in Japan, although the clinical application is limited compared with that in the United States. One reason for this limitation may be related to the difficulty of treatment planning and quality assurance (QA) for IMRT. It is not easy to carry out appropriate treatment planning with high-quality control of IMRT in daily practice. One of the reasons that IMRT has not been used more often in Japan is that there is a lack of medical physicists and dosimetrists in Japan, unlike the United States.

This book is an attempt to provide collected clinical evidence of IMRT with the appropriate advanced techniques of IMRT for clinicians and physicists. Several books on IMRT were published in the early 2000s. However, clinical evidence was scant at that time. In addition to the now-accumulated evidence for IMRT, the techniques for IMRT also have progressed. As an example, tumors and normal tissues move with time, and this movement may be clinically significant from second to second, day to day, week to week, or longer. It has been demonstrated that image-guided RT (IGRT) and/or adaptive RT (ART) are clinically advantageous for IMRT of these moving targets. Combined with a molecular imaging technique using PET/CT, IMRT based on molecular imaging will be soon available. This book covers these recent advances in IMRT.

In Part I, on foundations and techniques, the history, principles, quality assurance, treatment planning, radiobiology, IGRT, ART, and related topics of IMRT are presented. In Part II, on clinical application, several case studies including contouring and dose distribution with clinical results are described, following the description of indications and a review of clinical evidence for each tumor site.

While we were making plans to publish this book in 2013, our close friend Dr. K. Kian Ang, Professor of Radiation Oncology, Gilbert H. Fletcher Memorial Distinguished Chair, Department of Radiation Oncology, The University of Texas, MD Anderson Cancer Center, suddenly passed away. Dr. Ang was a great radiation oncologist, and produced significant achievements on RT for head and neck cancer, which included IMRT. The authors of this book are well recognized for their expertise in their respective fields both in Japan and the United States. In addition, all authors respected Dr. Ang very much and were greatly saddened by his sudden death. Thus, we decided this book should be dedicated to the memory of Dr. K. Kian Ang.

We would like to acknowledge Ms. Tamaki Yamamoto and Ms. Yoko Arai, Editorial Department, Springer Japan KK, for their significant contribution in editorial work on this book. The publication of the book was supported in part by a Grant-in-Aid for Cancer Research (H23-009, H26-090) from the Ministry of Health, Labour and Welfare of Japan. Finally, we hope that the information contained in this book will serve as a valuable resource for daily practice for many radiation oncologists and medical physicists.

Osaka-Sayama, Osaka, Japan Yasumasa Nishimura, M.D., Ph.D
Houston, TX, USA Ritsuko Komaki, M.D., F.A.C.R., F.A.S.T.R.O.

Contents

Part I

Foundations and Techniques

History of IMRT

1

Michael D. Mills and Shiao Y. Woo

Keywords
IMRT history • IGRT • Tomotherapy

1.1 IMRT: General Overview and Early History

There are several published historical reviews of intensity-modulated radiation therapy (IMRT), and this brief review will not recapitulate ground admirably covered by others [1–3]. The purpose of this review is to provide an experiential narrative that documents the emergence of IMRT beginning with the treatment of the first IMRT patient.

Briefly, IMRT is an advanced process of radiation therapy used to treat malignant and nonmalignant diseases. IMRT uses special beam modifiers to vary or modulate the intensity of the radiation over the field of delivery. The purpose is to manipulate the beam so that when all the radiation delivery is considered, the dose conforms closely to the tumor or target volume within the patient. IMRT may use multiple radiation beams of varying sizes and varying intensities to irradiate a tumor with precision and accuracy. The radiation intensity of each part of the beam is controlled, and the beam shape may change or multiple beams used throughout each treatment. The goal of IMRT is to shape the radiation dose to avoid or reduce exposure of healthy tissue and limit the side effects of treatment while delivering a therapeutic dose to the cancer.

Three-dimensional conformal radiation therapy (3DCRT) is a technique where the beams of radiation used in treatment are shaped to match the tumor. 3DCRT

M.D. Mills, Ph.D. • S.Y. Woo, M.D., FACR (✉)
Department of Radiation Oncology, James Graham Brown Cancer Center, University of Louisville, 529 S Jackson St, Louisville, KY 40202, USA
e-mail: sywoo001@exchange.louisville.edu

Y. Nishimura, R. Komaki (eds.), *Intensity-Modulated Radiation Therapy*,
DOI 10.1007/978-4-431-55486-8_1

technology emerged during the 1980s as CT information became more widely available and special computer platforms were developed to model the depiction and deposition of radiation dose over a CT image template. Previously, radiation treatments matched the height and width of the tumor, meaning that substantial healthy tissue was exposed to the full strength of the radiation beams. Advances in imaging technology made it possible to visualize and treat the tumor more precisely. Conformal radiation therapy uses the CT image targeting information to focus precisely on the tumor while avoiding the healthy surrounding tissue. This exact targeting made it possible to use higher levels of radiation in treatment, which are more effective in shrinking and killing tumors.

While 3DCRT planning and delivery allows for accurate dose conformity to irregular shapes, there are still limitations in the corrections that could be made. As its name implies, intensity-modulated radiation therapy allows the modulation of the intensity or fluence of each radiation beam, so each field may have one or many areas of high-intensity radiation and any number of lower-intensity areas within the same field, thus allowing for greater control of the dose distribution with the target. Also, with IMRT, the radiation beam can be broken up into many "beamlets," and the intensity of each beamlet can be adjusted individually. By modulating both the number of fields and the intensity of radiation within each field, we have limitless possibilities to sculpt radiation dose. In some situations, this may safely allow a higher dose of radiation to be delivered to the tumor, potentially increasing the chance of a cure [4, 5].

The concept of inverse planning for intensity-modulated radiation therapy was first elucidated by Anders Brahme [6]. His article was the first that demonstrated intensity-modulated fields of radiation would lead to more conformal dose distributions that would spare normal tissue. The first article dealt with the problem of inverse planning a target with complete circular symmetry. The article demonstrated how to generate an annulus of uniform dose around a completely blocked central circle by rotating a modulated beam profile.

Brahme later showed how inverse planning conceptually reverses the CT imaging planning process [7]. With CT, radiation beams are analyzed using a computer to produce an image. With inverse planning, the image, which is the ideal prescription for three-dimensional dose deposition, is generated by the physician as an input problem into a computer. The computer determines the position, shape, and intensity of the radiation beams (or the beamlets) to produce this ideal 3D dose delivery, thus fulfilling the prescription. Thus, the physician begins with the (ideal prescription) image and ends with the IMRT radiation beams.

The clinical implementation of forward planning IMRT is relatively easy, because it is closely related to conventional planning. Conventional forward planning mostly depends on the geometric relationships between the tumor and nearby sensitive structures. Time and effort requirements for quality assurance, planning, and delivery are similar to the experiences obtained with conformal radiotherapy. Manual definition of the segments leads to intuitive choices of the segment shapes based on the beam's eye view option of the planning system. Forward planning IMRT continues to be useful for the breast [8] and head & neck [9].

In comparison, inverse planning is less dependent on the geometric parameters but more on specification of volumes of tumor targets and sensitive structures, as well as their dose constraints. Inverse planning is far less related to conventional radiotherapy because the segment shapes are not defined manually and the number of segments is usually considerably larger. There are on occasion complex clinical situations which require the use of many beam directions and segments. In these cases, inverse planning may be the more efficient strategy.

At this point in time, the spatial technology and physics-based solutions are more advanced than our biological technology. Although we have the capability to plan and calculate doses accurately to within millimeters (or better), we are limited in our ability to identify microscopic disease with such accuracy and precision. We are also limited by the difficulties of immobilizing a patient for the time duration of an entire IMRT treatment (typically 15–30 min). Both patients and tumors move consequent to voluntary movement and involuntary motion such as peristalsis and respiration. Additionally, patients often lose weight over the course of the treatment. This renders the dosimetry inaccurate, and at some point the planning process must be redone. The next direction in radiation oncology is to account for this movement and is being called four-dimensional (4D) conformal radiotherapy (CRT), a logical progression from 3DCRT. Researchers have recently developed megavoltage cone-beam CT (MVCBCT) for clinical use. MVCBCT will allow the reconstruction of the actual daily delivered dose based on the patient's anatomy in real time. This will lead to "adaptive radiotherapy," the modulation of prescription and delivery based on the actual daily delivered dose, as opposed to planned dose [10].

IMRT creates and delivers a designed and prescribed three-dimensional dose distribution within a patient. In principle, modulated beams of radiation can be projected from any direction; however, most IMRT technologies do not enjoy this level of flexibility. This discussion will only consider those technologies that use coplanar beam directions as these are commercially available. The beams may be narrow (0.5–2.0 cm) fan beams, modulated in one dimension; this delivery system is termed "tomotherapy." The other major delivery schema uses divergent cone beams modulated in two dimensions. The modulation may be performed using milled metal (usually brass) blocks. More commonly, the modulation is accomplished by moving the leaves of the multileaf collimator. This may occur while the leaves are stationary during the time the beam is on (step-and-shoot IMRT), while the leaves move while the beam is on (dynamic MLC IMRT), or while both the leaves and the gantry move with the beam on (volumetric modulated arc therapy). For each of these techniques, delivery consists of a series of beam configurations each associated with a specific linear accelerator gantry angle. For each beam angle, a coplanar modulated beam is projected toward the isocenter.

Consider the fan beam delivery technology discussed above. The narrow beam rotation of a series of fan beams around a patient generates a dose distribution within a slice. This is conceptually analogous to the slice thickness of a CT (computerized tomography) scanner; therefore, this delivery technique was called tomotherapy. The first 2D IMRT unit consisted of add-on hardware to a conventional linear accelerator; it was the NOMOS Peacock unit [11]. Later, a helical tomotherapy unit was introduced to the marketplace, the TomoTherapy Hi-Art unit [12]. Continuing the CT analogy, to get a three-dimensional image representation of a patient from a CT scanner, it is necessary to capture and present multiple image slices. These CT slices may consist of a series of serial slices or a continuous helical slice. In order to deliver the three-dimensional IMRT dose distribution in tomotherapy, it is necessary to irradiate the serial or helical series of slices with the patient being positioned along the axis of rotation. In CT scanning, the axial resolution is limited by the slice thickness; similarly, in tomotherapy the axial resolution of the dose distribution is limited by the slice thickness. The parallels between CT scanners and tomotherapy treatment machines are seen in the ongoing development of the technology. The first applications of tomotherapy are being delivered one slice at a time, requiring accurate indexing of the patient between slices. The more modern development of helical tomotherapy in which the patient is moved continuously through the rotating fan beam can be compared with spiral (strictly helical) CT. Tomotherapy is an excellent choice for certain patient disease presentations such as head & neck cancer and some thoracic malignancies.

1.2 Early Phase, 1992–2002

Although the concept of IMRT and early algorithm for planning were developed in Sweden [6, 7], clinical application did not begin until a fully integrated IMRT planning and delivery system, namely, the NOMOS Peacock system, was invented and commissioned in 1993 by the collaborated effort between NOMOS [13–15] and Baylor College of Medicine/the Methodist Hospital (Houston, TX, USA). After obtaining investigational device exemptions and protocol approval by Baylor's Investigational Review Board, the first patient with brain metastases was to have three brain tumors treated simultaneously using IMRT in September 1993. However, at the eve of the treatment day, an exhaustive quality assurance (QA) testing discovered a glitch in the MIMiC collimator. The treatment was therefore canceled. After modification of the sensors for the moving vanes of the MIMiC was done and correction of the glitch was achieved, the first patient with a recurrent retropharyngeal cancer was treated by the Peacock system in March 1994. The initial patients (adults and children) were those with tumors in the brain or head & neck region where rigid fixation of the head by the invasive "Talon" system was used. A detailed QA system was developed [16]. Comparison between IMRT plan and stereotactic radiosurgery plan or three-dimensional (3D) conformal plan was performed [17, 18]. Several institutions such as Tufts New England Medical Center, University of Connecticut, and University of Washington and a radiation oncology practice in Pittsburg and one in Phoenix soon began treating patients with the Peacock system.

The rollout of the NOMOS Peacock IMRT system occurred in the summer of 1995. There was a concern that this paradigm shift of treatment delivery could deliver less than a tumoricidal dose to a portion of the target each day, depending on the plan configuration. There was much discussion and debate at this conference about what was the optimal beam configuration for IMRT. There was general consensus that energy between 6 and 10 MV photons was acceptable, with the overall optimal energy being about 8 MV.

By 1996, investigators at the Memorial Sloan Kettering Cancer Center began IMRT treatment for prostate cancer with the Varian dynamic multileaf collimator [19]. By July 1997, over 500 patients in the USA have received IMRT using the Peacock system. Early results have been reported [20–23]. Institutions such as Stanford University; UC, San Francisco; University of Washington; and others (including a few in Europe) all started to investigate the clinical use of IMRT. It was gradually recognized that rigid immobilization with an invasive device was not necessary for the majority of patients. There was quite a controversy when the use of a rectal balloon for treatment of prostate cancer was introduced [24]. Nonetheless, many investigators eventually recognized the value, and later, rectal balloon was adopted for proton therapy of prostate cancer. The early reports of clinical results of IMRT were single institutional reports but provided tentative evidence of reduced toxicity with IMRT [24–27].

A few important lessons were learnt during this period:

1. Understanding 3D anatomy on computerized tomography (CT) and magnetic resonance (MR), as well as the locoregional pathway of the spread of cancers, is of paramount importance when deciding on targets for treatment planning.
2. Significant improvement on the understanding of partial volume organ-at-risk tolerance is needed.
3. Treating with IMRT most tumors at a particular site (rather than just the most challenging cases) would more rapidly improve the skill, quality, and safety of the entire radiation oncology team.
4. Using IMRT for the entire course of treatment would more likely achieve the benefit of IMRT than using IMRT as a "boost" following the conventional 2D treatment.
5. A rotational IMRT technique could improve the treatment plan over static IMRT fields by increasing the "degrees of freedom" for beam angles.
6. If rotational technique is used, noncoplanar beams in most instances do not significantly improve the dose distribution over coplanar beams.
7. The flattening filter in an accelerator is not necessary for IMRT.
8. A higher dose rate (than that in a standard accelerator), because of the large monitor units required for IMRT, is desirable.
9. Beam energy higher than 10 MV is not necessary.
10. An accelerated fractionation scheme could be delivered once a day without resorting to multiple fractions per day [20].
11. A rigorous quality assurance (QA) system is mandatory [16].
12. Although the planning algorithm is an "inverse" process, the choice between the coverage of the target by the prescribed dose and the dose constraint to the adjacent organ(s) at risk dictates the parameter input into the planning computer to start the iteration. This is still a "forward" and somewhat "experienced-based" process.

The technical transition from conformal radiotherapy to IMRT was not always smooth. This was substantially because the two treatment systems had their own processes and terminologies. There was not a smooth path of connection between them, nor were there common terms of evaluation. There were often trade-off decisions to be made. For prostate patients, do we accept a lower rectum dose at the expense of a higher integral dose to the pelvis? Do we need and desire the dose escalation IMRT allows? Does the added complexity and cost of IMRT planning, verification, and delivery justify itself in terms of benefit to the patient? Pretreatment dose verification of the isodose curve plan became an important part of the quality assurance process, and for good reason. During the initial years of implementation, the accuracy of the treatment plan was not trusted by both physicians and physicists. Many patients were delayed and plans redone before treatment could begin with assurance of quality and safety. Today, verification hardware and software tools have made the process much more efficient, and the overall quality of IMRT delivery has substantially improved. However, more progress must be made before we see common implementation of adaptive therapy solutions [28].

The clinical advantage of IMRT is the greater control of dose to the normal tissues, allowing greater escalation of dose to the tumor. Because the ratio of normal tissue dose to tumor dose is reduced to a minimum with the IMRT approach, higher and more effective radiation doses can safely be delivered to tumors with fewer side effects compared with conventional radiotherapy techniques. IMRT also has the potential to reduce treatment toxicity, even when doses are not increased. Currently, IMRT is being used most often to treat cancers of the prostate, lung, head & neck, and central nervous system. IMRT has also been used in limited situations to treat breast and thyroid, as well as in gastrointestinal, gynecologic malignancies and certain types of sarcomas. IMRT may also be beneficial for treating pediatric malignancies. Radiation therapy, including IMRT, stops cancer cells from dividing and growing, thus slowing or stopping tumor growth. In many cases, radiation therapy is capable of killing all of the cancer cells, thus shrinking or eliminating tumors.

It was widely felt and discussed that aggressive use of electrons, especially in the head & neck area, would no longer be necessary or justified with IMRT. Planning algorithms, it was felt, were adequate to control, optimize, and present the dose properly in regions of sharp dose gradient. With proper selection of control points, the dose distribution would match and exceed the best conventional forward plans that utilized electrons. Many future IMRT machines would be built without electron capability.

Both 2D and 3D IMRT delivery have been performed both with and without a flattening filter. It was recognized in the NOMOS Peacock 1995 meeting that IMRT could be easily if not optimally performed without a flattening filter in place. The TomoTherapy Hi-Art machine was designed without a flattening filter. Modern conventional linear accelerator machines are now offering the option of beam delivery without the flattening filter in place. This has the advantage of increased dose output, making the treatment times shorter. Also, there are some advantages of a simpler photon spectrum with consequently better modeling of the beam within the treatment planning system.

The NOMOS MIMiC unit is an add-on collimator to a conventional linear accelerator. It delivers radiation in a tomographic fashion, one arc at a time. Each arc radiates a slice with both entrance and exit dose. The slice thickness is determined by the beamlet width: 0.85 cm in the 1 cm MIMIC mode and 1.7 cm in the 2 cm mode. The table must be advanced manually between each arc, making this potentially a very time-consuming treatment. The distance of table advance is defined by the width of the beam from the MIMiC device. The resolution of beam intensity modulation in the direction of couch motion is defined by the beamlet width. The binary collimator leaves are driven by compressed air into the computer-determined configuration during each moment of the delivery arc. This sound is described as being like "popping corn."

During the early years of IMRT, far more patients were treated with the NOMOS Peacock equipment than with any other technology. Most of the pioneers in IMRT gained most of their early experience with NOMOS equipment. Most disease sites were considered to see if IMRT could offer any therapeutic gain. IMRT demonstrated special promise for head & neck, prostate, and CNS treatments. It was determined that by using a collimator that narrows the fan beam, termed the "beak," NOMOS equipment could be used for stereotactic radiosurgery [29]. Several centers used the NOMOS equipment to treat cranial patients using arc vertex fields, adding a degree of freedom in addition to coplanar IMRT delivery.

1.3 Later Phase, 2002–Present

By 2002, most radiation oncology centers in North America and Europe and a few in Asia have become familiar with IMRT. About a third of radiation oncologists were already users of IMRT, and about 90 % of nonusers were planning to implement IMRT within 3 years [30]. The first tomotherapy unit which could deliver rotational IMRT and perform megavoltage scanning like a helical CT scan was installed in the University of Wisconsin.

1.3.1 TomoTherapy Hi-Art Machine at the University of Wisconsin

The TomoTherapy Hi-Art unit was designed and developed at the University of Wisconsin, Madison. A linear accelerator was mounted on a CT gantry using a slip-ring configuration. This allowed the linear accelerator to move in an arc so that a fan beam could be delivered in a continuous helical motion while the table is also advanced continuously. The energy of the unit was 6 MV photons. The unit was also capable of imaging a patient by lowering the energy to about 2 MV and utilizing a special CT detector array. Thus, the patient could be first imaged and then treated without changing machines or position. The TomoTherapy Hi-Art was the first popular image-guided radiation therapy (IGRT) treatment unit [31].

1.3.2 Verification of IMRT by the NOMOS MIMiC and the Wisconsin Machine

Early on, it occurred to those developing IMRT technology that the IMRT plan could be delivered to an IMRT dose verification phantom. It was a simple matter first to execute the plan onto the image of the phantom stored in the treatment planning system. The purpose of the IMRT dose verification phantom is to verify dose distributions and absolute dose values produced by IMRT beams, either subbeams or total beams. The phantom may be loaded with film, ion chambers, or thermoluminescent dosimeters. The verification is performed by irradiating the IMRT verification phantom and by comparing the measured phantom values and the calculated values of the radiotherapy treatment planning system. If the values match within a certain preset dose and distance, the plan is considered valid. Almost all early IMRT dose validation was performed in this manner [32].

The first commercial treatment planning system (TPS) available with an inverse planning code for IMRT was originally known as Peacock Plan from the NOMOS Corporation. It uses a simulated annealing algorithm and was originally designed to plan for delivery with the MIMiC technique. The beam modeling algorithm was a relatively simple ray-tracing algorithm optimized for speed to arrive at an inverse planned solution on the computer technology available in the mid-1990s.

Corvus was developed by the NOMOS Corporation after recognizing the need for a more sophisticated and accurate computer algorithm that would calculate dose distributions not only for the MIMiC but also for conventional plans. This algorithm employed a finite-size pencil beam (FSPB) convolution algorithm, which it incorporated during the final optimization process. It also allowed conventional planning of patient volumes with the boost volume to be calculated using the MIMiC delivery system [11].

The TomoTherapy planning system was developed by the same team that produced the Pinnacle system, and the photon calculation algorithms are similar. The TomoTherapy planner uses a simplified optimization algorithm, but performs the final dose calculation with its version of the superposition convolution algorithm. This results in very accurate calculations when presented with heterogeneities such as lung or air cavities in the head & neck region. The workstation uses an array of processors to reduce greatly the optimization time required to optimize a plan. The TomoTherapy planner does not calculate dose for conventional linear accelerators. It is possible, however, to export the dose grid from the TomoTherapy planner to a conventional planning system using DICOM RT exchange. This would allow the summation of conventional plans with TomoTherapy plans on an external workstation [33].

Later, volume-based IMRT such as VMAT and RapidArc and cone-beam CT on accelerators were introduced [34]. The planning systems have continued to improve including the addition of Monte Carlo-based algorithms and class solutions which have emerged to render planning more efficient [35–37]. With the introduction of four-dimensional (4D) CT and techniques of motion management, IMRT for moving targets such as those in the lung and liver became feasible and increasingly used. The National Cancer Institute in the USA published guidelines for the use of IMRT in clinical trials [38], and cooperative groups in North America and Europe increasingly allowed the use of IMRT in certain protocols. In addition to large nonrandomized series on lung cancer, prostate cancer, head & neck cancer, and anal cancer [39–44], results of a few randomized studies became available to provide class 1 evidence of the benefit of IMRT in head & neck cancer and prostate cancer [45–47]. In 2006, Hall raised the issue of total body exposure to leakage radiation because of the high number of monitor units used in IMRT and the resultant possibility of an increase in second malignancy (especially in children) [48]. Since then, improvements in the design of the multileaf collimator for IMRT have occurred. In 2010, Varian introduced a flattening filter-free accelerator with significantly higher output of monitor units per minute. At present, radiation treatment facilities without IMRT are a distinct rarity in North America and Europe, and increasing proportion of patients is being treated with IMRT. Hypo-fractionated schedules are being introduced for many tumor sites. Stereotactic body irradiation for lung cancer, liver tumors, and vertebral metastasis most frequently utilizes the technique of IMRT.

1.4 Summary

It would be reasonable to state that the introduction of IMRT has "revolutionized" radiation therapy. IMRT has proven to be a "versatile" technology that could provide excellent radiation dose distribution and in some cases better tumor control and lesser side effects. Further improvement of therapeutic ratio would likely require another genre of radiation, such as protons. Randomized studies are under way to determine if proton therapy could indeed produce a better outcome than what the best photons could do with IMRT.

References

1. Webb S (2003) The physical basis of IMRT and inverse planning. Br J Radiol 76:678–689
2. Hong TS, Ritter MA, Tomé WA, Harari PM (2005) Intensity-modulated radiation therapy: emerging cancer treatment technology. Br J Cancer 92:1819–1824
3. Sterzing F, Stoiber EM, Nill S, Bauer H, Huber P, Debus J, Münter MW (2009) Intensity modulated radiotherapy (IMRT) in the treatment of children and Adolescents – a single institution's experience and a review of the literature. Radiat Oncol 4:37
4. Stein J, Bortfeld T, Dorschel B, Schlegel W (1994) Dynamic x-ray compensation for conformal radiotherapy by means of multileaf collimation. Radiother Oncol 32:163–173
5. Bortfeld T, Boyer AL, Schlegel W, Kahler DL, Waldron TJ (1994) Realization and verification of three-dimensional conformal radiotherapy with modulated fields. Int J Radiat Oncol Biol Phys 30:899–908
6. Brahme A, Roos JE, Lax I (1982) Solution of an integral equation encountered in rotation therapy. Phys Med Biol 27:1221–1229
7. Brahme A (1988) Optimization of stationary and moving beam radiation therapy techniques. Radiother Oncol 12:129–140

8. Morganti AG et al (2011) Forward planned intensity modulated radiotherapy (IMRT) for whole breast postoperative radiotherapy. Is it useful? When? J Appl Clin Med Phys [S.l.] 12(2). ISSN 15269914. Available at: http://www.jacmp.org/index.php/jacmp/article/view/3451/2207. Accessed 20 Feb 2014. doi:10.1120/jacmp.v12i2.3451
9. Ferreira BC, do Carmo Lopes M, Mateus J, Capela M, Mavroidis P (2010) Radiobiological evaluation of forward and inverse IMRT using different fractionations for head and neck tumors. Radiat Oncol 5:57
10. Li X, Wang X, Li Y, Zhang X (2011) A 4D IMRT planning method using deformable image registration to improve normal tissue sparing with contemporary delivery techniques. Radiat Oncol 6:83
11. Sternick ES (1997) The theory and practice of intensity modulated radiation therapy. Advanced Medical Publishing, Madison
12. Mackie TR, Holmes T, Swerdloff S, Reckwerdt P, Deasy JO, Yang J et al (1993) Tomotherapy: a new concept for the delivery of conformal radiotherapy using dynamic collimation. Med Phys 20:1709–1719
13. Carol M, Grant WH III, Bleier AR, Kania AA, Targovnik HS, Butler EB, Woo SY (1996) The field-matching problem as it applies to the peacock three-dimensional conformal system for intensity modulation. Int J Radiat Oncol Biol Phys 34:183–187
14. Carol MP (1997) IMRT: where we are today. In: Sternick ES (ed) The theory and practice of intensity modulated radiation therapy. Advanced Medical Publishing, Madison, pp 17–36
15. Carol MP (1997) Where we go from here: one person's vision. In: Sternick ES (ed) The theory and practice of intensity modulated radiation therapy. Advanced Medical Publishing, Madison, pp 243–252
16. Grant W III, Butler EB, Woo SY, Targovnik H, Campbell RC, Carol MP (1994) Quality assurance of brain treatments with the Peacock treatment system. Radiology 193(Suppl):228
17. Woo SY, Grant W III, Bellezza D, Grossman R, Gildenberg P, Carpenter LS, Carol M, Butler EB (1996) A comparison of intensity modulated conformal therapy with a conventional external beam stereotactic radiosurgery system for the treatment of single and multiple intracranial lesions. Int J Radiat Oncol Biol Phys 35:593–597
18. Woo SY, Butler B, Grant WH (1997) Clinical experience: benign tumors of the CNS and head and neck tumors. In: Sternick ES (ed) The theory and practice of intensity modulated radiation therapy. Advanced Medical Publishing, Madison, pp 195–198
19. Ling CC, Burman C, Chui CS, LoSasso T, Mohan R, Spirou S et al (1997) Implementation of photon IMRT with dynamic MLC for the treatment of prostate cancer. In: Sternick ES (ed) Intensity modulated radiation therapy. Advanced Medical Publishing, Madison, pp 219–228
20. Butler EB, Teh BS, Grant WH, Uhl BM, Kuppersmith RB, Chiu JK, Donovan DT, Woo SY (1999) SMART (Simultaneous modulated accelerated radiation therapy) boost: a new accelerated fractionation schedule for the treatment of head and neck cancer with intensity modulated radiotherapy. Int J Radiat Oncol Biol Phys 45:21–32
21. Dawson LA, Anzai Y, Marsh L et al (2000) Local-regional recurrence pattern following conformal and intensity modulated RT for head and neck cancer. Int J Radiat Oncol Biol Phys 46:1117–1126
22. Teh BS, Mai WY, Augspurger ME, Uhi BM, McGary J, Dong L, Grant WH 3rd, Lu HH, Woo SY, Carpenter LS, Chiu JK, Butler EB (2001) Intensity modulated radiation therapy (IMRT) following prostatectomy: more favorable acute genitourinary toxicity profile compared to primary IMRT for prostate cancer. Int J Radiat Oncol Biol Phys 49:465–472
23. Kuppersmith RB, Teh BS, Donovan DT, Mai WY, Chiu JK, Woo SY, Butler EB (2000) The use of intensity modulated radiotherapy for the treatment of extensive and recurrent juvenile angiofibroma. Int J Pediatr Otorhinolaryngol 52:261–268
24. Teh BS, Mai W, Uhl BM, Augspurger ME, Uhl BM, McGary J, Dong L, Grant WH, Lu HH, Woo SY, Carpenter LS, Chiu JK, Butler EB (2001) Intensity-modulated radiation therapy (IMRT) for prostate cancer with the use of a rectal balloon for prostate immobilization: acute toxicity and dose-volume analysis. Int J Radiat Oncol Biol Phys 49(3):705–712

25. Mundt AJ, Roeske JC, Lujan AE et al (2001) Initial clinical experience with intensity modulated whole-pelvis radiation therapy in women with gynecologic malignancies. Gynecol Oncol 82:456–463
26. Huang E, Teh BS, Strother DR, Davis QG, Chiu JK, Lu HH, Carpenter LS, Mai WY, Chintagumpala MM, South M, Grant WH 3rd, Butler EB, Woo SY (2002) Intensity-modulated radiation therapy for pediatric medulloblastoma: early report on the reduction of toxicity. Int J Radiat Oncol Biol Phys 52(3):599–605
27. Zelefsky MJ, Fuks Z, Happersett L et al (2000) Clinical experience with intensity modulated radiation therapy (IMRT) in prostate cancer. Radiother Oncol 55:241–249
28. Kupelian P, Meyer J (2011) Image-guided, adaptive radiotherapy of prostate cancer: toward new standards of radiotherapy practice. In: Meyer JL (ed) IMRT, IGRT, SBRT – advances in the treatment planning and delivery of radiotherapy, ed 2, rev and ext. Front Radiat Ther Oncol. Basel, Karger, vol 43, pp 344–368
29. Zinkin HD, Rivard MJ, Mignano JE, Wazer DE (2004) Analysis of dose conformity and normal-tissue sparing using two different IMRT prescription methodologies for irregularly shaped CNS lesions irradiated with the beak and 1-cm mimic collimators. Int J Radiat Oncol Biol Phys 59(1):285–292
30. Mell LK, Roeske JC, Mundt AJ (2003) A survey of intensity-modulated radiation therapy use in the United States. Cancer 98:204–211
31. Mackie TR, Kapatoes J, Ruchala K, Lu W, Wu C, Olivera G, Forrest L, Tome W, Welsh J, Jeraj R, Harari P, Reckwerdt P, Paliwal B, Ritter M, Keller H, Fowler J, Mehta M (2003) Image guidance for precise conformal radiotherapy. Int J Radiat Oncol Biol Phys 56:89–105
32. Low DA, Mutic S, Dempsey JF, Gerber RL, Bosch WR, Perez CA, Purdy JA (1999) Quantitative dosimetric verification of an IMRT planning and delivery system. Radiother Oncol 49:305–316
33. Reckwerdt PJ, Mackie TR, Balog J, McNutt TR (1997) Three dimensional inverse treatment optimization for tomotherapy. In: Leavitt DD, Starkschall G (eds) Proceedings in 12th international conference on the use of computers in radiation therapy (Salt Lake City, Utah, May 1997). Medical Physics Publishing, Madison, pp 420–422
34. Yu CX, Tang G (2011) Intensity-modulated arc therapy: principles, technologies and clinical implementation. Phys Med Biol 56(5):R31–R54
35. Krieger T, Sauer OA (2005) Monte Carlo- versus pencil-beam-/collapsed-cone-dose calculation in a heterogeneous multi-layer phantom. Phys Med Biol 50:859–868
36. Fragoso M, Wen N, Kumar S, Liu D, Ryu S, Movsas B, Munther A, Chetty IJ (2010) Dosimetric verification and clinical evaluation of a new commercially available Monte Carlo-based dose algorithm for application in stereotactic body radiation therapy (SBRT) treatment planning. Phys Med Biol 55:4445–4464
37. Weksberg DC, Palmer MB, Vu KN, Rebueno NC, Sharp HJ, Luo D, Yang JN, Shiu AS, Rhines LD, McAleer MF, Brown PD, Chang EL (2012) Generalizable class solutions for treatment planning of spinal stereotactic body radiation therapy. Int J Radiat Oncol Biol Phys 84(3):847–853
38. National Cancer Institute (NCI) (2005) National Cancer Institute guidelines for the use of intensity-modulated radiation therapy in clinical trials. NCI, Bethesda. Available at: http://atc.wustl.edu/home/NCI/IMRT_NCI_Guidelines_v4.0.pdf
39. Jiang ZQ, Yang K, Komaki R, Wei X, Tucker SL, Zhuang Y, Martel MK, Vedam S, Balter P, Zhu G, Gomez D, Lu C, Mohan R, Cox JD, Liao Z (2012) Long-term clinical outcome of intensity-modulated radiotherapy for inoperable non-small cell lung cancer: the MD Anderson experience. Int J Radiat Oncol Biol Phys 83(1):332–339
40. Spratt DE, Pei X, Yamada J, Kollmeier MA, Cox B, Zelefsky MJ (2013) Long-term survival and toxicity in patients treated with high-dose intensity modulated radiation therapy for localized prostate cancer. Int J Radiat Oncol Biol Phys 85(3):686–692
41. Marta GN, Silva V, de Andrade Carvalho H, de Arruda FF, Hanna SA, Gadia R, da Silva JL, Correa SF, Vita Abreu CE, Riera R (2013) Intensity-modulated radiation therapy for head and

neck cancer: systematic review and meta-analysis. Radiother Oncol 110(1):9–15, pii: S0167-8140(13)00616-6. doi:10.1016/j.radonc2013.11.101
42. Beadle BM, Liao KP, Elting LS, Buchholz TA, Ang KK, Garden AS, Guadagnolo BA (2014) Improved survival using intensity-modulated radiation therapy in head and neck cancers: a SEER-Medicare analysis. Cancer 120(5):702–710. doi:10.1992/cncr.28372
43. Chen AM, Daly ME, Farwell DG, Vazquez E, Courquin J, Lau DH, Purdy JA (2014) Quality of life among long-term survivors of head and neck cancer treated by intensity-modulated radiotherapy. J Am Med Assoc Otolaryngol Head Neck Surg 140(2):129–133
44. Call JA, Prendergast BM, Jensen LG, Ord CB, Goodman KA, Jacob R, Mell LK, Thomas CR Jr, Jabbour SK, Miller RC (2014) Intensity-modulated radiation therapy for anal cancer: results from a multi-institutional retrospective cohort study. Am J Clin Oncol (Epub ahead of print)
45. Veldeman L, Madani I, Hulstaert F et al (2008) Evidence behind use of intensity-modulated radiotherapy: a systematic review of comparative clinical studies. Lancet Oncol 9(4):367–375. Erratum in: Lancet Oncol 9(6):513
46. Michalski JM, Yan Y, Watkins-Bruner D, Bosch WR, Winter K, Galvin JM, Bahary JP, Morton GC, Parliament MB, Sandler HM (2013) Preliminary toxicity analysis of 3-dimensional conformal radiation therapy versus intensity modulated radiation therapy on the high-dose arm of the radiation therapy oncology group 0126 prostate cancer trial. Int J Radiat Oncol Biol Phys 87(5):932–938
47. Gandhi AK, Sharma DN, Rath GK, Julka PK, Subramani V, Sharma S, Manigandan D, Lavirai MA, Kumar S, Thulkar S (2013) Early clinical outcomes and toxicity of intensity modulated versus conventional pelvic radiation therapy for locally advanced cervix carcinoma: a prospective randomized study. Int J Radiat Oncol Biol Phys 87(3):542–548
48. Hall EJ, Wuu CS (2003) Radiation-induced second cancers: the impact of 3D-CDT and IMRT. Int J Radiat Oncol Biol Phys 56(1):83–88

Principles of IMRT

2

Laurence E. Court, Peter Balter, and Radhe Mohan

Keywords
IMRT • VMAT • IMRT treatment planning • IMRT treatment delivery

2.1 Introduction

This chapter describes the basic principles of intensity-modulated radiation therapy (IMRT), providing the necessary background for subsequent clinical chapters. We start by describing the IMRT treatment planning and delivery process, introducing the wide range of different approaches and technologies currently in clinical use. Then, other topics important in the implementation of IMRT are discussed, including quality assurance (QA), facility design, and respiratory motion management. The chapter closes by reviewing some potential advantages and challenges of IMRT.

2.2 Treatment Planning

IMRT relies on many of the same tools for imaging, dose calculations, plan evaluation, QA, and delivery as conventional treatments do. However, some significant differences exist, particularly in the planning and treatment delivery processes. The following sections describe the workflow for the entire IMRT process, from the viewpoint of patients and clinic staff.

L.E. Court, Ph.D. (✉) • P. Balter, Ph.D. • R. Mohan, Ph.D.
Department of Radiation Physics, Unit 94, The University of Texas MD Anderson Cancer Center, 1515 Holcombe Blvd, Houston, TX 77030, USA
e-mail: lecourt@mdanderson.org

Y. Nishimura, R. Komaki (eds.), *Intensity-Modulated Radiation Therapy*,
DOI 10.1007/978-4-431-55486-8_2

2.2.1 Imaging and Delineation

The first step in the IMRT process, in common with conventional conformal RT, is to obtain images of the patient and delineate the targets and relevant normal tissues on those images. The primary type of imaging used for target delineation and dose calculation is computed tomography (CT), although other imaging modalities such as positron emission tomography (PET) and magnetic resonance imaging (MRI) can also be used. The various volumes that form the skeleton or outline of the treatment plan are described by the International Commission on Radiation Units and Measurements [52–54], and their clinical application is discussed in detail elsewhere in this book. The main volumes to be considered are the gross tumor volume (GTV), which is the gross demonstrable extent and location of the tumor; the clinical target volume (CTV), which includes the tissue that may contain subclinical malignant disease; and the planning target volume (PTV), which is the CTV after geometric expansion to account for uncertainties in the planning and treatment process. Other treatment plan components are the organs at risk (OARs), which are the normal tissues that can suffer radiation damage during treatment; the planning organ-at-risk volume, which is analogous to the PTV, but applied to normal tissues; and the remaining volume at risk, which describes uncontoured parts of the patient. These components are all important in creating an IMRT plan, and they are discussed further in other chapters. IMRT planning also involves the use of "dose-shaping" or "dummy" structures (sometimes called "pseudostructures"), which are nonanatomic structures created by treatment planners to guide optimization of the IMRT plans. One example of such a structure is a ring created around the target, to which the planner sets constraints that keep the dose to this region low. Planners may also add structures to which dose must be reduced after the first plan iteration to cover regions where the dose is too high, such as in a normal tissue or in the target itself. Additional concepts, such as the volume to account for respiratory motion (internal target volume [ITV] [54]), are also important for treating disease at particular anatomic sites (as described in other chapters).

The use of these volume definitions is not unique to IMRT. The quality and accuracy of the delineation of targets and normal tissues, however, require particular attention in IMRT, as this information is the basis for the creation of treatment fluences by inverse-planning algorithms [40]. Structures must be consistent from slice to slice to produce smooth 3D structures. Clinicians or treatment planners must also take care to "clean up" all structures created during the planning process. For example, inadvertent volumes, such as those created if the user accidentally pressed a mouse button when the cursor is not where they wanted it to be, must be removed. Such volumes may represent only a single point that may not be apparent visually, but they can become serious issues when the inverse-planning algorithm attempts to design the fluence that confers dose to them.

2.2.2 Number and Configuration of Radiation Beams

After the necessary structures have been delineated or contoured, the next step in the IMRT process is placement of the treatment beams, including the choice of the number of beams. This step requires determining the treatment isocenter, which may already have been set during the acquisition of scans during treatment simulation. Standard practice for isocenter placement varies among clinics. In some cases, the isocenter is placed in the center of the primary target (e.g., the center of the prostate), but in other cases, the isocenter is placed in a generic location (e.g., the anterior edge of C2 for head and neck tumors). As is true for conventional treatments, shifts may be apparent between the marks on the patient (from the treatment simulation) and the actual treatment isocenter. With IMRT (and unlike most conventional treatments), the isocenter is not necessarily within the treatment volume at all; rather, it may be placed so as to aid image-guided radiation therapy (IGRT) or to avoid geometric restrictions during treatment delivery. For example, when relatively wide targets are to be treated, common practice is to try to place the isocenter so as to minimize the number of adjacent fields needed to cover the entire target (in IMRT, the width of the fields is limited by the length of the multileaf collimator [MLC], as described later in this chapter).

In most planning systems, beams are positioned manually by the treatment planners, although automatic beam placement is also possible [66, 125]. Factors to be considered in beam placement include normal tissue location (e.g., we prefer to minimize beams that pass unnecessarily through the contralateral lung) and the desire to minimize treatment time. Thus, although using more beams provides more degrees of freedom for optimizing the plan, excessive numbers of beams should be avoided because of the additional time needed for treatment (and for plan optimization). Another common practice is to avoid using noncoplanar beams or multiple isocenters. The exact clinical trade-offs (target coverage, normal tissue dose, and treatment time) depend on the clinical situation, but in most cases, the appropriate number of beams is between 7 and 9 [99, 127]. Of course, IMRT may not be delivered as a series of individual beams, but rather may be delivered during a gantry rotation (as in tomotherapy or volume-modulated arc therapy [VMAT]), as described further below.

2.2.3 Treatment Plan Objectives

Once the beam configuration has been determined, the next step is to determine the treatment plan objectives—in other words, the doses that represent the intended treatment, such as target dose and coverage and normal tissue doses. In many cases, these doses come from templates, with standard objectives used for a given clinical site, although the doses can be edited based on individual patient prescriptions or anatomic characteristics. The constraints on those doses may be hard or soft and may be based on dose, dose-volume, or dose-response (e.g., predicted probability of tumor response), as described below. The desired objectives as specified are often not achieved because the optimizing software tries to balance the requirements of various structures. Thus, specified objectives may be quite different from what is desired. Specifying the objectives, in combination with suitable "penalties," usually leads to an acceptable approximation of what is desired. Penalties and specified objectives are often achieved through experience and may vary among institutions.

An important part of IMRT planning that is hidden from the user is how the inverse-planning algorithm quantifies how well the treatment plan (dose distribution) meets the planners' objectives. The functions used for this task are summarized below.

2.2.3.1 Hard and Soft Constraints

The constraints that the optimization algorithm attempts to meet can be either "hard" or "soft." A hard constraint is one that the treatment plan must meet. For example, the intensities in the fluence cannot be negative in value. In another example, the maximum dose to the spinal cord must not exceed 45 Gy. If the final dose distribution results in a spinal cord dose that exceeds this hard constraint, then the algorithm may automatically scale down the entire dose distribution. A soft constraint is one that could be violated, although violations may incur a penalty. For example, the mean dose to the parotid could be a soft constraint, reflecting our level of understanding of the radiobiology of the parotid and the clinical compromises needed when treating patients. That is, we would like to minimize the dose to the parotid, and we know that maintaining the mean dose at less than 26 Gy will produce less toxicity. However, we also know that a slightly higher dose would be acceptable, and ultimately, we want to treat the tumor and may be willing to sacrifice parotid function to do so. The level of "softness" of a constraint is controlled by the planner by increasing or decreasing its relative weight or penalty.

2.2.3.2 Dose- and Dose-Volume-Based Objective Functions

A simple objective function for optimizing dose distributions could be expressed as a sum of the squares of the differences of desired and computed dose at each point in the volume of interest. Each tissue could be assigned a different weight (or penalty) such that it contributes differently to the overall objective function. For tumors, dose increases (hot spots) and decreases (cold spots) may be important. For normal tissues, only dose increases would be considered. This simple dose-based approach is generally considered insufficient in practice. Radiobiologically, the response of both tumors and normal tissues to radiation is a function of the volume of the tissue that receives each level of dose—hence the common use of dose-volume histograms (DVHs) to assess the quality of radiation therapy plans. DV-based objectives are the most common approach used in IMRT optimization. For each normal structure, the DV constraint can be expressed as the volume of that structure that is allowed to receive a certain dose or higher. Typically, several DV objectives are used for each normal structure. For targets, a constraint is also included that describes the acceptable *minimum* dose to a certain volume, for example, the minimum dose to 95 % of the PTV. Further, as noted above, each constraint is also assigned a weight that reflects how much it will contribute to the overall objective function.

Notably, the DV constraints that treatment planners often set for plan optimization are not necessarily the same as those they are aiming for in the final treatment plan. Despite extensive and ongoing research into developing treatment plan optimization engines, treatment planning is still an "art" in that treatment planners are often required to "trick" or "massage" the optimization engine to obtain the optimal plan. The process is achieved by varying the DV constraints, varying the relative penalties of the different constraints, and adding dummy structures to help force dose either away from or toward certain areas. Thus, although optimization of a treatment plan is nominally automatic, the experience of the treatment planner is important in determining objectives, priorities, dummy tissues, and beam angles. One approach to mitigate the need for such experience-based artistry is the use of class solutions, which provide a systematic way to plan treatments for specific sites that is consistent and robust. Use of class solutions is particularly promising in the many situations in which objectives conflict [65, 120, 128].

One disadvantage of DV constraints is that each constraint may describe only a single point on the DVH curve. Use of multiple DV constraints can reduce this, and some planning systems actually allow the planner to draw the optimal DVH and then use that to guide the optimization. However, multiple DVHs could, in general, lead to the same dose-response, and another DVH in the space of DVHs of one structure may be more helpful to other structures. Thus, specifying the so-called optimal DVH may not be the ideal solution. One approach to help overcome the limitations of DV-based optimization is to supplement this process with dose-response-based constraints, such as constraints that are based on calculations of tumor control probability, normal tissue complication probability, or equivalent uniform dose [89, 132]. Constraints such as these have the potential advantage of including treatment response in the optimization. Notably, however, in many cases, the treatment response of the irradiated tissues is not well understood.

2.2.4 Optimization of Intensity Distribution

Once the beam configuration and plan objectives are established, the optimum intensity distribution for each beam can be determined. This is achieved through an iterative optimization process as follows. Each radiation ray (beamlet) is traced from the source of radiation through the patient. Only rays passing through the target plus a small margin are considered. The dose at each voxel in the patient is calculated for an initial set of weights for each individual beamlet, and the resulting dose distribution is then used to calculate an objective function that describes how close the current dose distribution is to the goals set by the treatment planner. The effect of a change in the weight of each individual ray or beamlet is then calculated, with the weight increased, decreased, or left the same depending on whether the change would be favorable for the patient. Mathematically, these changes in ray weight are determined from the gradient of the objective function with respect to the ray weight. Because improvements in the treatment plan are a result of changes in many rays from many beams, only small changes in ray weight may be permitted in any one iteration. This iterative process then continues until no further improvement occurs, at which point the optimization is assumed to have converged on the optimal solution. Most of the optimization algorithms used for IMRT planning use variations of gradient techniques. The large search space of RT plans can contain many local minima [20, 35, 131, 138], and alternative optimization approaches such as simulated annealing can reduce the probability of getting trapped in a nonoptimal local minima [11]. However, this is not a significant issue in clinical practice [67, 131]. Another common practice in clinical treatment planning (depending on the capabilities of the planning system) is to "massage" the optimization in real time by adjusting the constraints and weights as the optimization progresses.

At this point, the optimized intensity modulation must be converted to a deliverable field. Typically, this involves first determining the MLC sequence that will achieve a fluence as close to the optimal fluence as possible, given the physical constraints of the delivery mechanism (including the radiographic properties of the MLCs). The details of this sequencing process are described later in this chapter. Notably, in some situations the optimal fluence and the actual deliverable fluence are sufficiently different that the final dose distribution is compromised.

One approach used in some treatment planning systems to overcome this issue is to directly include the MLC constraints in the optimization process. For example, in direct aperture optimization, only MLC aperture shapes that satisfy the mechanical constraints of the MLC system are considered [1, 33, 108]. In this approach, the final plan typically uses fewer segments (apertures) than other approaches. In other systems, deliverable dose distributions are fed back into the optimizer to further adjust intensity distributions and the resulting leaf positions so that the optimized and deliverable dose distributions are essentially identical.

2.2.5 Dose Calculation

During inverse planning, the dose distribution is recalculated many times. Some compromise between dose accuracy and speed of the dose calculation is necessary because, in general, the faster an algorithm is, the less accurate it is and vice versa. If a fast, inaccurate dose calculation algorithm is used during the optimization, then the final dose calculation (calculated with an accurate algorithm) may well not be the same as the one calculated with the inaccurate algorithm, and it may not even be the optimal solution. Several solutions have been developed to minimize this issue. One approach is to start with a less accurate, fast algorithm to get close to the final solution and then carry out the final iterations using a slower, more accurate algorithm [81, 110]. The less accurate algorithm may be a simple pencil-beam algorithm that may not accurately model the effect of the delivery hardware (e.g., MLCs), as described below. The impact of this approach depends on the anatomic characteristics of the area being treated (e.g., significant heterogeneity in tissue density in the lungs) and the complexity of the treatment plan. The graphics processing unit recently emerged as an option for reducing the processing time for IMRT optimization and dose calculation [34, 45, 55, 76, 101]. The accuracy of dose calculations in the buildup region, especially with many tangential fields, is particularly important in IMRT, especially for anatomically complex areas such as the head and neck, and additional care is needed when commissioning the treatment planning system [27].

2.2.6 Treatment Plan Evaluation

IMRT dose distributions are usually very conformal, but they can also be very complex and are different from dose distributions in conventional RT. As is true for conventional RT, DVHs are useful tools for summarizing and comparing treatment plans. Unlike conventional RT, the need to review treatment field beam's eye view (including block shape) is usually not important in IMRT, with some notable exceptions (e.g., ensuring that beams do not travel across the top of the shoulders in patients being treated for head and neck cancer). Instead, the complex dose distributions, and clinical compromises that occur near normal tissues, underscore the importance of careful review of the dose distribution for each CT slice.

2.2.7 Special Planning Considerations

Some of the more common planning considerations experienced by clinicians are summarized here. Additional details on IMRT planning for tumors at various sites are given elsewhere in this book.

2.2.7.1 Targets in the Buildup Region

Severe skin reactions, reported in some patients treated with IMRT, can be caused by a variety of factors, including the use of immobilization masks or IGRT couches (both of which can have a "blousing" effect), multiple tangential fields (IMRT typically consists of many fields, many of which are tangential to the patient, unlike traditional treatments), inappropriate strategies during IMRT inverse planning (e.g., including the skin in the PTV expansion), and the inability of the treatment planning system to accurately calculate dose in the buildup regions [23, 27, 28, 44, 63, 118], particularly when the treatment targets are close to the patient's skin. Strategies used to mitigate these effects include delineating the skin as a sensitive structure (and applying a maximum dose constraint during optimization) and pulling the PTV back several millimeters from the body surface; however, care must be taken to avoid unintended consequences such as reduced target coverage [28].

2.2.7.2 Overlap Regions and Pseudostructures

Target volumes (PTVs) will often overlap with critical normal tissues, creating a potential conflict between target objectives and normal tissue constraints. For example, the PTV in head and neck treatments often overlaps with the parotid or other nearby structures. Various solutions to this potential dilemma have been proposed, including creating dummy (pseudo) structures with no overlap or implementing a priority system in the optimization. As noted previously, pseudostructures (structures that are not necessarily related to specific anatomic structures) are widely used in IMRT planning. Examples include ring structures created around target structures to help force the optimization to minimize dose to surrounding structures; structures created in regions that the planners expect, from experience, that the optimization process may deposit excess dose; and structures created based on isodose lines after initial optimization to remove unwanted high-dose regions (which could be in the target or normal tissues).

2.2.7.3 Hybrid IMRT Approaches

Treatment plans do not have to be constructed only for IMRT or only for VMAT, and many treatment centers combine IMRT and static treatments in therapy for breast cancer [73] or thoracic cancer (cancer of the lung or esophagus). In such cases, hybrid techniques typically concurrently combine static fields (~2/3 of the dose) and IMRT or VMAT fields (~1/3 of the dose) [17, 74]. Potential advantages of this type of treatments include a reduction in the volume of lung exposed to low doses.

2.3 Treatment Delivery

2.3.1 IMRT Delivery Hardware

Several hardware approaches are used to deliver IMRT; the most common involve rotating multileaved slits and moving MLCs and are described below. For the sake of completeness, we also briefly describe the use of compensators and jaws-only IMRT, although these approaches are rarely used for clinical purposes.

2.3.1.1 Compensators

Physical compensators (or compensating filters or modulators) can be used to create complex x-ray fluence distributions. The advantages of physical compensators include not requiring MLCs, with their attendant requirements for commissioning and maintenance (although these benefits are countered by issues related to the accuracy of machining and compensator placement). Other advantages include the finer resolution that is possible with compensators, the less complicated QA, and the lack of interplay effects (interactions between a moving radiation aperture and a moving target) [90], although interplay can be reduced with appropriate planning approaches [24–26]. Similarly, although some of the complexities involved in calculating dose for complex fluences created with MLCs (e.g., transmission, interleaf leakage, and tongue-and-groove effects) do not exist with physical compensators, other issues must be considered such as the effects of beam hardening and scatter from the filter [40]. Currently, at least one company in the United States is creating patient-specific compensators for IMRT (.decimal), and some users create their own [90]. The use of compensators in modern RT is extremely rare.

2.3.1.2 Rotating Multileaved Slit Approaches

The delivery of radiation using a rotating multileaved slit that produces an intensity-modulated fan beam is called tomotherapy [41]. Radiation can be delivered as a series of axial slices, where the patient is translated discretely through the linear accelerator (LINAC) between slices, or in a helical form, where the patient is translated continuously through the LINAC as the LINAC gantry rotates around the patient. These approaches can be considered analogous to axial and helical CT scans. Much of the initial experience with IMRT involved use of an axial tomotherapy system called MIMiC (Nomos) [40]. MIMiC was a binary MLC system, with two banks of pneumatically controlled opposing leaves arranged to give a fan beam of radiation parallel to the rotation of the LINAC. The intensity of the fan beam is modulated by controlling how long each leaf blocks the fan beam. The MIMiC system was an after-market add-on system that allowed centers to add IMRT capabilities to existing LINACs. Its successor still offers similar options (nomosStat; Best nomos, http://www.nomos.com/pdf/nomoSTAT_Bro_03.pdf).

Although axial tomotherapy has remained an after-market add-on system, helical tomotherapy was developed as a purpose-built system, called the TomoTherapy Hi-Art (Tomotherapy Inc). As is true for axial tomotherapy, helical tomotherapy has a fan beam parallel to the gantry rotation plane. In helical tomotherapy, the couch translates through the gantry as the gantry rotates. The pitch (couch movement/fan-beam width) is typically 0.2–0.5.

2.3.1.3 Multileaf Collimators

The vast majority of modern IMRT delivery systems use MLCs, small, individually motorized leaves that can be used to shape or modulate the intensity of the treatment field. Several basic approaches incorporate MLCs into the treatment unit [12], with the MLCs either taking the place of one of the LINAC adjustable jaw pairs or being positioned below the jaws. Common designs have between 10 and 60 opposed leaf pairs, with the width of the MLC in the beam's eye view at the isocenter plane between 2 mm and 1 cm, depending on the manufacturer and model.

The ability of MLCs to shape fields (or segments of IMRT fields) depends on several aspects of their physical design and control mechanism. These include maximum leaf travel (determined by the length of the MLC leaves), maximum field size perpendicular to the MLCs (MLC width × number of MLCs), and whether the leaves on one side can interdigitate with neighboring leaves on the opposite side. These are all important considerations; for example, a machine with small MLCs (such as those used for stereotactic applications) may not be able to cover sufficient length to treat head and neck tumors or large lung tumors.

The x-ray properties of the MLC can also have a significant effect on the dose distribution. Leakage of radiation through the MLCs is much more important in IMRT than in conventional RT because radiation is delivered with narrow openings of the moving leaves of MLCs, and so leakage contributes more to the target dose. For the same reason, scatter from the MLCs is also more important in IMRT than in conventional RT [68, 69].

Another important design consideration is the cross-sectional shape of the MLCs, which is complex because leaves must incorporate divergence in the direction perpendicular to their travel, and adjacent MLCs must overlap to minimize interleaf transmission. Details of this overlap are very important in IMRT, as the exposed stepped sides (known as the tongues) may block or scatter radiation, leading to underdosing the target [40]. This effect can be significant, with reported underdoses as large as 10–25 % [105, 116, 124], as shown in Fig. 2.1. Inclusion of this so-called tongue-and-groove effect in dose calculation algorithms is difficult, so leaf-sequencing algorithms are often designed to minimize this effect (rather than including it in the dose calculation), although some investigators have included this effect in the actual optimization stage [105].

The leaf end shape is also important. It can either be straight, in which case the collimator moves along the circumference of a circle, with the ends of the leaves always remaining along the divergent x-ray beam, or rounded, for designs in which the MCL moves perpendicular to the beam central axis. When MLCs have rounded ends, an offset of 0.4–1.1 mm is present between the edge of the radiation field and the nominal location of the MLC leaf, depending on leaf design, beam energy, and distance from central axis [40]. The effect of this offset must be included in the treatment planning system dose calculations.

The accuracy and precision of MLC positioning are also important. In conventional conformal RT, MLCs are used to define the aperture of the treatment beam, thereby conforming it to the treatment target. When used in this way, an uncertainty in the leaf position of 1–2 mm may be acceptable, because an uncertainty of this size (typically small compared with the total aperture size) has only minimal effects on the radiation output. However, in IMRT the situation is very different. First, the segments can be quite narrow (<1 cm), and uncertainties of only a few tenths of a millimeter can cause errors of several percentage points in delivered dose. Further, the cumulative dose distribution in IMRT comprises contributions from many

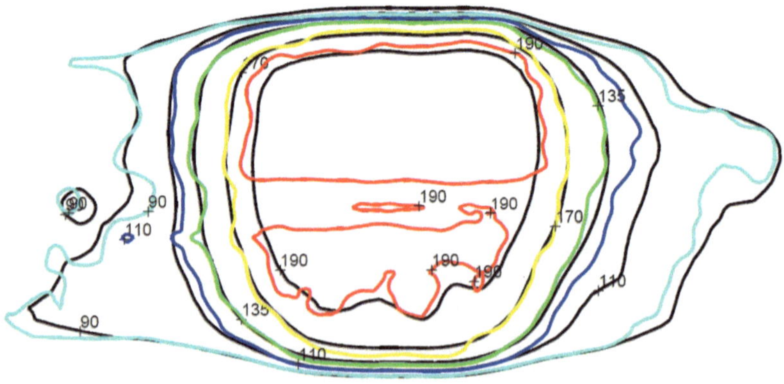

Fig. 2.1 Example of an IMRT case in which the tongue-and-groove effect resulted in a line of reduced dose through the target. *Black isodose lines* show the calculated dose distribution; *colored isodose lines*, the results of film-based IMRT quality assurance

segments. The beam edges move to many different locations during the treatment (i.e., not just at the edge of the target as is the case in conformal treatments), so it is essential that their positional accuracy is maintained to better than a millimeter. Without this level of accuracy, the contributions of the different segments may not sum correctly [40].

2.4 Volume-Modulated Arc Therapy

VMAT is a form of IMRT in which the treatment is delivered in one or more dynamically modulated arcs [5, 16, 93, 94, 104]. As the gantry rotates, the MLCs move, giving a different aperture shape for each angle of the gantry. The rate of rotation of the gantry and the LINAC dose rate can both be modulated during treatment to give the required delivered dose for each gantry angle. The quality of the planned dose distributions that can be achieved is equivalent to those that can be achieved with other forms of IMRT. The plan quality depends on achievable modulation, which, in turn, depends on the gantry speed, number of arcs, or both. The main advantage of VMAT is that the entire treatment can be completed quickly. For example, a typical treatment of two complete 360° arcs, with different couch rotations for each arc, takes less than 2.5 min. This advantage is significant, especially for a busy clinic, and thus we expect that VMAT will become the IMRT delivery technique of choice for most treatments.

2.4.1 Leaf Sequencing

As noted above, some optimization algorithms do not consider the physical characteristics or limitations of the delivery system when calculating the optimal intensity distribution. This optimal intensity is then used to create the MLC leaf positions (leaf positions as a function of time/monitor units [MUs]) that will deliver a fluence that is as close as possible to the optimized distribution.

In step-and-shoot multifield IMRT, modulated delivery is achieved with multiple static MLC segments, with each segment having its own aperture shape and weight (MU). The leaf-sequencing algorithm first coverts the optimized intensity distribution to discrete levels, which are then converted into separate MLC segments (Fig. 2.2). The ideal algorithm will create an MLC sequence for which the summation of all the segments gives a delivered fluence that is close to the optimized fluence, uses the minimum number of segments, and may also minimize the MLC motion between segments. In general, the agreement between the optimized and delivered intensity distribution increases as the number of intensity levels is increased. This process has been shown to increase the target coverage, but it also results in an increase in the number of MLC segments (MLC shapes). Because the beam is turned off as the MLCs move between segments, this can significantly affect the treatment delivery time. The advantage of step-and-shoot delivery is that factors such as MLC speed and dose rate are less important, so the IMRT delivery

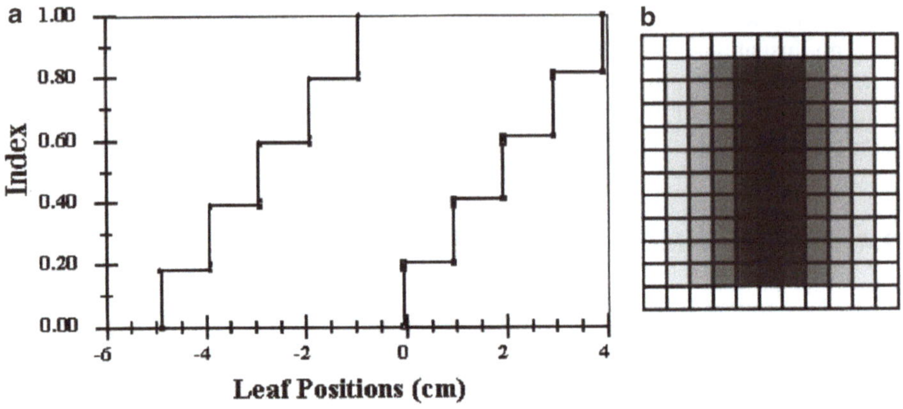

Fig. 2.2 (**a**) Leaf trajectory as a function of dose index for a step-and-shoot IMRT delivery. (**b**) is the resulting fluence map (From Xia and Verhey [133])

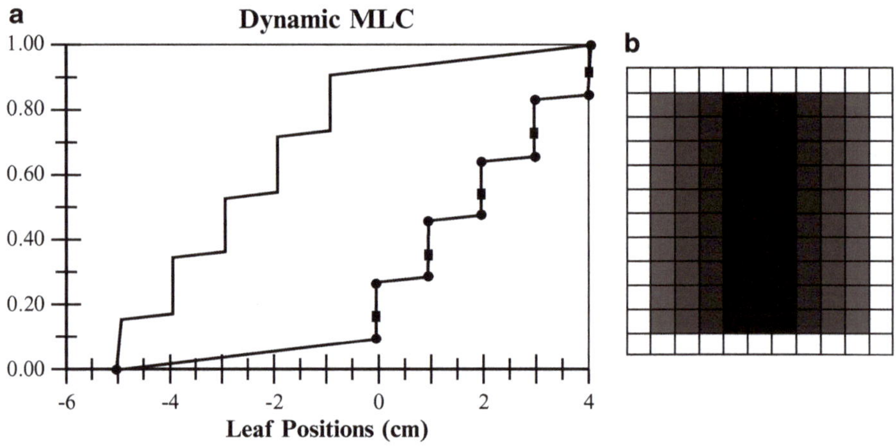

Fig. 2.3 (**a**) Leaf trajectories as a function of dose index for dynamic multileaf collimator (MLC) delivery. (**b**) is the resulting intensity fluence (From Xia and Verhey [133])

is possible with a less advanced treatment machine. Also, importantly, step-and-shoot IMRT typically requires less MUs than dynamic IMRT.

In sliding-window IMRT delivery, the MLCs move across the target volume while the radiation is on [8, 13, 40, 113]. The size of the gap between opposing MLCs and the speed of the MLCs are constantly changing. The dose rate may also be adjusted. Conceptually, the amount of radiation received by a point within the target is proportional to the number of MUs delivered while the system is in the open gap. When the two opposing leaves are far apart, the delivered dose is high; when they are closer together, the delivered dose is reduced (see Fig. 2.3). To account for unexpected variations in dose rate, the position of the MLCs is indexed to the delivered MUs rather than to time. Advantages of sliding-window IMRT include faster delivery than step-and-shoot IMRT, reduced numbers of MUs, and potentially reduced wear and tear on the MLC mechanism (because motion is mono-directional).

The delivered and ideal fluence can differ for several reasons, including overly complex ideal fluence distributions [80] and practical limitations related to the leaf design (transmission, non-divergent leaf end design, leaf scatter) [99]. Although these limitations reflect the choice of the planning and delivery system, treatment planners can take steps to minimize them. For example, planners should take care not to push the IMRT optimization excessively, as can happen when extreme values of the weights are used for the DV constraints. Some planning systems also allow the planner to control how smooth the ideal fluence will be. One way to monitor fluence complexity is to ensure that the MU per beam is not unusually high. Commonly, the agreement between ideal and deliverable MU decreases as the MU increases. Also, the complex MLC patterns that high-MU fields require can be more difficult for the MLC controller to deliver, resulting in unwanted delays at the time of treatment.

2.4.2 Jaws-Only IMRT

As described above, with MLC-based IMRT, MLCs are used to create many irregular shapes that are superimposed to create a complex fluence pattern. Complex fluences can also be created from many rectangular segments created by the LINAC jaws alone [38, 41]. The main advantage of jaws-only IMRT is the lack of additional complexity and expense of an MLC, which could allow IMRT to be achieved at lower cost. One possible application of this approach could be a low-cost LINAC for low- and middle-income countries, where affordable, reliable RT equipment is desperately needed [47–49]. In modern radiation therapy centers, however, IMRT is dominated by MLC-based delivery.

2.4.3 Image-Guided IMRT

IMRT alone can achieve impressive dose distributions, reducing toxicity to normal tissues. However, the high conformality that can be achieved with IMRT increases the need for image guidance (i.e., IGRT). Moreover, realizing these planned dose distributions over a treatment course lasting days or weeks requires highly accurate patient setup, particularly when taking advantage of tight dose distributions could lead to use of margins as small as 3–5 mm [30]. Many approaches are used for IGRT, including orthogonal (or stereotactic) kilovoltage or megavoltage x-ray imaging and CT imaging (cone beam or CT on rails), and MRI-guided treatments are only a few years away [57, 83, 102, 109]. Specifics of IGRT for IMRT are discussed elsewhere in this book.

2.5 Intrafraction Motion and IMRT

Use of IMRT to treat tumors in regions of the body that experience involuntary intrafraction motion (e.g., tumors in the lung or liver that move with respiration or tumors in the lower abdomen that move with the passing of bowel gas or other involuntary bodily functions) has been controversial for two reasons, namely, the potential for geometric miss and interplay effects between the motion of the tumor and the motion of the machine (gantry, collimator, and MLC) used to create the modulation pattern [10, 107, 136]. Both of these concerns can be managed by appropriate imaging and plan design.

The issue of geometric miss should be addressed by design of appropriate margins (in the same way that all geometric uncertainties are addressed) and the IGRT process (understanding the relationship between the imaging surrogate and the actual target and realizing that this relationship can be influenced by the motion management technique that is chosen). Motion can also be minimized by gating, abdominal compression, or other approaches [56]. Any residual motion should be carefully evaluated and included in the treatment margins.

The issue of the interplay effect has been extensively studied, and although extremely large dose deviations are theoretically possible, such deviations are generally not found in the MLC sequences of real clinical cases. Even when the interplay effect does cause dose deviations from day to day, these deviations average out after a few fractions [26]. Notably, however, treatment planners could potentially create an extremely complicated, overmodulated plan for which the interplay effect can become important. In situations where possible interplay effects are a concern, the dosimetric errors caused by the interplay effect can be reduced by reducing the dose rate [25]. This works because the longer treatment times result in more opportunities for the effects to average out. For the same reason, the interplay effect is not expected to be a significant clinical issue with stereotactic ablative RT (where the doses are high and treatment times are long). Another planning technique shown to reduce the impact of interplay effect for VMAT plans is to use several arcs instead of a single arc [26]. To minimize any interplay effects when moving targets are treated with IMRT (or VMAT), treatment planners should take care to not overmodulate the treatment plan and to use multiple arcs. If these approaches are not possible, a reduced dose rate can be considered.

2.6 Quality Assurance Specific to IMRT

2.6.1 Commissioning and Routine Machine Quality Assurance

Rigorous commissioning of the processes for IMRT planning and delivery is absolutely essential. Nevertheless, the Radiological Physics Center, an imaging and radiation core QA facility based at MD Anderson Cancer Center, has reported that as many as 28 % of institutions failed to meet even the loose criteria of ±7 % dose accuracy or 4 mm distance to agreement in a high gradient when subjecting a head and neck phantom to IMRT [51]. This is a rather frightening statistic, given that these institutions were obtaining credentials for implementing clinical trials involving IMRT (and presumably thought their planning and delivery process was adequate to treat patients). Although some of the failures resulted from incorrect phantom setup, other reasons include incorrect output factors in the treatment planning system, incorrect CT-to-density conversion, and inadequacies in beam modeling at the leaf ends. The commissioning process involves careful measurement of any physical parameters that the treatment planning system may need (e.g., MLC transmission), evaluation of the mechanical and radiation characteristics of the delivery system (e.g., MLC leaf positioning accuracy) [58], and end-to-end tests. Many of the tolerances (for commissioning or routine QA of IMRT equipment) are different than for non-IMRT machines. For example, step-and-shoot IMRT can involve segments with few MUs; the dose per MU, as well as flatness and symmetry of the beam, should be checked throughout the range of MUs used for IMRT [40, 58].

In addition to measuring individual characteristics of different parts of the delivery process, end-to-end tests are important as well. The American Association for Physicists in Medicine created a series of tests for IMRT commissioning that are designed to represent common clinical treatments. These tests include the measurement of point dose and also dose planes assessed by using the gamma criterion of 3 %/3 mm, the most prevalent standard for acceptance testing and QA [87]. Nine centers (all of which passed the Radiological Physics Center's phantom irradiation) planned, delivered, measured, and analyzed these tests, and the findings were used to create confidence levels for use as reference by other institutions attempting the same tests. Notably, however, despite the common acceptance of 3 %/3 mm as a standard [3], other criteria can be used; moreover, the 3 %/3 mm standard may not be appropriately "tight" for commissioning or for patient-specific QA [22]. Some have proposed a DVH-based metric as the final goal [15].

2.6.2 Patient-Specific Quality Assurance

Each plan in IMRT can be highly complex, and completing patient-specific QA before a patient's treatment is begun is common practice [3, 39, 40, 46, 103, 111]. This process verifies the ability of the treatment planning system to calculate the dose accurately for this patient's plan (which can be done with a secondary dose calculation software package, as used in conventional RT) and the ability of the delivery system to accurately deliver the dose. Typically, the QA process involves comparing a dose plane delivered to a regular phantom with the dose calculated by the treatment planning system for the same geometry [87]. However, correlation can be lacking between conventional IMRT QA passing rates and actual dose errors in anatomic regions of interest [88]. For example, plans can pass planar IMRT QA but still have relatively large dose errors to some of the patient's anatomy. For example, Kruse and colleagues concluded that gamma analysis on a per-plane basis for a set of highly modulated head and neck plans was insensitive for detecting calculational errors [60]. In a separate study, McKenzie and others found that some devices were relatively insensitive for detecting failing plans [75]. That said, these QA approaches should not be discounted, as they do ensure that no large errors in dose calculation are present. Products are now available that include a calculation of the dose to the patient's anatomy (rather than just to a dose plane in a phantom), although these products are still relatively new [22, 86, 91, 126]. Details on how to commission IMRT QA equipment and processes, equipment choices, and QA criteria are available elsewhere [39, 40, 78].

2.6.3 Process Quality Assurance

As indicated previously, the delivered dose distribution is less likely to match the planned distribution if the plan is excessively complex. Modeling MLCs tends to be more important for complex plans (e.g., the tongue-and-groove effect), and the interplay effect is also more pronounced for complex plans, underscoring the desirability of avoiding these complex plans. Complexity can be quantified in a variety of ways, including average distance between opposing MLCs, but the easiest is probably to quantify the number of MUs for the total plan. Plan MUs depend on a variety of factors, including treatment planning system, IMRT approach (step-and-shoot vs. sliding window), and expectations of the clinicians (tighter constraints or many iterations during the planning process will invariably result in plans with higher MUs). In the example shown in Fig. 2.4, a single, relatively inexperienced treatment planner was responsible for all plans that involved more than 2,000 MUs. Interestingly, this planner's plans tended to have more broken-up isodose lines as well as higher MUs, indicating that optimization engine was being overworked, which probably led to constraints being weighted too strongly. About a year after these data were obtained, the expected plan MUs had dropped significantly.

2.7 Facility Design for IMRT

The number of monitor units required for IMRT plans is much higher than that needed for conformal RT plans (with the exception of VMAT), leading to a significant increase in radiation "leakage." This leakage should be accounted for in shielding calculations by using the so-called IMRT factor, the ratio of average MU per unit prescribed absorbed dose needed for IMRT and the MU per unit prescribed dose for conventional treatments [84, 85]. The IMRT factor is a function of treatment site (being lower for simple plans like breast, higher for more complex plans like head and neck), delivery mode (higher for sliding window than for step-and-shoot IMRT), and treatment planning system (e.g., efficiency of the MLC sequencer). Given the wide range in published values (from 2 to 10 or more), the shielding designer should ensure the use of appropriate values, erring on the conservative

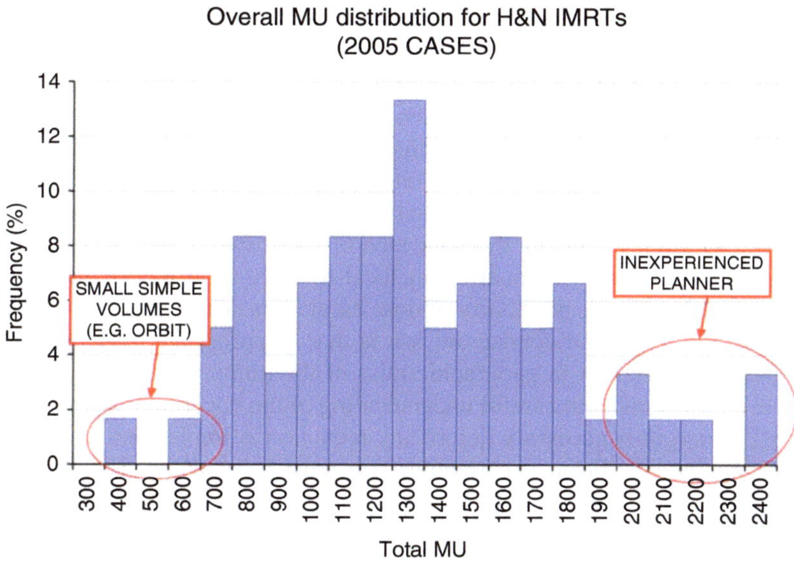

Fig. 2.4 Frequency distribution of total plan monitor units (MUs) in sliding-window IMRT for head and neck treatment at a single institution (These data were obtained in 2005, relatively early in the institution's experience with IMRT; in subsequent years, the total numbers of MUs were lower)

side. An additional consideration when designing a new treatment room is that IMRT tends to require less high-energy beams [95, 112, 119], in which case the high-energy workload may be reduced.

Tomotherapy, which is a special case of IMRT, has a narrow fan beam that is only a few centimeters long; thus, the primary barrier can be much narrower than that required by a conventional treatment unit. The IMRT factor for these units is relatively high, meaning that the secondary shielding barrier may have to be thicker than for conventional treatments [85]. The secondary shielding barrier requirements for scattered radiation are the same in tomotherapy and IMRT as for conventional treatments.

2.8 Advantages and Challenges of IMRT

The ability of IMRT to shape the dose around the target, thereby minimizing dose to adjacent normal structures, is significant, especially for individual patients. However, IMRT also has some limitations. In many cases, these limitations can be mitigated by careful clinical implementation (guided by awareness of the limitations) and future technology developments. Some of the advantages and disadvantages of IMRT are noted briefly below.

2.8.1 Higher Conformality/Margin Reduction

The high degree of conformality possible with IMRT can significantly improve normal tissue toxicity [4, 100, 106, 115, 129, 137], as discussed elsewhere throughout this volume. Combining this high degree of conformality with IGRT offers the potential for further reducing toxicity while maintaining local-regional control [64]. One group successfully reduced PTV margins for patients with head and neck cancer from 5 to 3 mm, reducing the incidence of gastrostomy tube dependence and esophageal stricture but without affecting local-regional control [19]. Image-guided IMRT may also offer the possibility of dose escalation, which can improve control of some tumors [117]. Of course, the risk is that inappropriate reduction in treatment margins will result in geographic miss of the tumor. Little clinical data are available on this issue, but careful understanding of the uncertainties accounted for in the PTV margin is necessary before any reduction is implemented. Even IGRT can still involve significant uncertainties that must be considered, particularly in target delineation [18, 77, 97, 130] and the actual extent of microscopic disease [123], but also in interfraction and intrafraction motion/deformation [121, 122].

2.8.2 Treatment Errors

Error rates in RT have been reported to be less than 1 % per fraction [42, 43, 70]. Moreover, the error rate is often reduced as new technology is introduced. For example, one group showed that the introduction of MLCs reduced error rates relative to "low-technology" machines without MLCs [72]. Error rates with IMRT have also been reported to be lower than those with three-dimensional/conventional RT [71]. One of the main reasons for the lower error rates is that IMRT usually does not involve the use of accessories (e.g., blocks, electron cones), the incorrect use of which is one of the main source of errors in conventional RT. Other reasons are that patients who require urgent treatment (with correspondingly rushed planning) tend to be treated with techniques other than IMRT. Presumably, the extensive patient-specific QA that is carried out for IMRT patients is also important in reducing error rates.

Even though the introduction of new technology tends to reduce error rates, the types of errors change when new technology is introduced. In one analysis, the most common errors with IMRT were found to be related to incorrect data entry (to the record-and-verify system) compared with conventional treatments, for which accessory and setup errors were more common [71]. As integration of the planning system, record-and-verify system, and treatment delivery system improves, the probability of such errors should be progressively reduced. Several IMRT treatment errors have had devastating consequences [50], including a well-publicized case in which a series of computer errors resulted in a patient being treated with MLCs parked in the open position instead of moving across the field to modulate the x-ray intensity. In addition to technology failures, treatment errors can also occur because of the different data needed to commission treatment planning systems for IMRT. For example, very small fields are possible in IMRT. If these are measured incorrectly (e.g., by using too large a detector), this can result in incorrect treatments [50]. Similarly, the radiation characteristics of MLCs (e.g., transmission) make a larger contribution to IMRT treatments than for conventional treatments, so incorrect entry of these parameters into the treatment planning system can result in incorrect dose calculations. Thus, as is true with any new technology, the clinical implementation of IMRT must be approached cautiously, with an understanding of the risks, full consideration of the workflow/processes, and appropriate staffing levels, training, tools, and techniques. Further guidance on safety considerations in IMRT is available elsewhere [82].

2.8.3 Unanticipated Clinical Consequences

IMRT dose distributions can be quite complex and are unusual in comparison with dose distributions from the pre-IMRT era. For example, depending on the technique or treatment site, high doses may be close to normal tissues, and large volumes of normal tissue may be exposed to low doses. Some of the dose-response data used clinically may have been based on patients who were not treated with IMRT, and care is needed when using such data to evaluate dose distributions from IMRT plans to avoid unanticipated clinical consequences. An example of such consequences was described by Allen and colleagues, who found that IMRT for mesothelioma led to an unexpectedly high rate of fatal pneumonitis [2]. This case highlighted the need for extreme care when applying DVH constraints to new clinical treatment techniques [2, 59].

2.8.4 Out-of-Field Dose and Secondary Malignancies

Patients treated with RT may be at increased risk of developing secondary malignancies caused by radiation outside of the treatment volume (i.e., out-of-field dose) [7, 14, 61]. For example, before the introduction of IMRT, Brenner and colleagues found the absolute risk of secondary malignancies caused by out-of-field dose to be 1.4 % among patients surviving more than 10 years after treatment [14]. Sources of out-of-field dose include photon leakage (proportional to MUs), radiation scattered from the collimators (also related to MUs), radiation scattered within the patient (proportional to target dose) [114], and neutrons, which are produced mostly through high-energy photons interacting with high-Z materials (e.g., tungsten or lead) [62]. Contributions of these factors depend on photon energy as well as distance from the target, with the former MU-dependent sources being most importantly distant from the field edge. Although the higher MUs required for IMRT mean that the risk of secondary malignancy is unavoidably higher, this increase can be minimized to some extent by the choice of IMRT approach (e.g., dynamic IMRT delivery vs. step and shoot) and energy (see Kry et al. [61]).

2.8.5 Concerns About Interplay Effects

The dose delivered to a tumor can vary from day to day when IMRT (whether step and shoot, dynamic, or VMAT) is used to treat moving tumors [9, 12, 29, 37, 107]. This variation results from the interplay between motion of the tumor and motion (or changing aperture shape/position) of the delivery system. As previously discussed, this effect can generally be minimized by careful planning that avoids overly complex plans. That is, for dynamic IMRT, small MLC separation and fast MLC motion should be avoided; for step-and-shoot IMRT, small segments with small MUs should be avoided; and for VMAT, more than one arc should be used. Interplay effects in complex treatment plans can be avoided by reducing the dose rate (and thus reducing the MLC speed and allowing more averaging over the respiratory cycle). In all but the most complex situations, dose deviations average out after several (<5) fractions [26]. Use of IMRT to treat moving targets is discussed in greater detail elsewhere in this volume.

2.8.6 Changes in Workflow

Another potential disadvantage of IMRT is the increased effort needed to create and check each treatment plan. Many factors contribute to the overall time needed to prepare an IMRT plan, including contouring of many more structures than for 3D conformal RT, plan optimization, and phantom planning (for patient-specific QA); the main factor is probably the time spent waiting for the treating physicians to provide target volumes [32]. The need for patient-specific IMRT QA before the first treatment can also add significantly to the effort in preparing a plan for treatment [79, 96]. Selection of the treatment planning system and other tools is important here; for example, some planning systems have better tools for manipulating structures (e.g., cleanup, processing, Boolean operations to combine anatomic structures or to exclude overlapping regions) than others [32]. Similarly, dose calculation times can vary widely between planning systems. Radiation oncology researchers and equipment manufacturers are investing considerable effort in developing segmentation, treatment planning, and general work management tools to improve the workflow of IMRT plan preparation, including reducing variations between users [36, 92, 134, 135], so these potential hurdles to the smooth integration of IMRT are being addressed.

The introduction of IMRT can also make some positive contributions to workflow. For example, forward planning can be extremely difficult. In some cases, only the best treatment planners can create conformal treatment plans that meet the clinical goals of target coverage, minimal dose hotspots, and acceptable normal tissue doses. The process of inverse planning, in which the plan is automatically created based on user-defined dose objectives, can significantly simplify this process. This reduction in planning time with IMRT has been noted by several groups [1, 73]. Also, because IMRT typically uses treatment beams from more directions than most conformal plans, the choice of beam angle is less important. Together, this means that at least in some cases, IMRT treatment planning is actually easier than planning for conventional treatments. Ongoing developments in multi-criteria optimization and automated plan optimization are further leading to greater planning efficiency and improved consistency and quality of IMRT plans. Multi-criteria optimization techniques allow users to navigate a space of multiple plans favoring individual anatomic structures to trade off competing clinical objectives; automated plan optimization automatically adjusts specified objectives and beam configuration to achieve an improved dose distribution.

2.8.7 Treatment Time

When IMRT was first introduced, the "beam-on" treatment time was significantly longer for IMRT than for treatment that involved conventional static fields. However, this time increase varies widely depending on the delivery technique [6]. Although this disadvantage still exists for some forms of IMRT, the relatively recent clinical introduction of VMAT, in which treatment is delivered with 1 or more arcs around the patient, means that IMRT treatments are much faster than the original forms [21, 31, 93, 98]. In fact, in cases in which conventional treatment would have included electron and photon fields, VMAT treatments are faster than conventional treatments.

2.9 Summary

We have described the basic principles of IMRT planning and delivery, together with the associated advantages and challenges that accompany its use in clinical settings. Subsequent chapters in this volume describe the use of IMRT for tumors at various anatomic sites in further detail.

References

1. Ahunbay EE, Chen GP, Thatcher S, Jursinic PA, White J, Albano K, Li XA (2007) Direct aperture optimization-based intensity-modulated radiotherapy for whole breast irradiation. Int J Radiat Oncol Biol Phys 67(4):1248–1258
2. Allen AM, Czerminska M, Jänne PA, Sugarbaker DJ, Bueno R, Harris JR, Court L, Baldini EH (2006) Fatal pneumonitis associated with intensity-modulated radiation therapy for mesothelioma. Int J Radiat Oncol Biol Phys 65(3):640–645
3. Basran PS, Woo MK (2008) An analysis of tolerance levels in IMRT quality assurance procedures. Med Phys 35(6):2300–2307
4. Beadle BM, Liao K-P, Elting LS, Buchholz TA, Ang KK, Garden AS, Guadagnolo BA (2014) Improved survival using intensity-modulated radiation therapy in head and neck cancers: a SEER-Medicare analysis. Cancer 120(5):702–710
5. Bedford JL (2009) Treatment planning for volumetric modulated arc therapy. Med Phys 36(11):5128–5138
6. Bohsung J, Gillis S, Arrans R, Bakai A, De Wagter C, Knöös T, Mijnheer BJ, Paiusco M, Perrin BA, Welleweerd H, Williams P (2005) IMRT treatment planning – a comparative

inter-system and inter-centre planning exercise of the ESTRO QUASIMODO group. Radiother Oncol 76(3):354–361
7. Boice JD Jr, Day NE, Andersen A, Brinton LA, Brown R, Choi NW, Clarke EA, Coleman MP, Curtis RE, Flannery JT (1985) Second cancers following radiation treatment for cervical cancer. An international collaboration among cancer registries. J Natl Cancer Inst 74(5):955–975
8. Bortfeld T, Boyer AL, Schlegel W, Kahler DL, Waldron TJ (1994) Realization and verification of three-dimensional conformal radiotherapy with modulated fields. Int J Radiat Oncol Biol Phys 30(4):899–908
9. Bortfeld T, Jiang SB, Rietzel E (2004) Effects of motion on the total dose distribution. Semin Radiat Oncol 14(1):41–51
10. Bortfeld T, Jokivarsi K, Goitein M, Kung J, Jiang SB (2002) Effects of intra-fraction motion on IMRT dose delivery: statistical analysis and simulation. Phys Med Biol 47(13):2203–2220
11. Bortfeld T, Schlegel W (1993) Optimization of beam orientations in radiation therapy: some theoretical considerations. Phys Med Biol 38(2):291–304
12. Boyer AL, Biggs P, Galvin J, Klein EE, LoSasso T, Low D, Mah K, Yu C (2001) Basic applications of multileaf collimators. American Association of Physicists in Medicine, Madison
13. Boyer AL, Yu CX (1999) Intensity-modulated radiation therapy with dynamic multileaf collimators. Semin Radiat Oncol 9(1):48–59
14. Brenner DJ, Curtis RE, Hall EJ, Ron E (2000) Second malignancies in prostate carcinoma patients after radiotherapy compared with surgery. Cancer 88(2):398–406
15. Bresciani S, Di Dia A, Maggio A, Cutaia C, Miranti A, Infusino E, Stasi M (2013) Tomotherapy treatment plan quality assurance: the impact of applied criteria on passing rate in gamma index method. Med Phys 40(12):121711
16. Bzdusek K, Friberger H, Eriksson K, Hårdemark B, Robinson D, Kaus M (2009) Development and evaluation of an efficient approach to volumetric arc therapy planning. Med Phys 36(6):2328–2339
17. Chan OSH, Lee MCH, Hung AWM, Chang ATY, Yeung RMW, Lee AWM (2011) The superiority of hybrid-volumetric arc therapy (VMAT) technique over double arcs VMAT and 3D-conformal technique in the treatment of locally advanced non-small cell lung cancer – a planning study. Radiother Oncol 101(2):298–302
18. Chao KSC, Bhide S, Chen H, Asper J, Bush S, Franklin G, Kavadi V, Liengswangwong V, Gordon W, Raben A, Strasser J, Koprowski C, Frank S, Chronowski G, Ahamad A, Malyapa R, Zhang L, Dong L (2007) Reduce in variation and improve efficiency of target volume delineation by a computer-assisted system using a deformable image registration approach. Int J Radiat Oncol Biol Phys 68(5):1512–1521
19. Chen BAM, Yu Y, Daly ME, Farwell DG, Benedict S, Purdy JA (2013) Long-term experience with reduced planning target volume margins for patients treated by intensity-modulated radiotherapy with daily image-guidance for head and neck cancer. Head Neck 36(12):1766–1772. doi:10.1002/hed.23532
20. Chui CS, Spirou SV (2001) Inverse planning algorithms for external beam radiation therapy. Med Dosim 26(2):189–197
21. Clivio A, Fogliata A, Franzetti-Pellanda A, Nicolini G, Vanetti E, Wyttenbach R, Cozzi L (2009) Volumetric-modulated arc radiotherapy for carcinomas of the anal canal: a treatment planning comparison with fixed field IMRT. Radiother Oncol 92(1):118–124
22. Coleman L, Skourou C (2013) Sensitivity of volumetric modulated arc therapy patient specific QA results to multileaf collimator errors and correlation to dose volume histogram based metrics. Med Phys 40(11):111715
23. Court L, Urribarri J, Makrigiorgos M (2010) Carbon fiber couches and skin sparing. J Appl Clin Med Phys 11(2):3241
24. Court L, Wagar M, Berbeco R, Reisner A, Winey B, Schofield D, Ionascu D, Allen AM, Popple R, Lingos T (2010) Evaluation of the interplay effect when using RapidArc to treat targets moving in the craniocaudal or right-left direction. Med Phys 37(1):4–11

25. Court L, Wagar M, Bogdanov M, Ionascu D, Schofield D, Allen A, Berbeco R, Lingos T (2011) Use of reduced dose rate when treating moving tumors using dynamic IMRT. J Appl Clin Med Phys 12(1):3276
26. Court LE, Seco J, Lu X-Q, Ebe K, Mayo C, Ionascu D, Winey B, Giakoumakis N, Aristophanous M, Berbeco R, Rottman J, Bogdanov M, Schofield D, Lingos T (2010) Use of a realistic breathing lung phantom to evaluate dose delivery errors. Med Phys 37(11):5850–5857
27. Court LE, Tishler R, Xiang H, Allen AM, Makrigiorgos M, Chin L (2008) Experimental evaluation of the accuracy of skin dose calculation for a commercial treatment planning system. J Appl Clin Med Phys 9(1):2792
28. Court LE, Tishler RB (2007) Experimental evaluation of the impact of different head-and-neck intensity-modulated radiation therapy planning techniques on doses to the skin and shallow targets. Int J Radiat Oncol Biol Phys 69(2):607–613
29. Court LE, Wagar M, Ionascu D, Berbeco R, Chin L (2008) Management of the interplay effect when using dynamic MLC sequences to treat moving targets. Med Phys 35(5):1926–1931
30. Court LE, Wolfsberger L, Allen AM, James S, Tishler RB (2008) Clinical experience of the importance of daily portal imaging for head and neck IMRT treatments. J Appl Clin Med Phys 9(3):2756
31. Cozzi L, Dinshaw KA, Shrivastava SK, Mahantshetty U, Engineer R, Deshpande DD, Jamema SV, Vanetti E, Clivio A, Nicolini G, Fogliata A (2008) A treatment planning study comparing volumetric arc modulation with RapidArc and fixed field IMRT for cervix uteri radiotherapy. Radiother Oncol 89(2):180–191
32. Das IJ, Moskvin V, Johnstone PA (2009) Analysis of treatment planning time among systems and planners for intensity-modulated radiation therapy. J Am Coll Radiol 6(7):514–517
33. De Gersem W, Claus F, De Wagter C, Van Duyse B, De Neve W (2001) Leaf position optimization for step-and-shoot IMRT. Int J Radiat Oncol Biol Phys 51(5):1371–1388
34. de Greef M, Crezee J, van Eijk JC, Pool R, Bel A (2009) Accelerated ray tracing for radiotherapy dose calculations on a GPU. Med Phys 36(9):4095–4102
35. Deasy JO (1997) Multiple local minima in radiotherapy optimization problems with dose-volume constraints. Med Phys 24(7):1157–1161. decimal. http://www.dotdecimal.com/. Accessed 10 Feb 2014
36. Deeley MA, Chen A, Datteri RD, Noble J, Cmelak A, Donnelly E, Malcolm A, Moretti L, Jaboin J, Niermann K, Yang ES, Yu DS, Dawant BM (2013) Segmentation editing improves efficiency while reducing inter-expert variation and maintaining accuracy for normal brain tissues in the presence of space-occupying lesions. Phys Med Biol 58(12):4071–4097
37. Duan J, Shen S, Fiveash JB, Popple RA, Brezovich IA (2006) Dosimetric and radiobiological impact of dose fractionation on respiratory motion induced IMRT delivery errors: a volumetric dose measurement study. Med Phys 33(5):1380–1387
38. Earl MA, Afghan MKN, Yu CX, Jiang Z, Shepard DM (2007) Jaws-only IMRT using direct aperture optimization. Med Phys 34(1):307–314
39. Ezzell GA, Burmeister JW, Dogan N, LoSasso TJ, Mechalakos JG, Mihailidis D, Molineu A, Palta JR, Ramsey CR, Salter BJ, Shi J, Xia P, Yue NJ, Xiao Y (2009) IMRT commissioning: multiple institution planning and dosimetry comparisons, a report from AAPM Task Group 119. Med Phys 36(11):5359–5373
40. Ezzell GA, Galvin JM, Low D, Palta JR, Rosen I, Sharpe MB, Xia P, Xiao Y, Xing L, Yu CX (2003) Guidance document on delivery, treatment planning, and clinical implementation of IMRT: report of the IMRT Subcommittee of the AAPM Radiation Therapy Committee. Med Phys 30(8):2089–2115
41. Fenwick JD, Tome WA, Soisson ET, Mehta MP, Rock Mackie T (2006) Tomotherapy and other innovative IMRT delivery systems. Semin Radiat Oncol 16(4):199–208
42. Ford EC, Gaudette R, Myers L, Vanderver B, Engineer L, Zellars R, Song DY, Wong J, Deweese TL (2009) Evaluation of safety in a radiation oncology setting using failure mode and effects analysis. Int J Radiat Oncol Biol Phys 74(3):852–858

43. Fraass BA, Lash KL, Matrone GM, Volkman SK, McShan DL, Kessler ML, Lichter AS (1998) The impact of treatment complexity and computer-control delivery technology on treatment delivery errors. Int J Radiat Oncol Biol Phys 42(3):651–659
44. Hadley SW, Kelly R, Lam K (2005) Effects of immobilization mask material on surface dose. J Appl Clin Med Phys 6(1):1–7
45. Hissoiny S, Ozell B, Despres P (2010) A convolution-superposition dose calculation engine for GPUs. Med Phys 37(3):1029–1037
46. Howell RM, Smith IP, Jarrio CS (2008) Establishing action levels for EPID-based QA for IMRT. J Appl Clin Med Phys 9(3):2721
47. IAEA (2003) A silent crisis: cancer treatment in developing countries. IAEA
48. IAEA (2013) AGaRT: The Advisory Group on increasing access to Radiotherapy Technology in low and middle income countries. IAEA, Vienna
49. IAEA (2014) IAEA: programme of action for cancer therapy. http://cancer.iaea.org/agart.asp. Accessed 17 Feb 14
50. IAEA (2014) Prevention of accidental exposure in radiation therapy. https://rpop.iaea.org/RPOP/RPoP/Content/AdditionalResources/Training/1_TrainingMaterial/AccidentPreventionRadiotherapy.htm. Accessed 31 Mar 2014
51. Ibbott GS, Followill DS, Molineu HA, Lowenstein JR, Alvarez PE, Roll JE (2008) Challenges in credentialing institutions and participants in advanced technology multi-institutional clinical trials. Int J Radiat Oncol Biol Phys 71(1 Suppl):S71–S75
52. ICRU Report 83: prescribing, recording, and reporting photon-beam intensity-modulated radiation therapy (IMRT) (2010). J ICRU 10(1):NP. doi:10.1093/jicru/ndq002
53. International Commission on Radiation Units and Measurements (1993) Prescribing, recording, and reporting photon beam therapy, vol 50, ICRU report. International Commission on Radiation Units and Measurements, Bethesda
54. International Commission on Radiation Units and Measurements (1999) Prescribing, recording, and reporting photon beam therapy, vol 62, ICRU report. International Commission on Radiation Units and Measurements, Bethesda
55. Jacques R, Taylor R, Wong J, McNutt T (2010) Towards real-time radiation therapy: GPU accelerated superposition/convolution. Comput Methods Prog Biomed 98(3):285–292
56. Keall PJ, Mageras GS, Balter JM, Emery RS, Forster KM, Jiang SB, Kapatoes JM, Low DA, Murphy MJ, Murray BR, Ramsey CR, Herk MBV, Vedam SS, Wong JW, Yorke E (2006) The management of respiratory motion in radiation oncology report of AAPM Task Group 76. Med Phys 33(10):3874–3900
57. Kitamura K, Court LE, Dong L (2003) Comparison of imaging modalities for image-guided radiation therapy (IGRT). Nihon Igaku Hoshasen Gakkai Zasshi 63(9):574–578
58. Klein EE, Hanley J, Bayouth J, Yin F-F, Simon W, Dresser S, Serago C, Aguirre F, Ma L, Arjomandy B, Liu C, Sandin C, Holmes T (2009) Task Group 142 report: quality assurance of medical accelerators. Med Phys 36(9):4197–4212
59. Komaki R, Liao Z, Liu H, Tucker S, Rice D (2006) Fatal pneumonitis associated with intensity-modulated radiation therapy for mesothelioma. In regard to Allen et al (Int J Radiat Oncol Biol Phys 2006. 65:640–645). Int J Radiat Oncol Biol Phys 66(5):1595–1596
60. Kruse JJ (2010) On the insensitivity of single field planar dosimetry to IMRT inaccuracies. Med Phys 37(6):2516–2524
61. Kry SF, Salehpour M, Followill DS, Stovall M, Kuban DA, White RA, Rosen II (2005) The calculated risk of fatal secondary malignancies from intensity-modulated radiation therapy. Int J Radiat Oncol Biol Phys 62(4):1195–1203
62. Kry SF, Salehpour M, Followill DS, Stovall M, Kuban DA, White RA, Rosen II (2005) Out-of-field photon and neutron dose equivalents from step-and-shoot intensity-modulated radiation therapy. Int J Radiat Oncol Biol Phys 62(4):1204–1216
63. Lee N, Chuang C, Quivey JM, Phillips TL, Akazawa P, Verhey LJ, Xia P (2002) Skin toxicity due to intensity-modulated radiotherapy for head-and-neck carcinoma. Int J Radiat Oncol Biol Phys 53(3):630–637

64. Liao ZX, Komaki RR, Thames HD Jr, Liu HH, Tucker SL, Mohan R, Martel MK, Wei X, Yang K, Kim ES, Blumenschein G, Hong WK, Cox JD (2010) Influence of technologic advances on outcomes in patients with unresectable, locally advanced non-small-cell lung cancer receiving concomitant chemoradiotherapy. Int J Radiat Oncol Biol Phys 76(3):775–781
65. Likhacheva A, Palmer M, Du W, Brown PD, Mahajan A (2012) Intensity modulated radiation therapy class solutions in Philips Pinnacle treatment planning for central nervous system malignancies: standardized, efficient, and effective. Pract Radiat Oncol 2(4):e145–e153
66. Liu HH, Jauregui M, Zhang X, Wang X, Dong L, Mohan R (2006) Beam angle optimization and reduction for intensity-modulated radiation therapy of non-small-cell lung cancers. Int J Radiat Oncol Biol Phys 65(2):561–572
67. Llacer J, Deasy JO, Portfeld TR, Solberg TD, Promberger C (2003) Absence of multiple local minima effects in intensity modulated optimization with dose-volume constraints. Phys Med Biol 48(2):183–210
68. Lorenz F, Nalichowski A, Rosca F, Killoran J, Wenz F, Zygmanski P (2008) An independent dose calculation algorithm for MLC-based radiotherapy including the spatial dependence of MLC transmission. Phys Med Biol 53(3):557–573
69. Lorenz F, Nalichowski A, Rosca F, Kung J, Wenz F, Zygmanski P (2007) Spatial dependence of MLC transmission in IMRT delivery. Phys Med Biol 52(19):5985–5999
70. Macklis RM, Meier T, Weinhous MS (1998) Error rates in clinical radiotherapy. J Clin Oncol 16(2):551–556
71. Margalit DN, Chen Y-H, Catalano PJ, Heckman K, Vivenzio T, Nissen K, Wolfsberger LD, Cormack RA, Mauch P, Ng AK (2011) Technological advancements and error rates in radiation therapy delivery. Int J Radiat Oncol Biol Phys 81(4):e673–e679
72. Marks LB, Light KL, Hubbs JL, Georgas DL, Jones EL, Wright MC, Willett CG, Yin FF (2007) The impact of advanced technologies on treatment deviations in radiation treatment delivery. Int J Radiat Oncol Biol Phys 69(5):1579–1586
73. Mayo CS, Urie MM, Fitzgerald TJ (2005) Hybrid IMRT plans – concurrently treating conventional and IMRT beams for improved breast irradiation and reduced planning time. Int J Radiat Oncol Biol Phys 61(3):922–932
74. Mayo CS, Urie MM, Fitzgerald TJ, Ding L, Lo YC, Bogdanov M (2008) Hybrid IMRT for treatment of cancers of the lung and esophagus. Int J Radiat Oncol Biol Phys 71(5):1408–1418
75. McKenzie E, Balter P, Jones J, Followill D, Stingo F, Pulliam K, Kry S (2013) SU-E-T-158: evaluation of the sensitivities of patient specific IMRT QA dosimeters. Med Phys 40(6):240
76. Men C, Gu X, Choi D, Majumdar A, Zheng Z, Mueller K, Jiang SB (2009) GPU-based ultrafast IMRT plan optimization. Phys Med Biol 54(21):6565–6573
77. Michalski JM, Lawton C, El Naqa I, Ritter M, O'Meara E, Seider MJ, Lee WR, Rosenthal SA, Pisansky T, Catton C, Valicenti RK, Zietman AL, Bosch WR, Sandler H, Buyyounouski MK, Ménard C (2010) Development of RTOG consensus guidelines for the definition of the clinical target volume for postoperative conformal radiation therapy for prostate cancer. Int J Radiat Oncol Biol Phys 76(2):361–368
78. Mijnheer B (2006) Guidelines for the verification of IMRT: the ESTRO QUASIMODO project. Radiother Oncol 81:S173–S173
79. Miles EA, Clark CH, Urbano MTG, Bidmead M, Dearnaley DP, Harrington KJ, A'Hern R, Nutting CM (2005) The impact of introducing intensity modulated radiotherapy into routine clinical practice. Radiother Oncol 77(3):241–246
80. Mohan R, Arnfield M, Tong S, Wu Q, Siebers J (2000) The impact of fluctuations in intensity patterns on the number of monitor units and the quality and accuracy of intensity modulated radiotherapy. Med Phys 27(6):1226–1237
81. Mohan R, Wu Q, Wang X, Stein J (1996) Intensity modulation optimization, lateral transport of radiation, and margins. Med Phys 23(12):2011–2021

82. Moran JM, Dempsey M, Eisbruch A, Fraass BA, Galvin JM, Ibbott GS, Marks LB (2011) Safety considerations for IMRT: executive summary. Med Phys 38(9):5067–5072
83. Murphy MJ (2012) Kilovoltage radiography for robotic linac IGRT. In: Bourland JD (ed) Image-guided radiation therapy, Imaging in medical diagnosis and therapy. CRC Press, Boca Raton, pp 147–155
84. Mutic S, Low DA, Klein EE, Dempsey JF, Purdy JA (2001) Room shielding for intensity-modulated radiation therapy treatment facilities. Int J Radiat Oncol Biol Phys 50(1):239–246
85. NCRP (2005) NCRP report 151: structural shielding design and evaluation for megavoltage X- and gamma-ray radiotherapy facilities. National Council on Radiation Protection and Measurements, Bethesda
86. Nelms BE, Opp D, Robinson J, Wolf TK, Zhang G, Moros E, Feygelman V (2012) VMAT QA: measurement-guided 4D dose reconstruction on a patient. Med Phys 39(7):4228–4238
87. Nelms BE, Simon JA (2007) A survey on planar IMRT QA analysis. J Appl Clin Med Phys 8(3):2448
88. Nelms BE, Zhen H, Tome WA (2011) Per-beam, planar IMRT QA passing rates do not predict clinically relevant patient dose errors. Med Phys 38(2):1037–1044
89. Niemierko A (1997) Reporting and analyzing dose distributions: a concept of equivalent uniform dose. Med Phys 24(1):103–110
90. Oguchi H, Obata Y (2009) Commissioning of modulator-based IMRT with XiO treatment planning system. Med Phys 36(1):261–269
91. Olch AJ (2012) Evaluation of the accuracy of 3DVH software estimates of dose to virtual ion chamber and film in composite IMRT QA. Med Phys 39(1):81–86
92. Olsen LA, Robinson CG, He GR, Wooten HO, Yaddanapudi S, Mutic S, Yang D, Moore KL (2014) Automated radiation therapy treatment plan workflow using a commercial application programming interface. Pract Radiat Oncol 4(6):358–367
93. Otto K (2008) Volumetric modulated arc therapy: IMRT in a single gantry arc. Med Phys 35(1):310–317
94. Papp D, Unkelbach J (2014) Direct leaf trajectory optimization for volumetric modulated arc therapy planning with sliding window delivery. Med Phys 41(1):011701
95. Pasler M, Georg D, Wirtz H, Lutterbach J (2011) Effect of photon-beam energy on VMAT and IMRT treatment plan quality and dosimetric accuracy for advanced prostate cancer. Strahlenther Onkol 187(12):792–798
96. Pawlicki T, Yoo S, Court LE, McMillan SK, Rice RK, Russell JD, Pacyniak JM, Woo MK, Basran PS, Shoales J, Boyer AL (2008) Moving from IMRT QA measurements toward independent computer calculations using control charts. Radiother Oncol 89(3):330–337
97. Persson GF, Nygaard DE, Hollensen C, Munck af Rosenschold P, Mouritsen LS, Due AK, Berthelsen AK, Nyman J, Markova E, Roed AP, Roed H, Korreman S, Specht L (2012) Interobserver delineation variation in lung tumour stereotactic body radiotherapy. Br J Radiol 85(1017):e654–e660
98. Popescu CC, Olivotto IA, Beckham WA, Ansbacher W, Zavgorodni S, Shaffer R, Wai ES, Otto K (2010) Volumetric modulated arc therapy improves dosimetry and reduces treatment time compared to conventional intensity-modulated radiotherapy for locoregional radiotherapy of left-sided breast cancer and internal mammary nodes. Int J Radiat Oncol Biol Phys 76(1):287–295
99. Popple RA, Fiveash JB, Brezovich IA (2007) Effect of beam number on organ-at-risk sparing in dynamic multileaf collimator delivery of intensity modulated radiation therapy. Med Phys 34(10):3752–3759
100. Portelance L, Chao KSC, Grigsby PW, Bennet H, Low D (2001) Intensity-modulated radiation therapy (IMRT) reduces small bowel, rectum, and bladder doses in patients with cervical cancer receiving pelvic and para-aortic irradiation. Int J Radiat Oncol Biol Phys 51(1):261–266
101. Pratx G, Xing L (2011) GPU computing in medical physics: a review. Med Phys 38(5):2685–2697

102. Raaymakers BW, Lagendijk JJ, Overweg J, Kok JG, Raaijmakers AJ, Kerkhof EM, van der Put RW, Meijsing I, Crijns SP, Benedosso F, van Vulpen M, de Graaff CH, Allen J, Brown KJ (2009) Integrating a 1.5 T MRI scanner with a 6 MV accelerator: proof of concept. Phys Med Biol 54(12):N229–N237
103. Ramsey C, Dube S, Hendee WR (2003) It is necessary to validate each individual IMRT treatment plan before delivery. Med Phys 30(9):2271–2273
104. Rao M, Yang W, Chen F, Sheng K, Ye J, Mehta V, Shepard D, Cao D (2010) Comparison of Elekta VMAT with helical tomotherapy and fixed field IMRT: plan quality, delivery efficiency and accuracy. Med Phys 37(3):1350–1359
105. Salari E, Men C, Romeijn HE (2011) Accounting for the tongue-and-groove effect using a robust direct aperture optimization approach. Med Phys 38(3):1266–1279
106. Samuelian JM, Callister MD, Ashman JB, Young-Fadok TM, Borad MJ, Gunderson LL (2012) Reduced acute bowel toxicity in patients treated with intensity-modulated radiotherapy for rectal cancer. Int J Radiat Oncol Biol Phys 82(5):1981–1987
107. Seco J, Sharp GC, Turcotte J, Gierga D, Bortfeld T, Paganetti H (2007) Effects of organ motion on IMRT treatments with segments of few monitor units. Med Phys 34(3):923–934
108. Shepard DM, Earl MA, Li XA, Naqvi S, Yu C (2002) Direct aperture optimization: a turnkey solution for step-and-shoot IMRT. Med Phys 29(6):1007–1018
109. Shirato H, Ishikawa M, Shimizu S, Bengua G, Sutherland K, Onimaru R, Aoyama H (2012) Kilovoltage X-ray IMRT and IGRT. In: Bourland JD (ed) Image-guided radiation therapy, Imaging in medical diagnosis and therapy. CRC Press, Boca Raton, pp 131–146
110. Siebers JV, Tong S, Lauterbach M, Wu Q, Mohan R (2001) Acceleration of dose calculations for intensity-modulated radiotherapy. Med Phys 28(6):903–910
111. Smith JC, Dieterich S, Orton CG (2011) It is STILL necessary to validate each individual IMRT treatment plan with dosimetric measurements before delivery. Med Phys 38(2):553–555
112. Solaiappan G, Singaravelu G, Prakasarao A, Rabbani B, Supe SS (2009) Influence of photon beam energy on IMRT plan quality for radiotherapy of prostate cancer. Rep Pract Oncol Radiother 14(1):18–31
113. Stein J, Bortfeld T, Dorschel B, Schlegel W (1994) Dynamic X-ray compensation for conformal radiotherapy by means of multi-leaf collimation. Radiother Oncol 32(2):163–173
114. Stovall M, Blackwell CR, Cundiff J, Novack DH, Palta JR, Wagner LK, Webster EW, Shalek RJ (1995) Fetal dose from radiotherapy with photon beams: report of AAPM Radiation Therapy Committee Task Group No. 36. Med Phys 22(1):63–82
115. Sulman EP, Schwartz DL, Le TT, Ang KK, Morrison WH, Rosenthal DI, Ahamad A, Kies M, Glisson B, Weber R, Garden AS (2009) IMRT reirradiation of head and neck cancer – disease control and morbidity outcomes. Int J Radiat Oncol Biol Phys 73(2):399–409
116. Sykes JR, Williams PC (1998) An experimental investigation of the tongue and groove effect for the Philips multileaf collimator. Phys Med Biol 43(10):3157–3165
117. Takeda K, Takai Y, Narazaki K, Mitsuya M, Umezawa R, Kadoya N, Fujita Y, Sugawara T, Kubozono M, Shimizu E, Abe K, Shirata Y, Ishikawa Y, Yamamoto T, Kozumi M, Dobashi S, Matsushita H, Chida K, Ishidoya S, Arai Y, Jingu K, Yamada S (2012) Treatment outcome of high-dose image-guided intensity-modulated radiotherapy using intra-prostate fiducial markers for localized prostate cancer at a single institute in Japan. Radiat Oncol (Lond) 7:105
118. Thomas SJ, Hoole AC (2004) The effect of optimization on surface dose in intensity modulated radiotherapy (IMRT). Phys Med Biol 49(21):4919–4928
119. Tyagi A, Supe SS, Sandeep S, Singh MP (2010) A dosimetric analysis of 6 MV versus 15 MV photon energy plans for intensity modulated radiation therapy (IMRT) of carcinoma of cervix. Rep Pract Oncol Radiother 15(5):125–131
120. van der Est H, Prins P, Heijmen BJM, Dirkx MLP (2012) Intensity modulated radiation therapy planning for patients with a metal hip prosthesis based on class solutions. Pract Radiat Oncol 2(1):35–40
121. van der Wielen GJ, Mutanga TF, Incrocci L, Kirkels WJ, Vasquez Osorio EM, Hoogeman MS, Heijmen BJM, de Boer HCJ (2008) Deformation of prostate and seminal vesicles relative to intraprostatic fiducial markers. Int J Radiat Oncol Biol Phys 72(5):1604–1611.e1603

122. van Kranen S, van Beek S, Rasch C, van Herk M, Sonke J-J (2009) Setup uncertainties of anatomical sub-regions in head-and-neck cancer patients after offline CBCT guidance. Int J Radiat Oncol Biol Phys 73(5):1566–1573
123. van Loon J, Siedschlag C, Stroom J, Blauwgeers H, van Suylen R-J, Knegjens J, Rossi M, van Baardwijk A, Boersma L, Klomp H, Vogel W, Burgers S, Gilhuijs K (2012) Microscopic disease extension in three dimensions for non-small-cell lung cancer: development of a prediction model using pathology-validated positron emission tomography and computed tomography features. Int J Radiat Oncol Biol Phys 82(1):448–456
124. Wang X, Spirou S, LoSasso T, Stein J, Chui CS, Mohan B (1996) Dosimetric verification of intensity-modulated fields. Med Phys 23(3):317–327
125. Wang X, Zhang X, Dong L, Liu H, Wu Q, Mohan R (2004) Development of methods for beam angle optimization for IMRT using an accelerated exhaustive search strategy. Int J Radiat Oncol Biol Phys 60(4):1325–1337
126. Watanabe Y, Nakaguchi Y (2013) 3D evaluation of 3DVH program using BANG3 polymer gel dosimeter. Med Phys 40(8):082101
127. Webb S (1994) Optimizing the planning of intensity-modulated radiotherapy. Phys Med Biol 39(12):2229–2246
128. Weksberg DC, Palmer MB, Vu KN, Rebueno NC, Sharp HJ, Luo D, Yang JN, Shiu AS, Rhines LD, McAleer MF, Brown PD, Chang EL (2012) Generalizable class solutions for treatment planning of spinal stereotactic body radiation therapy. Int J Radiat Oncol Biol Phys 84(3):847–853
129. Welsh J, Gomez D, Palmer MB, Riley BA, Mayankkumar AV, Komaki R, Dong L, Zhu XR, Likhacheva A, Liao Z, Hofstetter WL, Ajani JA, Cox JD (2011) Intensity-modulated proton therapy further reduces normal tissue exposure during definitive therapy for locally advanced distal esophageal tumors: a dosimetric study. Int J Radiat Oncol Biol Phys 81(5):1336–1342
130. White EA, Brock KK, Jaffray DA, Catton CN (2009) Inter-observer variability of prostate delineation on cone beam computerised tomography images. Clin Oncol 21(1):32–38
131. Wu Q, Mohan R (2002) Multiple local minima in IMRT optimization based on dose-volume criteria. Med Phys 29(7):1514–1527
132. Wu Q, Mohan R, Niemierko A, Schmidt-Ullrich R (2002) Optimization of intensity-modulated radiotherapy plans based on the equivalent uniform dose. Int J Radiat Oncol Biol Phys 52(1):224–235
133. Xia P, Verhey LJ (2001) Delivery systems of intensity-modulated radiotherapy using conventional multileaf collimators. Med Dosim 26(2):169–177
134. Yang J, Amini A, Williamson R, Zhang L, Zhang Y, Komaki R, Liao Z, Cox J, Welsh J, Court L, Dong L (2013) Automatic contouring of brachial plexus using a multi-atlas approach for lung cancer radiation therapy. Pract Radiat Oncol 3(4):e139–e147
135. Yang J, Beadle BM, Garden AS, Gunn B, Rosenthal D, Ang K, Frank S, Williamson R, Balter P, Court L, Dong L (2014) Auto-segmentation of low-risk clinical target volume for head and neck radiation therapy. Pract Radiat Oncol 4(1):e31–e37
136. Yu CX, Jaffray DA, Wong JW (1998) The effects of intra-fraction organ motion on the delivery of dynamic intensity modulation. Phys Med Biol 43(1):91–104
137. Zelefsky MJ, Fuks Z, Hunt M, Yamada Y, Marion C, Ling CC, Amols H, Venkatraman ES, Leibel SA (2002) High-dose intensity modulated radiation therapy for prostate cancer: early toxicity and biochemical outcome in 772 patients. Int J Radiat Oncol Biol Phys 53(5):1111–1116
138. Zhang X, Liu H, Wang X, Dong L, Wu Q, Mohan R (2004) Speed and convergence properties of gradient algorithms for optimization of IMRT. Med Phys 31(5):1141–1152

Radiobiology for IMRT

3

Yuta Shibamoto, Chikao Sugie, Hiroyuki Ogino, and Natsuo Tomita

Keywords
Intermittent radiation • Sublethal damage repair • Reoxygenation • Linear–quadratic model • Biologically effective dose

3.1 Introduction

Intensity-modulated radiation therapy (IMRT) has established its role in the definitive treatment of various cancers. Owing to the excellent dose distribution in the target volume and sparing of normal tissues, higher doses than those can be delivered with the conventional technique can be administered safely, and clinical data showing the advantage of IMRT are accumulating; our own experiences with prostate cancer have been published recently [1, 2]. While advantages of IMRT are evident upon physical grounds, a few radiobiological issues remain unresolved regarding the evaluation of radiation doses employed in this new treatment modality. One of the issues is regarding the prolonged beam delivery time. In conventional radiotherapy, photon beams from one portal are irradiated continuously, and radiation delivery time in daily treatment is usually within 1–2 min. In contrast, IMRT takes a much longer time, ranging from 3 to 20 min or even longer for one treatment session. In IMRT, segments in target volumes receive intermittent irradiation even

Y. Shibamoto (✉) • C. Sugie
Department of Radiology, Nagoya City University Graduate School of Medical Sciences, 1 Kawasumi, Mizuho-cho, Mizuho-ku, Nagoya 467-8601, Japan
e-mail: yshiba@med.nagoya-cu.ac.jp

H. Ogino
Nagoya Proton Therapy Center, Nagoya City West Medical Center, Nagoya 462-8505, Japan

N. Tomita
Department of Radiation Oncology, Aichi Cancer Center, Nagoya 464-8681, Japan

Y. Nishimura, R. Komaki (eds.), *Intensity-Modulated Radiation Therapy*,
DOI 10.1007/978-4-431-55486-8_3

during one fixed-portal beam delivery, so the situation is quite complicated. From a radiobiologic point of view, it is questioned whether the radiation dose delivered with such intermissions is equivalent to that administered without breaks, since it is well known that sublethal damage repair (SLDR) occurs when intervals are set between two radiation doses [3, 4]. To date, several studies have been conducted to address this issue, and we review the results and summarize our previous studies on this issue in the first part of this article.

Another issue is regarding the evaluation of different fractionation schedules and conversion of radiation doses using mathematical models. With the development of high-precision radiotherapy, various fractionation schedules become available, because irradiation to the normal tissues can be minimized. For IMRT, while most institutions use a conventional daily dose of 2 Gy or a slightly higher daily dose, some investigators use much higher daily doses up to 8 Gy with a much fewer fraction number [5]. Trends toward the use of a high dose per fraction are more evident in intensity-modulated stereotactic radiotherapy [6, 7]. To evaluate the treatment outcome, comparison among different fractionation schedules is necessary. For this purpose, many clinicians use the linear–quadratic (LQ) formalism. However, it has been questioned whether the LQ model is really applicable to high-dose-per-fraction treatment [8, 9]. Therefore, evaluation of the reliability of LQ formalism in the high-dose range and, if inadequate, proposal of the method to correct the error or alternative models are important issues in clinical radiation biology. In the latter part of this article, we review recent works on this issue, including our own studies. The main part of this article has been published elsewhere [10], which is reproduced after updating, with permission from the publisher.

3.2 Biological Effects of Intermittent Radiation Delivery

3.2.1 SLDR During Intermittent Radiation Exposure in Cultured Cells

First, we show the results of our own experiments. We have conducted four series of laboratory studies regarding the biological effects of intermittent irradiation [10]. In the first study, the effects of fractionated doses delivered at intervals of a few minutes were evaluated in EMT6 mouse mammary sarcoma and SCCVII mouse squamous cell carcinoma cells [11]. These two cell lines were employed throughout the series of experiments, and their characteristics were described in detail previously [12, 13]. In experiments where 8 Gy was given in 2 fractions, SLDR was observed when the interval was 2 min or longer in EMT6 cells and when it was 3 min or longer in SCCVII cells. In the next experiment where 8 Gy was given in 5 fractions at intervals of 1–5 min, significant SLDR was observed when the interval was 2 min or longer in both cell lines (Fig. 3.1). When the interval was 5 min, 8 Gy in 5 fractions corresponded to 7.38 Gy in a single fraction in EMT6 cells and 7.29 Gy in SCCVII cells.

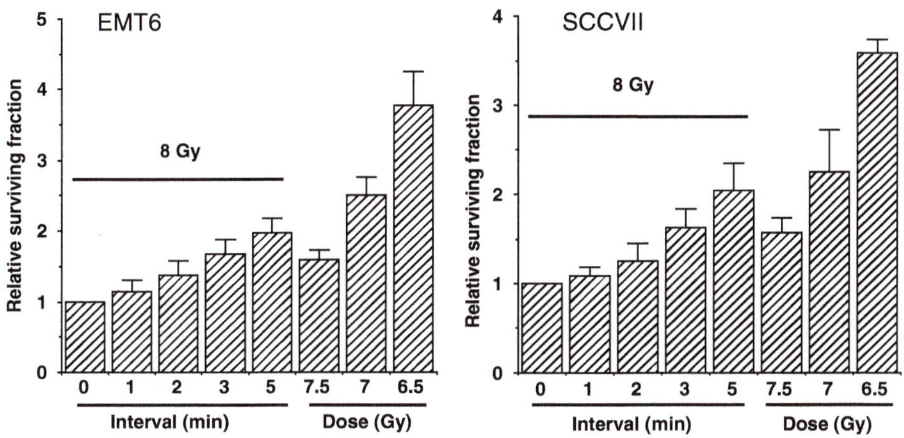

Fig. 3.1 Relative surviving fractions of EMT6 and SCCVII cells after 8 Gy given without a break or in 5 fractions at various intervals. Cell survival after continuous 8-Gy irradiation was regarded as 1. *Bars* represent SD (Reproduced from Ref. [10] with permission from the publisher)

Furthermore, the effects of 2 Gy given in 5 or 10 fractions at intervals of 0.5–5 min were estimated in EMT6 cells using the cytokinesis-block micronucleus assay, which is a sensitive assay to evaluate radiation effects at low doses [13, 14]. In the 5-fraction experiment, the micronucleus frequency decreased significantly as compared to single 2-Gy irradiation when the interval was 2 min or longer (Fig. 3.2). When the interval was 5 min, 5 fractions of 0.4 Gy corresponded to a single dose of 1.72 Gy. In the 10-fraction experiment, the micronucleus frequency decreased when the interval was 1 min or longer. With an interval of 3 min each, 10 fractions of 0.2 Gy corresponded to a single dose of 1.76 Gy.

To summarize this in vitro study, it was concluded that dose-modifying factors of 1.08–1.16 need to be considered when the total irradiation time is 20–30 min. However, further in vivo study is considered necessary to extrapolate this result to clinical situations.

3.2.2 Influence of the Fraction Dose and Number and Dose Rate on the Biological Effect

The next in vitro study was conducted to investigate the effects of intermittent irradiation like IMRT [15]. A total dose of 8 Gy was given to EMT6 and SCCVII cells in 2, 5, 10, 20, and 40 fractions within a fixed period of 15, 30, or 46 min, and the effects were compared with continuous 8-Gy irradiation given at a dose rate of 1.55 Gy/min or at reduced dose rates over 15, 30, or 46 min. The 20- and 40-fraction schedules would be closer to the IMRT situation than other irradiation schedules. When the total radiation time was 15 min, there were no differences in cell survival among the fractionation schedules, but when the period was 30 or 46 min, the radiation effect tended to decrease with an increase in the fraction number up to 20

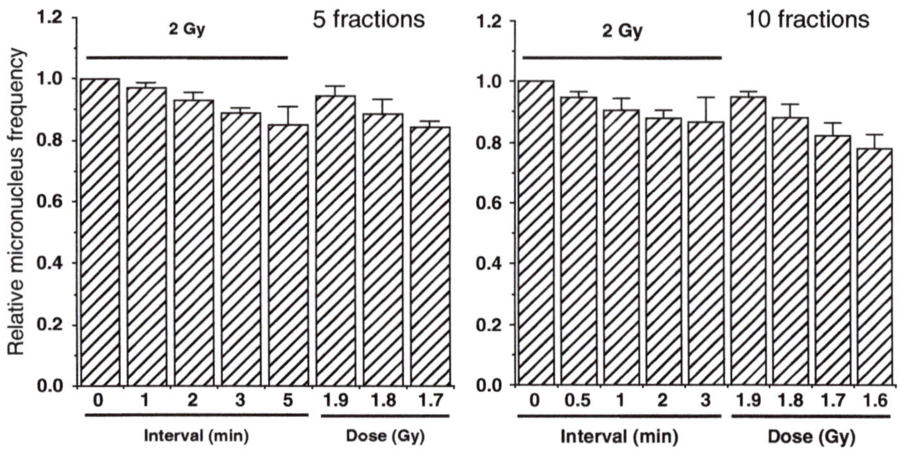

Fig. 3.2 Relative micronucleus frequency in EMT6 cells after 2 Gy given without a break or in 5 or 10 fractions at various intervals. The micronucleus frequency after continuous 2-Gy irradiation was regarded as 1. *Bars* represent SD (Reproduced from Ref. [10] with permission from the publisher)

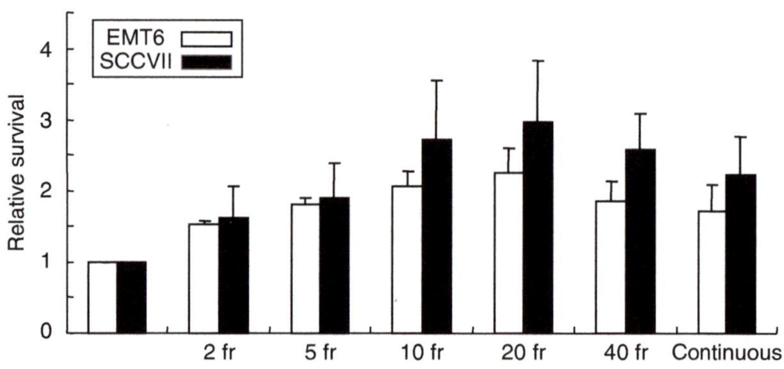

Fig. 3.3 Relative surviving fractions of EMT6 and SCCVII cells after 8 Gy given in 2–40 fractions and prolonged continuous irradiation given over 46 min. The control group received a single dose of 8 Gy over 5.3 min. Cell survival of the control group was regarded as 1. *Bars* represent SD (Reproduced from Ref. [10] with permission from the publisher. Regarding differences between groups, see Ref. [15])

fractions (Fig. 3.3). Two-fraction irradiation yielded the greatest effect among the fractionated radiation groups. Continuous low-dose-rate irradiation had a greater effect than 20- or 40-fraction irradiation. Implications regarding the clinical application of these results are complicated; nevertheless, this study showed that biological effects could differ with the fractionation schedule even when the total radiation time and dose are identical. Judging from the in vitro study, intermittent irradiation as used in IMRT seems to be less effective than continuous irradiation, and to minimize the decrease of biological effects, total irradiation time should be kept as short as possible.

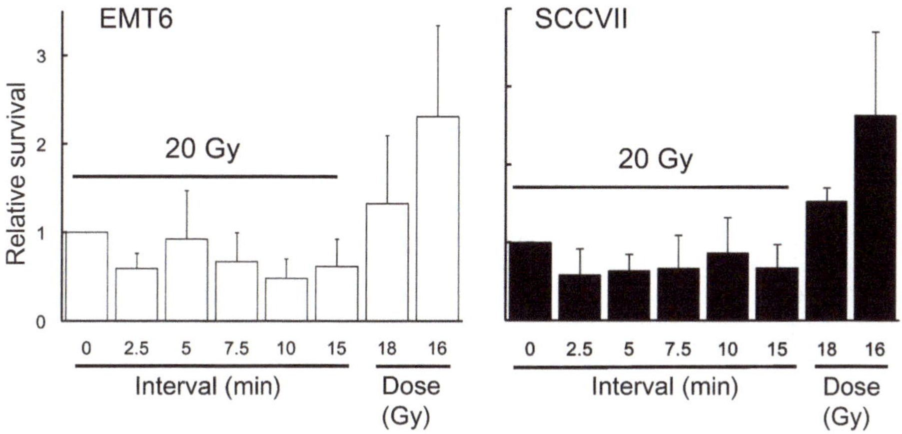

Fig. 3.4 Relative surviving fractions of EMT6 and SCCVII cells irradiated in vivo at 16, 18, or 20 Gy without a break or 20 Gy in 5 fractions at various intervals. Cell survival after continuous 20-Gy irradiation was regarded as 1. *Bars* represent SD Reproduced from Ref. [10] with permission from the publisher)

3.2.3 Effects of Intermittent Irradiation on Murine Tumors In Vivo

The effect of prolonged radiation delivery was also studied in vivo [16]. EMT6 and SCCVII tumors were transplanted into the hind legs of Balb/c and C3H/HeN mice, respectively. When subcutaneous tumors grew to 1 cm in their longest diameter, the mice received 20 Gy in 2, 5, or 10 fractions at various intervals. Within 24 h from the first irradiation, the tumors were excised, minced, and enzymatically disaggregated into single cells. Then, cell survival was assessed using a colony assay. Figure 3.4 shows the results of a 5-fraction experiment. Contrary to the in vitro data, no decrease in radiation effects was observed; instead, by placing 2.5-, 7.5-, 10-, or 15-min intervals for EMT6 tumors and 2.5-, 5-, 7.5-, or 15-min intervals for SCCVII tumors, the effect became stronger. Similar results were obtained in 10-fraction experiments. It was speculated that SLDR in vivo might be counterbalanced or overweighed by other phenomena such as reoxygenation. Therefore, we investigated reoxygenation in SCCVII tumors during intervals of several minutes in the next study.

3.2.4 Reoxygenation Shortly After Irradiation and SLDR In Vivo in the Absence of Reoxygenation

Using 1-cm-diameter SCCVII tumors transplanted into C3H mice, reoxygenation at 0–15 min after a 13-Gy dose was investigated [17]; the hypoxic fraction was measured at 0, 2.5, 5, 10, and 15 min after 13 Gy using a paired survival curve assay. At given times, the irradiated mice were divided into alive and dead groups and received a second irradiation with 15 Gy. Cell survival in the two groups was compared to

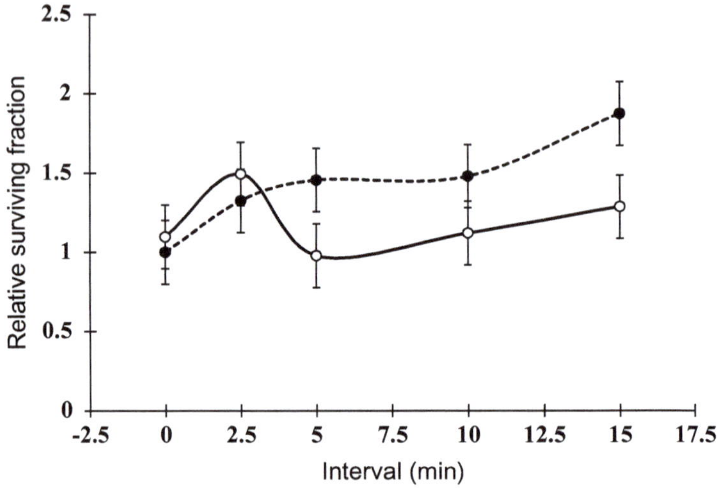

Fig. 3.5 Relative surviving fractions of SCCVII tumor cells after a priming dose of 13 Gy and a second dose of 15 Gy given at 0–15 min intervals to air-breathing (○) or dead (●) mice. The surviving fraction in the dead group that received the second dose immediately after the priming dose was regarded as 1. The hypoxic fraction is given by the surviving fraction of tumor cells in air-breathing mice divided by that in dead mice at respective time points. *Bars* represent SE (Reproduced from Ref. [10] with permission from the publisher)

assess the hypoxic fraction. As shown in Fig. 3.5, the hypoxic fraction was 100 % at 0 and 2.5 min after the end of the first irradiation, but, at 5 min, it fell to 67 % (95 % confidence interval, 41–93 %). Thus, reoxygenation was observed at 5 min after irradiation. It was suggested that rapid reoxygenation could compensate for SLDR in vivo.

To investigate the effect of intermittent irradiation under conditions of restricted reoxygenation, 1-cm-diameter SCCVII tumors in the hind legs of C3H mice were irradiated with the leg fixed using adhesive tape. This procedure was considered to increase the hypoxic fraction and restrict reoxygenation [18]. Figure 3.6 compares the growth delay of SCCVII tumors irradiated with 20 Gy, 25 Gy, or 5 fractions of 5 Gy given at 3-, 6-, or 10-min intervals. The effect of radiation decreased by imposing intervals of 3–7 min; the effect of 25 Gy given in 5 fractions was between that of 20 Gy and that of 25 Gy delivered continuously. Therefore, it was suggested that the effects of intermittent radiation in vivo decrease due to SLDR when reoxygenation is restricted.

3.2.5 Other Laboratory Studies on the Biological Effects of Intermittent Irradiation

Classically, Elkind and his coworkers [3, 4] were the first to report the SLDR phenomenon. In their experiments, a significant increase in cell survival was observed

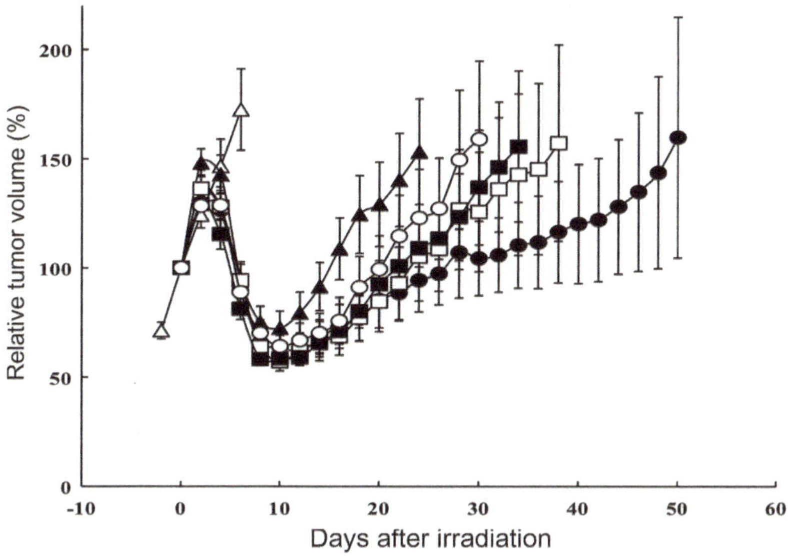

Fig. 3.6 Relative volumes of SCCVII tumors after 20 (▲) or 25 (●) Gy given as a single dose or 5 fractions of 5 Gy given at intervals of 3 (○), 6 (■), or 10 (□) minutes. △: no irradiation. Tumor volumes before irradiation in each group were regarded as 100 %. *Bars* represent SE (Reproduced from Ref. [10] with permission from the publisher)

when intervals of 30 min or longer were set between two radiation doses. However, they never investigated shorter intervals. With the development of radiotherapy techniques, it has become necessary to investigate the influence of radiation interruptions of shorter than 30 min.

After the 1990s, Benedict et al. [19] attempted to estimate dose correction factors for stereotactic radiosurgery using U-87MG cells in vitro. In their experiments, the effect of radiation decreased with prolongation of the treatment time, and the correction factor of 0.02–0.03 Gy/min was proposed when a total dose of 6–18 Gy was given. This indicates that when the treatment time prolongs by 30 min, 8 Gy would correspond to approximately 7.1–7.4 Gy delivered continuously, giving dose-modifying factors of 1.08–1.13. These results appear to agree with our own. Mu et al. [20] conducted an in vitro study with V79 cells using much more complicated fractionation schedules than those we employed and compared the surviving fraction ratios between the continuous and prolonged delivery of radiation with those estimated by biological models derived from the LQ model. Their conclusion was that the biological models underestimated the effect of prolonging the fraction time when a total dose of 2 Gy was fractionated. Therefore, estimation of the influence of prolonging the treatment time using biological models alone seems to be insufficient in clinical practice. More recently, Zheng et al. [21] investigated the impact of prolonged fraction delivery times simulating IMRT on two cultured nasopharyngeal carcinoma cell lines. The fraction delivery time was 15, 36, or 50 min. The dose-modifying factor for a fraction dose of 2 Gy was 1.05 when the delivery time was 15 min, but it increased to 1.11 or 1.18 when the time prolonged to 50 min. They emphasized, however, that these results do not necessarily hold in vivo. Moiseenko et al. [22] obtained results similar to those of the abovementioned studies and suggested that DNA repair underlies the increase in cell survival observed when dose delivery is prolonged, based on measurement of the retention of gammaH2AX, a measure of the lack of DNA damage repair.

Moiseenko et al. [23] investigated the correlation between the magnitude of the loss of effect brought about by prolonged radiation delivery and the α/β ratio in three cell lines. When their results were projected to a 30-fraction treatment, the dose deficit to bring cell survival to the same level was 4.1 Gy in one line, but it was as large as 24.9 and 31.1 Gy in the other two lines. The dose deficit did not relate to the α/β ratio of the three cell lines. On the other hand, Zheng et al. [24] also investigated the issue in two hepatocellular carcinoma cell lines, and a significant decrease in cell survival due to prolonged fraction delivery was observed in one line with an α/β ratio of 3.1 Gy but not in another with an α/β ratio of 7.4 Gy. Therefore, the relationship with the α/β ratio remains unclear and requires further investigation.

All these results indicate that SLDR certainly takes place when radiation delivery is prolonged or given intermittently in daily stereotactic irradiation and IMRT settings. However, it should be noted that these results were obtained using in vitro single cells. Until recently, there have been no in vivo studies except for our own ones, but other studies have been published. The results of a study by Wang et al. [25] agree with our own; when C57BL mice bearing Lewis lung cancer were irradiated under conditions of limited reoxygenation, intermittent radiation delivery led to a significant reduction in the biological effects. The study by Jiang et al. [26] also showed a similar result. However, more in vivo investigations appear to be warranted in the near future. Our study suggests that SLDR in vivo can be counterbalanced by reoxygenation. In tumors that reoxygenate rapidly, therefore, the adverse effects of prolonging the radiation delivery time may be none or negligible. However, little is known about the reoxygenation of tumors in humans, so this issue is also an important topic to be investigated in the future to elucidate the effect of intermittent or prolonged radiation delivery in clinical practice.

3.3 Applicability of the LQ Model to High-Dose-per-Fraction Radiotherapy

3.3.1 Current Controversy

To compare different fractionation schedules, the LQ formalism ($n_2 d_2 / n_1 d_1 = (1 + d_1 / [\alpha / \beta]) / (1 + d_2 / [\alpha / \beta])$ (where d_1 and d_2 are fractional doses and n_1 and n_2 are fraction numbers) and the biologically effective dose (BED) derived from the LQ model (BED = $D(1 + d/[\alpha/\beta])$, where D is the total dose and d is the fractional dose) are often used because of their convenience and simplicity [10, 27]. While LQ formalism is useful for conversion between relatively low radiation doses as used in conventional radiotherapy, it has been suggested that it is not applicable to higher daily doses or smaller fraction numbers [9, 10]. However, many clinicians have used LQ formalism to convert hypofractionated doses to single doses in their publications [28, 29], and many have used BED to evaluate the doses of stereotactic irradiation [30, 31]. To further complicate the issue, some investigators, in contrast, claim that the LQ model is applicable to stereotactic irradiation [32, 33]. The ground for the latter group is somewhat limited in that the existing clinical data do not significantly deviate from those expected from LQ model calculations, and their data do not necessarily indicate that the LQ model fits best to the high-dose data. Since clinical data usually contain many errors, experimental evaluation of the reliability of the LQ model in single-fraction and hypofractionated radiation schedules appears to be important and desirable.

3.3.2 Cell Survival Data for the Reliability of the LQ Model at High Doses per Fraction

The theoretical basis behind the LQ model not being applicable with high doses per fraction is that dose-survival curves for cultured cells cannot be fitted well by the LQ model in high-dose ranges. This has been pointed out for a long time; in the pioneering work of Puck and Markus [34] who established the colony formation assay, the high-dose region of the dose-survival curve was apparently straight in HeLa cells. Therefore, the LQ model, with which the cell survival curve continues to bend downward at high doses, does not seem to fit the actual curves at high doses. Joiner and Bentzen [10] stated that extrapolations by the LQ model beyond 5–6 Gy per fraction are likely to lack clinically useful precision. More recently, Garcia et al. [35] investigated the compatibility of the LQ model regarding dose-survival curves of 4 cell lines in broad dose ranges. In the 4 lines, the LQ model did not fit the curves at very high dose ranges that were >7.5, 9.5, 11.5, or 13 Gy depending on the cell line. Therefore, the inadequacy of the LQ model at high doses was clearly demonstrated.

In a previous study, our group investigated the reliability of LQ formalism in converting hypofractionated doses (in 2–5 fractions) to single doses in single cells and spheroids in culture [36]. That study showed that LQ formalism is inadequate in doing so; the equivalent single doses for the hypofractionated doses calculated by LQ formalism were apparently lower than the equivalent single doses actually measured. LQ formalism underestimated the effect of fractionated irradiation. The magnitudes of errors were 6–19 % for 2- or 3-fraction schedules in cultured V79 and EMT6 single cells and 18–30 % for 2- to 5-fraction schedules in V79 spheroids. Since the reoxygenation of hypoxic tumor cells takes place in in vivo tumors between respective fractions [17, 37, 38], the compatibility of LQ formalism to single and hypofractionated radiation regimens was also investigated using murine tumors in the subsequent study.

Using EMT6 tumors, the applicability of LQ formalism for converting hypofractionated doses (in 2–5 fractions) to single doses was evaluated [39], as in the previous in vitro study. Again, the use of LQ formalism produced large errors; the equivalent single doses for the hypofractionated doses calculated from LQ formalism were much lower than the equivalent single doses actually measured. The magnitudes of errors were larger than those seen in the in vitro study; they were 21–31 % for 2- or 3-fraction schedules and 27–42 % for 4- or 5-fraction schedules. The possible larger discrepancy in in vivo tumors as compared to in vitro single cells and spheroids was considered to be largely due to the reoxygenation of hypoxic tumor cells during intervals between fractions in the hypofractionated groups. This study clearly showed that LQ formalism is inadequate for high-dose-per-fraction radiotherapy, especially in in vivo tumors.

To further evaluate the appropriateness of the BED concept in hypofractionated irradiation, we compared 2- to 5-fraction irradiation schedules simultaneously in the EMT6 tumors in Balb/mice [39]. Total doses of 18–30 Gy were given in 2–5 fractions to the tumor-bearing mice at 4-h intervals, and tumor cell survival was assessed employing an in vivo–in vitro colony assay, as in the previous experiment. Tumor cell survival was plotted against the total dose and BED3.5. In the in vitro cell survival determination conducted along with the in vivo experiment, the α/β ratio of the cell line was 3.5 Gy, so BED3.5 was adopted as a substitute for "BED10" often used clinically to represent the tumor response. Figure 3.7 shows tumor cell survival plotted against the total dose and BED3.5. Respective dose–response curves almost overlapped when cell survival was plotted against actual radiation doses. However, the curves tended to shift downward by increasing the fraction number when cell survival was plotted against BED3.5. If the BED concept is correct, the respective cell survival curves would overlap on this figure. Thus, it seems that BED is inadequate for use in this dose-per-fraction range, especially for tumors.

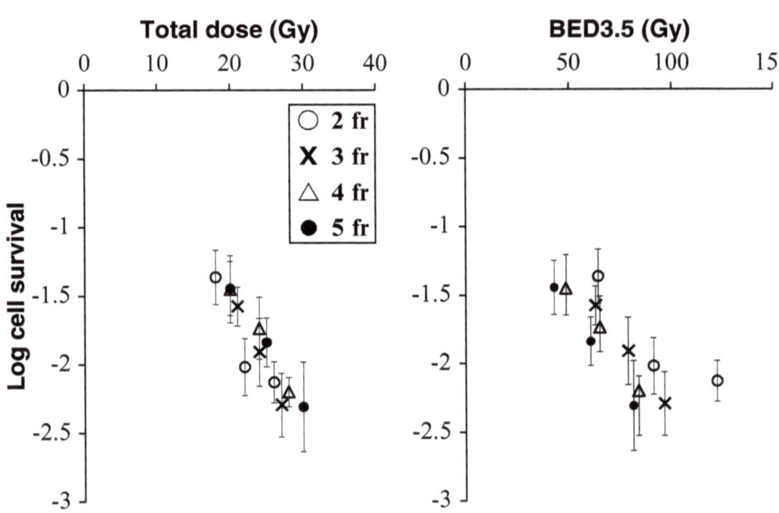

Fig. 3.7 Surviving fractions of EMT6 cells in vivo after 2-fraction (○), 3-fraction (X), 4-fraction (△), or 5-fraction (●) irradiation plotted against the total radiation dose and BED3.5. *Bars* represent SE (Reproduced from Ref. [10] with permission from the publisher)

The total dose reflected the actual effect (tumor cell survival) more accurately than BED in this experiment. The calculated BED tended to become larger than expected from the actual effects when the fraction number decreased. Thus, BED tends to overestimate the actual biologically effective dose with increasing radiation doses.

3.3.3 Normal Tissue Response Data for the Reliability of the LQ Model at High Doses per Fraction

The reliability of the LQ model can also be evaluated based on normal tissue data. In classic radiobiology studies, raw data for various normal tissue responses from animal and human studies were presented as a series of dose–response curves [40–43]. Measured responses were plotted against total radiation doses for each schedule. From horizontal cuts, isoeffect doses could be read off, and these isoeffect doses could then be plotted as a log dose against the log number of fractions or log fraction size. Since the isoeffect curves are concave downward, it is difficult to determine any particular slope for the curves. Instead, the isoeffect curves can be plotted as the reciprocal total dose as a function of the dose per fraction [43]. This reciprocal total dose or F_e plot was elaborated to estimate the α/β ratio of normal tissues [40]. When the normal tissue response data fall in a straight line on this F_e plot, the LQ model is considered to be appropriate. The isoeffect curves for most normal tissues were linear in the dose range of 1–8 Gy [44], suggesting that the LQ model is adequate in this range of dose per fraction. Brenner [33] found that the isoeffect curves for the rat spinal cord response, mouse skin reaction, and murine intestinal damage could be visually fitted with straight lines in the dose range between 0 and 25 Gy and insisted that the LQ model is applicable throughout this dose range. However, statistical validation of the linearity was not performed. Later, Astrahan [45] analyzed the data for various normal tissues in more detail and found that the LQ formula closely fitted the curve for the late reaction of the mouse spinal cord for fractions up to about 10 Gy. However, the data for cervical vascular damage did not fit the LQ model but fitted the LQL (linear–quadratic–linear) model, which is stated later. Fowler et al. [46] suggested that for certain epithelial tissues, the LQ model may be applicable up to 23 Gy per fraction.

These observations are somewhat contradictory and confusing, but the discrepancy may be, in part, explained by the α/β ratio for the normal tissue responses. The applicability of the LQ model may not simply depend on the absolute dose per fraction; for a tissue with a large α/β ratio, its applicability may be extended to a higher dose region. This is the case with epithelial tissues that usually have an α/β ratio of around 10 Gy. Since the α/β ratio represents the dose at which cell killing from linear (α) and quadratic (β) components of the LQ formula is equal, the LQ model holds around the dose level of the α/β ratio. However, with the increase in the dose, the β-cell kill component dominates in the LQ model, from which actual cell survival data have been shown to deviate. This deviation appears to become evident in the dose range over twofold the α/β ratio [35]. From these considerations, it may be said that the model is applicable up to a radiation dose approximately twofold the α/β ratio.

Recently, Borst et al. [47] analyzed radiation pneumonitis data in patients undergoing stereotactic body radiotherapy. Various fractionation schedules were employed ranging from 35 Gy in 4 fractions to 60 Gy in 8 fractions. They tried to correlate the mean lung dose with the occurrence of radiation pneumonitis. They found that the data were best fitted by the LQ model with an α/β ratio of 3 Gy. Although the prescribed dose per fraction was 7.5–2 Gy, the mean lung dose per fraction is usually much lower, so it may not be surprising that the LQ model fitted their mean lung dose data.

3.3.4 Other Alternatives to the LQ Model

Since it is becoming clearer that LQ formalism is not adequate for stereotactic irradiation, other models have been proposed one after another. These include the universal survival curve model [48], LQL model [49] (or modified LQ model [50]), and generalized LQ (gLQ) model [51]. The universal survival curve model hybridizes the LQ model for low doses and the classic multitarget model ($S=1-\left(1-e^{-D/D_0}\right)^n$, where S is the surviving fraction, D is the dose, D_0 is a parameter that determines the final slope of the survival curve, and n is the y-intercept of the asymptote) [52] for high doses beyond a single transition dose (D_T). Hence, the concept is relatively simple. The LQL model derived from a mechanism-based lethal–potentially lethal model [53] has a mechanistic basis. Although the equations for the LQL model are more complex, cell survival curves extend nearly linearly in a high-dose range, as compared to the LQ model [49]. Therefore, the applicability of the universal survival curve model and LQL model to a high-dose region may be similar. The most recently proposed gLQ model takes SLDR and the conversion of sublethal damage to lethal damage during irradiation into account; the model is designed to cover any dose delivery patterns. All of these newer models seem to fit better than the LQ model in the high-dose range. We have also evaluated how the LQ and other models fit experimental data. In an in vitro study, the classic multitarget model and the repairable–conditionally repairable model tended to fit better than the LQ model at high doses [54]. In the near future, it is desirable for an optimal model to be established for clinical use in high-dose-per-fraction radiotherapy. However, it should be noted that these models are generally applicable to the normal tissue response, especially late damage, and not to tumors, since none of these models take the reoxygenation phenomenon, as well as cell cycle effects, host immune effects, and effects on vascular/stromal elements, into account. When the overall treatment time becomes longer than that conventionally used, a factor deriving from repopulation should also be considered [55, 56]. In future studies, models that incorporate these factors as well as reoxygenation should be developed in order to use the models for in vivo tumor responses to high-dose-per-fraction irradiation and more conventional radiotherapy.

Acknowledgments This work was supported in part by Grants-in-Aid for Scientific Research from the Japanese Ministry of Education, Culture, Sports, Science and Technology (20591501, 23591846).

References

1. Takemoto S, Shibamoto Y, Ayakawa S et al (2012) Treatment and prognosis of patients with late rectal bleeding after intensity-modulated radiation therapy for prostate cancer. Radiat Oncol 7:87
2. Manabe Y, Shibamoto Y, Sugie C et al (2014) Toxicity and efficacy of three dose-fractionation regimens of intensity-modulated radiation therapy for localized prostate cancer. J Radiat Res 55(3):494–501
3. Elkind MM, Sutton H (1960) Radiation response of mammalian cells grown in culture. I. Repair of X-ray damage in surviving Chinese hamster cells. Radiat Res 13:556–593
4. Elkind MM, Sutton-Gilbert H, Moses WB et al (1965) Radiation response of mammalian cells grown in culture. V. Temperature dependence of the repair of X-ray damage in surviving cells (aerobic and hypoxic). Radiat Res 25:359–376
5. Iuchi T, Hatano K, Narita Y et al (2006) Hypofractionated high-dose irradiation for the treatment of malignant astrocytomas using simultaneous integrated boost technique by IMRT. Int J Radiat Oncol Biol Phys 64:1317–1324
6. Kim MJ, Yeo SG, Kim ES et al (2013) Intensity-modulated stereotactic body radiotherapy for stage I non-small cell lung cancer. Oncol Lett 5:840–844
7. Boda-Heggemann J, Mai S, Fleckenstein J et al (2013) Flattening-filter-free intensity modulated breath-hold image-guided SABR (Stereotactic ABlative Radiotherapy) can be applied in a 15-min treatment slot. Radiother Oncol 109:505–509
8. Kirkpatrick JP, Meyer JJ, Marks LB (2008) The linear-quadratic model is inappropriate to model high dose per fraction effects in radiosurgery. Semin Radiat Oncol 18:240–243
9. Joiner MC, Bentzen SM (2009) Fractionation: the linear-quadratic approach. In: Joiner M, van der Kogel A (eds) Basic clinical radiobiology. Hodder Arnold, London, pp 102–119
10. Shibamoto Y, Otsuka S, Iwata H, Tomita N et al (2012) Radiobiological evaluation of the radiation dose as used in high-precision radiotherapy: effect of prolonged delivery time and applicability of the linear-quadratic model. J Radiat Res 53:1–9
11. Shibamoto Y, Ito M, Sugie C et al (2004) Recovery from sublethal damage during intermittent exposures in cultured tumor cells: implications for dose modification in radiosurgery and IMRT. Int J Radiat Oncol Biol Phys 59:1484–1490
12. Shibamoto Y, Yukawa Y, Tsutsui K et al (1986) Variation in the hypoxic fraction among mouse tumors of different types, sizes, and sites. Jpn J Cancer Res 77:908–915
13. Shibamoto Y, Streffer C, Fuhrmann C et al (1991) Tumor radiosensitivity prediction by the cytokinesis-block micronucleus assay. Radiat Res 128:293–300
14. Shibamoto Y, Streffer C, Sasai K et al (1992) Radiosensitization efficacy of KU-2285, RP-170 and etanidazole at low radiation doses: assessment by in vitro cytokinesis-block micronucleus assay. Int J Radiat Biol 61:473–478
15. Ogino H, Shibamoto Y, Sugie C et al (2005) Biological effects of intermittent radiation in cultured tumor cells: influence of fraction number and dose per fraction. J Radiat Res 46:401–406
16. Sugie C, Shibamoto Y, Ito M et al (2006) The radiobiological effect of intermittent radiation exposure in murine tumors. Int J Radiat Oncol Biol Phys 64:619–624
17. Tomita N, Shibamoto Y, Ito M et al (2008) Biological effect of intermittent radiation exposure in vivo: recovery from sublethal damage versus reoxygenation. Radiother Oncol 86:369–374
18. Shibamoto Y, Sasai K, Abe M et al (1987) The radiation response of SCCVII tumor cells in C3H/He mice varies with the irradiation conditions. Radiat Res 109:352–354
19. Benedict SH, Lin PS, Zwicker RD et al (1997) The biological effectiveness of intermittent irradiation as a function of overall treatment time: development of correction factors for linac-based stereotactic radiotherapy. Int J Radiat Oncol Biol Phys 37:765–769
20. Mu X, Löfroth PO, Karlsson M et al (2003) The effect of fraction time in intensity modulated radiotherapy: theoretical and experimental evaluation of an optimisation problem. Radiother Oncol 68:181–187

21. Zheng XK, Chen LH, Wang WJ et al (2010) Impact of prolonged fraction delivery times simulating IMRT on cultured nasopharyngeal carcinoma cell killing. Int J Radiat Oncol Biol Phys 78:1541–1547
22. Moiseenko V, Banáth JP, Duzenli C et al (2008) Effect of prolonging radiation delivery time on retention of gammaH2AX. Radiat Oncol 3:18
23. Moiseenko V, Duzenli C, Durand RE et al (2007) In vitro study of cell survival following dynamic MLC intensity-modulated radiation dose delivery. Med Phys 34:1514–1520
24. Zheng XK, Chen LH, Yan X et al (2005) Impact of prolonged fraction dose-delivery time modeling intensity-modulated radiation therapy on hepatocellular carcinoma cell killing. World J Gastroenterol 11:1452–1456
25. Wang X, Xiong XP, Lu J et al (2011) The in vivo study on the radiobiologic effect of prolonged delivery time to tumor control in C57BL mice implanted with Lewis lung cancer. Radiat Oncol 6:4
26. Jiang L, Xiong XP, Hu CS et al (2013) In vitro and in vivo studies on radiobiological effects of prolonged fraction delivery time in A549 cells. J Radiat Res 54:230–234
27. Withers HR, Thames HD Jr, Peters LJ et al (1983) A new isoeffect curve for change in dose per fraction. Radiother Oncol 1:187–191
28. Wulf J, Baier K, Mueller G et al (2005) Dose-response in stereotactic irradiation of lung tumors. Radiother Oncol 77:83–87
29. Milano MT, Katz AW, Schell MC et al (2008) Descriptive analysis of oligometastatic lesions treated with curative-intent stereotactic body radiotherapy. Int J Radiat Oncol Biol Phys 72:1516–1522
30. Onishi H, Shirato H, Nagata Y et al (2011) Stereotactic body radiotherapy (SBRT) for operable stage I non-small-cell lung cancer: can SBRT be comparable to surgery? Int J Radiat Oncol Biol Phys 81:1352–1358
31. Takeda A, Sanuki N, Kunieda E et al (2009) Stereotactic body radiotherapy for primary lung cancer at a dose of 50 Gy total in five fractions to the periphery of the planning target volume calculated using a superposition algorithm. Int J Radiat Oncol Biol Phys 73:442–448
32. Guckenberger M, Klement RJ, Allgaeuer M et al (2013) Applicability of the linear-quadratic formalism for modeling local tumor control probability in high dose per fraction stereotactic body radiotherapy for early stage non-small cell lung cancer. Radiother Oncol 109:13–20
33. Brenner DJ (2008) The linear-quadratic model is an appropriate methodology for determining iso-effective doses at large doses per fraction. Semin Radiat Oncol 18:234–239
34. Puck TT, Marcus PI (1956) Action of X-rays on mammalian cells. J Exp Med 103:653–666
35. Garcia LM, Leblanc J, Wilkins D et al (2006) Fitting the linear-quadratic model to detailed data sets for different dose ranges. Phys Med Biol 51:2813–2823
36. Iwata H, Shibamoto Y, Murata R et al (2009) Estimation of errors associated with use of linear-quadratic formalism for evaluation of biologic equivalence between single and hypofractionated radiation doses: an in vitro study. Int J Radiat Oncol Biol Phys 75:482–488
37. Shibamoto Y, Kitakabu Y, Murata R et al (1994) Reoxygenation in the SCCVII tumor after KU-2285 sensitization plus single or fractionated irradiation. Int J Radiat Oncol Biol Phys 29:583–586
38. Murata R, Shibamoto Y, Sasai K et al (1996) Reoxygenation after single irradiation in rodent tumors of different types and sizes. Int J Radiat Oncol Biol Phys 34:859–865
39. Otsuka S, Shibamoto Y, Iwata H et al (2011) Compatibility of the linear-quadratic formalism and biologically effective dose concept to high-dose-per-fraction irradiation in a murine tumor. Int J Radiat Oncol Biol Phys 81:1538–1543
40. Douglas BG, Fowler JF (1976) The effect of multiple small doses of x rays on skin reactions in the mouse and a basic interpretation. Radiat Res 66:401–426
41. van der Kogel AJ (1985) Chronic effects of neutrons and charged particles on spinal cord, lung, and rectum. Radiat Res 8(Suppl):S208–S216
42. Peck JW, Gibbs FA (1994) Mechanical assay of consequential and primary late radiation effects in murine small intestine: alpha/beta analysis. Radiat Res 138:272–281

43. Fowler JF (1984) Total doses in fractionated radiotherapy – implications of new radiobiological data. Int J Radiat Biol 46:103–120
44. Fowler JF (1989) The linear-quadratic formula and progress in fractionated radiotherapy. Br J Radiol 62:679–694
45. Astrahan M (2008) Some implications of linear-quadratic-linear radiation dose-response with regard to hypofractionation. Med Phys 35:4161–4172
46. Fowler JF, Tomé WA, Fenwick JD et al (2004) A challenge to traditional radiation oncology. Int J Radiat Oncol Biol Phys 60:1241–1256
47. Borst GR, Ishikawa M, Nijkamp J et al (2010) Radiation pneumonitis after hypofractionated radiotherapy: evaluation of the LQ(L) model and different dose parameters. Int J Radiat Oncol Biol Phys 77:1596–1603
48. Park C, Papiez L, Zhang S et al (2008) Universal survival curve and single fraction equivalent dose: useful tools in understanding potency of ablative radiotherapy. Int J Radiat Oncol Biol Phys 70:847–852
49. Guerrero M, Carlone M (2010) Mechanistic formulation of a linear-quadratic-linear (LQL) model: split-dose experiments and exponentially decaying sources. Med Phys 37:4173–4181
50. Guerrero M, Li XA (2004) Extending the linear-quadratic model for large fraction doses pertinent to stereotactic radiotherapy. Phys Med Biol 49:4825–4835
51. Wang JZ, Huang Z, Lo SS et al (2010) A generalized linear-quadratic model for radiosurgery, stereotactic body radiation therapy, and high-dose rate brachytherapy. Sci Transl Med 2:39ra48
52. Butts JJ, Katz R (1967) Theory of RBE for heavy ion bombardment of dry enzymes and viruses. Radiat Res 30:855–871
53. Curtis SB (1986) Lethal and potentially lethal lesions induced by radiation – a unified repair model. Radiat Res 106:252–270
54. Iwata H, Matsufuji N, Toshito T et al (2013) Compatibility of the repairable-conditionally repairable, multi-target and linear-quadratic models in converting hypofractionated radiation doses to single doses. J Radiat Res 54:367–373
55. Scott OC (1990) Mathematical models of repopulation and reoxygenation in radiotherapy. Br J Radiol 63:821–823
56. Nakamura K, Brahme A (1999) Evaluation of fractionation regimen in stereotactic radiotherapy using a mathematical model of repopulation and reoxygenation. Radiat Med 17:219–225

Treatment Planning of IMRT for Head and Neck Malignancies

4

Toru Shibata

Keywords
Intensity-modulated radiation therapy (IMRT) • Treatment planning • Optimization

4.1 Introduction

Intensity-modulated radiation therapy (IMRT) is an advanced form of conformal irradiation techniques that can generate highly optimized dose distributions with steep dose gradients [1]. Studies have demonstrated the benefits of IMRT for improving target coverage and decreasing the doses delivered to the organ at risk (OAR), particularly in clinical situations where concave targets surround adjacent OARs [2–4]. In prostate cancer, concave absorbed-dose distributions have enabled the establishment of dose escalation strategies without increasing rectal complications [5]. In head and neck cancers, this approach enables for a greater sparing of various normal tissue structures such as the salivary glands, optic nerves, temporal lobes, auditory apparatus, brainstem, and pharyngeal constrictor muscles while also improving tumor control [4, 6–8].

In this chapter, the process for treatment planning of IMRT will be discussed from a practical perspective. This process consists of several phases, as follows: (1) delineation of the target volume and various normal tissue structures on three-dimensional (3D) images, (2) optimization of the radiation dose distribution using inverse planning, (3) leaf-sequence generation followed by calculation of dose distributions, and (4) evaluation of the planning results. Additional rounds of

T. Shibata, M.D., Ph.D. (✉)
Department of Radiation Oncology, Kagawa University Hospital,
1750-1 Ikenobe, Miki-cho, Kita-gun, Kagawa 761-0793, Japan
e-mail: tshibata@med.kagawa-u.ac.jp

Y. Nishimura, R. Komaki (eds.), *Intensity-Modulated Radiation Therapy*,
DOI 10.1007/978-4-431-55486-8_4

iterative optimization would be required with constraint modifications that define the objective function until satisfactory planning goals have been achieved.

The International Commission on Radiation Units and Measurements (ICRU) addressed the issue of the volume definition and dose specification for radiotherapy by publishing the ICRU Report 50 in 1993 and its supplement 62 in 1999 [9, 10]. The most recent ICRU Report 83 (ICRU 83) described updated information to standardize the techniques and procedures for prescribing, recording, and reporting of IMRT using growing technological advances [11]. It should be noted that the most protocols used in previous or ongoing clinical trials may be still based on previous reports. Therefore, the definition of a consistent volume and dose specification for treatment planning of IMRT should be translated along with ICRU 83. The descriptions in this chapter have been adequately adapted according to the recent changes in the recommendations.

4.2 Delineation of Target Volumes, Critical Structures, and the Regions of Interest at Risk

In inverse planning of IMRT, radiation oncologists must explicitly define what is to be treated and what is to be avoided. The careful selection and accurate delineation of target volumes and structures is essential because errors in this process could not be recovered by the subsequent optimizations and may often result in the marginal recurrences of the tumors and/or severe late radiation toxicity to the adjacent structures after IMRT. The proper delineation of target volumes and OARs to achieve optimal dose distributions requires experience in anatomical and diagnostic imaging and a thorough understanding of the spread patterns of the disease. Consensus guidelines have been published and updated for target volume delineation to help standardize treatment planning of IMRT at specific tumor sites [12–18]. Nevertheless, standards for target definition, dose specification, and normal tissue constraints for IMRT are still evolving.

During treatment planning and reporting processes, several volumes related to the tumor and normal organs are typically defined as follows: (1) the gross tumor volume (GTV), (2) clinical target volume (CTV), (3) planning target volume (PTV), (4) OAR, (5) planning organ-at-risk volume (PRV), (6) internal target volume (ITV), (7) treated volume (TV), and (8) remaining volume at risk (RVR).

4.2.1 Gross Tumor Volume (GTV)

According to ICRU 83 [11], GTV is defined as the gross demonstrable extent and location of the tumor. GTV may consist of a primary tumor, metastatic regional lymph nodes, or distant metastasis. Typically, different GTVs are delineated for primary tumors and regional nodes. However, if primary lesions could not be differentiated from adjacent nodal lesions, a single GTV encompassing both the primary tumor and the nodes may be delineated. Such situations are often observed in advanced nasopharyngeal carcinomas (NPCs) that have infiltrated into the retropharyngeal space, including possible nodal lesions. Combinations of various anatomical and/or functional imaging modalities can be useful for the delineation of GTV.

4.2.1.1 GTV Delineation Using Anatomical Imaging

Typically, anatomical imaging modalities such as computed tomography (CT) and magnetic resonance imaging (MRI) have been used to contour the boundary of GTV. Among these, contrast-enhanced CT is used most commonly for treatment planning. It reveals the precise anatomical findings of primary tumors and involved nodes. In addition, CT enables the radiation dose to be calculated and the generation of the digitally reconstructed images necessary for treatment verifications.

In general, MRI provides better anatomical visualizations than CT because of its excellent soft tissue contrast. In cases of advanced NPC with the skull base involvement (Fig. 4.1c), MRI is superior to CT for determining tumor extension and the involvement of the skull base, cavernous sinus, and parapharyngeal and retropharyngeal spaces. In particular, gadolinium-enhanced and/or fat-saturated T1-weighted images with 3D views are useful for GTV delineation, as shown in Fig. 4.2a. A previous study demonstrated that, compared with CT, MRI-based target volumes were significantly larger, more irregularly shaped, and did not always include the CT targets in NPCs [19]. Other studies showed that CT–MRI image fusion improved the delineation of the target volume and decreased variability between observers in head and neck or prostate cancers [20, 21]. Therefore, the combined use of MRI and CT is desirable for target determination at most tumor sites. In addition, diffusion-weighted MRI may significantly improve nodal staging in comparison with CT or routine MRI [22]. Despite various imaging modalities, physical examination is essential for the accurate determination of GTV. For example, base-of-tongue cancers could be examined easily by palpation, and the precise mucosal extension of oropharyngeal and laryngeal cancers could be revealed only on visual inspection or using fiber-optic endoscopy.

For prostate IMRT, MRI is superior to CT for defining of the prostate apex for targeting and the erectile tissues for sparing purposes [23–26]. Furthermore, the location of a dominant lesion in the prostate could be well predicted using multiple MRI parameters, including diffusion-weighted, dynamic contrast-enhanced, T2-mapping, and 3D spectroscopic MR images for high-dose intraprostatic boosting using IMRT [27, 28].

4.2.1.2 GTV Delineation Using Functional Imaging

^{18}Fluorodeoxyglucose positron emission tomography (FDG-PET) has been applied increasingly to radiation treatment planning to increase the accuracy of GTV delineation besides the staging work-up. FDG-PET may be expected to visualize the "true" pathological tumor volume or reflect the "actual" tumor burden depending on the levels of glucose metabolism. Interestingly, several studies showed that PET-based GTV (PET-GTV) was often smaller than the CT- or MRI-based GTVs [29–31]. Daisne et al. showed that the macroscopic GTVs in surgical specimens were even smaller than GTVs obtained using FDG-PET, suggesting that all imaging

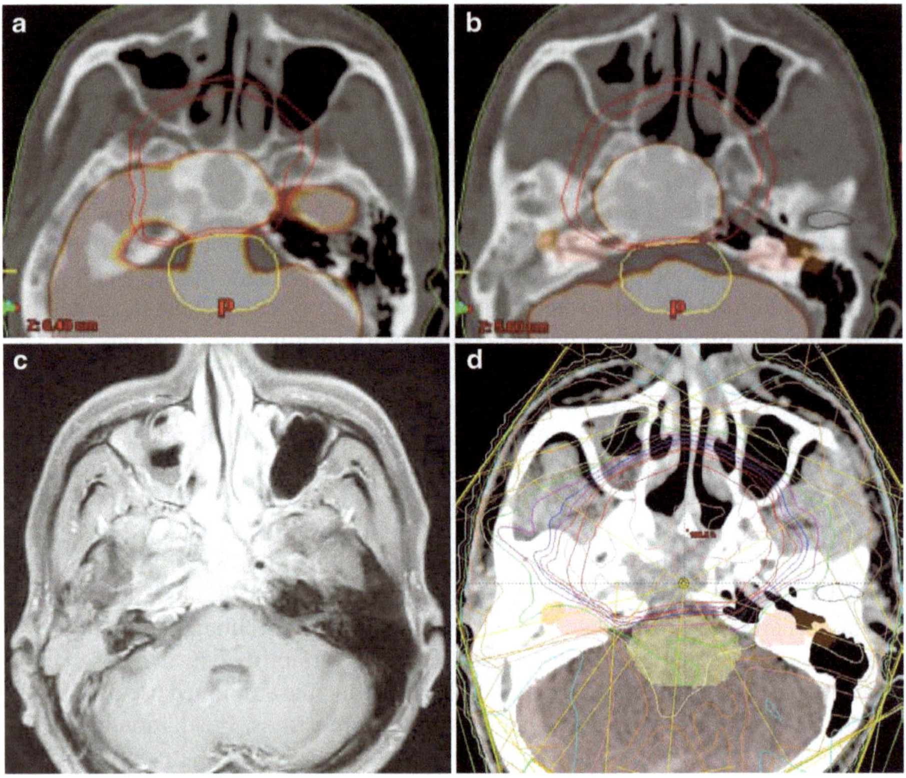

Fig. 4.1 Delineation of the target volumes using FDG-PET/CT and MRI. An advanced nasopharyngeal carcinoma showing extensive infiltrations to the skull base and the clivus. (**a**) and (**b**) The FDG-PET/CT fusion images for delineation of CTV/PTV. The physiological uptake in the brain considerably disturbed the determination of the extent of tumor near the skull base region. (**c**) The T1-weighted MRI well visualized the tumor infiltration to the skull base and even into the right carotid canal as the gadolinium-enhanced areas. In this case, MRI is the most useful for delineation of CTV/PTV in combination with CT in bone window. (**d**) The dose distributions in an initial plan of two-step IMRT. Because of overlapping of the PTV and the ventral portion of the brainstem, CTV-to-PTV margins at this region were reduced up to 2 mm

modalities overestimate GTV. However, when examined in detail, all imaging modalities underestimated the actual tumor extent, particularly in terms of the small mucosal and submucosal infiltration of pharyngolaryngeal squamous cell carcinomas in specimens [29]. Therefore, the importance of clinical examinations for determining GTVs should be emphasized.

Unlike CT and MR images, in which the tumor has well-defined margins, it should be noted that the extent and the size of PET-GTV is highly dependent on the display windowing. Different studies have suggested various methods to determine the outline of PET-GTV, such as using the threshold of 2.5–4 of the standardized uptake value (SUV), 20–50 % of the maximal SUV, or the multiple-threshold methods [20, 31–33]. However, a standardized technique for contouring PET-GTV is

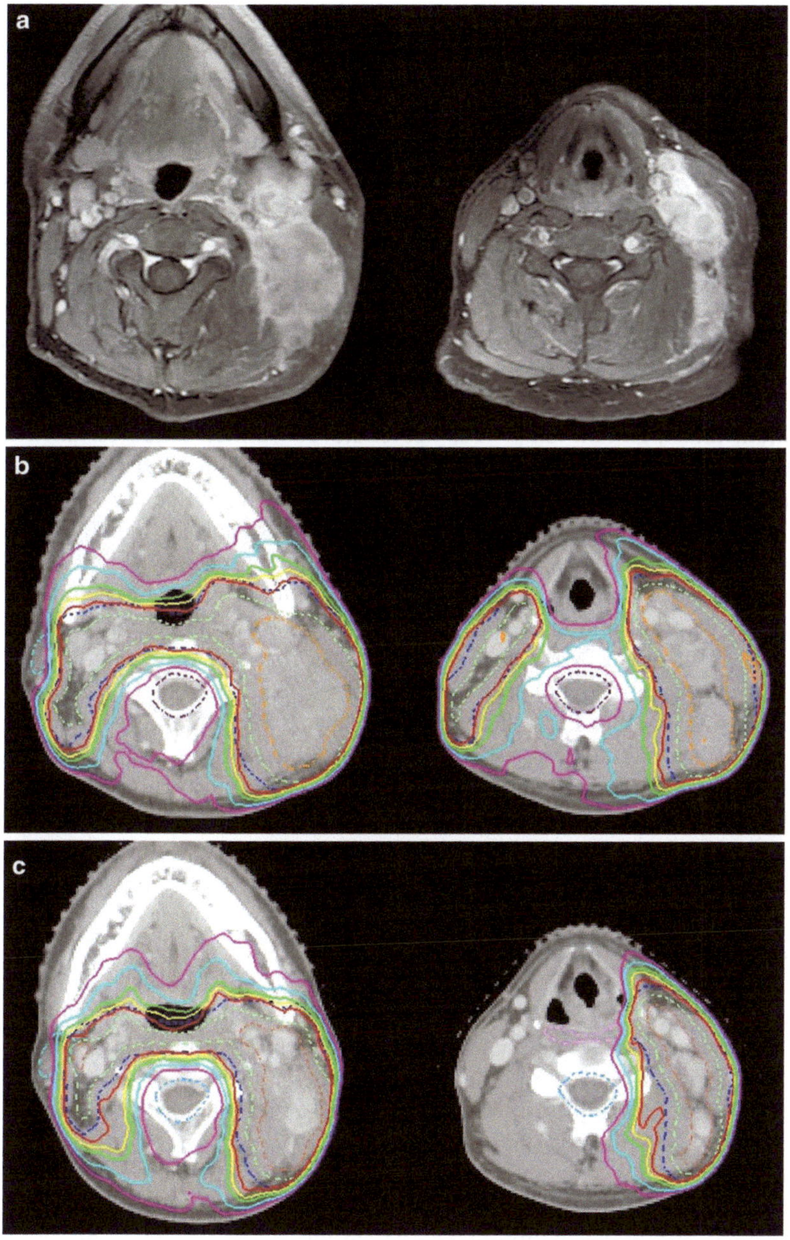

Fig. 4.2 A case of nasopharyngeal carcinoma with lymph node metastasis. (**a**) The T1-weighted MRI with gadolinium enhancement clearly demonstrated the extensive extracapsular invasions of the bulky nodal metastasis in a case of nasopharyngeal cancer before IMRT. (**b**) The initial plan to the large field included both the clinically detectable tumors and the elective regions. (**c**) The boost plan to the high-dose boost volume on the CT at the dose of 36 Gy. Please note the marked volume changes of both the lymph node volumes and the body contours during IMRT with concurrent chemotherapy. Thus, a two-step IMRT planning is a useful means for an adaptive planning strategy

still not available, and a universal standard may be impossible to establish because FDG uptakes can vary among individual tumors depending on their histological type and differentiation grade and the expression of metabolic enzymes. Moreover, PET-GTV delineation can be challenging because certain areas can be FDG avid, particularly in head and neck regions such as the tonsils and the base of the tongue, skeletal muscles, the thyroid gland, the brain, and parotid glands (Fig. 4.1a). Despite these limitations, some authors have attempted to validate the automatic segmentation of GTVs on FDG-PET using objective and user-independent methods. For example, Gregoire et al. [16] reported the use of automatically delineated PET-GTV in a dose escalation trial. In contrast, Eisbruch et al. proposed using a composite of CT-GTV and PET-GTVs [34, 35]. This proposal has been supported by several studies suggesting that local recurrences are not always detected within PET-GTVs [35, 36]. Therefore, FDG-PET should not be used as the sole modality for GTV definition, and high-risk GTV volumes could be defined based on all available information from the findings of CT, PET/CT, MRI, and physical examination.

Functional imaging using PET with various tracers could be also used to define subvolumes within GTV depending on specific biological factors. A partial dose escalation with IMRT may be attractive if we could consistently define certain subvolumes that are at a high risk of failure due to tumor hypoxia, high clonogen density, or the cell proliferation rate [37]. For example, the use of ^{18}F-fluoromisonidazole (FMISO) PET-guided IMRT dose painting was proposed to potentially overcome the radioresistance of hypoxia in head and neck cancers [38]. However, hypoxia imaging using FMISO-PET showed significant spatial variability when repeated sessions were performed either before or during therapy; this finding suggests that reoxygenation and dynamic changes occur at the locations of tumor hypoxia [39, 40]. Thus, GTV segmentation methods using functional imaging are appealing but are still being investigated.

4.2.2 Clinical Target Volume (CTV)

As described in ICRU 83, CTV is the volume of tissue that contains a demonstrable GTV and/or subclinical malignant disease with a certain probability of occurrence that is considered relevant for therapy. However, there is no general consensus regarding what probability is considered relevant for therapy. Based on clinical judgment, a possibility of microscopic disease typically over 5–10 % would be assumed to require treatment. The concepts of subclinical malignant disease suggest the possibility of both microscopic extension outside GTV and the potential involvement of regional lymph nodes, which are usually undetectable by clinical examinations or using any imaging modalities.

CTV may be created practically by adding margins (5–10 mm) to GTV, although these margins should be tailored based on anatomical and pathological considerations. Particularly, in head and neck tumors, a thorough understanding of the complex regional anatomy is required, including knowledge of the cervical compartments and fasciae that are barriers to tumor extension. Different head and neck primary tumors also exhibit a variety of spread patterns and biological behaviors. Therefore, the delineation of CTV is currently based on clinical experience and oncological considerations. Consensus guidelines and recommendations [12–18] have been proposed to help standardize the target volume selection and delineation of head and neck tumors. A 2013 update by Gregiore et al. [14] clearly illustrated several node groups not considered previously, including the supraclavicular, retroauricular, occipital, buccal, and parotid nodes. They have also shown that translation from the node levels to CTV delineation may need some adjustments. In terms of the risk of extracapsular extensions (ECEs), two previous studies reported that ECEs were limited to within 5 mm of nodes <3 cm at their largest diameter in 96 % of cases and were always within less than 10 mm [41] or that ECEs were always within 8 mm range [42]. In cases with larger lymph nodes, CTV delineation may need to consider that expansion from the nodal GTV may be limited by the surrounding structures, such as the sternocleidomastoid muscle, the paraspinal muscles, or the parotid gland. The proper implementation of these guidelines in daily practice will require common sense and good knowledge of anatomy and oncology, which could result in a reduction in treatment variations.

4.2.3 Planning Target Volume (PTV)

PTV is a geometrical concept that was introduced for treatment planning and evaluation. It is a recommended tool to shape dose distributions to ensure that the prescribed dose will actually be delivered to all parts of CTV at a clinically acceptable probability despite geometrical uncertainties such as organ motion and setup variations. The delineation of GTV and CTV is based on anatomical and oncological considerations and is independent of the irradiation technique used. In contrast, the delineation of PTV is dependent on the technique used and is part of the treatment prescription.

ICRU 83 [11] emphasized that the delineation of the primary PTV margin should not be compromised even in cases when PTV encroaches or overlaps with other PTVs, OARs, or PRVs. Developments in treatment planning software now make it possible to manage dose distributions using priority rules during optimization. Alternatively, PTV could be subdivided into regions with different dose prescriptions. Such methods may be often applicable to IMRT optimization for head and neck cancers. In some cases, PTV extends close to or even outside the patient's skin because of tumor invasion or added margins. Most dose computation algorithms cannot accurately compute the absorbed dose in buildup regions. To overcome this limitation, ICRU 83 provided possible solutions such as PTV subdivisions and relaxation of the absorbed-dose objectives for planning.

4.2.4 Organs at Risk (OAR)

OAR or critical normal structures are tissues that, if irradiated, could suffer significant morbidity and thus may influence treatment planning and/or the absorbed-dose prescription. In principle, all nontarget tissues could be OARs. However, normal tissues that are considered typically as OARs depend on the location of CTV and/or the prescribed absorbed dose.

From a functional perspective, OARs have been divided conceptually into "serial" and "parallel" organs [43]. In serial OARs such as the spinal cord, the integrity of each functional subunit with a serial architecture is critical to organ function, and the inactivation of any individual subunit results in dysfunction of the whole organ. In parallel OARs such as the parotid gland, independent functional subunits are organized with a parallel architecture, and it is assumed that no complication will occur until a sufficiently large number of functional subunits have been eliminated, in so-called volume effects. This concept of tissue organization is operationally useful for determining dose–volume constraints. In principle, for serial OARs showing a threshold binary response, the maximum dose in a given volume is typically the best predictor of complications. In contrast, for parallel OARs showing graded absorbed-dose responses, the mean dose or the volume that receives excessive doses of a defined value has been used to predict complications [44].

For establishing normal tissue complication probability (NTCP) models, Emami et al. reported the tolerance dose for irradiation of one-third, two-thirds, or the whole of various organs [45]. Burman et al. provided fits to Emami's consensus dose–volume data using a Lyman model [46]. In addition, Kutcher et al. proposed a dose–volume histogram (DVH) reduction algorithm analysis that enabled the extrapolation of Emami's constraints to any dose distribution [47]. These mathematical methods amounted to a common formula that introduced the generalized equivalent uniform dose (EUD). The resulting Lyman–Kutcher–Burman model remains the most widely used NTCP model, although it is not always the best model that should be considered [48–50]. Recently, the Qualitative Analyses of Normal Tissue Effects in the Clinic (QUANTEC) has provided a critical overview of current knowledge of the quantitative dose–response and dose–volume relationships for clinically relevant normal tissue endpoints, and QUANTEC provided practical guidelines that enabled radiation oncologists to reasonably categorize toxicity risk based on dose–volume parameters or the results of models for a number of treatment sites [51, 52]. Information provided by QUANTEC is not currently ideal, and continued adjustments and validations should be necessary after the release of further clinical data regarding dose–volume modeling [53–56].

4.2.5 Planning Organ-at-Risk Volume (PRV)

Similar to PTV, uncertainties and variations regarding the position of OAR during treatment must be considered to avoid serious complications. For this reason, margins have to be added to OARs to compensate for these uncertainties and variations using similar principles as those used for PTV, which defines PRV.

A margin for serial OAR is much more critical than that for parallel OAR. It should be noted that the delineation of PTV and PRV often results in one or more overlapping regions. To ensure a sufficient sparing of normal tissue, the use of priority rules in the planning system for PTV or PRV can be subdivided into regions with different dose constraints. Nevertheless, ICRU recommends that the absorbed dose should be reported in the full PRV and PTV [11].

4.2.6 Treated Volume (TV)

As described in ICRU 83 [11], TV is the volume of the tissue enclosed within the specific isodose envelope that receives the absorbed dose specified by the radiation oncology team as being appropriate to achieve tumor eradication or palliation, within the boundaries of acceptable complications. According to ICRU, $D_{98\,\%}$ (the absorbed dose for 98 % of PTV) could be selected to determine TV. It is important to identify the shape, size, and position of TV in relation to PTV because it can provide information to evaluate the causes of local recurrences inside or outside the PTV.

4.2.7 Remaining Volume at Risk (RVR)

In ICRU 83, RVR is defined operationally as the difference between the volume enclosed by the external contour of the patient and that of CTVs and OARs on the slices that have been imaged. RVR is important for the treatment optimization process. If RVR is not specifically determined and the dose objectives are not imposed accordingly, the planning systems for IMRT will then extensively seek any possible solutions to achieve the desired dose distribution according to the planning aims. It is possible that this could result in the delivery of unacceptably high doses elsewhere in RVR and that steep dose gradients may not be produced between PTV and RVR. To avoid these difficulties, the planning aims should be applied to RVR. In addition, the absorbed dose in RVR may be useful for estimating the risk of late adverse effects, such as carcinogenesis. Therefore, evaluating the absorbed doses in RVR is particularly important in younger patients who are expected to have a long lifespan.

4.3 IMRT Treatment Planning and Optimization

4.3.1 IMRT Optimization Process

Compared with conventional RT, the main features of IMRT optimization include the use of mathematical objective functions defined by a set of dose constraints and their priorities and the use of a computer-based algorithm to seek the optimal solution in an iterative manner.

IMRT optimization consists of the following processes: (1) defining constraints and establishing an objective function, (2) computing the beamlet weights or the beam segment shapes and weights, (3) computing the absorbed-dose distribution, and then (4) evaluating if the results meet the optimization requirement. If necessary, this then leads to additional rounds of iterative optimization with constraint modifications that define the objective function. Finally, the radiation oncologist makes a decision regarding the acceptability of the plan.

4.3.2 Optimization and Objective Functions

4.3.2.1 Planning Aims and Prescription

In treatment planning for IMRT, the distribution of the absorbed dose to any specified volume including PTVs, PRVs, OARs, and TV could be prioritized and tailored using an iterative process referred to as optimization. In ICRU 83, the "planning aims" are defined as the dosimetric goals when starting the IMRT treatment planning. Parameters such as the desired absorbed doses and planning constraints should be described in planning protocols when initiating the planning process. In contrast, "prescription" represents the finally accepted set of values based on the modified "planning aims" via the optimization process. The use of multiple dose–volume objective functions for each defined volume can lead to increased precision in the planning aims. Analysis of treatment outcome in terms of tumor control and normal tissue toxicity as a function of the absorbed-dose and dose distribution provides important information regarding desirable dose–volume constraints, such as D_{V} (the absorbed dose in fraction V of the volume) and V_{D} (the volume receiving at least the absorbed dose D). During the optimization process, the priority of one constraint over another and/or the priority of one volume over another are specified by parameters that quantitatively weigh the set of priorities. The priority ranking is a clinical decision that may be described in a clinical protocol. The constraints can then be modified iteratively to achieve an acceptable goal with an optimal intensity pattern.

4.3.2.2 Dose and Dose–Volume Objective Functions

To achieve the desired absorbed-dose distribution, a set of constraints is specified for selected target volumes and normal structures. In general, the objective functions for IMRT are established to achieve absorbed-dose homogeneity within PTVs and reduce the absorbed dose in PRVs or OARs according to the dose and dose–volume criteria for each structure and its relative importance. The main reason for the use of such criteria in the IMRT planning is that dose coverage of volumes can be determined explicitly from DVH and could be better controlled via the optimization process.

As shown in Fig. 4.3, the maximum (D_{max}) and the minimum (D_{min}) dose values for PTVs have been generally specified as dose-based objective functions to limit both the hot and cold spots, respectively. In addition, for PRVs of critical OARs, a high priority is that the D_{max} values should not exceed the tolerance limits. The dose–volume metrics D_{min} ($=D_{100\,\%}$) or D_{max} ($=D_{0\,\%}$) values represent those at a single or small number of voxels with the lowest or the highest absorbed doses, respectively. These values may not be accurately computed because they are highly

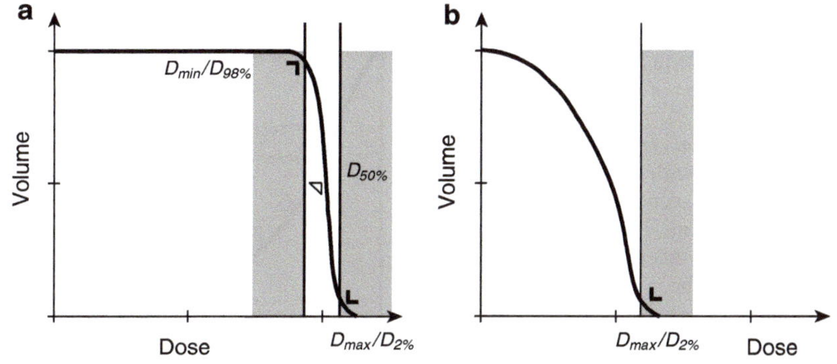

Fig. 4.3 Objective functions in dose-based optimization of the IMRT planning. (**a**) Application of minimal (D_{min}) and maximal dose (D_{max}) constraints to achieve, as much as possible, the dose homogeneity in the PTV. (**b**) Application of maximal dose (D_{max}) constraints with the high-priority factor in the PRV of the critical structure such as the spinal cord to prevent the dose delivery beyond the threshold. According to the ICRU 83, $D_{98\,\%}$ and $D_{2\,\%}$ are preferable instead of D_{min} and D_{max}, respectively. $D_{50\,\%}$ should be reported for evaluation of the PTV dosing

sensitive to the resolution of the calculation and accuracy of PTV. Therefore, ICRU 83 recommends that $D_{100\,\%}$ or $D_{0\,\%}$ should be replaced by the near-minimum, $D_{98\,\%}$, or the near-maximum, $D_{2\,\%}$, values, respectively, which can be determined more accurately [11]. It is also recommended that other values, such as $D_{95\%}$, may also be used but should not replace $D_{98\,\%}$. In addition, the median absorbed dose, specified by $D_{50\,\%}$, should be reported because it is considered to correspond best to the previously defined dose at the ICRU reference point. The clinical relevance of the lowest doses can depend on their position within PTV. In cases close to the edge of PTV, a low-dose region may have less clinical relevance than one that is well within PTV. Therefore, it is important to not rely solely on DVH for evaluation but to instead carefully inspect the dose distributions slice by slice to ensure adequate dose coverage for PTV.

For parallel OARs, mainly dose–volume-based criteria with relative priorities are applied for optimization, as shown in Fig. 4.4. Their dose–volume constraints are specified as V_D. If the sole criterion is specified for OAR, any DVH curves that cross through this particular point obviously meet such a criterion (Fig. 4.5). To avoid such limitations, multiple dose–volume criteria with their relative weights for a single structure can be given, or even the entire DVH could be defined, which enables the shape of the cumulative DVH for that structure to be controlled.

Optimization may be performed using physical (or dose- and dose–volume-based) rather than biological (or dose–response-based) criteria, such as tumor control probability (TCP), NTCP, and EUD. Dose–response-based optimization is particularly attractive because IMRT provides a dose-painting strategy for some biological targets according to functional imaging for metabolic status, hypoxia, and cell proliferation.

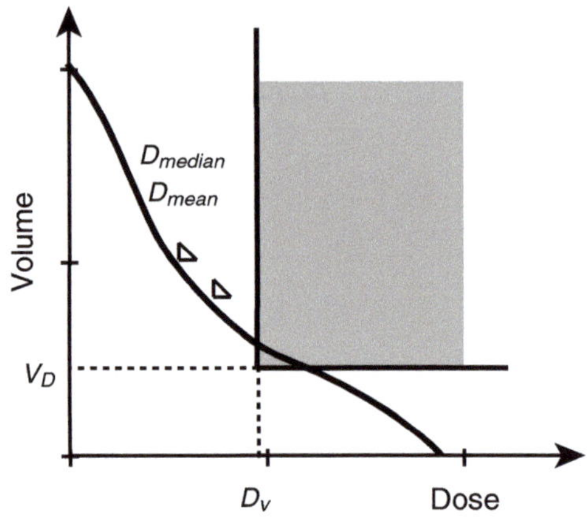

Fig. 4.4 Objective functions in dose–volume-based optimization of the IMRT planning. Application of the dose–volume-based constraint at the point (D_V, V_D) in the structure with a large volume effect such as the parallel-like organs. DVH constraint indicated that no more than V_D % of the volume should receive a dose of D_V Gy. D_{median} and D_{mean} values should be reported for evaluation

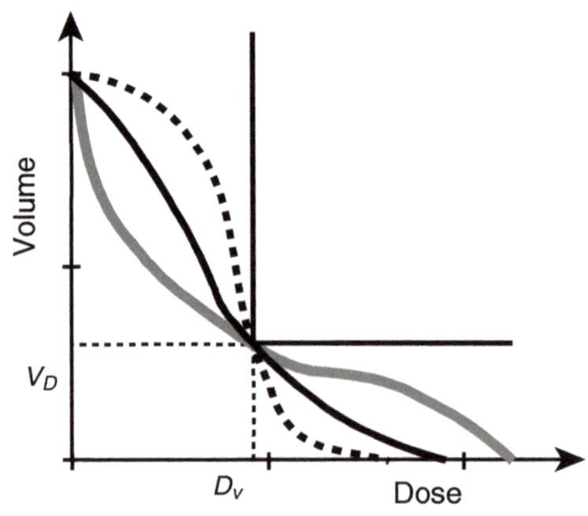

Fig. 4.5 Representation of limitations of the DVH for the structure with a large volume effect such as the parallel-like organs. Three DVH curves producing the different effects will fit for one dose–volume-based constraint (at D_V and V_D of the volume). Application of multiple constraints may well control the shape of the DVH as desired

4.3.2.3 Iterative Optimization Using a Computer-Based Algorithm

A computer-based algorithm for inverse planning can incorporate "hard" constraints that cannot be altered or "soft" constraints that could be changed. Hard constraints restrict the process to only solutions that are feasible; examples include physical constraints such as forbidding negative beam intensities and restrictions on the beam size or direction. Solutions that are based on hard constraints are not the "best," but can be feasible if they merely satisfy all hard constraints. Instead, "soft" constraints can better help achieve clinical goals. Such constraints can include specifications such as dose–volume uniformity, other dose–volume criteria, or the absorbed-dose limits to various structures. The optimization algorithm searches the parameter space seeking to achieve an objective function for the generation of an optimized solution in combination with soft and hard constraints. It should be noted that hard constraints bind the solution space, whereas soft constraints define a global minimum for a given objective function.

Optimization algorithms used to search for the possible solutions could be classified into the following two broad categories: deterministic and stochastic methods. In deterministic methods, the same solution will always be found when the same setup and initial conditions are used; they do not contain any random elements. Examples of deterministic methods include linear least squares and gradient descents [57]. A commonly used gradient descent in IMRT is nonlinear least-squares minimization [58], which is also called the steepest descent method. This algorithm follows the negative gradient of the objective function, similar to a downhill skier who could always go down the steepest slopes. Therefore, it iteratively finds a value for the objective function that is smaller than the last, getting closer to the minimum using relatively few steps [44, 59–61]. Compared with stochastic methods, the main advantage of the gradient descent and other deterministic methods is speed. However, if multiple minima are expected, the gradient descent searches then cannot escape from a local minimum of the objective function without finding the global minimum [62].

In inverse planning for IMRT, one of the most widely used stochastic algorithms is simulated annealing. Annealing is a physical process involving the heating and controlled cooling of a material being crystallized. This process is simulated during optimization of treatment plans. Higher temperatures indicate larger sizes of the random steps within the search space and even control the probability of "hill climbing." In other words, many iteration steps may result in the temporal acceptance of a worse treatment plan than the previous one, because of the random principle involved [63, 64]. Generally, stochastic algorithms allow plans to escape from the local minima and continue to explore the solution space to reach the global minimum if there is unlimited time to search the parameter space. Because time is restricted for treatment planning, the iterative search needs to be stopped either manually or automatically if its progress has slowed. The computed value of the objective function is usually displayed graphically to help decide when to stop the search. In practice, stochastic and gradient descent methods could be combined.

If the cost function depends only on the dose-based objectives with the aim of one dose per volume, local minima do not exist. In contrast, dose–volume-based objectives can cause local minima. Multiple local minima can exist when optimizing beam angles and numbers and during optimization using biological models (dose–response-based), in which even different dose distributions can result in the same response probability. If multiple minima exist with a dose–volume-based optimization, it is difficult to know whether a given solution is optimal. In such cases, one must attempt to discover ways to force inverse planning systems into different parts of the solution space by changing the initial conditions.

4.3.2.4 Leaf Segmentation, Forward Dose Calculation, and Plan Evaluation

Dose constraints are assigned to both target volumes and OARs, and inverse optimization is then performed to identify the individual weights of a large number of beamlets. The computer adjusts the fluences of these beamlets according to the required planning dose objectives. After optimization, the intensity profiles need to be converted into delivery instructions for the specific system used. For multileaf collimator (MLC) systems, a leaf-segmentation or leaf-sequencing algorithm determines a series of "deliverable" MLC shapes and/or movements to best replicate the desired dose pattern. It is required that MLC should move at the maximum allowable speed in each segment, thereby minimizing the treatment time. Both transmission through MLCs and the effect of their rounded edges should be considered. Therefore, the original intensity profiles cannot be converted with complete fidelity using these processes, and so the final leaf motion is converted back into a deliverable intensity profile for subsequent forward calculations of absorbed dose and monitor unit (MU).

Once the forward absorbed-dose calculation has been completed using the deliverable profiles, the plan is evaluated using standard methods including planar dose distributions, DVHs, and radiobiological indices such as TCP and NTCP. If the dose distribution does not achieve the planning aims, the optimization parameters, optimization-only structures, or beams are adjusted, and the process is repeated. The choice between plan improvement and plan acceptance is often based on tradeoffs among conflicting aims. As such, the prescription could be considered acceptable but be different from the original planning aim.

4.3.2.5 Beamlet Optimization and Aperture-Based Optimization

For both stochastic and deterministic strategies, there are two general frameworks for IMRT optimization: beamlet optimization [64–66] and aperture-based optimization [67]. Both strategies use the same criteria (constraints and objective functions). For beamlet optimization, each field is discretized into a grid of beamlets with modulated intensities. Beamlet-based optimization, which was pioneered for serial TomoTherapy (TomoTherapy Inc., Madison, WI, USA), is used to design dynamic MLC IMRT because it can achieve relatively highly modulated intensity patterns. After the leaf-sequencing step, calculating the final absorbed dose and MU could result in a significantly different absorbed-dose distribution than that suggested by the optimized plan; this discrepancy is called "convergence error" [68]. The cause of the discrepancy is partly because a simpler model of energy deposition is often used for the optimization algorithm than for the final dose calculations to accelerate the process. Because the accuracy of the final dose calculations is most important, it considers the limitations of MLC delivery that are not accounted for by the optimizer, as described above.

Aperture-based optimization takes into account the limitations of MLC during each optimization step, so that the segmentation step is eliminated. Instead, the best set of aperture shapes (and their relative weights) is directly sought to deliver the intensity pattern without explicit discretization of the field into a grid of beamlets. This approach is used for segmental MLC and intensity-modulated arc therapy (IMAT). If the same dose calculation algorithm is used for iterative optimization of apertures as that for the final absorbed-dose calculation, this method will then also avoid convergence errors. The automated aperture-based optimization procedure, in addition to new segments, could modify the boundaries of previous segments to better avoid normal tissue or improve the target dose homogeneity.

4.3.2.6 Methods of Delivery of IMRT

IMRT can be delivered using a conventional linear accelerator (linac) equipped with an MLC system designed to shape field apertures. This type of IMRT is delivered at fixed gantry angles by delivering multiple field segments statically (known as step-and-shoot IMRT) or by having the leaf pairs slide across the field dynamically (called dynamic or sliding window IMRT) [69]. The TomoTherapy HI-ART System [70] is a rotational IMRT approach that uses a binary collimator. Recently, IMAT was developed as a form of rotational IMRT [71, 72].

As the complexity of the IMRT plan increases, there is a dramatic increase in the number of MUs required to deliver IMRT [73, 74]. This increase in MU is associated with increased treatment time and a greater leakage of radiation from MLCs, which can increase the total body dose and the risk of secondary cancers [75]. It is also associated with increased dosimetric uncertainty. Therefore, attempts should be taken to reduce such complexity, including setting intensity limits and placing penalties on the cost function [76].

IMAT has greater flexibility in shaping the dose distribution than conventional IMRT. Studies have shown that IMAT could reduce treatment time and provide better plan quality in terms of target dose coverage and the sparing of normal tissue, as it uses fewer MUs per fraction as compared with fixed-field IMRT [77–79].

4.4 Solutions to the Pitfalls Associated with IMRT Planning Optimization

The clinical situations described below are often encountered during IMRT planning: (1) overlapping volumes and conflicting absorbed-dose objectives and (2) planned absorbed dose in the buildup region and in a PTV extending outside the body contour.

4.4.1 Overlapping Volumes with Conflicting Optimization Criteria

The concepts of PTV and PRV have been widely adopted to compensate for the geometric uncertainties and variations in the position of targets and structures. Adding margins around the contoured volumes can often result in significant overlaps between different volumes, including PTVs and PRVs. A conflict can always occur because the planning aims at overlapping volumes that lack a common desired dose range, which results in a highly heterogeneous dose distribution with unacceptably cold and hot spots in PTVs and PRVs or even in RVR.

In cases of advanced NPCs with extensive infiltration to the skull base or the clivus (Fig. 4.1), PTV frequently encroaches or overlaps with the PRV of the brainstem. Some previous protocols have suggested that the margin could be as small as 1 mm in these situations [80]. In addition to prostate IMRT planning, one may shrink the CTV-to-PTV margin in the anterior–posterior (AP) direction to spare the rectum [5]. Generally, the setup uncertainty could be larger in the AP direction than in other directions during prostate IMRT. However, ICRU 83 recommends that the primary PTV margins should not be compromised in such manners.

An inverse planning algorithm cannot automatically judge to make much or less of parts of PTVs. In such cases, parts of PTV must be separated or prioritized specifically. ICRU 83 also suggests at least two different methods to resolve conflicts in the overlapping volumes [11]. One method is based on subdividing the volumes. Different dose–volume criteria are applied to individual subdivisions, including overlapping and nonoverlapping volumes. In the second method, the dose–volume constraints are relaxed for one or more of the contoured volumes that overlap. Both methods intend to achieve the same goal: a "controlled underdosage" for parts of PTV, a "controlled overdosage" for parts of PRV, or both. By changing the importance of the constraints, it is possible to achieve an underdosage in PTV to an overdosage in PRV, or anything between these extremes. Both methods require priority ranking according to a clinical decision that should be specified in the planning protocols. These methods enable the delineation of PTV and PRV to be maintained without compromise. For example, PTV often overlaps the parotid gland during head and neck IMRT (Fig. 4.6). Different dose–volume constraints could be applied separately to the overlapping and nonoverlapping volumes to manipulate target coverage at a higher priority than parotid sparing. The doses prescribed for the whole parotid gland should then be calculated to evaluate or predict functional sparing.

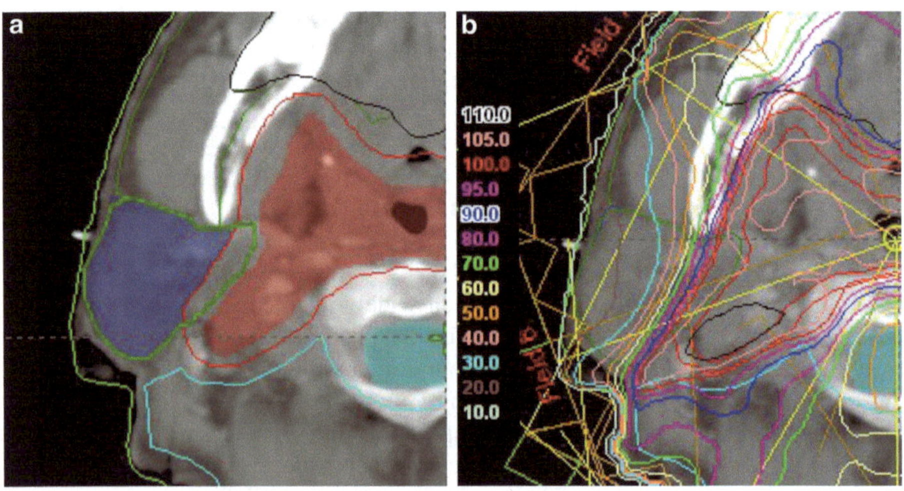

Fig. 4.6 A solution in IMRT optimization for the overlapping of the PTV and the parotid gland. (**a**) A planning conflict may occur in the case that the contour of right parotid gland overlaps with the PTV. (**b**) One may prefer to choose a solution of controlled overdosing to the overlapping parts of the parotid to maintain the whole PTV coverage with higher priority (Reprinted from Jpn J Clin Radiol 2013: 58(5), with permission)

4.4.2 PTV in the Buildup Region or Outside the Body Contour

In some clinical situations such as in head and neck cancers, CTV is located very close to the surface of the skin. By adding the CTV-to-PTV margin in a 3D concentric manner, parts of PTV could extend into the buildup region or even outside the body contour. This may be an extreme analogy of the above-mentioned overlapping volumes. It is known that the doses calculated in most dose computation algorithms are often inaccurate and lower than the doses delivered to the buildup regions. During IMRT optimization, such an algorithm could cause convergence errors by increasing the absorbed dose in parts of PTV within these regions, which often leads to unacceptably heterogeneous absorbed-dose distributions in PTV or regions of high absorbed doses elsewhere [68]. Instead, removing the part of PTV outside the body contour or bordering PTV expansion at the skin contour avoids irrelevant optimization of the absorbed dose in air; however, this is not an ideal solution.

ICRU 83 suggests some solutions for these limitations [11]. If CTV is very close to the skin in the buildup region, underdosage is a real and unacceptable problem that is better solved by adding a bolus, as in conventional radiotherapy. It is better to use the bolus for the CT scan, so that it is accurately represented in the planning. If underdosage in the buildup region is clinically acceptable, two different methods could be followed, as mentioned above for overlapping volumes: PTV subdivision or relaxation of the absorbed-dose objectives for planning.

4.5 Designs for IMRT Planning Protocols

4.5.1 Simultaneous Integrated Boost (SIB) and Sequential IMRT Strategies

Because IMRT can achieve excellent dose homogeneity within target volumes and conformally avoid adjacent normal tissues, it permits a variety of treatment designs with different doses and fractionations toward the gross tumor targets and subclinical target areas. Such strategies, known as the SIB technique, are commonly used to perform differential dose painting (66–74 Gy to the gross disease and 50–60 Gy to the subclinical disease) for each treatment fraction throughout the entire radiation course, particularly in head and neck tumors. In contrast, standard fractionation could also be used to design IMRT strategies with an initial plan (46–50 Gy to a large field) and a separated boost plan (20–24 Gy to the boost volume) for up to 7 weeks, which are known as sequential (or two-step) IMRT techniques.

Some authors reported that the SIB strategy not only produces a superior dose distribution but is also an easier, more efficient, and a less error-prone way to plan and deliver IMRT, because it uses the same plan for the entire treatment course [6, 81–83]. Because each target volume receives different doses per fraction in SIB, the prescribed nominal dose and fraction must be adjusted appropriately.

The standard dose per fraction could be chosen for the gross disease using SIB, which may be a lower dose per fraction for the subclinical disease or in elective areas. At the Mallinckrodt Institute of Radiology, SIB IMRT for the definitive treatment of head and neck cancers prescribed 70 Gy in 35 fractions at 2 Gy per fraction to the gross disease, 63 Gy at 1.8 Gy per fraction to the adjacent soft tissue and the nodal volume at a high risk, and 56 Gy at 1.6 Gy per fraction to the elective nodal regions. Chao et al. have shown that this strategy is well tolerated with concurrent chemotherapy [3].

The standard 2 Gy per fraction could also be selected for lower-dose target volumes, which would require a higher dose per fraction, 2.5 Gy or more, at the gross targets. At Virginia Commonwealth University, an IMRT protocol using accelerated fractionation with SIB was performed for locally advanced head and neck squamous cell carcinomas. Total doses of 68.1, 70.8, and 73.8 Gy were delivered in 30 fractions to GTV with escalating doses per fraction (2.27, 2.36, and 2.46 Gy, respectively). CTV was defined as the tissue within 1 cm around GTV (at a high risk of subclinical disease) that received 60 Gy in 30 fractions of 2.0 Gy. The elective volume received 54 Gy in 30 fractions of 1.8 Gy. Lauve et al. reported that 70.8 Gy in 30 fractions of 2.36 Gy was the maximum tolerated dose deliverable [84].

In the Radiation Therapy Oncology Group (RTOG) 0022 trial, patients with early oropharyngeal cancers were treated using SIB IMRT that delivered 66 Gy in 30 fractions of 2.2 Gy to the PTV of the gross disease, 60 Gy to the high-risk regions, and 54 Gy to the subclinical disease regions. Eisbruch et al. [85] reported that this moderately accelerated hypofractionated IMRT without chemotherapy was feasible and that it achieved high tumor control rates and reduced salivary toxicity.

In the RTOG 0225 phase II trial, NPCs were treated using SIB IMRT with the delivery of 70 Gy in 33 fractions of 2.12 Gy to the PTV of the gross disease and 59.4 Gy to the subclinical regions with or without chemotherapy. Lee et al. reported that this SIB regimen was feasible with or without chemotherapy and this regimen produced excellent results with a 90 % LRPF rate [80].

In a previous study, Mohan et al. [83] suggested that SIB IMRT resulted in a superior dose distribution compared with sequential IMRT and that it permitted the delivery of graded doses to various targets with different risks while sparing normal tissues to the greatest extent possible. Evaluating the isoeffect using the linear-quadratic model revealed that the biological effective dose to normal tissues outside the targets using accelerated SIB strategies was lower; however, normal tissues embedded within or adjacent to target volumes were assumed to receive a similar dose to that received by the tumor and, therefore, were at higher risks compared with IMRT with standard fractionation. However, long-term data are not yet available for these approaches regarding the late effects on normal tissues, including the muscle, blood vessels, and nerves embedded within target volumes, or regarding concerns with the concurrent use of chemotherapy.

Sequential IMRT, such as two-step IMRT [86, 87], may be a useful adaptive strategy and may even have a lower-dose conformity than SIB. It is known that measurable anatomic changes occur during fractionated IMRT for head and neck cancers. These changes in body contour, target volumes, and OARs may become significant during the second half of treatment (after 3–4 weeks) and could potentially have a detrimental effect on the dose distribution. Therefore, it is reasonable to develop an adaptive RT scheme that takes into account such treatment-related anatomical and volumetric changes [88]. Nishimura et al. previously revealed excellent overall survival and the locoregional control rates without xerostomia using this IMRT method to treat NPCs [86, 87].

4.5.2 An Experience of a Japanese Multi-institutional Trial Using Two-Step IMRT Planning

The Radiation Therapy Study Group of the Japan Clinical Oncology Group (JCOG) is currently conducting the first multi-institutional trial using IMRT for locoregionally advanced NPC. In this phase II study (JCOG 1015) [89], the treatment regimen is concurrent chemoradiotherapy with two-step IMRT (70 Gy in 35 fractions over 47 days). This two-step IMRT was designed to deliver an initial large-field IMRT plan of 46 Gy in 23 fractions to the gross targets and elective regions followed by a boost IMRT plan of 24 Gy that was confined to the gross and high-risk target volumes (Figs. 4.2 and 4.7). We assumed that this adaptive IMRT scheme may be desirable for undifferentiated or poorly differentiated NPC, which is known to be radiosensitive and to show rapid tumor shrinkage during IMRT with concurrent chemotherapy. Even with such advantages, this method requires a longer time to perform two separate planning steps including contouring, inverse planning with more difficulty, and repeated quality assurance procedures compared with SIB without a replanning step.

Because this is the first multi-institutional trial using IMRT in Japan, dry-run tests were performed prior to the initiation of the trial to generate IMRT plans using the same CT image data sets together with MRI and FDG-PET scans of two NPC model cases according to the protocol requirements of the individual participating institutions. During the initial survey, there were large interinstitutional variations in the contours of target volumes and OARs and the planning results (data not shown). To standardize the contouring rules among institutions, we have developed a consensus-based atlas for this particular scheme and have finished repeated dry-run tests, so that all participants were trained for this two-step IMRT planning. Such procedures have been recognized as useful for establishing standardized IMRT planning in a multi-institutional setting. There is now a strict quality control and quality assurance program in place, including a dosimetry audit, dry run, and individual case review using the ITC Remote Review Tool. Therefore, we now await the results of this prospective trial using two-step IMRT.

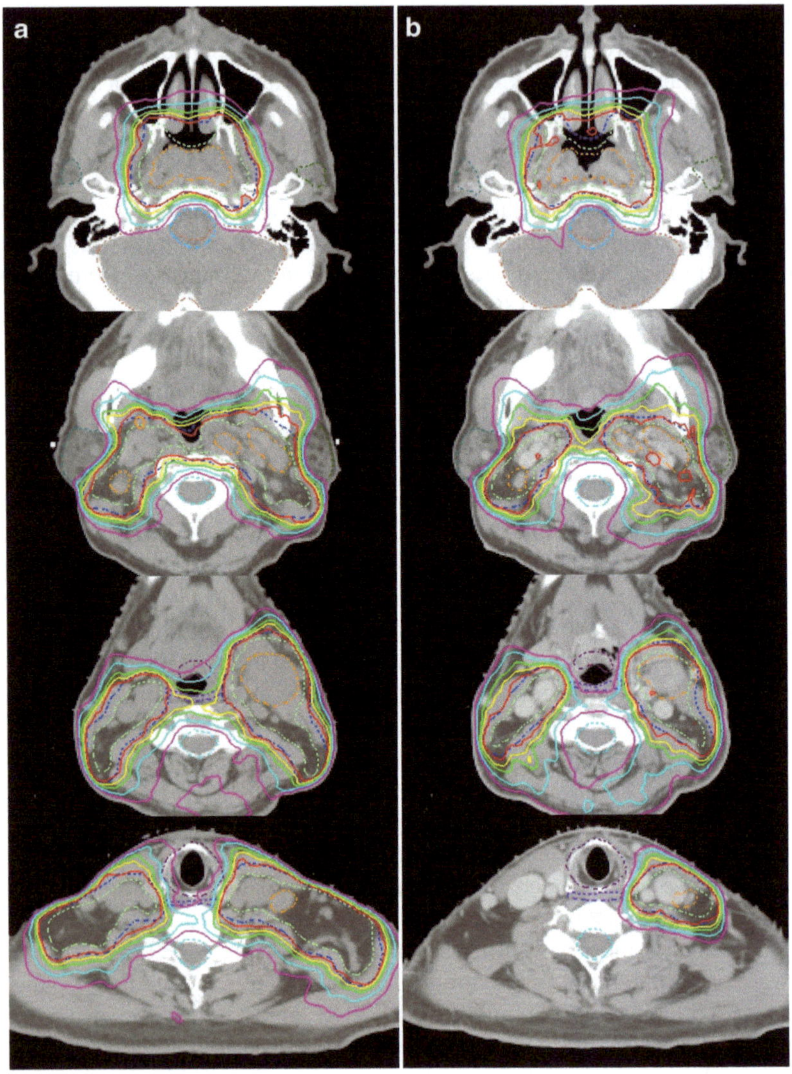

Fig. 4.7 An example of a two-step IMRT planning in JCOG 1015 protocol. (**a**) The initial plan of 46 Gy in 23 fractions to the large field included both the clinically detectable tumors and the elective regions. (**b**) The boost plan of 24 Gy in 12 fractions up to 70 Gy in total to the high-dose boost volume

Acknowledgments This study was supported in part by a Grant-in-Aid for Cancer Research (H23-009) from the Ministry of Health, Labour and Welfare of Japan and by a Grant-in-Aid for Scientific Research (26461892) from Japan Society for the Promotion of Science.

References

1. Brahme A (1987) Design principles and clinical possibilities with a new generation of radiation therapy equipment. A review. Acta Oncol 26(6):403–412
2. Burman C, Chui CS, Kutcher G, Leibel S, Zelefsky M, LoSasso T, Spirou S, Wu Q, Yang J, Stein J, Mohan R, Fuks Z, Ling CC (1997) Planning, delivery, and quality assurance of intensity-modulated radiotherapy using dynamic multileaf collimator: a strategy for large-scale implementation for the treatment of carcinoma of the prostate. Int J Radiat Oncol Biol Phys 39(4):863–873
3. Chao KS, Deasy JO, Markman J, Haynie J, Perez CA, Purdy JA, Low DA (2001) A prospective study of salivary function sparing in patients with head-and-neck cancers receiving intensity-modulated or three-dimensional radiation therapy: initial results. Int J Radiat Oncol Biol Phys 49(4):907–916
4. Wu Q, Manning M, Schmidt-Ullrich R, Mohan R (2000) The potential for sparing of parotids and escalation of biologically effective dose with intensity-modulated radiation treatments of head and neck cancers: a treatment design study. Int J Radiat Oncol Biol Phys 46(1):195–205
5. Zelefsky MJ, Fuks Z, Happersett L, Lee HJ, Ling CC, Burman CM, Hunt M, Wolfe T, Venkatraman ES, Jackson A, Skwarchuk M, Leibel SA (2000) Clinical experience with intensity modulated radiation therapy (IMRT) in prostate cancer. Radiother Oncol 55(3):241–249
6. Chao KS, Majhail N, Huang CJ, Simpson JR, Perez CA, Haughey B, Spector G (2001) Intensity-modulated radiation therapy reduces late salivary toxicity without compromising tumor control in patients with oropharyngeal carcinoma: a comparison with conventional techniques. Radiother Oncol 61(3):275–280
7. Eisbruch A, Ten Haken RK, Kim HM, Marsh LH, Ship JA (1999) Dose, volume, and function relationships in parotid salivary glands following conformal and intensity-modulated irradiation of head and neck cancer. Int J Radiat Oncol Biol Phys 45(3):577–587
8. Lee N, Xia P, Quivey JM, Sultanem K, Poon I, Akazawa C, Akazawa P, Weinberg V, Fu KK (2002) Intensity-modulated radiotherapy in the treatment of nasopharyngeal carcinoma: an update of the UCSF experience. Int J Radiat Oncol Biol Phys 53(1):12–22
9. ICRU (1993) Prescribing, recording, and reporting photon beam therapy, ICRU report 50. International Commission on Radiation Units and Measurements, Bethesda
10. ICRU (1999) Prescribing, recording, and reporting photon beam therapy (Supplement to ICRU report 50), ICRU report 62. International Commission on Radiation Units and Measurements, Bethesda
11. ICRU (2010) ICRU report 83: prescribing, recording, and reporting photon-beam Intensity-Modulated Radiation Therapy (IMRT). J ICRU 10(1):1–106
12. Chao KS, Wippold FJ, Ozyigit G, Tran BN, Dempsey JF (2002) Determination and delineation of nodal target volumes for head-and-neck cancer based on patterns of failure in patients receiving definitive and postoperative IMRT. Int J Radiat Oncol Biol Phys 53(5):1174–1184
13. Eisbruch A, Foote RL, O'Sullivan B, Beitler JJ, Vikram B (2002) Intensity-modulated radiation therapy for head and neck cancer: emphasis on the selection and delineation of the targets. Semin Radiat Oncol 12(3):238–249
14. Gregoire V, Ang K, Budach W, Grau C, Hamoir M, Langendijk JA, Lee A, Le QT, Maingon P, Nutting C, O'Sullivan B, Porceddu SV, Lengele B (2014) Delineation of the neck node levels for head and neck tumors: a 2013 update. DAHANCA, EORTC, HKNPCSG, NCIC CTG, NCRI, RTOG, TROG consensus guidelines. Radiother Oncol 110(1):172–181
15. Gregoire V, Coche E, Cosnard G, Hamoir M, Reychler H (2000) Selection and delineation of lymph node target volumes in head and neck conformal radiotherapy. Proposal for standardizing terminology and procedure based on the surgical experience. Radiother Oncol 56(2):135–150
16. Gregoire V, Daisne JF, Geets X, Levendag P (2003) Selection and delineation of target volumes in head and neck tumors: beyond ICRU definition. Rays 28(3):217–224

17. Gregoire V, Eisbruch A, Hamoir M, Levendag P (2006) Proposal for the delineation of the nodal CTV in the node-positive and the post-operative neck. Radiother Oncol 79(1):15–20
18. Hartford AC, Palisca MG, Eichler TJ, Beyer DC, Devineni VR, Ibbott GS, Kavanagh B, Kent JS, Rosenthal SA, Schultz CJ, Tripuraneni P, Gaspar LE (2009) American Society for Therapeutic Radiology and Oncology (ASTRO) and American College of Radiology (ACR) Practice Guidelines for Intensity-Modulated Radiation Therapy (IMRT). Int J Radiat Oncol Biol Phys 73(1):9–14
19. Emami B, Sethi A, Petruzzelli GJ (2003) Influence of MRI on target volume delineation and IMRT planning in nasopharyngeal carcinoma. Int J Radiat Oncol Biol Phys 57(2):481–488
20. Geets X, Daisne JF, Arcangeli S, Coche E, De Poel M, Duprez T, Nardella G, Gregoire V (2005) Inter-observer variability in the delineation of pharyngo-laryngeal tumor, parotid glands and cervical spinal cord: comparison between CT-scan and MRI. Radiother Oncol 77(1):25–31
21. Rasch C, Steenbakkers R, van Herk M (2005) Target definition in prostate, head, and neck. Semin Radiat Oncol 15(3):136–145
22. Dirix P, Vandecaveye V, De Keyzer F, Op de Beeck K, Poorten VV, Delaere P, Verbeken E, Hermans R, Nuyts S (2010) Diffusion-weighted MRI for nodal staging of head and neck squamous cell carcinoma: impact on radiotherapy planning. Int J Radiat Oncol Biol Phys 76(3):761–766
23. Hentschel B, Oehler W, Strauss D, Ulrich A, Malich A (2011) Definition of the CTV prostate in CT and MRI by using CT-MRI image fusion in IMRT planning for prostate cancer. Strahlenther Onkol 187(3):183–190
24. Perna L, Fiorino C, Cozzarini C, Broggi S, Cattaneo GM, De Cobelli F, Mangili P, Di Muzio N, Calandrino R (2009) Sparing the penile bulb in the radical irradiation of clinically localised prostate carcinoma: a comparison between MRI and CT prostatic apex definition in 3DCRT, Linac-IMRT and Helical Tomotherapy. Radiother Oncol 93(1):57–63
25. Rasch C, Barillot I, Remeijer P, Touw A, van Herk M, Lebesque JV (1999) Definition of the prostate in CT and MRI: a multi-observer study. Int J Radiat Oncol Biol Phys 43(1):57–66
26. Wachter S, Wachter-Gerstner N, Bock T, Goldner G, Kovacs G, Fransson A, Potter R (2002) Interobserver comparison of CT and MRI-based prostate apex definition. Clinical relevance for conformal radiotherapy treatment planning. Strahlenther Onkol 178(5):263–268
27. Riches SF, Payne GS, Desouza NM, Dearnaley D, Morgan VA, Morgan SC, Partridge M (2014) Effect on therapeutic ratio of planning a boosted radiotherapy dose to the dominant intraprostatic tumour lesion within the prostate based on multifunctional MR parameters. Br J Radiol 87(1037):20130813
28. van Lin EN, Futterer JJ, Heijmink SW, van der Vight LP, Hoffmann AL, van Kollenburg P, Huisman HJ, Scheenen TW, Witjes JA, Leer JW, Barentsz JO, Visser AG (2006) IMRT boost dose planning on dominant intraprostatic lesions: gold marker-based three-dimensional fusion of CT with dynamic contrast-enhanced and 1H-spectroscopic MRI. Int J Radiat Oncol Biol Phys 65(1):291–303
29. Daisne JF, Duprez T, Weynand B, Lonneux M, Hamoir M, Reychler H, Gregoire V (2004) Tumor volume in pharyngolaryngeal squamous cell carcinoma: comparison at CT, MR imaging, and FDG PET and validation with surgical specimen. Radiology 233(1):93–100
30. Geets X, Daisne JF, Tomsej M, Duprez T, Lonneux M, Gregoire V (2006) Impact of the type of imaging modality on target volumes delineation and dose distribution in pharyngo-laryngeal squamous cell carcinoma: comparison between pre- and per-treatment studies. Radiother Oncol 78(3):291–297
31. Paulino AC, Koshy M, Howell R, Schuster D, Davis LW (2005) Comparison of CT- and FDG-PET-defined gross tumor volume in intensity-modulated radiotherapy for head-and-neck cancer. Int J Radiat Oncol Biol Phys 61(5):1385–1392
32. Ciernik IF, Dizendorf E, Baumert BG, Reiner B, Burger C, Davis JB, Lutolf UM, Steinert HC, Von Schulthess GK (2003) Radiation treatment planning with an integrated positron emission

and computer tomography (PET/CT): a feasibility study. Int J Radiat Oncol Biol Phys 57(3):853–863
33. Okubo M, Nishimura Y, Nakamatsu K, Okumura M, Shibata T, Kanamori S, Hanaoka K, Hosono M (2010) Radiation treatment planning using positron emission and computed tomography for lung and pharyngeal cancers: a multiple-threshold method for [(18)F]fluoro-2-deoxyglucose activity. Int J Radiat Oncol Biol Phys 77(2):350–356
34. Wang D, Schultz CJ, Jursinic PA, Bialkowski M, Zhu XR, Brown WD, Rand SD, Michel MA, Campbell BH, Wong S, Li XA, Wilson JF (2006) Initial experience of FDG-PET/CT guided IMRT of head-and-neck carcinoma. Int J Radiat Oncol Biol Phys 65(1):143–151
35. Soto DE, Kessler ML, Piert M, Eisbruch A (2008) Correlation between pretreatment FDG-PET biological target volume and anatomical location of failure after radiation therapy for head and neck cancers. Radiother Oncol 89(1):13–18
36. Madani I, Duthoy W, Derie C, De Gersem W, Boterberg T, Saerens M, Jacobs F, Gregoire V, Lonneux M, Vakaet L, Vanderstraeten B, Bauters W, Bonte K, Thierens H, De Neve W (2007) Positron emission tomography-guided, focal-dose escalation using intensity-modulated radiotherapy for head and neck cancer. Int J Radiat Oncol Biol Phys 68(1):126–135
37. Gregoire V, Haustermans K (2009) Functional image-guided intensity modulated radiation therapy: integration of the tumour microenvironment in treatment planning. Eur J Cancer 45(Suppl 1):459–460
38. Lee NY, Mechalakos JG, Nehmeh S, Lin Z, Squire OD, Cai S, Chan K, Zanzonico PB, Greco C, Ling CC, Humm JL, Schoder H (2008) Fluorine-18-labeled fluoromisonidazole positron emission and computed tomography-guided intensity-modulated radiotherapy for head and neck cancer: a feasibility study. Int J Radiat Oncol Biol Phys 70(1):2–13
39. Nehmeh SA, Lee NY, Schroder H, Squire O, Zanzonico PB, Erdi YE, Greco C, Mageras G, Pham HS, Larson SM, Ling CC, Humm JL (2008) Reproducibility of intratumor distribution of (18)F-fluoromisonidazole in head and neck cancer. Int J Radiat Oncol Biol Phys 70(1):235–242
40. Thorwarth D, Eschmann SM, Paulsen F, Alber M (2007) A model of reoxygenation dynamics of head-and-neck tumors based on serial 18F-fluoromisonidazole positron emission tomography investigations. Int J Radiat Oncol Biol Phys 68(2):515–521
41. Ghadjar P, Simcock M, Schreiber-Facklam H, Zimmer Y, Grater R, Evers C, Arnold A, Wilkens L, Aebersold DM (2010) Incidence of small lymph node metastases with evidence of extracapsular extension: clinical implications in patients with head and neck squamous cell carcinoma. Int J Radiat Oncol Biol Phys 78(5):1366–1372
42. Apisarnthanarax S, Elliott DD, El-Naggar AK, Asper JA, Blanco A, Ang KK, Garden AS, Morrison WH, Rosenthal D, Weber RS, Chao KS (2006) Determining optimal clinical target volume margins in head-and-neck cancer based on microscopic extracapsular extension of metastatic neck nodes. Int J Radiat Oncol Biol Phys 64(3):678–683
43. Withers HR, Taylor JM, Maciejewski B (1988) Treatment volume and tissue tolerance. Int J Radiat Oncol Biol Phys 14(4):751–759
44. Bortfeld T (1999) Optimized planning using physical objectives and constraints. Semin Radiat Oncol 9(1):20–34
45. Emami B, Lyman J, Brown A, Coia L, Goitein M, Munzenrider JE, Shank B, Solin LJ, Wesson M (1991) Tolerance of normal tissue to therapeutic irradiation. Int J Radiat Oncol Biol Phys 21(1):109–122
46. Burman C, Kutcher GJ, Emami B, Goitein M (1991) Fitting of normal tissue tolerance data to an analytic function. Int J Radiat Oncol Biol Phys 21(1):123–135
47. Kutcher GJ, Burman C, Brewster L, Goitein M, Mohan R (1991) Histogram reduction method for calculating complication probabilities for three-dimensional treatment planning evaluations. Int J Radiat Oncol Biol Phys 21(1):137–146
48. Daly ME, Luxton G, Choi CY, Gibbs IC, Chang SD, Adler JR, Soltys SG (2012) Normal tissue complication probability estimation by the Lyman-Kutcher-Burman method does not

accurately predict spinal cord tolerance to stereotactic radiosurgery. Int J Radiat Oncol Biol Phys 82(5):2025–2032
49. Gulliford SL, Partridge M, Sydes MR, Webb S, Evans PM, Dearnaley DP (2012) Parameters for the Lyman Kutcher Burman (LKB) model of Normal Tissue Complication Probability (NTCP) for specific rectal complications observed in clinical practise. Radiother Oncol 102(3):347–351
50. Semenenko VA, Li XA (2008) Lyman-Kutcher-Burman NTCP model parameters for radiation pneumonitis and xerostomia based on combined analysis of published clinical data. Phys Med Biol 53(3):737–755
51. Bentzen SM, Constine LS, Deasy JO, Eisbruch A, Jackson A, Marks LB, Ten Haken RK, Yorke ED (2010) Quantitative Analyses of Normal Tissue Effects in the Clinic (QUANTEC): an introduction to the scientific issues. Int J Radiat Oncol Biol Phys 76(3 Suppl):S3–S9
52. Marks LB, Yorke ED, Jackson A, Ten Haken RK, Constine LS, Eisbruch A, Bentzen SM, Nam J, Deasy JO (2010) Use of normal tissue complication probability models in the clinic. Int J Radiat Oncol Biol Phys 76(3 Suppl):S10–S19
53. Appelt AL, Vogelius IR, Farr KP, Khalil AA, Bentzen SM (2014) Towards individualized dose constraints: adjusting the QUANTEC radiation pneumonitis model for clinical risk factors. Acta Oncol 53(5):605–612
54. Beetz I, Steenbakkers RJ, Chouvalova O, Leemans CR, Doornaert P, van der Laan BF, Christianen ME, Vissink A, Bijl HP, van Luijk P, Langendijk JA (2014) The QUANTEC criteria for parotid gland dose and their efficacy to prevent moderate to severe patient-rated xerostomia. Acta Oncol 53(5):597–604
55. Lee TF, Fang FM (2013) Quantitative analysis of normal tissue effects in the clinic (QUANTEC) guideline validation using quality of life questionnaire datasets for parotid gland constraints to avoid causing xerostomia during head-and-neck radiotherapy. Radiother Oncol 106(3):352–358
56. Moiseenko V, Wu J, Hovan A, Saleh Z, Apte A, Deasy JO, Harrow S, Rabuka C, Muggli A, Thompson A (2012) Treatment planning constraints to avoid xerostomia in head-and-neck radiotherapy: an independent test of QUANTEC criteria using a prospectively collected dataset. Int J Radiat Oncol Biol Phys 82(3):1108–1114
57. Morse PM, Feshbach H (1953) Methods of theoretical physics. McGraw-Hill, New York/London
58. Bortfeld T, Schlegel W (1993) Optimization of beam orientations in radiation therapy: some theoretical considerations. Phys Med Biol 38(2):291–304
59. Holmes T, Mackie TR (1994) A comparison of three inverse treatment planning algorithms. Phys Med Biol 39(1):91–106
60. Spirou SV, Chui CS (1998) A gradient inverse planning algorithm with dose-volume constraints. Med Phys 25(3):321–333
61. Xing L, Chen GT (1996) Iterative methods for inverse treatment planning. Phys Med Biol 41(10):2107–2123
62. Deasy JO (1997) Multiple local minima in radiotherapy optimization problems with dose-volume constraints. Med Phys 24(7):1157–1161
63. Mageras GS, Mohan R (1993) Application of fast simulated annealing to optimization of conformal radiation treatments. Med Phys 20(3):639–647
64. Webb S (1992) Optimization by simulated annealing of three-dimensional, conformal treatment planning for radiation fields defined by a multileaf collimator: II. Inclusion of two-dimensional modulation of the x-ray intensity. Phys Med Biol 37(8):1689–1704
65. Bortfeld T, Burkelbach J, Boesecke R, Schlegel W (1990) Methods of image reconstruction from projections applied to conformation radiotherapy. Phys Med Biol 35(10):1423–1434
66. Holmes T, Mackie TR, Simpkin D, Reckwerdt P (1991) A unified approach to the optimization of brachytherapy and external beam dosimetry. Int J Radiat Oncol Biol Phys 20(4):859–873
67. Shepard DM, Earl MA, Li XA, Naqvi S, Yu C (2002) Direct aperture optimization: a turnkey solution for step-and-shoot IMRT. Med Phys 29(6):1007–1018

68. Jeraj R, Keall P (2000) The effect of statistical uncertainty on inverse treatment planning based on Monte Carlo dose calculation. Phys Med Biol 45(12):3601–3613
69. Intensity Modulated Radiation Therapy Collaborative Working Group (2001) Intensity-modulated radiotherapy: current status and issues of interest. Int J Radiat Oncol Biol Phys 51(4):880–914
70. Mackie TR, Holmes T, Swerdloff S, Reckwerdt P, Deasy JO, Yang J, Paliwal B, Kinsella T (1993) Tomotherapy: a new concept for the delivery of dynamic conformal radiotherapy. Med Phys 20(6):1709–1719
71. Yu CX (1995) Intensity-modulated arc therapy with dynamic multileaf collimation: an alternative to tomotherapy. Phys Med Biol 40(9):1435–1449
72. Ling CC, Zhang P, Archambault Y, Bocanek J, Tang G, Losasso T (2008) Commissioning and quality assurance of RapidArc radiotherapy delivery system. Int J Radiat Oncol Biol Phys 72(2):575–581
73. Mohan R, Arnfield M, Tong S, Wu Q, Siebers J (2000) The impact of fluctuations in intensity patterns on the number of monitor units and the quality and accuracy of intensity modulated radiotherapy. Med Phys 27(6):1226–1237
74. Craft D, Suss P, Bortfeld T (2007) The tradeoff between treatment plan quality and required number of monitor units in intensity-modulated radiotherapy. Int J Radiat Oncol Biol Phys 67(5):1596–1605
75. Kry SF, Salehpour M, Followill DS, Stovall M, Kuban DA, White RA, Rosen II (2005) The calculated risk of fatal secondary malignancies from intensity-modulated radiation therapy. Int J Radiat Oncol Biol Phys 62(4):1195–1203
76. Broderick M, Leech M, Coffey M (2009) Direct aperture optimization as a means of reducing the complexity of intensity modulated radiation therapy plans. Radiat Oncol 4:8
77. Kjaer-Kristoffersen F, Ohlhues L, Medin J, Korreman S (2009) RapidArc volumetric modulated therapy planning for prostate cancer patients. Acta Oncol 48(2):227–232
78. Rao M, Yang W, Chen F, Sheng K, Ye J, Mehta V, Shepard D, Cao D (2010) Comparison of Elekta VMAT with helical tomotherapy and fixed field IMRT: plan quality, delivery efficiency and accuracy. Med Phys 37(3):1350–1359
79. Verbakel WF, Cuijpers JP, Hoffmans D, Bieker M, Slotman BJ, Senan S (2009) Volumetric intensity-modulated arc therapy vs. conventional IMRT in head-and-neck cancer: a comparative planning and dosimetric study. Int J Radiat Oncol Biol Phys 74(1):252–259
80. Lee N, Harris J, Garden AS, Straube W, Glisson B, Xia P, Bosch W, Morrison WH, Quivey J, Thorstad W, Jones C, Ang KK (2009) Intensity-modulated radiation therapy with or without chemotherapy for nasopharyngeal carcinoma: radiation therapy oncology group phase II trial 0225. J Clin Oncol 27(22):3684–3690
81. Dogan N, King S, Emami B, Mohideen N, Mirkovic N, Leybovich LB, Sethi A (2003) Assessment of different IMRT boost delivery methods on target coverage and normal-tissue sparing. Int J Radiat Oncol Biol Phys 57(5):1480–1491
82. Lamers-Kuijper E, Heemsbergen W, van Mourik A, Rasch C (2011) Sequentially delivered boost plans are superior to simultaneously delivered plans in head and neck cancer when the boost volume is located further away from the parotid glands. Radiother Oncol 98(1):51–56
83. Mohan R, Wu Q, Manning M, Schmidt-Ullrich R (2000) Radiobiological considerations in the design of fractionation strategies for intensity-modulated radiation therapy of head and neck cancers. Int J Radiat Oncol Biol Phys 46(3):619–630
84. Lauve A, Morris M, Schmidt-Ullrich R, Wu Q, Mohan R, Abayomi O, Buck D, Holdford D, Dawson K, Dinardo L, Reiter E (2004) Simultaneous integrated boost intensity-modulated radiotherapy for locally advanced head-and-neck squamous cell carcinomas: II – clinical results. Int J Radiat Oncol Biol Phys 60(2):374–387
85. Eisbruch A, Harris J, Garden AS, Chao CK, Straube W, Harari PM, Sanguineti G, Jones CU, Bosch WR, Ang KK (2010) Multi-institutional trial of accelerated hypofractionated intensity-modulated radiation therapy for early-stage oropharyngeal cancer (RTOG 00–22). Int J Radiat Oncol Biol Phys 76(5):1333–1338

86. Nishimura Y, Nakamatsu K, Shibata T, Kanamori S, Koike R, Okumura M, Suzuki M (2005) Importance of the initial volume of parotid glands in xerostomia for patients with head and neck cancers treated with IMRT. Jpn J Clin Oncol 35(7):375–379
87. Nishimura Y, Shibata T, Nakamatsu K, Kanamori S, Koike R, Okubo M, Nishikawa T, Tachibana I, Tamura M, Okumura M (2010) A two-step intensity-modulated radiation therapy method for nasopharyngeal cancer: the Kinki University experience. Jpn J Clin Oncol 40(2):130–138
88. Nishi T, Nishimura Y, Shibata T, Tamura M, Nishigaito N, Okumura M (2013) Volume and dosimetric changes and initial clinical experience of a two-step adaptive intensity modulated radiation therapy (IMRT) scheme for head and neck cancer. Radiother Oncol 106(1):85–89
89. Ishikura S, Ito Y, Hiraoka M (2011) JCOG Radiation Therapy Study Group: history and achievements. Jpn J Clin Oncol 41(11):1241–1243

IGRT for IMRT

5

Hidenobu Tachibana and Tetsuo Akimoto

Keywords
IMRT • IGRT • Image guidance • High-precision radiation therapy

5.1 Introduction

Recent technological developments, including of computing and image processing technologies, have played a vital role in the introduction of modern techniques of radiation therapy, such as intensity-modulated radiation therapy (IMRT) and stereotactic body radiation therapy (SBRT), which enable us to deliver high doses of radiation more precisely as compared to that possible in conventional radiation therapy. For improving the clinical outcomes of treatment of locoregional tumors, it is indispensable to increase the treatment intensity without eliciting an increase in the severity or incidence of treatment-related complications. Among the various radiation therapy techniques available currently, IMRT is the most sophisticated. Treatment plans of IMRT are based on a set of computed tomography (CT) images of patients obtained in the treatment position, and delineations of the gross tumor volume (GTV) and clinical target volume (CTV) are carried out based on information obtained from clinical diagnostic imaging. In the delineation of the target volume, precise determination of the tumor extent is important, and image fusion or detailed diagnostic information regarding the extent of tumors, including from advanced imaging modalities such as multi-slice CT, magnetic resonance image (MRI), and functional imaging, must be gathered during the process of treatment planning for IMRT. For this reason, the imaging quality and optimal selection of imaging modalities are essential for

H. Tachibana, Ph.Ds. • T. Akimoto, M.D., Ph.D. (✉)
Division of Radiation Oncology and Particle Therapy, National Cancer Center Hospital East, 6-5-1 Kashiwanoha, Kashiwa, Chiba 277-8577, Japan
e-mail: takimoto@east.ncc.go.jp

Y. Nishimura, R. Komaki (eds.), *Intensity-Modulated Radiation Therapy*,
DOI 10.1007/978-4-431-55486-8_5

IMRT. In addition, application of imaging modalities for IMRT is also important, because geometrical errors and positional gaps between the planned target localization and actual target localization would compromise the clinical outcomes due to insufficient target coverage and/or overdose to the surrounding normal tissues. Proper management of target localization can be performed by image guidance using several imaging systems and is directly related to the concept of treatment margins such as the width of the GTV, CTV, setup margin (SM), or internal margin (IM). Image-guided radiation therapy (IGRT), which refers to radiation therapy under image guidance, is aimed at reducing these margins without compromising the success of local control or the clinical outcomes.

In this section, we summarize the concept of IGRT, the importance of image guidance in IMRT, the advantages and disadvantages of IGRT technologies and the systems used for application of these technologies in clinical practice, and the concept of treatment margins in IGRT. The process of radiation therapy has always been more or less image-guided, and introduction of the concept of IGRT has made us to move from patient-oriented positioning toward target volume-based positioning. Therefore, proper understanding of the concepts and techniques are indispensable for successful implementation of modern radiation therapy techniques.

5.2 Why Is Image Guidance Needed for IMRT?

As described in the Introduction section, IMRT has been applied for patients with various primary sites of cancer, including prostate cancer, head and neck cancer, brain tumor, lung cancer, etc. and enables delivery of higher doses to the target organs or tumors while keeping the doses to the surrounding normal tissues and organs at relatively low levels. In fact, dose-escalated IMRT has been demonstrated to yield excellent clinical outcomes without increasing the incidence of late complications, such as rectal bleeding, in patients with localized prostate cancer. In the radiotherapeutic management of head and neck cancer also, IMRT has been shown to yield satisfactory local control while sparing organs at risk, such as the parotid glands, thereby leading to a reduction in the frequency of late radiation morbidities such as xerostomia. However, excellent dose distribution per se is insufficient to obtain satisfactory clinical outcomes, partly because interfractional organ motion or deformation of the target tumors or organs caused by respiratory motion or compression due to expansion of the surrounding organs could cause positional gaps between the planned dose distribution to the targets and the actual irradiation area. These organ motions can be attributable to an unexpectedly insufficient dose to the target tumor or excessive dose to the surrounding normal tissues. The major problems that result from such positional errors are an increased probability of impaired local control or development of unexpected complications resulting from an excessive radiation dose to normal tissues. Therefore, reduction and/or management of positional errors are indispensable to maximize the physical and clinical advantages of IMRT. Image guidance is an effective approach for the management of positional

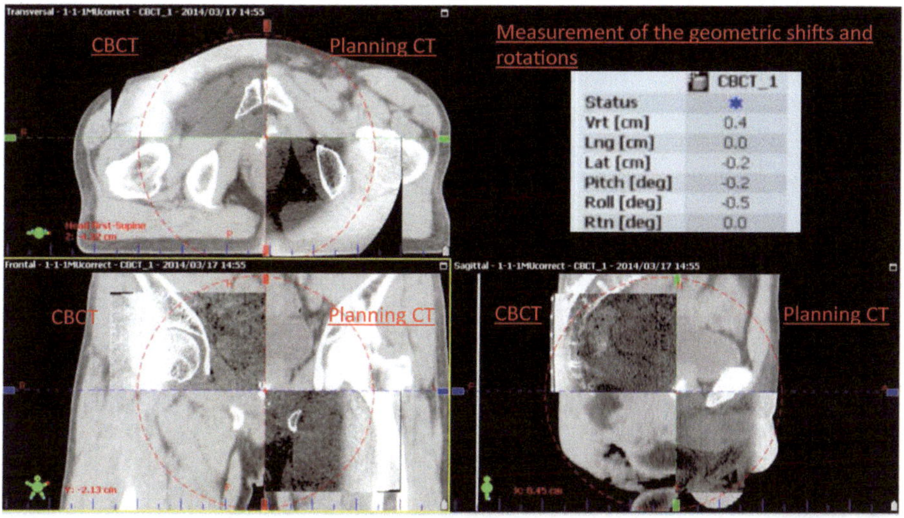

Fig. 5.1 Image registration measuring geometric variation between treatment planning and actual positioning prior to the start of treatment (blended images of planning CT and kV-CBCT images are shown)

errors. Through the process of actual procedures of daily treatments, we summarize the importance of image guidance.

Based on the treatment planning and dose evaluation, daily treatment is given. The patient lies down on a treatment couch, and radiation therapists or technologists position the patient in accordance to tattoos or markers on the patient's skin surface that almost represent the position of the isocenter of the X-ray beams. These processes are common for conventional radiation therapy, including 2-dimensional or 3-dimensional (2D or 3D) radiation therapy and IMRT. Treatments that are performed after checking the marker position of the skin alone are classified as radiation therapy without image guidance (non-IGRT). In contrast, recent technological advancements enable us to precisely evaluate the position or posture of the patients daily through various modalities of image guidance. For IGRT, several imaging modalities can be used to acquire the information on the patient's body position or the position of the target organs, including bone structures, soft tissues, or a tumor itself. In daily treatment, the distance and rotation of the patient's position from the planned position are checked, and positional gaps are evaluated by using image registration of the treatment planning attached software programs (Fig. 5.1). In order to correct the positional gap or differences, the treatment couch is moved, and the patient is repositioned appropriately before the start of the daily treatment.

Ideally, treatment beams should be directed only to the volume of the tumor without any margins such as setup margin, CTV or PTV margins; however, this is unrealizable, mainly because the patient's movements, including physiological movements and unexpected movements of the tumor and patient, are unavoidable (Fig. 5.2). In addition, changes in the patient's body contour could also be expected during the

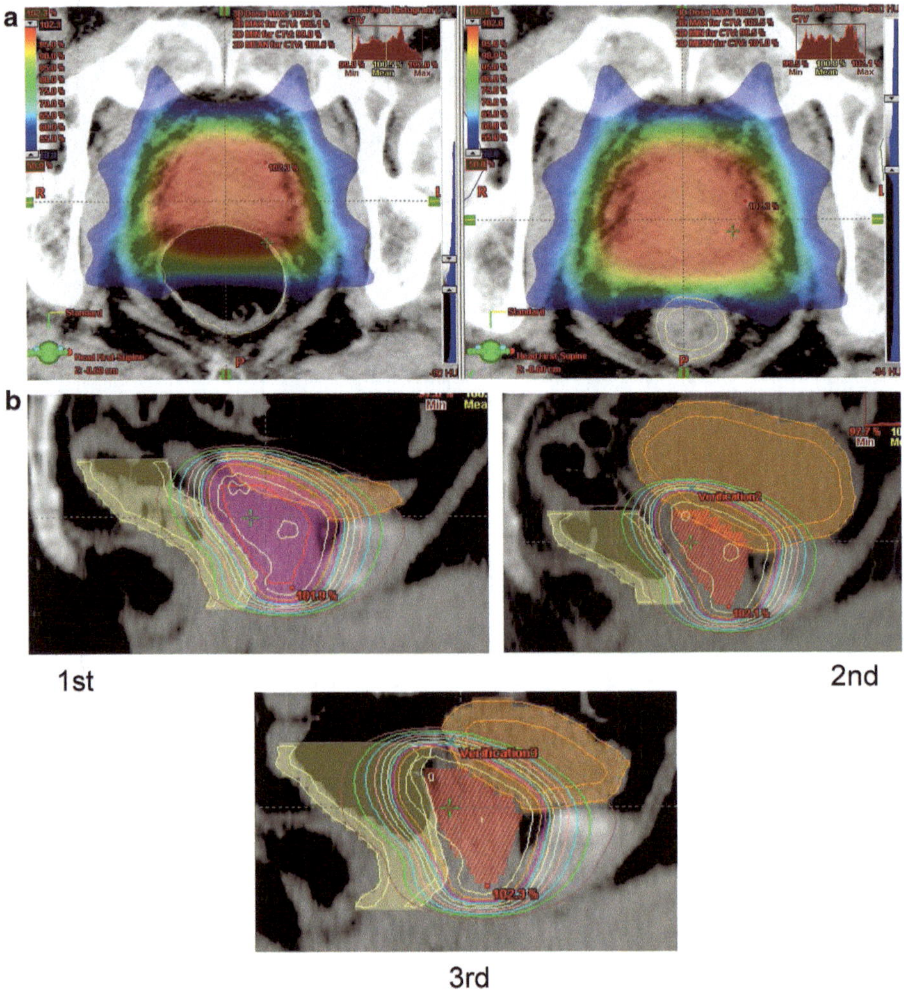

Fig. 5.2 (**a**) Differences in irradiated volumes according to the expansion of the rectum in the patients with prostate cancer. (**b**) Differences in irradiated volumes according to the deformation of the prostate gland and expansion of the rectum

course of treatment, due to body weight loss, etc. Hence, it is necessary to add optimal margins surrounding the target tumor or organs, indicating that the treatment volume becomes larger than the actual tumor volume. The treatment planning can be done precisely to cover the volume of the tumor while keeping the dose to the normal tissues within acceptable level by evaluating the dose-volume histogram. However, the treatment plan is only a snapshot of a moment, no matter how precisely it is prepared. Therefore, it is necessary to keep in mind the discrepancies and limitations of treatment planning and the problems of daily treatment. In other words, the balance between tumor coverage and sparing of normal tissues is a trade-off, and IGRT provides an optimal and effective solution toward realizing ideal radiation therapy.

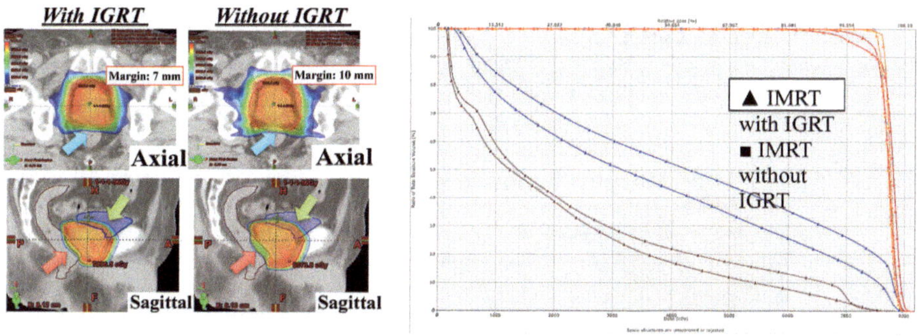

Fig. 5.3 Target coverage according to different width of the PTV margins. Compared to a situation shown in the *left*, the larger margin can provide safer for tumor coverage (prostate) during the treatment irradiation; however, organ at risk (rectum) receive more dose and it may increase the risk of side effect

IGRT is an effective tool for direct visual evaluation of the target tumor location in relation to the location of the landmark structures, such as bones, close to the target, which represent the target location or position. In addition, online correction of 3D positional gaps is also possible based on optimal image guidance. Thus, IGRT enables coverage of the treatment volume with less extra margins or volume even after providing for organ motion or patient setup errors, indicating that tumor can be covered by higher dose while allowing the surrounding organs to be spared (Fig. 5.3).

The laser-setup technique is still used for non-IGRT. The tattoos and the markers on the patient's skin as an indication of the isocenter are movable by 3–5 mm; therefore, the markers on the skin are not necessarily an accurate representation of the intended target position. However, the radiation therapists position the patient with the use of the correspondence between the point of the tattoos/markers and the position of the laser indicating the irradiation isocenter. Thus, extra margin for the uncertainties must be added in the treatment planning. This means that the setup margin should be increased for obtaining sufficient coverage of the target volume.

The dose distributions of IMRT with or without image guidance for prostate cancer are demonstrated in Fig. 5.4. The method used for image guidance is kV-cone-beam computed tomography (kV-CBCT). In the treatment plan made without image guidance, an additional 3 mm margin around the PTV is provided for as compared to the PTV margin in the treatment plan made with image guidance. Comparison between the treatment plans made with or without image guidance shows that utilization of image guidance allows the dose to the rectum and urinary bladder to be reduced, especially in terms of the area receiving less than 50 Gy, due to the effect of the reduced margin. Figure 5.5 shows a comparison of the dose volume histogram (DVH) between the treatment plans made with or without image guidance. The volume of the rectum receiving 60 Gy (V60 Gy) in the treatment plan with image guidance was 8 % and that without image guidance was 13 %. Similarly, V60 Gy for the urinary bladder in the plan with image guidance was 25 % and that without image guidance was 35 %. As described above, the utilization of image

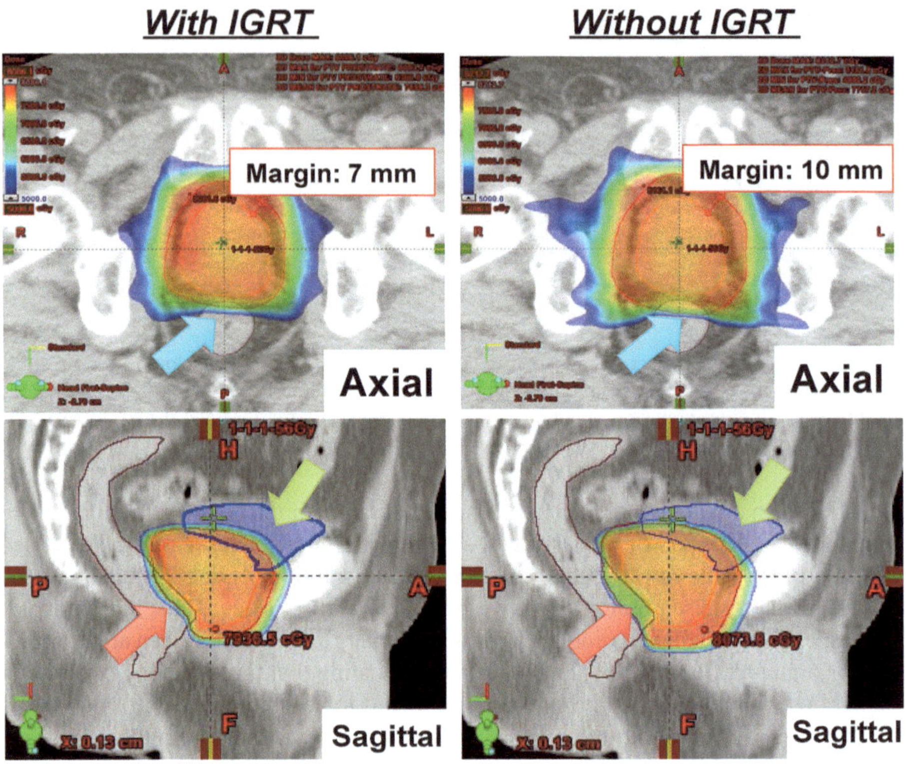

Fig. 5.4 Dose distributions in IMRT with and without image guidance for localized prostate cancer

guidance contributes to reducing the PTV margin, resulting in dose sparing to the organs at risk. On the other hand, complicated dose distribution, such as concave-shaped dose distribution, and the higher-dose gradient of IMRT are susceptible to geometric uncertainties as compared to the case in conventional radiation therapy techniques. We need to pay attention to the balance between the benefit of obtaining a reduced margin and the geometric uncertainties of IMRT.

Furthermore, the accuracy and effectiveness of image guidance are influenced by the imaging modalities used. The contrast and resolution of images differ according to the imaging modality used, and low contrast or insufficient resolution cause difficulties in the image registration, resulting in discrepancies between the distance in the treatment plan and the actual treatment positions. Therefore, medical physicists must perform quality assurance for the imaging devices, the computer program used for registration, and the accuracy of the couch table movement. It is indispensable for radiation oncologists and radiation technologists to understand the characteristics and limitations of these imaging devices.

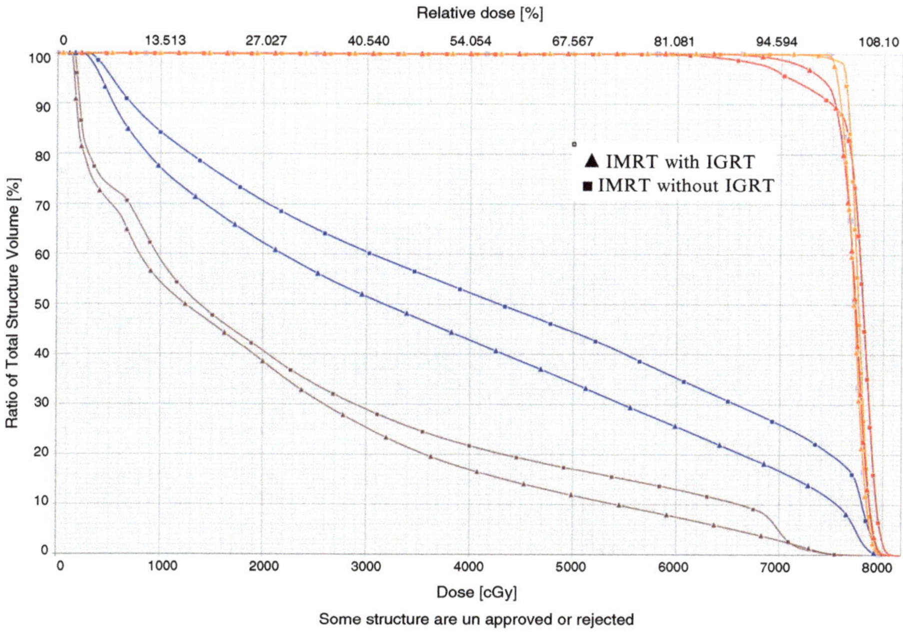

Fig. 5.5 Comparison of the DVH between the treatment plans prepared with and without image guidance for localized prostate cancer

Currently, 2D kV imaging and 3D kV-CBCT are commonly available clinically as imaging modalities. 2D kV images are easier to acquire than CBCT with less radiation exposure dose. The main advantages of CBCT are that it enables identification of both bony structures and soft tissues, in contrast to the planer X-ray images in which soft tissues are difficult to visualize [1]. Soft tissue structures, such as tumors in the prostate and lung, not only move away from the intended position relative to the isocenter but also change in shape during the course of treatment because of physical expansion of the intestines or respiratory movements. Therefore, volumetric CBCT registration would be better than 2D bony structure registration [2–5].

5.3 Image Guidance System for IGRT

5.3.1 Megavoltage (MV) Imaging and MV Cone-Beam Computed Tomography (CBCT)

Electronic portal imaging devices (EPIDs) generate 2D images for portal verification of the radiation field as well as verification of the patient positioning [6]. There are a variety of EPID types, including liquid-filled (Li-Fi) EPID, camera-based EPID, and amorphous silicon (a-Si) EPID. Among these, a-Si EPID is the most commonly used in clinical practice [7, 8]. The EPID is equipped with the treatment machine where the therapy beam is along with (Fig. 5.6). In regard to the

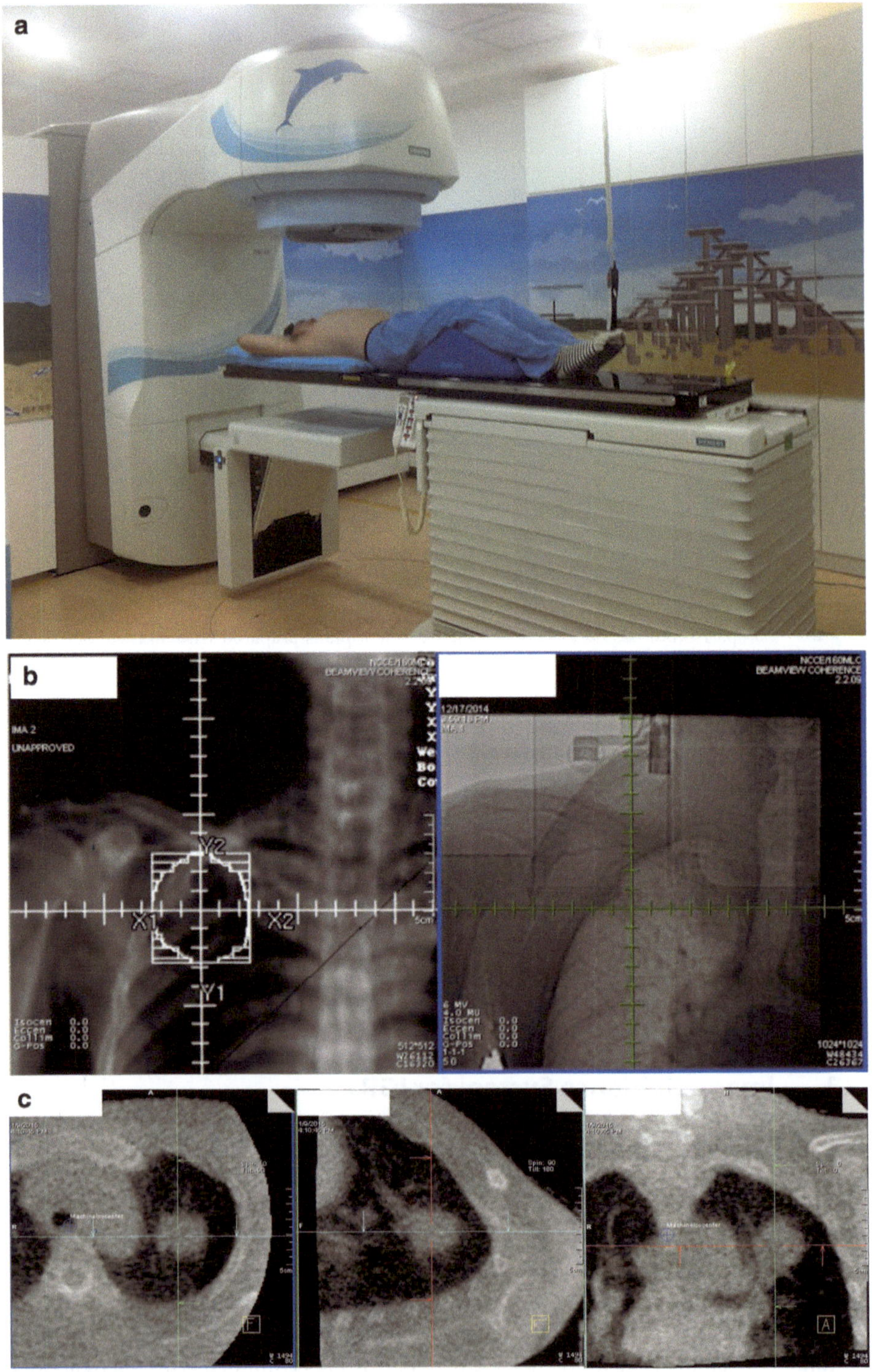

Fig. 5.6 2D MV image (**b**) and 3D MV-CBCT image (**c**) using a Linac with an on-board MV imaging system (**a**)

advantages and disadvantages of using EPIDs for image guidance, the contrast resolution of an EPID is lower than that of kV imaging [9]; however, the accuracy in matching the EPID images is usually excellent, reported to be on the order of 1 mm or 1° from a phantom study performed using (semi-) automated matching software [5, 10, 11]. On the other hand, large intra- and interobserver variations exist in the interpretation of images in image registration using an EPID [12–15]. Usage of an EPID can provide positional information regarding whether the target volume is properly covered or not even during treatment with respiratory or physiological movements of target tumors or organs.

MV-CBCT imaging can generate 3D images using the EPID. During rotational irradiation of the MV beam using relatively low exposure, projection images can be acquired automatically and the CBCT images subsequently reconstructed. The main disadvantages of using MV beam energy for the acquisition of volumetric images are the degraded quality of the images and increased radiation exposure of the patients due to increased Compton scatters. An important benefit of MV-CBCT is the high resolution of the acquired images despite the presence of dense metal objects such as hip prostheses or high-Z dental enamels [16]. When MV-CBCT was applied clinically, the absorbed dose in the irradiated area was about 5–9 cGy according to the anatomical site, thickness of the patient's body or the irradiated volume, and the maximum dose was between 9 and 17 cGy [17]. In MV-CBCT images, the bony anatomy and 3D boundaries between the soft tissues and bony structures can be easily identified, and image registration can also be carried out easily with low-dose exposure. However, the dose required for identifying soft tissue boundaries, such as those of the prostate gland, urinary bladder, rectum, etc., would increase [17].

5.3.2 kV Imaging and kV-CBCT

2D kV imaging using an on-board imager installed in the treatment machine is an effective method to evaluate setup errors precisely and adjust the patient's position properly. The kV imaging system consists of a conventional X-ray tube and amorphous silicon X-ray detectors mounted orthogonally to the treatment beam axis (Fig. 5.7). The X-ray tube and detectors are retractable during treatment. The advantages of kV imaging over MV imaging for the correction of setup errors include not only lower dose exposure but also smaller interobserver variability, which allows improved reproducibility of the patient's setup [18].

kV-CBCT imaging is performed during rotational irradiation of a kV beam and simultaneous acquisition of projection images. The kV-CBCT images are reconstructed from the projection images produced by the retractable X-ray tube and the detectors; the image contrast in kV-CBCT is generally superior to that in MV-CBCT [19, 20]. CBCT has the advantage of having the ability to detect soft tissues, which enables improved accuracy of setup of the patients, accompanying rapid 3D image registration, as compared to MV-CBCT. However, streak artifacts caused by high-Z materials are found in kV-CBCT images because of significant photon attenuation as compared to the case in MV-CBCT images.

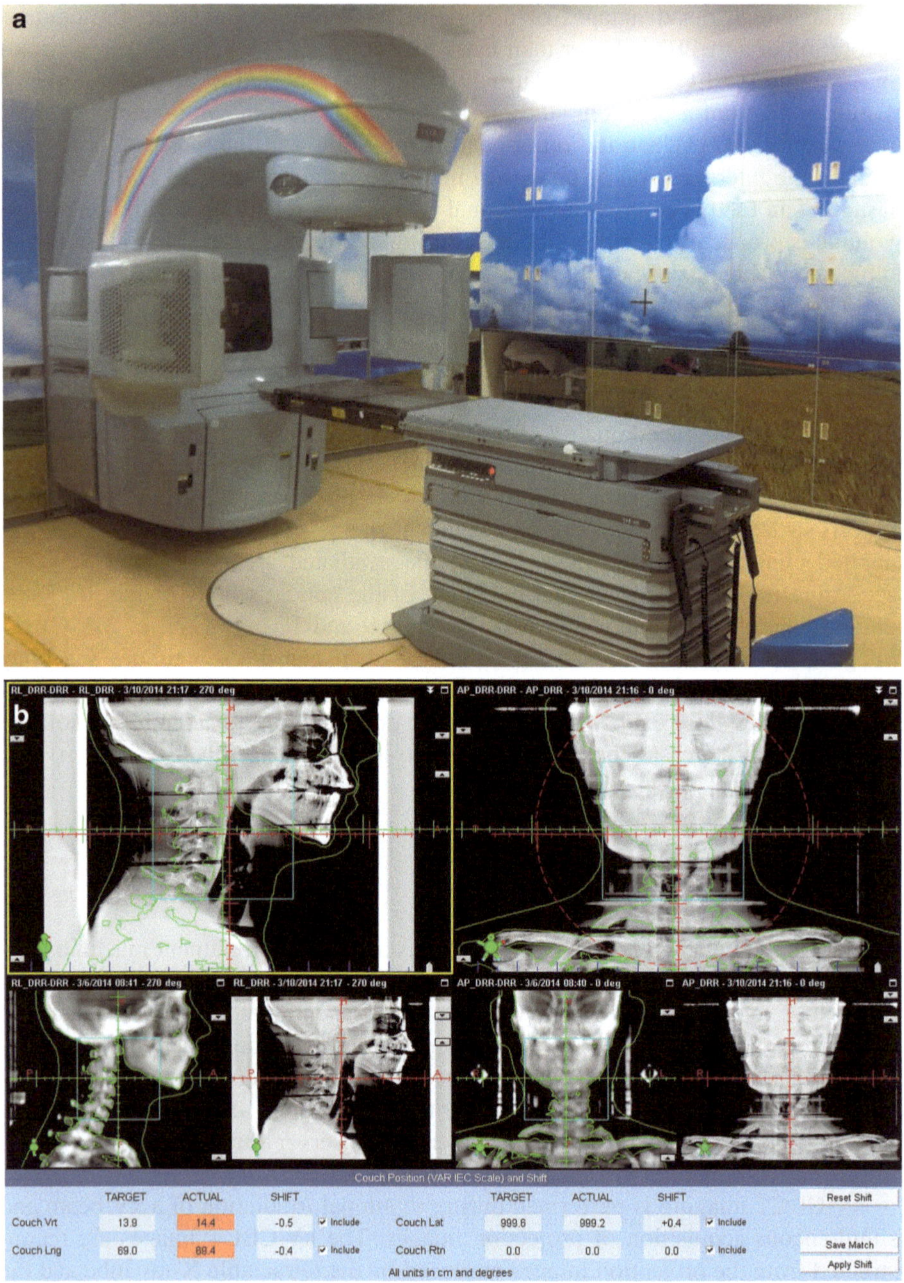

Fig. 5.7 2D kV image (**b**) and 3D kV-CBCT image (**c**) using a Linac with an on-board kV imaging system (**a**)

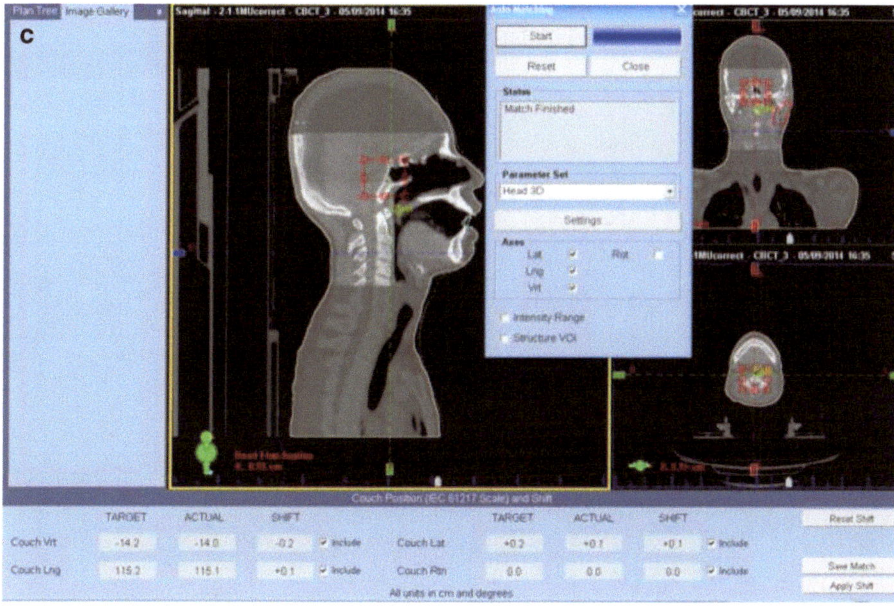

Fig. 5.7 (continued)

5.3.3 CT-on-Rails

CT-on-rails is defined as a diagnostic CT scanner installed in the treatment room for the purpose of verification of patient positioning (Fig. 5.8). The gantry of the CT-on-rails scanner can move across the patient, instead of the couch moving. The CT-on-rails scanner moves in line with the couch orientation and yields CT images, while the patient remains in the treatment position before or after the treatment. Diagnostic-quality CT images can be obtained by CT-on-rails, because the CT scanner installed in the CT-on-rails was originally developed for diagnostic radiology. Therefore, usage of CT-on-rails as a tool for image guidance may be expected to improve the accuracy of detection of soft tissue lesions, including the target tumors or organs, and reduce the interobserver variations in image registrations [21, 22]. However, several disadvantages of CT-on-rails have also been pointed out. From the structural basis of CT-on-rails, the CT scanner and the treatment machine need to be operated independently, because the CT-on-rails does not share the same gantry. These systematic limitations may cause the uncertainties in the following: (1) the patient couch position after a rotation, (2) the precision of the CT coordinates, and (3) alignment of contours with structures in the CT images. In addition, there is sometimes a vital time lag between the actual treatment and the image acquisitions for image guidance, resulting in unexpected movement of the patients during the time lag and impairment of the setup accuracy [23].

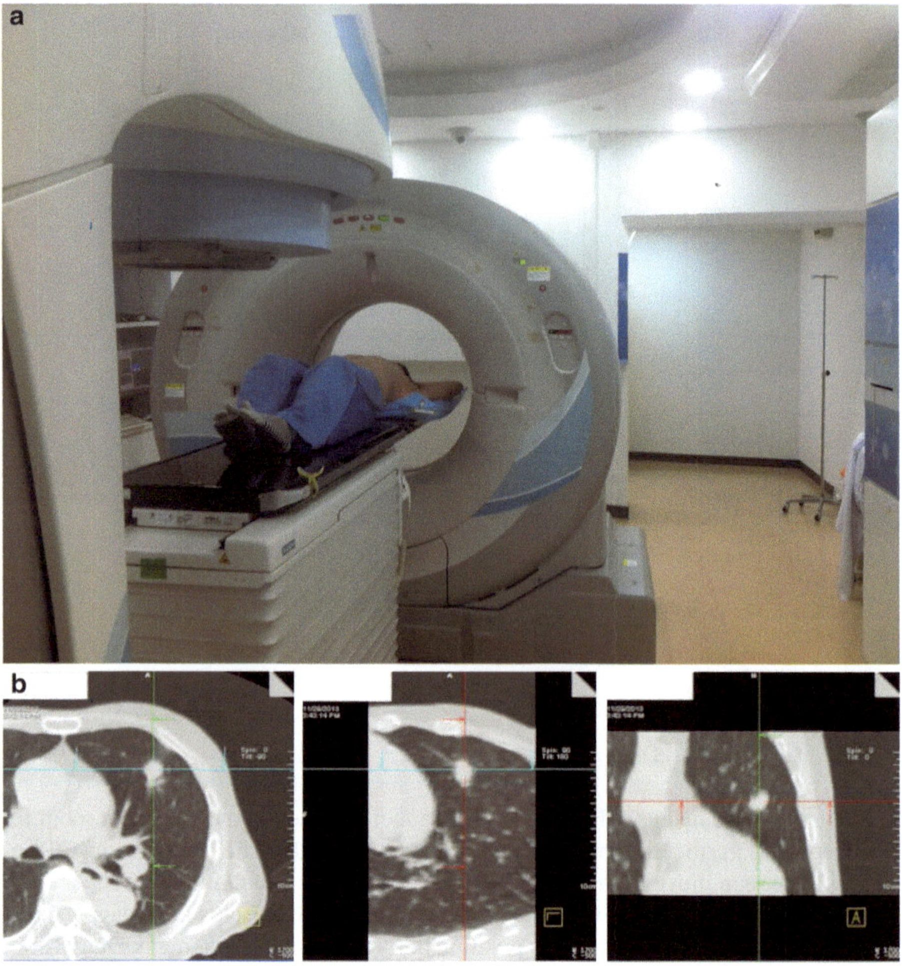

Fig. 5.8 3D CT image (**b**) using CT-on-rails system (**a**)

5.3.4 2D kV Stereoscopic Imaging

2D kV stereoscopic imaging allows measurement of 3D geometric variations with 2 kV X-ray tubes and the corresponding X-ray detectors in which the two X-ray beam axes are crossed at a certain point (Fig. 5.9). Two kV 2D images are acquired from a patient, and subsequently, two sets of 2D coordinates of geometric variations from the two images can be converted to the 3D variation.

Here, we introduce three commercially available systems. The CyberKnife system consists of a compact linear accelerator mounted on a robotic arm. The manipulator arm is controlled to direct the radiation beams to the region of the beam intersection of the two orthogonal X-ray imaging systems integrated to provide image guidance during the treatment. The ExacTrac system uses the optical

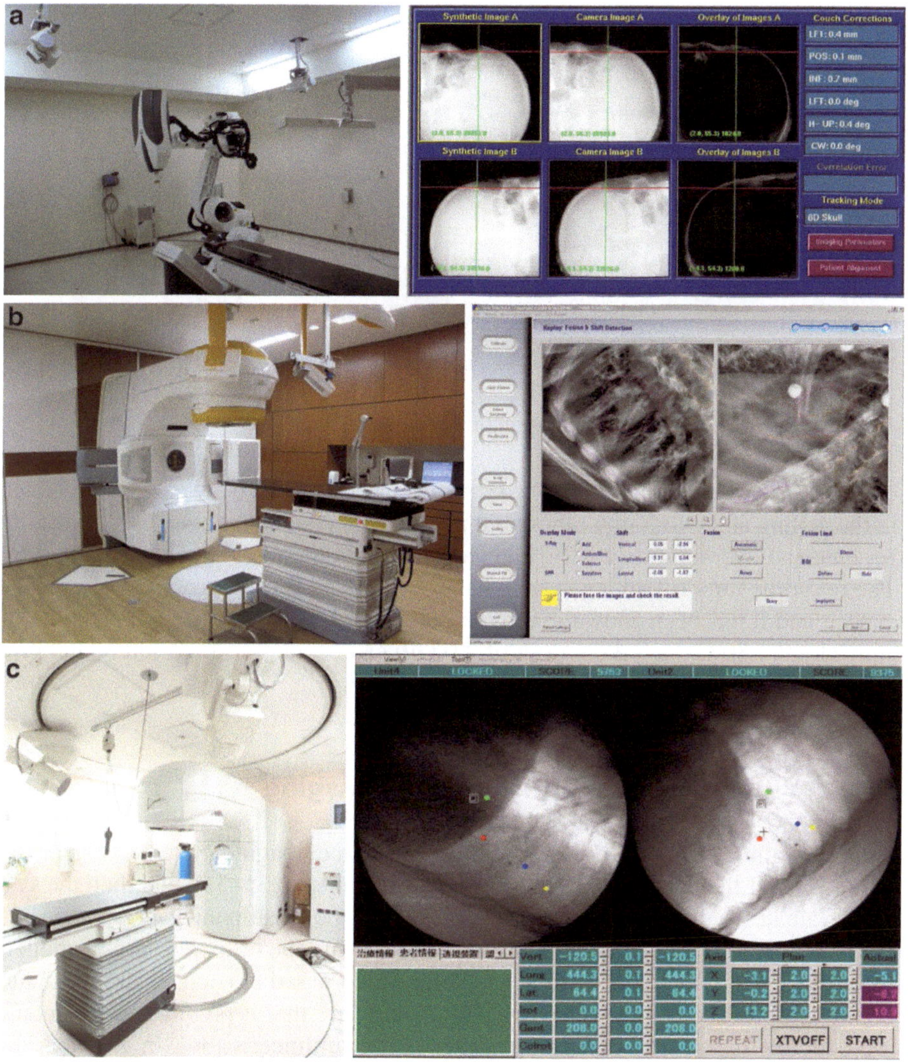

Fig. 5.9 2D kV stereoscopic imaging systems. (**a**) The CyberKnife system. (**b**) The Novalis Linac system with the ExacTrac imaging system. (**c**) Real-time tumor-tracking radiotherapy system

positioning system and two sets of kV imaging systems to enable patient positioning during the treatment. The optical positioning system is used for the initial patient setup as well as the couch motion control. In addition, the system can manage the patient's respiration and gated beam irradiation. The X-ray imaging system is used for verification of the patient's position and readjustment using the internal anatomy or implanted fiducial markers on the images. The accuracy of positioning using the optical guidance is on the order of submillimeters [24]. The real-time tumor-tracking system was designed to track the targets using fiducial markers implanted in or

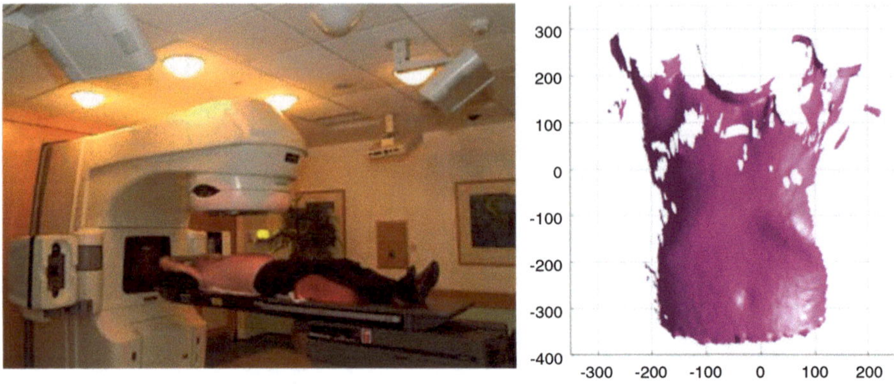

Fig. 5.10 Video-based 3D imaging system (http://www.visionrt.com/)

close to the tumors in real time. The system consists of two X-ray cameras mounted in the ceiling and two X-ray tubes on the floor. The system has been used clinically for lung and liver tumors [25–29] and reduced tumor motion to less than 5 mm [26]. The real-time tumor-tracking system has been applied for the treatment of prostate cancer, and it has been shown that hypofractionated IMRT (2.5 Gy × 28 fr = 70 Gy, equivalent to 80 Gy at a daily dose of 2.0 Gy) can be administered safely and provides a reasonable biochemical control rate [30].

5.3.5 Video-Based 3D Optical Surface Imaging

During treatment, the video-based 3D optical surface imaging system can generate 3D models of the patient surface using photogrammetry [31–34], and then the system evaluates and quantitates the difference in the positional gap between the planned 3D models of the patient as the reference and the observed surface models [35] (Fig. 5.10). The major advantages of the video-based 3D optical surface imaging system, as demonstrated in a phantom study, are that it provides accurate and reproducible image guidance on the order of submillimeters [36]. In addition, the optical surface imaging system allows for continual monitoring of the patient surface without any radiation exposure during the actual treatment. In regard to its clinical applicability, the video-based 3D optical surface imaging system is especially useful as an image guidance system for radiation therapy of breast cancer, because the breast surface is nonrigid, and alignment of the 3D breast surface achieved by using this system yields greater breast correspondence than that obtained with laser or portal imaging systems [31].

5.3.6 Ultrasound (US) Imaging

US images of the target tumors or organs are acquired before the start of treatment and the integrated computer software program overlays the planned contours of the

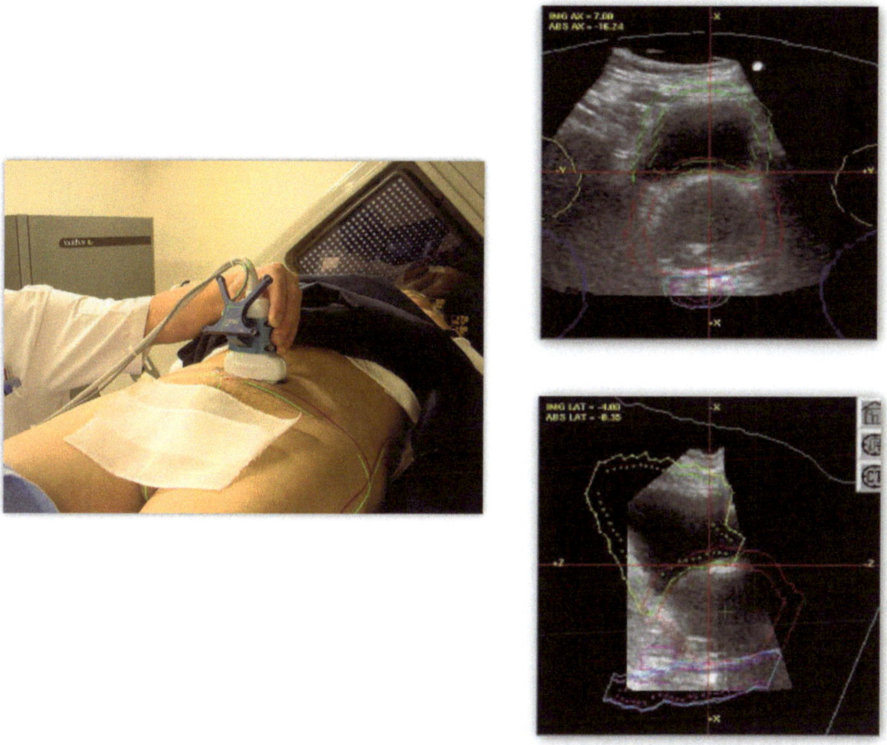

Fig. 5.11 US imaging system

target tumors or organs such as the prostate gland, which are delineated at the treatment planning, over the acquired US images (Fig. 5.11). Subsequently, adjustments of the patient's position are performed to align the contours to the targets while monitoring the actual target position on the US image. The advantages of US imaging as an image guidance system are as follows: (1) no radiation exposure, (2) noninvasive image guidance, (3) excellent visualization of soft tissues, (4) relatively rapid acquisition of images, and (5) low cost, including the entire system. The first commercially available US imaging system was the BAT system (North American Scientific, USA), which can provide 3D information of the target based on two US images. Subsequently, a 3D US imaging system was developed (SonArray, Zmed Inc., USA) so as to overcome the limitations of the former system. Currently, the US imaging system is mostly used for image guidance in radiotherapeutic approaches for localized prostate cancer. The disadvantage of US image guidance is that it is partly dependent on the user's experience and expertise, resulting in high inter-user variability and reduced accuracy relative to CBCT guidance with or without fiducial markers. The US imaging systems that are currently in use clinically calculate the required shift, but do not allow for automated correction of the treatment couch. This results in a compromise of accuracy and prolongation of the treatment time. Other limitations of the US imaging system are forced movement or deformation of

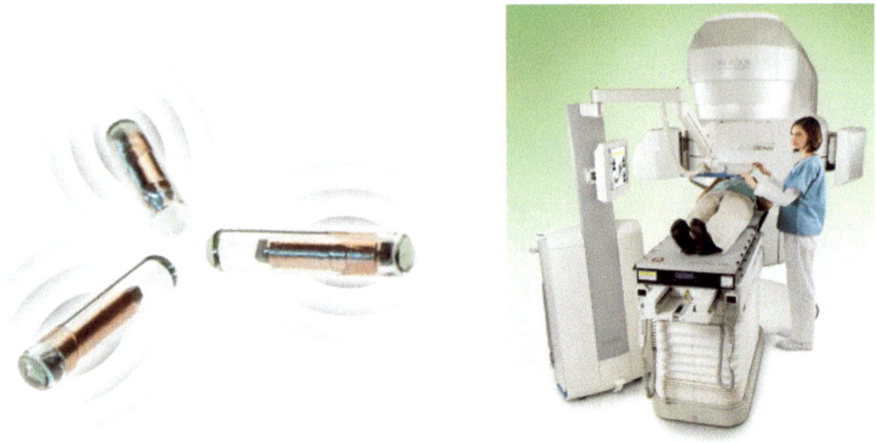

Fig. 5.12 Electromagnetic tracking system (http://www.varian.com/asjp/oncology/imaging_solutions/calypso/)

the target tumors or organs by the external pressure of the US probe during the acquisition of images and also non-visualizability of the target tumors or organs located behind bony structures [37].

5.3.7 Electromagnetic Tracking Systems

The Calypso system (Varian Medical Systems, USA) can track the locations of target tumors or organs in real time by using tiny electromagnetic transponders implanted within the tumor before the treatment (Fig. 5.12). Upon excitement of the system, the transponders yield radiofrequency waves, and the detection of radiofrequency waves by the receiver activates analysis of the location of the transponders. The transponders are implanted by an invasive method; however, the system provides real-time tracking of the target during the treatment with no radiation dose delivered for imaging. The system has been demonstrated in some studies to show interfraction localization within 2 mm of the X-ray-based positioning [38, 39].

5.3.8 Magnetic Resonance Imaging (MRI)

MRI-based radiation therapy systems include MRI imaging and treatment beam delivery in the same room, and both radiation therapy and monitoring of the target location in 2D/3D can be performed in real time using MRI simultaneously (Fig. 5.13). The main advantage of MRI for image guidance is that it provides high contrast of soft tissues, yielding clear detection of the boundary between the target tumor and the surrounding normal tissues, and the modality does not entail radiation exposure. However, the clinical usefulness of a linear accelerator equipped with an

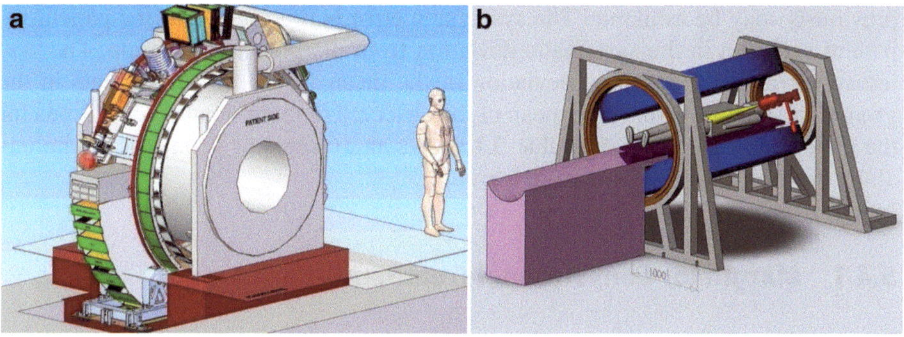

Fig. 5.13 Radiotherapy machine equipped with an MR imaging system. (**a**) System from the University Medical Center Utrecht, Netherlands (http://medicalphysicsweb.org/cws/article/research/51279). (**b**) System from the Cross Cancer Institute, the University of Alberta (http://www.mp.med.ualberta.ca/linac-mr/)

MRI imaging system is still under investigation [40–49]. The cobalt irradiation system with MRI (Viewray, USA) is already available commercially.

5.4 Concept of Margin in IGRT

The PTV covers the CTV plus a margin, considering geometric uncertainties. Larger margins are associated with unnecessarily higher doses to normal organs surrounding the target; however, an insufficient margin would result in a less-than-optimal dose to the CTV. Thus, an adequate PTV margin setting is essential to obtain satisfactory outcomes of radiation therapy.

Patients' geometric displacements correspond to the differences between a reference image, such as digital reconstructed radiograph (DRR) or CT simulation images, and a comparative image acquired pretreatment and posttreatment, such as with EPID, CBCT, etc. In the other words, the geometric displacements can be represented as deviations between the intended treatment plan and the geometry of the actually irradiated area. The displacement can be divided into two types of errors: systematic and random errors. Systematic errors, denoted as Σ, refer to differences in the mean irradiation geometry in fractionated treatment from the geometry in the treatment plan. Random errors, represented by σ, may occur by chance and may be due to day-to-day variations, including organ motions and patient movements during the treatment. The margin computation methods have been reported previously [50, 51]. This textbook introduces a margin recipe from J. C. Stroom [51].

$$\text{Margin} = 2.0\Sigma + 0.7\delta$$

The margin computation is retrospectively performed from the results of the patients' treatments. When planning treatment for a particular patient, the geometrical errors that the patient will experience in future are not known. However, the measured uncertainty data in a group of similar patients who have been treated previously may be available. The systematic error is represented as the patient-to-patient variation in the systematic deviation from the planning situation. The systematic error is the standard deviation of the mean values for the patients in the group. The random error is the mean of the observed random standard deviations for the patients in the group [5]. Table 5.1 shows an example of geometric deviations along the x-axis for the patients in a group.

5.4.1 Margin for Conventional Treatment

In conventional treatment, there is no geometric correction prior to each treatment by image guidance. In the example, a 4.1 mm expansion to the CTV is needed for the margin from the result of Table 5.1.

5.4.2 Margin for IGRT

When image guidance (so-called online correction, in which the patient is repositioned by image acquisition, image registration, and treatment couch shift pretreatment) is applied, the geometric deviation of the systematic error and the random error on the day seems to be zero, and subsequently, the margin may be reduced to zero. However, the pretreatment images do not contain the information on the intrafractional organ motion from the start to the end of the irradiation. Furthermore, the treatment couch shift has some uncertainties, and interobserver discrepancies may occur during the IGRT. The margin can be on the order of submillimeters and be close to zero, but can never be zero.

5.5 Clinical Examples of IGRT for Tumor Sites

In this section, we summarize the clinical usefulness and problems that need to be resolved in the application of IMRT with image guidance for tumors at several different primary sites.

5.5.1 Lesions of the Central Nervous System (CNS)

IGRT in frameless radiation therapy such as stereotactic radiosurgery (SRS) or stereotactic radiation therapy (SRT) has a significant role in the treatment of CNS lesions, including primary brain tumors and brain metastases. Minimizing normal tissue toxicity by generating sharp dose gradients, which must be directed accurately to the tumor volume, needs safe and effective SRS or SRT. The implementation of image-guided stereotactic localization using optical image guidance or stereoscopic X-ray imaging can support frameless SRS and/or SRT. In a study of patients undergoing frameless SRS with an image guidance system, the geometric

Table 5.1 Results of margin calculation using geometric deviations in five patients receiving RT with and without image guidance

		Day 1			Day 2			Day 3			Day 4			Day 5				
Patient	IGRT	Pre-	Post-	Total	Pre-	Post-	Total	Pre-	Post-	Total	Pre-	Post-	Total	Pre-	Post-	Total	Mean	SD
No. 1	w/o	2.7	N/A	2.7	3.6	N/A	3.6	−4.2	N/A	−4.2	1.7	N/A	1.7	2.5	N/A	2.5	**1.3**	**3.1**
	w/z	0.0	0.5	0.5	0.0	1.6	1.6	0.0	−0.2	−0.2	0.0	−1.2	−1.2	0.0	1.5	1.5	**0.4**	**1.2**
No. 2	w/o	−2.0	N/A	−2.0	0.6	N/A	0.6	−1.4	N/A	−1.4	−2.0	N/A	−2.0	−2.6	N/A	−2.6	**−1.5**	**1.2**
	w/z	0.0	−0.5	−0.5	0.0	0.7	0.7	0.0	0.2	0.2	0.0	0.2	0.2	0.0	0.5	0.5	**0.2**	**0.5**
No. 3	w/o	−2.7	N/A	−2.7	−2.3	N/A	−2.3	−3.1	N/A	−3.1	−2.0	N/A	−2.0	1.3	N/A	1.3	**−1.8**	**1.8**
	w/z	0.0	0.0	0.0	0.0	0.0	0.0	0.0	0.5	0.5	0.0	−0.3	−0.3	0.0	0.8	0.8	**0.2**	**0.5**
No. 4	w/o	1.3	N/A	1.3	2.9	N/A	2.9	−3.3	N/A	−3.3	3.1	N/A	3.1	−1.7	N/A	−1.7	**0.5**	**2.8**
	w/z	0.0	0.6	0.6	0.0	0.5	0.5	0.0	0.2	0.2	0.0	−0.4	−0.4	0.0	0.5	0.5	**0.3**	**0.4**
No. 5	w/o	−1.9	N/A	−1.9	−2.0	N/A	−2.0	0.0	N/A	0.0	1.8	N/A	1.8	−1.6	N/A	−1.6	**−0.7**	**1.6**
	w/z	0.0	0.0	0.0	0.0	0.0	0.0	0.0	0.2	0.2	0.0	−0.3	−0.3	0.0	−0.2	−0.2	**−0.1**	**0.2**

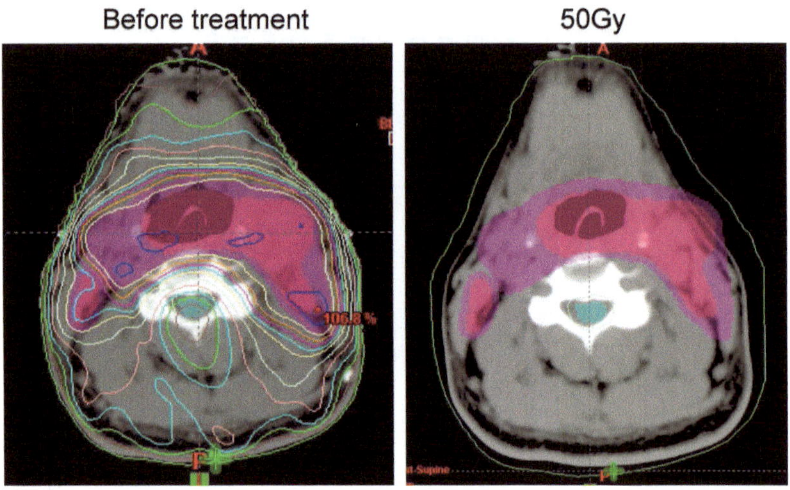

Fig. 5.14 Dose distribution was affected by deformation of the body contour due to weight loss and tumor shrinkage during the course of IMRT

deviations were approximately 0.2, 0.3, and 0.3 mm in the left-right (LR), superior-inferior (SI), and anterior-posterior (AP) directions, respectively. The margins for residual errors after correction in the LR, SI, and AP directions were determined to be 0.8 mm [52].

5.5.2 Head and Neck Cancer

Radiation therapy plays a vital role in the treatment of early and locally advanced head and neck cancer, because head and neck cancers usually develop from important organs for speech or swallowing. Therefore, organ and/or function preservation is an important factor while selecting the most appropriate treatment modalities. Among the various treatment modalities, radiation therapy with or without chemotherapy has been established as a radical and effective treatment approach for head and neck cancer due to the excellent preservation of the QOL of the patients after treatment. However, late radiation-related morbidities such as xerostomia or dysgeusia caused by radiation therapy, especially 2D or conventional radiation therapy, are important issues that need to be resolved [53]. Under these circumstances, IMRT has attracted attention as an effective and useful approach to resolve the aforementioned issues [54]. The confined and complicated dose distribution of IMRT is especially effective for sparing organs at risk such as the parotid glands, spinal cord, etc. However, changes in the target volume and shape occur due to tumor shrinkage or weight loss, indicating that image guidance that ensures target coverage would be indispensable during the course of radiation therapy (Fig. 5.14).

5.5.3 Breast Cancer

Establishment of the clinical usefulness of application of IGRT for breast cancer is at an early stage; however, several recent studies have shown the usefulness of image guidance to overcome the variability of the target location due to soft tissue motion, breathing, and variability of the patient's setup. An increasing number of recent studies have described that image guidance can improve the accuracy of beam delivery and reduce the margin of normal organs exposed in radiation therapy for breast cancer [55–59]. In a study of accelerated partial breast irradiation (APBI) using IMRT, gold markers were implanted around the surgical cavity and detected by port film [55]. The image guidance with gold markers resulted in improved accuracy of beam delivery in 79 % of the treatment sessions with an average shift of 6 mm in each 3D direction and enabled reduction of the margin size from 1 cm to 5 mm. A study focusing on comparison of the accuracy of different image guidance approaches in patients undergoing APBI [56] demonstrated that kV imaging was superior to surface imaging using 3D video, kV imaging of the chest wall, or laser alignment of skin surface markers. In a study evaluating the role of CBCT, CBCT was shown to be advantageous for visualizing the soft tissues at the tumor site [57, 58]. The use of CBCT provides additional positioning benefit beyond kV/MV imaging, and CBCT guidance can reduce targeting errors as compared with the usage of skin landmarks, at an average of approximately 1–1.5 mm in each 3-dimensional direction, which could yield an average reduction of the margin width from 8.8 mm to 3.6 mm.

5.5.4 Lung Cancer

Lung cancer is the leading cause of cancer death in men and women in the United States and in men in Japan. Among lung cancers, non-small cell lung cancer (NSCLC) accounts for nearly 80 % of all the cases, and only a quarter of all patients with NSCLC are suitable candidates for surgical resection at the time of diagnosis. In patients with stage I/II disease, surgical resection is the standard treatment option, and radiation therapy such as SBRT or particle therapy is indicated for those who cannot tolerate surgical resection due to comorbidities or refusal to undergo surgery [59]. Among patients who are not suitable candidates for surgery, those with locally advanced NSCLC require concurrent chemoradiation therapy (CRT).

Radiation therapy for lung cancer, especially NSCLC, has evolved from 2D to 3D radiation therapy due to the advances in radiation therapy technologies. The treatment planning and precision of target delineation have also progressed, matching in pace with the progress of the treatment machines. Tumor motion due to respiration is also an important issue, and this affects the concept of the margins for the target. For this reason, 4-dimensional CT (4DCT) is recommended for the process of treatment planning; however, organ motion differs greatly among patients [60]. Hence, organ motion during the course of radiation therapy should be considered individually.

5.5.5 Liver Tumor

Even though surgical resection remains the gold standard for the management of primary (hepatocellular carcinoma, HCC) and metastatic liver disease, radiation therapy, including stereotactic body radiation therapy (SBRT), IMRT, and particle therapy, is currently an option for the treatment of patients who are not suited for surgery or invasive procedures. High-dose SBRT yields local control rates of 70–80 %, which may improve not only the survival but also the QOL [61]. Patients with HCC often have preexisting liver dysfunction, which may serve as a limitation against application of radiation therapy due to the excessive dose to the normal liver. Therefore, application of image guidance is effective to spare the normal liver by focusing the dose on the target tumors.

5.5.6 Stomach Neoplasms

Indication of radiation therapy for stomach tumors is still limited, and one of the promising indications is mucosa-associated lymphoid tissue (MALT) lymphoma of the stomach. Similar to the case for primary tumors at other sites, conformal dose delivery and selective sparing of critical structures such as the kidneys, liver, and spinal cord are important. In addition, the size and shape of the stomach change easily as compared to those of solid organs such as the liver. Aggarwal A et al. carried out a 3D evaluation of the displacement of the gastric remnant during adjuvant IGRT-IMRT and demonstrated large variations in the gastric remnant volume during the course of radiation therapy [62]. In this context, image guidance would be useful for the treatment of gastric neoplasms.

5.5.7 Prostate Cancer

The incidence of prostate cancer has been increasing in the United States and also in Asian countries, including Japan, mainly because of the widespread use of the PSA screening test. Conventionally, surgery is the standard of care for localized prostate cancer; however, modern radiation therapy techniques have been established as an effective treatment modality of first choice, since the results of clinical trials of dose escalation or combined treatment with androgen ablation have revealed that radiation therapy with or without androgen ablation is as effective as surgery. However, it has also been demonstrated that treatment-related complications increase as the total dose of radiation is increased [63]. This trend is prominent especially in cases treated by the conventional radiotherapeutic approach. Clinical trials of IMRT as a radiotherapeutic approach for localized prostate cancer have revealed improved clinical outcomes without increase in the frequency of late complications such as rectal bleeding [64]. In these advancements, the importance of image guidance has been recognized [65].

5.5.8 Uterine Cervix

Among pelvic malignant neoplasms other than prostate cancer, uterine cervical cancer is a good indication for IMRT, because radiation therapy combined with concurrent chemotherapy is associated with severe (grade 3–4) gastrointestinal and genitourinary late toxicity rates of 6–23 % [66], leading to distressing lifelong symptoms, including malabsorption, incontinence, and fistulae. The irradiated volume of the conventional 4-field "box" technique was much larger than the CTV, resulting in excessive doses to normal tissues. IMRT for cervical cancer enables highly conformal dose delivery to target volumes such as the uterine cervix, parametrium, and pelvic lymph nodes, with lower doses to adjacent organs at risk such as the small intestine, rectum, and urinary bladder [67–72]. Pelvic organs, especially the intestines and urinary bladder, are likely to show positional and volumetric changes during the course of radiation therapy. Therefore, usage of image guidance for IMRT of uterine cervical cancer is advantageous, in that it reduces the geometric uncertainties during the treatment.

5.5.9 Spine Lesions

Radiation therapy has been established as a palliative treatment for bone metastases because of its effectiveness in yielding pain relief and improving the neurologic symptoms associated with bone metastases. The standard fractionation schema for bone metastases is 30 Gy in 10 fractions or 20 Gy in 5 fractions. Recent clinical studies have revealed that patients with oligometastases, including single bone metastases, sometimes survive for longer periods of time as compared to those with multiple metastases. Therefore, the optimal technique and effectiveness of high-dose irradiation for spine metastases have been vigorously studied, and IMRT and SBRT have been established as promising methods that can deliver high doses safely. However, since critical organs such as the spinal cord are located close to the metastatic lesions, ensuring precise irradiation is indispensable.

5.6 Radiation Exposure According to the Imaging Modalities Used for the Image Guidance

As described above, image guidance is an essential tool for patient positioning, target localization, and external beam alignment in IMRT. However, some imaging modalities are associated with higher radiation exposure levels of the patient; therefore, attention must be paid to the radiation exposure dose associated with the imaging modality used for the image guidance. Table 5.2 shows a summary of the radiation exposure dose according to the imaging modality used for the image guidance. In general, the radiation exposure level of a patient is higher with 3D imaging in MV-CBCT and kV-CBCT than with 2D kV and MV imaging. The dose to the skin is larger than the dose to the central part of the body because of the usage of the

Table 5.2 Radiation exposure from the imaging systems used for IGRT

IGRT system	Image acquisition	Patient dose per image
kV or MV 2D planer	2D	1–3 mGy [73]
MV-CBCT	3D	35–110 mGy (Siemens) [74]
		7–35 mGy (Tomotherapy) [75]
kV-CBCT	3D	30–60 mGy at surface [76]
		30 mGy at central [76]
		~23 mGy at surface [77]
		~16 mGy at central [77]
CT-on-rails	3D	10–50 mGy [73]
Stereoscopic kV imaging	2D	0.10–2.0 mGy (CyberKnife) [78]
		0.33–0.55 mGy (Novalis) [78]
		0.20–20 mGy (RTRT) [78]

kilovoltage X-ray beam. In the case of stereoscopic kV imaging, intermittent image acquisition, in which multiple images are generated using X-ray irradiations, is performed during treatment. Thus, the prolonged treatment time causes more radiation exposure, suggesting that the appropriate frequency of intermittent image acquisition should be determined and justified.

5.7 Summary

In this chapter, we have reviewed the role of image guidance in modern radiation therapy technologies, including IMRT, and discussed the advantages and disadvantages of the imaging modalities used for image guidance for various tumors from different primary sites. The following are the key points.

1. IMRT has been established as a standard radiotherapeutic approach for head and neck cancer, prostate cancer, thoracic malignancies, uterine cervical cancer, etc.
2. Treatment plan is only a snapshot of a moment, no matter how precisely it is prepared. Inter- and intrafractional organ motion must be considered.
3. Interfractional organ motion or deformation of the target tumors or organs caused by respiratory motion or compression due to expansion of the surrounding organs may cause positional gaps between the planned dose distribution area to the targets and the actual irradiation area, resulting in an unexpectedly insufficient dose to the target tumor and/or excessive dose to the surrounding normal tissues.
4. Image guidance is an essential tool to achieve effective and safe IMRT delivery, as it enables reduction of the geometric deviations that can occur between the intended treatment plan and the actual irradiated area.
5. Each of the imaging modalities used for imaging guidance has its own advantages and disadvantages.

6. Optimal margin setting is partly dependent on the balance of the risk between optimal target coverage and sparing of risk organs when IGRT is applied.
7. A multi-professional team comprising radiation oncologists, medical physicists and imaging technology experts responsible for the IGRT, should be established to ensure the quality assurance of IMRT and IGRT.

References

1. Zhang L, Garden AS, Lo J et al (2006) Multiple regions-of-interest analysis of setup uncertainties for head-and-neck cancer radiotherapy. Int J Radiat Oncol Biol Phys 64:1559–1569
2. O'Daniel JC, Dong L, Zhang L et al (2006) Dosimetric comparison of four target alignment methods for prostate cancer radiotherapy. Int J Radiat Oncol Biol Phys 66:883–891
3. Borst GR, Sonke JJ, Betgen A et al (2007) Kilo-voltage cone-beam computed tomography setup measurements for lung cancer patients; first clinical results and comparison with electronic portal-imaging device. Int J Radiat Oncol Biol Phys 68:555–561
4. Court LE, Dong L (2003) Automatic registration of the prostate for computed-tomography-guided radiotherapy. Med Phys 30:2750–2757
5. Bijhold J, Lebesque JV, Hart AA et al (1992) Maximizing setup accuracy using portal images as applied to a conformal boost technique for prostatic cancer. Radiother Oncol 24:261–271
6. Bijhold J, Gilhuijs KG, van Herk M (1992) Automatic verification of radiation field shape using digital portal images. Med Phys 19(4):1007–1014
7. Cremers F, Frenzel T, Kausch C et al (2004) Performance of electronic portal imaging devices (EPIDs) used in radiotherapy: image quality and dose measurements. Med Phys 31(5):985–996
8. Bailey DW, Kumaraswamy L, Bakhtiari M et al (2012) EPID dosimetry for pretreatment quality assurance with two commercial systems. J Appl Clin Med Phys 13(4):3736
9. Pesznyák C, Polgár I, Weisz C et al (2011) Verification of quality parameters for portal images in radiotherapy. Radiol Oncol 45(1):68–74
10. Hanley J, Lumley MA, Mageras GS et al (1997) Measurement of patient positioning errors in three-dimensional conformal radiotherapy of the prostate. Int J Radiat Oncol Biol Phys 37:435–444
11. Pouliot J, Lirette A (1996) Verification and correction of setup deviations in tangential breast irradiation using EPID: gain versus workload. Med Phys 23(8):1393–1398
12. Bissett R, Boyko S, Leszczynski K et al (1995) Radiotherapy portal verification: an observer study. Br J Radiol 68:165–174
13. Ezz A, Munro P, Porter AT et al (1992) Daily monitoring and correction of radiation field placement using a video-based portal imaging system: a pilot study. Int J Radiat Oncol Biol Phys 22:159–165
14. Herman MG, Abrams RA, Mayer RR (1994) Clinical use of on-line portal imaging for daily patient treatment verification. Int J Radiat Oncol Biol Phys 28:1017–1023
15. Perera T, Moseley J, Munro P (1999) Subjectivity in interpretation of portal films. Int J Radiat Oncol Biol Phys 45:529–534
16. Pouliot J, Bani-Hasemi A, Chen J et al (2005) Low-dose megavoltage cone-beam CT for radiation therapy. Int J Radiat Oncol Biol Phys 61:552–560
17. Gayou O, Parda DS, Johnson M et al (2007) Patient dose and image quality from mega-voltage cone beam computed tomography imaging. Med Phys 34(2):499–506
18. Pisani L, Lockman D, Jaffray D et al (2000) Setup error in radiotherapy: on-line correction using electronic kilovoltage and megavoltage radiographs. Int J Radiat Oncol Biol Phys 47(3):825–839

19. Langen KM, Meeks SL, Pouliot J (2008) Quality assurance of onboard megavoltage computed tomography imaging and target localization systems for on- and off-line image-guided radiotherapy. Int J Radiat Oncol Biol Phys 71(1 Suppl):S62–S65
20. Yoo S, Kim GY, Hammoud R et al (2006) A quality assurance program for the on-board imagers. Med Phys 33:4431–4447
21. Court LE, Dong L, Taylor N et al (2004) Evaluation of a contour-alignment technique for CT-guided prostate radiotherapy: an intra- and interobserver study. Int J Radiat Oncol Biol Phys 59:412–418
22. Langen KM, Zhang Y, Andrews RD et al (2005) Initial experience with megavoltage (MV) CT guidance for daily prostate alignments. Int J Radiat Oncol Biol Phys 62:1517–1524
23. Owen R, Kron T, Foroudi F et al (2009) Comparison of CT on rails with electronic portal imaging for positioning of prostate cancer patients with implanted fiducial markers. Int J Radiat Oncol Biol Phys 74:906–912
24. Bova FJ, Buatti JM, Friedman WA et al (1997) The University of Florida frameless high-precision stereotactic radiotherapy system. Int J Radiat Oncol Biol Phys 38:875–882
25. Jiang SB (2006) Technical aspects of image-guided respiration-gated radiation therapy. Med Dosim 31:141–151
26. Shirato H, Shimizu S, Kitamura K et al (2000) Four-dimensional treatment planning and fluoroscopic real-time tumor tracking radiotherapy for moving tumor. Int J Radiat Oncol Biol Phys 48:435–442
27. Shirato H, Shimizu S, Kunieda T et al (2000) Physical aspects of a real time tumor-tracking system for gated radiotherapy. Int J Radiat Oncol Biol Phys 48:1187–1195
28. Shimizu S, Shirato H, Ogura S et al (2001) Detection of lung tumor movement in real-time tumor-tracking radiotherapy. Int J Radiat Oncol Biol Phys 51:304–310
29. Shirato H, Seppenwoolde Y, Kitamura K et al (2004) Intrafractional tumor motion: lung and liver. Semin Radiat Oncol 14:10–18
30. Kitamura K, Shirato H, Shinohara N et al (2003) Reduction in acute morbidity using hypofractionated intensity-modulated radiation therapy assisted with a fluoroscopic real-time tumor-tracking system for prostate cancer: preliminary results of a phase I/II' study. Cancer J 9:268–276
31. Bert C, Metheany KG, Doppke KP et al (2006) Clinical experience with a 3D surface patient setup system for alignment of partial-breast irradiation patients. Int J Radiat Oncol Biol Phys 64:1265–1274
32. Brahme A, Nyman P, Skatt B (2008) 4D laser camera for accurate patient positioning, collision avoidance, image fusion and adaptive approaches during diagnostic and therapeutic procedures. Med Phys 35:1670–1681
33. Moore C, Lilley F, Sauret V et al (2003) Opto-electronic sensing of body surface topology changes during radiotherapy for rectal cancer. Int J Radiat Oncol Biol Phys 56:248–258
34. Schoffel PJ, Harms W, Sroka-Perez G et al (2007) Accuracy of a commercial optical 3d surface imaging system for realignment of patients for radiotherapy of the thorax. Phys Med Biol 52:3949–3963
35. Willoughby T, Lehmann J, Bencomo JA et al (2012) Quality assurance for nonradiographic radiotherapy localization and positioning systems: report of Task Group 147. Med Phys 39:1728–1747
36. Bert C, Metheany KG, Doppke K et al (2005) A phantom evaluation of a stereo-vision surface imaging system for radiotherapy patient setup. Med Phys 32:2753–2762
37. Fuss M, Wong A, Fuller CD et al (2007) Image-guided intensity-modulated radiotherapy for pancreatic carcinoma. Gastrointest Cancer Res 1:2–11
38. D'Amboise DJ, Buyout J, Chatty IJ et al (2012) Continuous localization technologies for radiotherapy delivery: report of the American Society for Radiation Oncology Emerging Technology Committee. Pract Radiat Oncol 2:145–150
39. Shah AP, Kupelian PA, Waghorn BJ et al (2013) Real-time tumor tracking in the lung using an electromagnetic tracking system. Int J Radiat Oncol Biol Phys 86(3):477–483

40. Dempsey J, Dionne B, Fitzsimmons J et al (2006) A real-time MRI guided external beam radiotherapy delivery system. Med Phys 33:2254
41. Fallone BG, Murray B, Rathee S et al (2009) First MR images obtained during megavoltage photon irradiation from a prototype integrated linac-MR system. Med Phys 36:2084–2088
42. Kirkby C, Stanescu T, Rathee S et al (2008) Patient dosimetry for hybrid MRI-radiotherapy systems. Med Phys 35:1019–1027
43. Lamey M, Yun J, Burke B et al (2010) Radio frequency noise from an MLC: a feasibility study of the use of an MLC for linac-MR systems. Phys Med Biol 55:981–994
44. Raaijmakers AJ, Raaymakers BW, Lagendijk JJ (2007) Experimental verification of magnetic field dose effects for the MRI-accelerator. Phys Med Biol 52:4283–4291
45. Raaijmakers AJ, Raaymakers BW, van der Meer S et al (2007) Integrating a MRI scanner with a 6 MV radiotherapy accelerator: impact of the surface orientation on the entrance and exit dose due to the transverse magnetic field. Phys Med Biol 52:929–939
46. Raaijmakers AJE, Raaymakers BW, Lagendijk JJW (2005) Integrating a MRI scanner with a 6 MV radiotherapy accelerator: dose increase at tissue-air interfaces in a lateral magnetic field due to returning electrons. Phys Med Biol 50:1363–1376
47. Raaymakers BW, Raaijmakers AJ, Kotte AN et al (2004) Integrating a MRI scanner with a 6 MV radiotherapy accelerator: dose deposition in a transverse magnetic field. Phys Med Biol 49:4109–4118
48. Santos DM, St Aubin J, Fallone BG et al (2012) Magnetic shielding investigation for a 6 MV in-line linac within the parallel configuration of a linac-MR system. Med Phys 39:788–797
49. St Aubin J, Santos DM, Steciw S et al (2010) Effect of longitudinal magnetic fields on a simulated in-line 6 MV linac. Med Phys 37:4916–4923
50. Stroom JC, Heijmen BJ (2002) Geometrical uncertainties, radiotherapy planning margins, and the ICRU-62 report. Radiother Oncol 64(1):75–83
51. Van Herk M, Remeijer P, Rasch C et al (2000) The probability of correct target dosage: dose-population histograms for deriving treatment margins in radiotherapy. Int J Radiat Oncol Biol Phys 47(4):1121–1135
52. Badakhshi H, Kaul D, Wust P et al (2013) Image-guided stereotactic radiosurgery for cranial lesions: large margins compensate for reduced image guidance frequency. Anticancer Res 33(10):4639–4643
53. Wang X, Hu C, Eisbruch A (2011) Organ-sparing radiation therapy for head and neck cancer. Nat Rev Clin Oncol 8(11):639–648
54. Lohia S, Rajapurkar M, Nguyen SA et al (2014) A comparison of outcomes using intensity-modulated radiation therapy and 3-dimensional conformal radiation therapy in treatment of oropharyngeal cancer. J Am Med Assoc Otolaryngol Head Neck Surg 140(4):331–337
55. Leonard CE, Tallhamer M, Johnson T et al (2010) Clinical experience with image-guided radiotherapy in an accelerated partial breast intensity-modulated radiotherapy protocol. Int J Radiat Oncol Biol Phys 76:528–534
56. Gierga D, Riboldi M, Turcotte J et al (2008) Comparison of target registration errors for multiple image-guided techniques in accelerated partial breast irradiation. Int J Radiat Oncol Biol Phys 70:1239–1246
57. White E, Cho J, Vallis K et al (2007) Cone beam computed tomography guidance for setup of patients receiving accelerated partial breast irradiation. Int J Radiat Oncol Biol Phys 68:547–554
58. Kim L, Wong J, Yan D (2007) On-line localization of the lumpectomy cavity using surgical clips. Int J Radiat Oncol Biol Phys 69:1305–1309
59. Grutters JP, Kessels AG, Pijls-Johannesma M et al (2010) Comparison of the effectiveness of radiotherapy with photons, protons and carbon-ions for non-small cell lung cancer: a meta-analysis. Radiother Oncol 95(1):32–40
60. Shirato H, Onimaru R, Ishikawa M et al (2012) Real-time 4-D radiotherapy for lung cancer. Cancer Sci 103(1):1–6

61. Nouhaud E, Créhange G, Cueff A et al (2013) Stereotactic Body radiation therapy for liver tumors with or without rotational intensity modulated radiation therapy. BMC Res Notes 6:492
62. Aggarwal A, Chopra S, Paul SN et al (2014) Evaluation of internal target volume in patients undergoing image-guided intensity modulated adjuvant radiation for gastric cancers. Br J Radiol 87(1033):20130583
63. Kuban DA, Tucker SL, Dong L et al (2008) Long-term results of the M. D. Anderson randomized dose-escalation trial for prostate cancer. Int J Radiat Oncol Biol Phys 70(1):67–74
64. Zelefsky MJ, Deasy JO (2013) Improved long-term outcomes with IMRT: is it better technology or better physics? Int J Radiat Oncol Biol Phys 87(5):867–868
65. Stephans KL, Xia P, Tendulkar RD et al (2010) The current status of image-guided external beam radiotherapy for prostate cancer. Curr Opin Urol 20(3):223–228
66. Mundt AJ, Mell LK, Roeske JC (2003) Preliminary analysis of chronic gastrointestinal toxicity in gynecology patients treated with intensity-modulated whole pelvic radiation therapy. Int J Radiat Oncol Biol Phys 56(5):1354–1360
67. Portelance L, Chao KS, Grigsby PW, Bennet H, Low D (2001) Intensity- modulated radiation therapy (IMRT) reduces small bowel, rectum, and bladder doses in patients with cervical cancer receiving pelvic and para-aortic irradiation. Int J Radiat Oncol Biol Phys 51(1):261–266
68. Cozzi L, Dinshaw KA, Shrivastava SK et al (2008) A treatment planning study comparing volumetric arc modulation with RapidArc and fixed field IMRT for cervix uteri radiotherapy. Radiother Oncol 89(2):180–191
69. Kidd EA, Siegel BA, Dehdashti F et al (2010) Clinical outcomes of definitive intensity-modulated radiation therapy with fluorodeoxyglucose-positron emission tomography simulation in patients with locally advanced cervical cancer. Int J Radiat Oncol Biol Phys 77(4):1085–1091
70. Hasselle MD, Rose B, Kochanski JD et al (2011) Clinical outcomes of intensity-modulated pelvic radiation therapy for carcinoma of the cervix. Int J Radiat Oncol Biol Phys 80(5):1436–1445
71. Forrest J, Presutti J, Davidson M, Hamilton P, Kiss A, Thomas G (2012) A dosimetric planning study comparing intensity-modulated radiotherapy with four-field conformal pelvic radiotherapy for the definitive treatment of cervical carcinoma. Clin Oncol 24(4):e63–e70
72. Renard-Oldrini S, Brunaud C, Huger S et al (2012) Dosimetric comparison between the intensity modulated radiotherapy with fixed field and Rapid Arc of cervix cancer. Cancer Radiother 16(3):209–214
73. Santos JDL, Popple R, Agazaryan N et al (2013) Image guided radiation therapy (IGRT) technologies for radiation therapy localization and delivery. Int J Radiat Oncol Biol Phys 87:33–45
74. Morin S, Gillis A, Chen J et al (2006) Megavoltage cone-beam CT: system description and clinical applications. Med Dosim 31:51–61
75. Meeks SL, Harmon JF, Langen KM et al (2005) Performance characterization of megavoltage computed tomography imaging on a helical tomotherapy unit. Med Phys 32:2673–2681
76. Wen N, Guan H, Hammoud R et al (2007) Dose delivered from Varian's CBCT to patients receiving IMRT for prostate cancer. Phys Med Biol 52:2267–2276
77. Islam MK, Purdie TG, Norrlinger BD et al (2006) Patient dose from kilovoltage cone beam computed tomography imaging in radiation therapy. Med Phys 33:1573–1582
78. Murphy MJ, Balter J, Balter S et al (2007) The management of imaging dose during image-guided radiotherapy: report of the AAPM Task Group 75. Med Phys 34:4041–4043

Adaptive Radiation Therapy in Intensity-Modulated Radiation Therapy for Head and Neck Cancer

6

Yasumasa Nishimura

Keywords
Adaptive radiation therapy • Intensity-modulated radiation therapy • Image-guided radiation therapy • Head and neck cancer

6.1 Introduction

The development and successful clinical use of intensity-modulated radiation therapy (IMRT) is a significant advance in radiation therapy (RT). The dose conformality possible with IMRT makes it particularly effective for cancer in complex anatomic regions such as the head and neck, because the clinical target volumes (CTVs) are contiguous to organs at risk, which for head and neck cancer includes the salivary glands, brainstem, and spinal cord. Two randomized clinical trials comparing IMRT with conventional RT for patients with early-stage nasopharyngeal cancer showed a significant benefit of IMRT with regard to salivary function and quality of life [1, 2]. Single-institution reports and multi-institutional prospective trials of IMRT for head and neck cancer have shown excellent rates of locoregional control and overall survival [3–6].

Because treatment-planning and quality assurance procedures for IMRT are time- and labor-intensive, most investigators use the initial IMRT plan, created during treatment simulation and planning, during the entire course of IMRT. However, significant anatomic changes during fractionated RT, such as shrinkage of the primary tumor or nodal masses and loss of body weight, have been reported during cancer treatment and are of particular importance in head and neck cancer [5–8]

Y. Nishimura, M.D., Ph.D. (✉)
Department of Radiation Oncology, Kinki University Faculty of Medicine,
377-2, Ohno-Higashi, Osaka-Sayama, Osaka 589-8511, Japan
e-mail: ynishi@med.kindai.ac.jp

Y. Nishimura, R. Komaki (eds.), *Intensity-Modulated Radiation Therapy*,
DOI 10.1007/978-4-431-55486-8_6

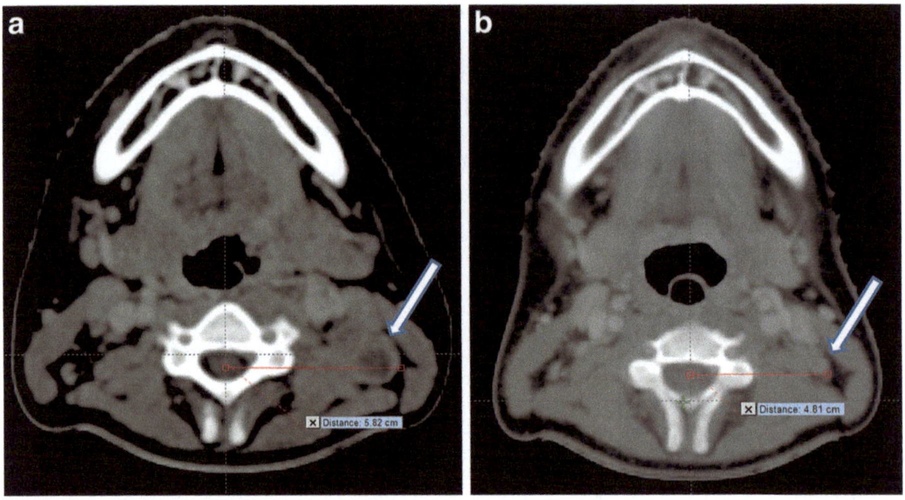

Fig. 6.1 (**a**) *Left upper* jugular lymph node metastasis (*white arrow*) from nasopharyngeal cancer was noted on CT before IMRT. (**b**) On the 25th day of IMRT (after delivery of 36 Gy in 18 fractions), the metastatic lymph node had shifted 1 cm medially because of body weight loss and tumor shrinkage

(Fig. 6.1). For example, parotid gland volume was shown in two separate analyses to decrease substantially during IMRT, to 74 % of the initial volume in one analysis and to 82 % in the other [8, 9]. Treatment-induced changes in body contour, target volumes, and organs at risk during IMRT can affect the dose distribution to the target volume and the organs at risk, which can lead to marginal recurrence [10–12] as well as delayed toxicity. For head and neck cancer, new masks may need to be created to compensate for body weight loss and tumor shrinkage during treatment, and treatment simulation may need to be repeated so that new, more accurate plans can be generated (replanning). The process for modifying treatment plans by using systematic feedback of such measurements is known as adaptive radiation therapy (ART) [13]. This chapter discusses the concept of ART and its necessity in the treatment of cancer in general and head and neck cancer in particular.

6.2 Concept of ART and Image-Guided RT

Tumors and normal tissues move with time, and this movement may be clinically significant from second to second, day to day, week to week, or longer, and the variation in their position can be considered to take place intra (within)- or inter (between)-treatment sessions. Each tumor site has its own movement characteristics. For example, thoracic tumors are predominantly affected by periodic respiratory motion, whereas head and neck tumors are affected more by gradual tumor shrinkage over time (Fig. 6.1). The position of prostate tumors can change from day to day owing to differences in bowel and bladder fillings, although additional momentary and irregular changes can also result from peristalsis. However, no tumor is immune from change in position, size, or movement, whether momentary or more gradual.

Substantial effort has been devoted in RT in reducing the effect of variations related to the treatment process, such as set-up error and geometric variations in volumes of targets and critical normal organs. To compensate for these variations (also called uncertainties), the use of predefined uniform margins around the CTV has been suggested. The term planning target volume (PTV) denotes a CTV that is volumetrically expanded to compensate for treatment set-up uncertainties [14]. Such expansions often present dosimetric trade-offs because they often overlap geographically with adjacent organs at risk. A favorable therapeutic ratio relies on accurate knowledge of set-up uncertainties, normal tissue dose tolerances, and delineation of treatment targets to safely minimize PTV expansions. However, radiation treatment planning with compensation strategies has not been customized to account for variations in individual patients over the course of treatment.

ART is an approach to correct for daily or weekly variations in tumor and normal tissue positioning or volume through online or offline modification of original RT plans [13, 15–17]. ART has two components: (1) means of detecting such changes and (2) means of intervention. Successful implementation of each of these components determines the overall success of the clinical application of ART. The first component of ART, detecting changes, is also known as image-guided RT (IGRT). Obtaining images from patients while they are in the treatment room can provide extensive datasets documenting movements and anatomic change of targets during and between RT sessions. The use of computed tomography (CT) in the treatment room can depict geometrically accurate soft tissue targets. In-room CT scanners, tomotherapy-based megavoltage CT, and gantry-mounted cone-beam CT are now all available to provide in-room three-dimensional (3D) imaging [16, 17]. Such imaging can reveal two types of variation: systematic, in which the mean of the observed positions is offset from the prescribed position, or random, which can be measured as daily changes from the mean. Systematic uncertainties are amenable to offline analyses of images acquired at treatment. However, random errors can be addressed only with daily online imaging and modification. Thus, IGRT can reduce set-up uncertainties and improve the management of organ motion, consequently allowing dose escalation, an improved therapeutic ratio, or both. Because the goal of IGRT is to correct for set-up errors and minimize the PTV margin, image guidance does not typically involve modification of the original treatment plan but rather may involve repositioning treatment fields (Fig. 6.2).

In contrast to IGRT, the intent of ART is to appropriately modify a radiation treatment plan to account for temporal changes in anatomy (Fig. 6.2). In theory, ART can occur on three different timescales: offline between fractions, online immediately before a fraction, and in real time during a treatment fraction [16, 17]. ART is closely linked to image-guidance processes because any volumetric images acquired for the IGRT procedure could also be used to monitor changes in anatomy and to design new plans. Ideally, in-room volumetric images could be sent to a

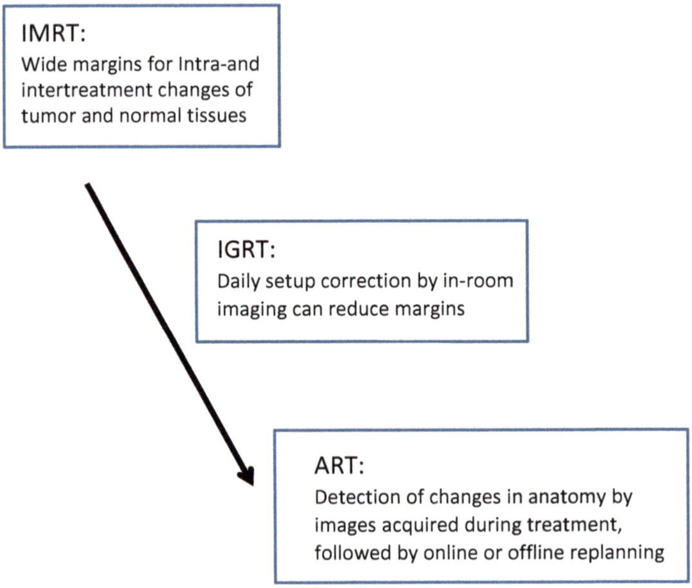

Fig. 6.2 The concept of ART in comparison with IMRT and IGRT

treatment-planning system, where a new treatment plan based on current anatomy could be generated via automated deformable image registration software and sent back to the therapy machine for delivery. The adapted plan could be either deployed immediately (online correction) or used for future treatments (offline correction).

Manual segmentation of treatment-planning images requires too much time and effort on behalf of treating physicians and staff to be practical for the routine use of ART. Deformable image registration for atlas-based autosegmentation is an effective alternative for serial adaptive replanning [18–21]. If the contour exists in one of the reference CT images, then deformable transformation can be used to transform reference contours onto the newly acquired CT images with minimal manual input. This approach is well suited to ART, given that the original plan can serve as the reference for this process. Manual recontouring can take hours, which would not be practical for online ART and would strain resources for the application of offline ART as well.

6.3 Volume and Dosimetric Changes During IMRT

Several studies have demonstrated that anatomic changes take place during IMRT with concurrent chemotherapy for head and neck cancer [7, 9, 20, 22–24]. In one of these studies, Barker et al. [7] evaluated 14 patients with head and neck cancer treated by an integrated CT-linear accelerator system that allows CT imaging at daily RT sessions; these investigators noted rapid weight loss and anatomic change in both tumors and at-risk organs after 3–4 weeks of RT. Specifically, the gross tumor volume (GTV) decreased throughout therapy at a median rate of 0.2 cm^3 per treatment day (range, 0.01–1.95 cm^3/day), and the GTV decreased at a median rate of 1.8 % per treatment day. The absolute volume loss was larger for large tumors and nodes. By the final day of treatment, the median relative loss of the GTV was approximately 70 % (range, 10–92 %), and the center of the mass was displaced by a median 3.3 mm (range, 0–17.3 mm). The volume of the parotid glands also decreased over time; at the end of treatment, the median loss was 28 % (range, 5.9–53.6 %), and the median medial shift was 3.1 mm (range, −0.3 to 9.9 mm) [7].

Wu et al. [20] retrospectively analyzed 11 patients with locally advanced head and neck cancer who had had one IMRT planning scan and 6 weekly helical CT scans. Each patient had 1–6 repeated plans ("replans") created based on the weekly helical CTs, and doses were accumulated on the planning CT. Although the cumulative doses to targets or the spinal cord, brainstem, and mandible were unchanged on the replans relative to the original plans, significant increases in parotid doses were observed without replanning. These investigators concluded that replanning would preserve sparing of only the parotid glands.

On the other hand, Hunter et al. [18] evaluated 18 patients with oropharyngeal cancer by daily cone-beam CT with clinical set-up alignment. In that study, differences between planned and delivered doses to the parotid glands were small relative to the standard deviations of the dose–salivary flow function, suggesting that ART is not likely to produce measurable improvements in salivary output in most cases.

Duma et al. [25] reported a dosimetric advantage of ART triggered exclusively by soft tissue changes detected on daily set-up helical megavoltage CT during helical tomotherapy for head and neck cancer. At those authors' institution, ART is routinely used for patients with head and neck cancer when (1) inspection or palpation reveals a change in soft tissues or the mask was loose or (2) the IGRT CT scan shows a soft tissue change >5 mm. Of 94 patients with head and neck cancer treated with helical tomotherapy, replanning was done for 11 patients (12 %) for whom soft tissue changes were >5 mm as identified by IGRT CT (i.e., none had clinically evident changes); replanning was done mostly at the end of the third week of treatment. In these patients, shrinkage of the body diameter by 1 cm did not affect the coverage of the PTV but translated into a slightly higher dose to the PTV itself and a higher delivered dose to the normal tissues outside the PTV.

At our institution, we adopted a two-step method for IMRT for head and neck cancer and are exploring its use as ART [8, 9, 11]. We obtain two sets of treatment-planning CT scans, the first before IMRT (CT-1) and the second during the third or fourth week of IMRT (CT-2); the second set of plans is used for an IMRT boost dose after 46–50 Gy has been given and the patient fitted with a new thermoplastic mask. We recently analyzed geometric and dosimetric changes between the first and second set of scans in 20 consecutive patients treated with this two-step IMRT process [8]. Twenty consecutive patients with pharyngeal cancer (10 nasopharyngeal, 6 oropharyngeal, and 4 hypopharyngeal) were treated with a two-step IMRT method in which 46–50 Gy was given in 23–25 fractions to the whole neck, followed by a boost dose to the high-risk CTV for a total dose of 60–70 Gy in 30–35 fractions (median dose, 70 Gy); patients also received concurrent cisplatin-based chemotherapy (80 mg/m^2 every 3 weeks). The spinal cord and parotid glands were recontoured without margins on both sets of CT scans, and the primary tumors (GTV-p) and the largest metastatic lymph node (GTV-n) were also recontoured without margins. On the CT-2 scans, the mean GTV-p volume had dropped to 37 % ± 24 % of the CT-1 plan ($P=0.002$) and the mean GTV-n volume to 48 % ± 37 % of the CT-1 value ($P=0.081$). The parotid gland volume also decreased to 82 % ± 12 % of the initial volume ($P<0.0001$), and the lateral surface of the parotid glands on CT-2 had shifted medially by an average of 4.2 ± 2.9 mm ($P<0.0001$) [8].

To analyze dosimetric changes, we created three sets of IMRT plans: Plan-1 was the actual initial IMRT plan (based on the pretreatment CT-1 scans), Plan-2 was the actual boost plan (based on the CT-2 scans), and Plan-3, which involved transferring the initial IMRT plan to the CT-2 scans and matching them for isocenter and bony alignment. Dose distribution plans were then recalculated to obtain dosimetric parameters of the recontoured target volumes and organs at risk, and variables were compared between Plan-3 and Plan-1 to evaluate the effects of anatomic changes on dosimetric outcomes. The replanning effects for dosimetric outcomes were compared for Plan-2 and Plan-3. Dose parameters were calculated for a total prescribed dose of 70 Gy for each plan. The mean doses to the parotid glands (D_{mean}) were 25.4 Gy in Plan-1, 20.0 Gy in Plan-2, and 30.3 Gy in Plan-3 [8]; these differences were significant for Plan-1 vs. Plan-3 (5.0 ± 5.1 Gy, or 120 %; $P<0.0001$) and for Plan-2 vs. Plan-3 (10.3 ± 3.6 Gy, or 66 %; $P<0.0001$). The corresponding doses to 2 % of the spinal cord (D_2) were 37.2 Gy in Plan-1, 36.7 Gy in Plan-2, and 39.1 Gy in Plan-3 (significant difference for Plan-3 vs. Plan-1 [1.9 ± 2.0 Gy or 105 %; $P=0.0003$] and marginally significant difference for Plan-2 vs. Plan-3 [2.4 ± 5.2 Gy or 94 %; $P=0.0507$]). These findings, plus those from another group showing the value of adaptive replanning for significant changes in the maximum dose to the spinal cord and dose to 50 % of the parotid glands (D_{50}) [22], lead us to conclude that our two-step IMRT method can effectively prevent increases in high-dose exposure of the spinal cord and parotid glands.

Figure 6.3 illustrates two sets of plans, the Plan-1 (lower left) based on the CT-1 scan (upper left) and the Plan-3 (lower right) based on the CT-2 scans (upper right), for a patient with nasopharyngeal cancer. Shrinkage of the neck diameter is evident on CT-2, and the left upper jugular lymph nodes (GTV-n; white arrow) have apparently regressed as well. On Plan-3 (lower right), the parotid glands had shifted so that, in the absence of replanning, they were now included in the 70 % dose region (blue color). Comparing Plan-1 with Plan-3 also indicated a slight (but significant) increase in D_{mean} to the GTV-p (0.6 ± 0.7 Gy, $P=0.0007$) and D_{98} to the GTV-p (0.8 ± 0.6 Gy, $P<0.0001$) but no changes in the D_{mean} and D_{98} of the GTV-n. These findings lead us to conclude that planned doses to GTV-p and GTV-n could be delivered by using the initial IMRT plan, without ART.

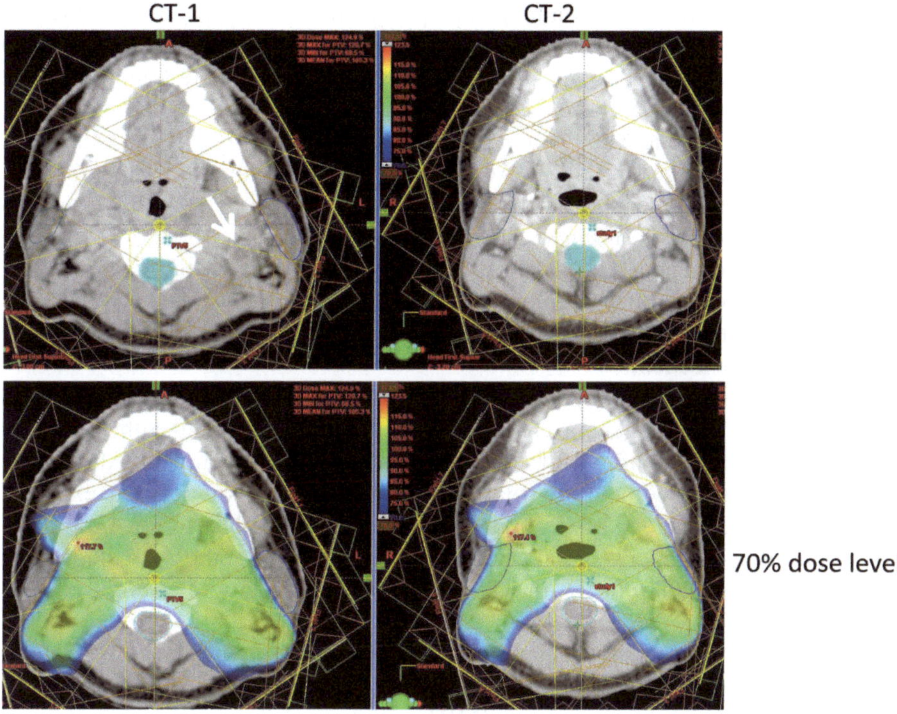

Fig. 6.3 *Top row*, treatment-planning CT scan obtained before IMRT (CT-1, *upper left*) and a second CT scan (CT-2, *upper right*) obtained after 38 Gy in 19 fractions for a patient with nasopharyngeal cancer. Shrinkage of the neck diameter is evident on CT-2, and the *left upper* jugular lymph nodes (GTV-n; *white arrow*) have apparently regressed as well. *Bottom row*, dose distributions on the initial IMRT plan on CT-1 (Plan-1, *lower left*) and the same plan transferred onto CT-2 (Plan-3, *lower right*). Most of the bilateral parotid glands would have been included in the 70 % dose level (*blue color*) in Plan-3

6.4 Clinical Studies of ART

The dosimetric advantage of ART in reducing the dose to various organs at risk such as the parotid glands and spinal cord is apparent. The substantial investments in equipment, staffing, and staff time required for widespread implementation of ART require demonstration that these dosimetric advantages will translate into clinical benefit. However, to date only a few studies have prospectively evaluated the clinical benefit of ART for patients with head and neck cancer.

In one such study, Schwartz et al. [19] described the use of ART for patients with advanced head and neck cancer in which a daily CT-guided setup and deformable image registration were used. Of the 22 patients enrolled, all required at least one round of replanning because of changes in CTV and normal tissues, and eight patients (36 %) required a second replanning. The median trigger point for the first adaptive plan was the 16th treatment fraction. For the eight patients with two replans, the median trigger points for the first replan were the 11th fraction and the 22nd fraction for the second replan. The elapsed interval from triggering in-room CT imaging to subsequent delivery of the prompted ART plan was 1.7 working days (median, 2 days; range, 1–4 days). At a median follow-up time of 31 months, disease control rates at 2 years were 100 % local and 95 % regional. Parotid dose sparing was improved with a single ART replanning by a mean of 0.6 Gy for the contralateral parotid and 1.3 Gy for the ipsilateral parotid relative to standard IGRT; for the eight patients with two replans, these values were 0.8 Gy and 4.1 Gy. Chronic toxicity in this study was considered encouraging, and the locoregional control rate was considered excellent.

Berwouts et al. [26] reported a phase I clinical trial of three-phase ART for head and neck cancer. Ten patients were enrolled, and scans were obtained before treatment, after 8 fractions, after 18 fractions, and at the end of treatment. All patients completed treatment without interruptions or severe acute toxicities. The extent of GTV reduction at the end of treatment was 72 %. At a median follow-up time of 13 months, 9 of 10 patients had no evidence of disease. The authors concluded that their three-phase ART process, in which currently available tools were used, was feasible.

In another clinical study, Zhao et al. [27] retrospectively evaluated the role of replanning in IMRT for nasopharyngeal cancer. Of 175 patients with nasopharyngeal cancer who had been treated with IMRT, 158 showed obvious anatomic changes including tumor shrinkage, nodal shrinkage, or weight loss before 20 fractions had been delivered; 33 of those patients had repeat CT scanning and replanning during the course of treatment, and their outcomes were compared with those of 66 control patients who had not had replanning who were matched in terms of clinical disease stage and type of anatomic changes. IMRT replanning improved the 3-year local progression-free survival rate for patients with T3–T4 tumors and also reduced late effects for patients with large lymph nodes (N2, N3). These investigators concluded that ART is recommended for patients with advanced (T3, T4 or N2, N3) nasopharyngeal cancer. Indeed, because patients with locally advanced nasopharyngeal cancer often present with large lymph nodes in the neck and because both primary tumors and neck lymph nodes regress rapidly with RT, ART is especially desirable for such disease. Lee et al. [5] described a phase I/II study of a two-step, two-plan method involving the use of IMRT with a simultaneous integrated boost (SIB) technique for 20 patients with nasopharyngeal cancer. The results of this study suggested that tumor regression of early-stage pharyngeal tumors may be sufficiently small as to render ART unnecessary; however, even patients with early-stage tumors showed notable changes in body weight and contour when the RT was given with chemotherapy. Therefore, we consider ART to be indicated for head and neck cancer that is to be treated by concurrent chemoradiation.

The two-step method of IMRT used at the author's institution, described earlier in this chapter, has been used to treat head and neck cancer beginning in 2000. This two-step method obviously requires more time for treatment planning and quality assurance than a single-step SIB method, which can provide several dose levels for

Table 6.1 Patient, tumor, and treatment characteristics

Characteristic	Value or no. of patients
Age, median, years	56 (range 14–81)
Sex	
Male	39
Female	11
Performance status score	
0	39
1	10
2	1
Tumor histology	
WHO type I (keratinizing SCC)	8
WHO type II (non-keratinizing SCC)	37
WHO type III (Lymphoepithelial)	5
TNM disease stage (UICC 7th, 2009)	
I	7
II	8
III	18
IVa	12
IVb	5
Total radiation dose	60–70 Gy in 2-Gy fractions (median 68 Gy)
Concurrent chemotherapy	
Cisplatin (80 mg/m^2), 1–3 cycles	45
Adjuvant chemotherapy	
Cisplatin (70 mg/m^2) plus fluorouracil (700 mg/m^2)×4–5 days	31

CTVs and GTVs simultaneously. Initially, it took 5 working days for treatment planning and its verification, and treatment of the IMRT plan was begun 7–10 days after CT simulation. Because daily or weekly IGRT was not possible at that time, a second treatment-planning CT (CT-2) for the boost IMRT plan was obtained for all patients with head and neck cancer during the third or fourth week of treatment. Our clinical findings with the use of this technique for nasopharyngeal cancer are summarized below [11].

Fifty patients with stage I–IVB nasopharyngeal cancer were treated with the two-step IMRT method from 2000 to 2010 (Table 6.1). For all patients, treatment-planning CT scans were obtained twice, once before and once during IMRT, which were given to a total dose of 60–70 Gy in 28–35 fractions (median dose, 68 Gy). Forty-five of these patients also received concurrent chemotherapy (1–3 cycles of cisplatin 80 mg/m^2 every 3 weeks) with IMRT (two patients were excluded because of advanced age [>76 years] and three for poor renal or cardiac function). Thirty-one patients received 1 or 2 courses of adjuvant chemotherapy (cisplatin 70 mg/m^2, fluorouracil 700 mg/m^2 over 4–5 days) after the IMRT.

At a median follow-up time of 55 months for surviving patients, overall survival rates for the 45 patients treated with concurrent chemotherapy were 90 % at 3 years and 84 % at 5 years (Fig. 6.4a); the corresponding locoregional control rates were 94 % and 89 % (Fig. 6.4b). Xerostomia was scored at 1–2 years after treatment for

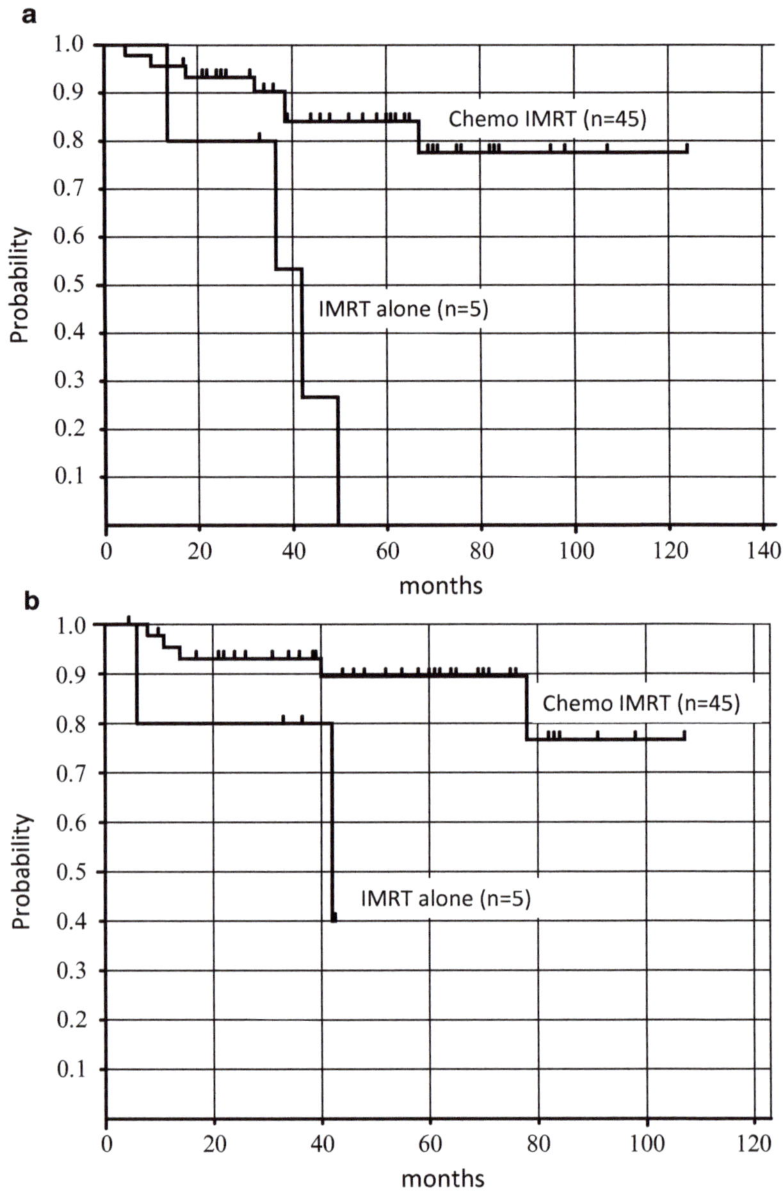

Fig. 6.4 (**a**) Overall survival and (**b**) locoregional control for 45 patients with nasopharyngeal cancer treated with IMRT and concurrent chemotherapy and five patients treated with IMRT alone

49 patients (excluding 1 patient who died at 3 months after treatment); at that time, 15 patients had grade 0, 22 had grade 1, 11 had grade 2, and 1 had grade 3 xerostomia. The authors concluded that this two-step IMRT approach was the simplest for ART, was effective for preventing xerostomia, and produced good locoregional control. These retrospective single-institution findings are now being extended in a prospective multi-institutional phase II trial of this two-step IMRT method for nasopharyngeal cancer by the Japan Clinical Oncology Group (JCOG 1015), with the goal of evaluating the feasibility and efficacy of this approach for nasopharyngeal cancer.

6.5 Future of ART

Online ART, i.e., that which involves in-room imaging of the target areas of interest and sending those images to a treatment-planning system, which creates a new treatment plan that accounts for changes in volume and anatomy and sends it back to the treatment machine for immediate delivery, could provide greater treatment precision but only at the cost of substantial increases in physician and staff effort and treatment time [16, 17]. Moreover, online ART is even more challenging than that because the patients must remain immobilized on the treatment couch while waiting for ART correction. As noted previously, most anatomic changes in head and neck cancer take place gradually over the first few weeks of treatment [7], and thus, real-time intervention is probably not needed, at least in the absence of an acute, unforeseen event (such as rapid disease progression). Therefore, in our opinion, offline ART (when the new treatment plan is delivered at some interval after its creation) seems to be the more practical approach for most patients with head and neck cancer [15, 17]. Even when daily or weekly IGRT is used, usually only one or two replans are necessary during 7 weeks of treatment for head and neck cancer. Many studies have shown that at least one replan is necessary for most patients, and about 36 % of patients will require two replans [19]. Even without weekly IGRT, ART can be done clinically by a planned second treatment CT in the third or fourth week of IMRT as described in our two-step IMRT method [11], because in most studies of volumetric changes, both primary tumors and metastatic lymph nodes regress within the first 4 weeks of treatment [7]. This makes ART easier to implement in the clinical treatment flow.

In the ART process, manual contouring of the CT images used for treatment is quite time- and labor-intensive and is also susceptible to intra- and interobserver variation. Hence, the development and clinical implementation of automated deformable image registration has facilitated the use of ART [18–21]. The replanning process remains the most time-consuming component of ART, leading several groups to study auto-replanning algorithms [16, 17]. Such algorithms are essential for online ART and may also reduce the workload requirements for offline ART.

Another technique that may be helpful in the ART process is ^{18}F-fluoromisonidazole positron emission tomography (F-MISO PET)/CT, a molecular imaging technique that can visualize hypoxic areas within tumors (Fig. 6.5) that can then be subjected

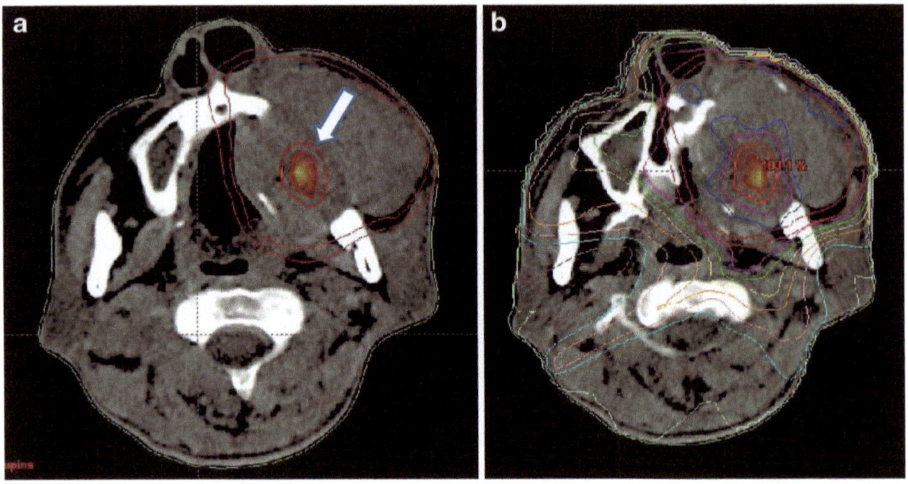

Fig. 6.5 (**a**) An ^{18}F-misonidazole positron emission tomography/computed tomography scans of a patient with advanced maxillary cancer. The hypoxic area (*white arrow*) is that in which the standardized uptake value of F-MISO is ≥1.60. (**b**) Dose distribution for an intensity-modulated radiation therapy plan with a simultaneous integrated boost for the patient shown in (**a**). In this plan, a total dose of 82.5 Gy in 33 fractions can be delivered to the hypoxic area, and a total of 66 Gy in 33 fractions is given simultaneously to the residual normoxic tumor volume

to dose escalation by "dose-painting" IMRT [28]. Although some reports have shown that F-MISO PET/CT-based dose painting can allow doses to hypoxic regions to be escalated to 78, 84, or even 105 Gy [29, 30], no clinical trials of this approach have been reported [28, 31]. A small series from the author's institution has documented changes in intratumoral uptake of F-MISO before and during RT [32]; in that study, uptake of ≥1.60 standardized uptake value (SUV) is taken to indicate a hypoxic area. However, clinical trials involving this technique may not be appropriate at this time for several reasons, among the reproducibility of intratumoral measurements of F-MISO [33]. Nehmeh et al. [34] showed that intratumoral distribution of F-MISO over a 3-day period without treatment was reproducible in only 6 of 13 human tumors; the other 7 tumors showed changes in F-MISO distribution after 3 days. The other factor complicating the use of this method is the reoxygenation of hypoxic areas during fractionated RT. In our study, both the SUV_{max} and the areas of F-MISO accumulation decreased in six of the eight tumors after delivery of approximately 20 Gy [32]. Another limitation of using PET in ART is the poor resolution of PET images. Discrepancies are common between PET images and the underlying microscopic reality represented by autoradiography [31, 35]. Therefore, dose escalation to hypoxic subvolumes of a tumor revealed by F-MISO PET/CT before RT seems inappropriate. However, if high-resolution F-MISO PET/CT becomes more reliable, and if online ART becomes more widely available in the near future, dose-painting IMRT for hypoxic subvolumes within tumors may become clinically feasible [17, 31].

6.6 Conclusions

Current ART strategies remain labor- and resource-intensive. However, several clinical studies of ART for head and neck cancer have demonstrated its feasibility and dosimetric advantages. With the development of IGRT, deformable image registration, and other related techniques, clinical application of ART has become feasible. Because the ultimate clinical effectiveness of ART in head and neck cancer remains undefined at this time, prospective clinical trials comparing IMRT, with or without ART, are warranted to clarify whether its dosimetric advantages translate into clinical benefits.

Acknowledgments This study was supported in part by a Grant-in-Aid for Cancer Research (H23-009, H26-090) from the Ministry of Health, Labour, and Welfare of Japan and by a Grant-in-Aid for Scientific Research (25461932) from the Ministry of Education, Culture, Sports, Science, and Technology, Japan.

Conflict of Interest Statement The author has no conflicts to declare.

References

1. Pow EH, Kwong DL, McMillan AS et al (2006) Xerostomia and quality of life after intensity-modulated radiotherapy vs. conventional radiotherapy for early-stage nasopharyngeal carcinoma: initial report on a randomized controlled clinical trial. Int J Radiat Oncol Biol Phys 66:981–991
2. Kam MK, Leung SF, Zee B et al (2007) Prospective randomized study of intensity-modulated radiotherapy on salivary gland function in early-stage nasopharyngeal carcinoma patients. J Clin Oncol 25:4873–4879
3. Lee N, Xia P, Quivey JM et al (2002) Intensity-modulated radiotherapy in the treatment of nasopharyngeal carcinoma: an update of the UCSF experience. Int J Radiat Oncol Biol Phys 53:12–22
4. Lee N, Harris J, Garden AS et al (2009) Intensity-modulated radiation therapy with or without chemotherapy for nasopharyngeal carcinoma: radiation therapy oncology group phase II trial 0225. J Clin Oncol 27:3684–3690
5. Lee SW, Back GM, Yi BY et al (2006) Preliminary results of a phase I/II study of simultaneous modulated accelerated radiotherapy for nondisseminated nasopharyngeal carcinoma. Int J Radiat Oncol Biol Phys 65:152–160
6. Wolden SL, Chen WC, Pfister DG, Kraus DH, Berry SL, Zelefsky MJ (2006) Intensity-modulated radiation therapy (IMRT) for nasopharynx cancer: update of the Memorial Sloan-Kettering experience. Int J Radiat Oncol Biol Phys 64:57–62
7. Barker JL Jr, Garden AS, Ang KK et al (2004) Quantification of volumetric and geometric changes occurring during fractionated radiotherapy for head-and-neck cancer using an integrated CT/linear accelerator system. Int J Radiat Oncol Biol Phys 59:960–970
8. Nishi T, Nishimura Y, Shibata T et al (2013) Volume and dosimetric changes and initial clinical experience of a two-step adaptive intensity modulated radiation therapy (IMRT) scheme for head and neck cancer. Radiother Oncol 106:85–89
9. Nishimura Y, Nakamatsu K, Shibata T et al (2005) Importance of the initial volume of parotid glands in xerostomia for patients with head and neck cancers treated with IMRT. Jpn J Clin Oncol 35:375–379

10. Cannon DM, Lee NY (2008) Recurrence in region of spared parotid gland after definitive intensity-modulated radiotherapy for head and neck cancer. Int J Radiat Oncol Biol Phys 70:660–665
11. Nishimura Y, Shibata T, Nakamatsu K et al (2010) A two-step intensity-modulated radiation therapy method for nasopharyngeal cancer: the Kinki University experience. Jpn J Clin Oncol 40:130–138
12. Schoenfeld GO, Amdur RJ, Morris CG et al (2008) Patterns of failure and toxicity after intensity-modulated radiotherapy for head and neck cancer. Int J Radiat Oncol Biol Phys 71:377–385
13. Yan D, Vicini F, Wong J et al (1997) Adaptive radiation therapy. Phys Med Biol 42:123–132
14. Suzuki M, Nishimura Y, Nakamatsu K et al (2006) Analysis of interfractional set-up errors and intrafractional organ motions during IMRT for head and neck tumors to define an appropriate planning target volume (PTV)- and planning organs at risk volume (PRV)-margins. Radiother Oncol 78:283–290
15. Grégoire V, Jeraj R, Lee JA, O'Sullivan B (2012) Radiotherapy for head and neck tumours in 2012 and beyond: conformal, tailored, and adaptive? Lancet Oncol 13:e292–e300
16. Schwartz DL, Dong L (2011) Adaptive radiation therapy for head and neck cancer-can an old goal evolve into a new standard? J Oncol 2011:1–13. pii: 690595. doi:10.1155/2011/690595
17. Schwartz DL (2012) Current progress in adaptive radiation therapy for head and neck cancer. Curr Oncol Rep 14(2):139–147
18. Hunter KU, Fernandes LL, Vineberg KA et al (2013) Parotid glands dose-effect relationships based on their actually delivered doses: implications for adaptive replanning in radiation therapy of head-and-neck cancer. Int J Radiat Oncol Biol Phys 87:676–682
19. Schwartz DL, Garden AS, Thomas J et al (2012) Adaptive radiotherapy for head-and-neck cancer: initial clinical outcomes from a prospective trial. Int J Radiat Oncol Biol Phys 83:986–993
20. Wu Q, Chi Y, Chen PY et al (2009) Adaptive replanning strategies accounting for shrinkage in head and neck IMRT. Int J Radiat Oncol Biol Phys 75:924–932
21. Zhen X, Yan H, Zhou L et al (2013) Deformable image registration of CT and truncated cone-beam CT for adaptive radiation therapy. Phys Med Biol 58(22):7979–7993
22. Ahn PH, Chen CC, Ahn AI et al (2011) Adaptive planning in intensity-modulated radiation therapy for head and neck cancers: single-institution experience and clinical implications. Int J Radiat Oncol Biol Phys 80:677–685
23. Bhide SA, Davies M, Burke K et al (2010) Weekly volume and dosimetric changes during chemoradiotherapy with intensity-modulated radiation therapy for head and neck cancer: a prospective observational study. Int J Radiat Oncol Biol Phys 76:1360–1368
24. Kuo YC, Wu TH, Chung TS et al (2006) Effect of regression of enlarged neck lymph nodes on radiation doses received by parotid glands during intensity-modulated radiotherapy for head and neck cancer. Am J Clin Oncol 29:600–605
25. Duma MN, Kampfer S, Schuster T et al (2012) Adaptive radiotherapy for soft tissue changes during helical tomotherapy for head and neck cancer. Strahlenther Onkol 188:243–247
26. Berwouts D, Olteanu LA, Duprez F et al (2013) Three-phase adaptive dose-painting-by-numbers for head-and-neck cancer: initial results of the phase I clinical trial. Radiother Oncol 107:310–316
27. Zhao L, Wan Q, Zhou Y et al (2011) The role of replanning in fractionated intensity modulated radiotherapy for nasopharyngeal carcinoma. Radiother Oncol 98:23–27
28. Hoeben BA, Bussink J, Troost EG et al (2013) Molecular PET imaging for biology-guided adaptive radiotherapy of head and neck cancer. Acta Oncol 52:1257–1271
29. Choi W, Lee SW, Park SH et al (2010) Planning study for available dose of hypoxic tumor volume using fluorine-18-labeled fluoromisonidazole positron emission tomography for treatment of the head and neck cancer. Radiother Oncol 97:176–182

30. Lee NY, Mechalakos JG, Nehmeh S et al (2008) Fluorine-18-labeled fluoromisonidazole positron emission and computed tomography-guided intensity-modulated radiotherapy for head and neck cancer: a feasibility study. Int J Radiat Oncol Biol Phys 70:2–13
31. Geets X, Grégoire V, Lee JA (2013) Implementation of hypoxia PET imaging in radiation therapy planning. Q J Nucl Med Mol Imag 57(3):271–282
32. Tachibana I, Nishimura Y, Shibata T et al (2013) A prospective clinical trial of tumor hypoxia imaging with 18F-fluoromisonidazole positron emission tomography and computed tomography (F-MISO PET/CT) before and during radiation therapy. J Radiat Res 54:1078–1084
33. Lin Z, Mechalakos J, Nehmeh S et al (2008) The influence of changes in tumor hypoxia on dose-painting treatment plans based on 18F-FMISO positron emission tomography. Int J Radiat Oncol Biol Phys 70:1219–1228
34. Nehmeh SA, Lee NY, Schroder H et al (2008) Reproducibility of intratumor distribution of 18F-fluoromisonidazole in head and neck cancer. Int J Radiat Oncol Biol Phys 70:235–242
35. Christian N, Lee JA, Bol A et al (2009) The limitation of PET imaging for biological adaptive-IMRT assessed in animal models. Radiother Oncol 91:101–106

Part II

Clinical Application

7 Brain Tumor: How Should We Manage Glioblastoma in the Era of IMRT?

Toshihiko Iuchi and Kazuo Hatano

Keywords
IMRT • Glioblastoma • Planning

7.1 Introduction

Glioblastoma (GBM) is the most frequently occurring primary brain tumor. The standard treatment for patients with GBM is tumor removal followed by radiation therapy (RT) with concurrent and adjuvant temozolomide (TMZ).

A randomized cooperative study has demonstrated the survival benefit of post-surgical RT [1], and other studies have demonstrated improved survival of patients treated at a total radiation dose of 60 Gy as compared with the lower treatment dose levels [2, 3]. Therefore, 60 Gy has become the standard treatment dose for this tumor. However, the survival time of patients with GBM after RT is still limited [4], and local failure is the predominant pattern of failure after RT. This suggests that more intensive RT targeting of the regional tumor will improve patient survival. On the other hand, higher dose delivery (>50 Gy) also runs the risk of cerebral necrosis, and this risk increases with increasing radiation dose [5]. Necrosis frequently causes neurological deterioration in patients, and whether or not the treatment dose should be escalated has been debated.

T. Iuchi (✉)
Division of Neurological Surgery, Chiba Cancer Center, 666-2 Nitona-cho, Chuo-ku, Chiba 260-0801, Japan
e-mail: tiuchi@chiba-cc.jp

K. Hatano
Division of Radiation Oncology, Tokyo Bay Advanced Imaging and Radiation Oncology Clinic MAKUHARI, Chiba, Japan

Y. Nishimura, R. Komaki (eds.), *Intensity-Modulated Radiation Therapy*,
DOI 10.1007/978-4-431-55486-8_7

Recently, innovations in irradiation technique, such as the three-dimensional radiation therapy and intensity-modulated radiation therapy (IMRT), have contributed to the precise delivery of the treatment dose to the target while reducing the dose to the surrounding normal brain tissue. To date, these advantages of IMRT have been utilized for three different purposes: to minimize RT-related toxicities by decreasing the dose to the surrounding functioning structures, to improve tumor control as a result of the delivery of increased treatment dose to the target, and to increase the biologically effective dose (BED) by increasing the dose per fraction.

In the subsequent sections, the advantages and the practical planning of IMRT for the treatment of GBM patients are discussed.

7.2 Biology of GBMs

GBM is the most malignant brain tumor. The aggressive and treatment-resistant nature of the GBM cells, such as their rapid growth, infiltrative characteristics, and hypoxic condition, should be considered in radiotherapy treatment planning.

7.2.1 Growth Speed

GBM is a fast-growing tumor. The potential tumor doubling time has been reported to be only 9–12 days [6, 7]. Repopulation during irradiation should not be ignored in the treatment of these fast-growing tumors. The dose loss per day induced by repopulation and its relationship to effective tumor doubling time and intrinsic radiosensitivity have been previously reported in a study involving in vitro analyses [8]. These data indicate the advantage of hypofractionated irradiation with high dose per fraction, which has become possible using highly conformal irradiation such as IMRT.

7.2.2 Infiltrative Ability

The spread of GBM cells beyond the enhancing tumor and into the perifocal edema has been reported after a pathological evaluation of biopsy samples [9]. A more precise evaluation of the distribution of neoplastic cells in autopsy cases has also revealed the infiltration of tumor cells into the low-density area on CT images and in some cases even beyond the low-density area [10]. The migration of GBM cells into the high-intensity area on T2-weighted MR images has also been reported [11]. These findings have suggested that the infiltrative ability of GBM cells and the infiltrating area should be encompassed in the clinical target volumes. However, in contrast to the tumor bulk, these areas contain both tumor cells and functioning neurons; differentiation of tumor cells from the functioning neurons, depending upon the differences in response to treatment, is required.

7.2.3 Hypoxia

Tumor hypoxia plays a significant role in radioresistance [12, 13]. The hypoxic condition of GBM is also well known and accepted as one of the causes of the radioresistant features of this tumor [14]. This solid tumor hypoxia can be classified into chronic and acute hypoxia. The large intercapillary distance that occurs as a result of rapid tumor growth causes chronic hypoxia, and unusual vascular conditions, high interstitial pressure, tortuous vessels, and bidirectional flow cause acute hypoxia [15]. The distribution of the acute hypoxic area may alter during fractionated RT, owing to changes in causative factors as a result of treatment [16]; in addition, reoxygenation during fractionated RT is suspected [17].

7.3 Advantages of IMRT in GBM Treatment

7.3.1 Prescribed Dose and Fractionation

With conventional irradiation techniques, accelerated hyperfractionation using smaller doses per fraction is one of the methods employed to increase the radiation effect while limiting injury to the surrounding normal tissue [18]. However, previous prospective trials of accelerated hyperfractionation have failed to improve the survival of patients with GBM [19, 20].

Hypofractionation, the other treatment approach for improved control of the regional tumor, has several advantages over conventional RT; there are increased cell death as a result of the higher doses per fraction used and a reduced repopulation effect as a consequence of the shortening of the overall treatment time. Shortened treatment time is also a significant benefit for patients and their families, because patients with GBM only have a limited survival time after the completion of treatment. Although there may also be a risk of enhanced radioresistance because of a reduction in reoxygenation during fractionated irradiation, hypofractionated irradiation has become a frequent choice in the treatment of GBM patients in the past decade (Table 7.1) [21–29]. At our institution, we performed hypofractionated IMRT, 68 Gy by 8 fractions during 10 days, and found the significant effect of this abbreviated treatment on the local control of GBMs [29]. As shown in Table 7.1, different schemes of hypofractionation have been reported, and doses/fraction and number of fractions are varied. This flexibility in fractionation is one of the advantages of IMRT, but it also makes it difficult to compare the outcomes between different treatment plans. Biologically effective dose (BED), which is calculated on the basis of the linear-quadratic (LQ) model, is commonly used to standardize the doses. However, the value of the α/β ratio that should be used for GBM is still debated, and the unreliability of the LQ model in the dose range over which the α/β ratio is used has been reported [30]. In spite of these remaining problems, hypofractionated high-dose IMRT contributes to the better control of the regional tumor [28, 29] without severe acute toxicity.

Table 7.1 Published trials of hypofractionated high-dose IMRT Dose delivery for the central region and tumor control

Author/study	PTV	Dose/fraction	Fractions	BED	CHT	MST
Floyd et al. [21] *n*=18	GTV	5.0 Gy	10	75 Gy	No	7 m
Sultanem et al. [22] *n*=25	GTV	3.0 Gy	20	78 Gy	No	9.5 m
Monjazeb et al. [23]	GTV+5 mm	2.5 Gy	28	88 Gy	No	13.6 m
		2.5 Gy	30	94 Gy		
n=21		2.5 Gy	32	100 Gy		
Tsien et al. [24]	GTV+5 mm	2.2 Gy	30	81 Gy	TMZ	20.1 m
n=38		2.4 Gy		89 Gy		
		2.5 Gy		94 Gy		
		2.6 Gy		98 Gy		
		2.7 Gy		103 Gy		
Panet-Raymond et al. [25] *n*=35	GTV	3.0 Gy	20	78 Gy	TMZ	14.4 m
Morganti et al. [26]	GTV+1.5 cm	2.4 Gy	25	74 Gy	TMZ	20 m
		2.5 Gy	25	78 Gy		
n=19		2.6 Gy	25	82 Gy		
Chen et al. [27]	GTV+5 mm	3.0 Gy	20	78 Gy	TMZ	16.2 m
n=19		4.0 Gy	20	84 Gy		
		5.0 Gy	20	90 Gy		
		6.0 Gy	20	96 Gy		
Reddy et al. [28] *n*=24	GTV+5 mm	6.0 Gy	10	96 Gy	TMZ	16.6 m
Iuchi et al. [29] *n*=46	GTV+5 mm	8.5 Gy	8	126 Gy	TMZ	20.0 m

Abbreviations: *PTV* planning target volume, *BED* biologically effective dose, *CHT* chemotherapy, *MST* median survival time, *GTV* gross tumor volume, *TMZ* temozolomide. α/β ratio of 10 Gy was employed to calculate BED in this table

7.3.2 Simultaneous Integrated Boost Technique

As mentioned above, GBM cells are highly infiltrative and have already spread widely into the surrounding brain at diagnosis. Although both the tumor bulk and infiltrating lesion should be included as treatment targets, their situations are different and a different treatment approach is required. The tumor bulk includes dens of tumor cells and few functioning neurons, which indicates the need for higher-dose RT and functional safety in this area. Tumor cells in this zone are suspected of being hypoxic, and dose escalation to overcome this problem is also expected. In contrast, the area of infiltration includes both tumor cells and functioning neurons, and differentiation of these cells owing to their different radiosensitivity is required. Tumor cells in this area are well oxygenated, and the required total radiation dose may be lower than that needed for the treatment of the tumor bulk. Less frequent distant failure after conventional RT also supports this approach; conventional dose delivery for the treatment of this area is desirable to spare the functioning neurons. Therefore, differing dose delivery is required to control both of these lesions. Using the simultaneous integrated boost technique, different doses can be delivered to layered targets. This technique is suitable for the treatment of infiltrative tumors including GBM. For GBM, two- or three-layered targets are usually settled, surrounding the enhancing tumor.

7.4 Immobilization System

The anatomic accuracy of IMRT depends on the reproducibility of the geometrical positioning of patients between the planning CT and treatment or during fractionated irradiations. Thermoplastic masks are widely used for noninvasive immobilization of patients during treatment (Fig. 7.1). However, we should be aware of the limitations of this immobilization system. Mask shrinkage after fabrication may have an effect on the immobilization of the patient, and flexibility during treatment may decrease reproducibility [31]. Head and shoulder masks, bite mouthpieces, and head and body cushions are employed to improve reproducibility, but the range of the interfractional displacement of the target has been reported as being 2.0–3.0 mm [32, 33]. This reproducibility error should be included in the margin surrounding the lesions during the settlement of the targets. Furthermore, the changes regarding the condition of the patient's head after craniotomy, such as the presence of edema in the face and scalp, subcutaneous hematoma, and cerebrospinal fluid (CSF) leakage, also have an effect on the accuracy of immobilization. Therefore, immobilization devices should be fabricated after improvement in these postsurgical changes. Indeed, such an approach would mean a delay in the initiation of postsurgical RT; however, an improved outcome for patients with GBM has been reported after the

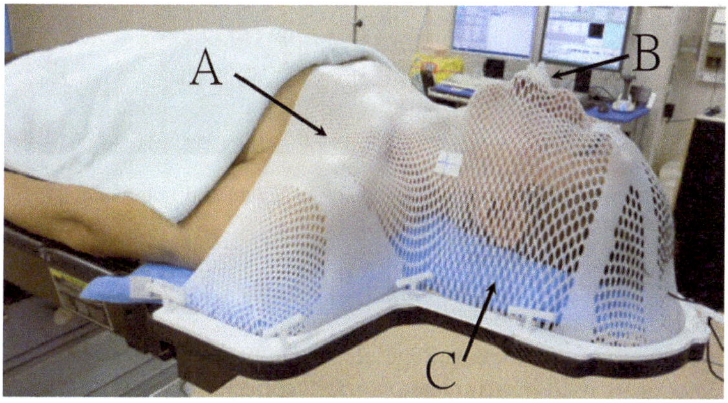

Fig. 7.1 Noninvasive immobilization system. Thermoplastic head and shoulder mask (*A*), bite mouthpieces (*B*), and head and body cushions (*C*) are used to improve the reproducibility of treatment

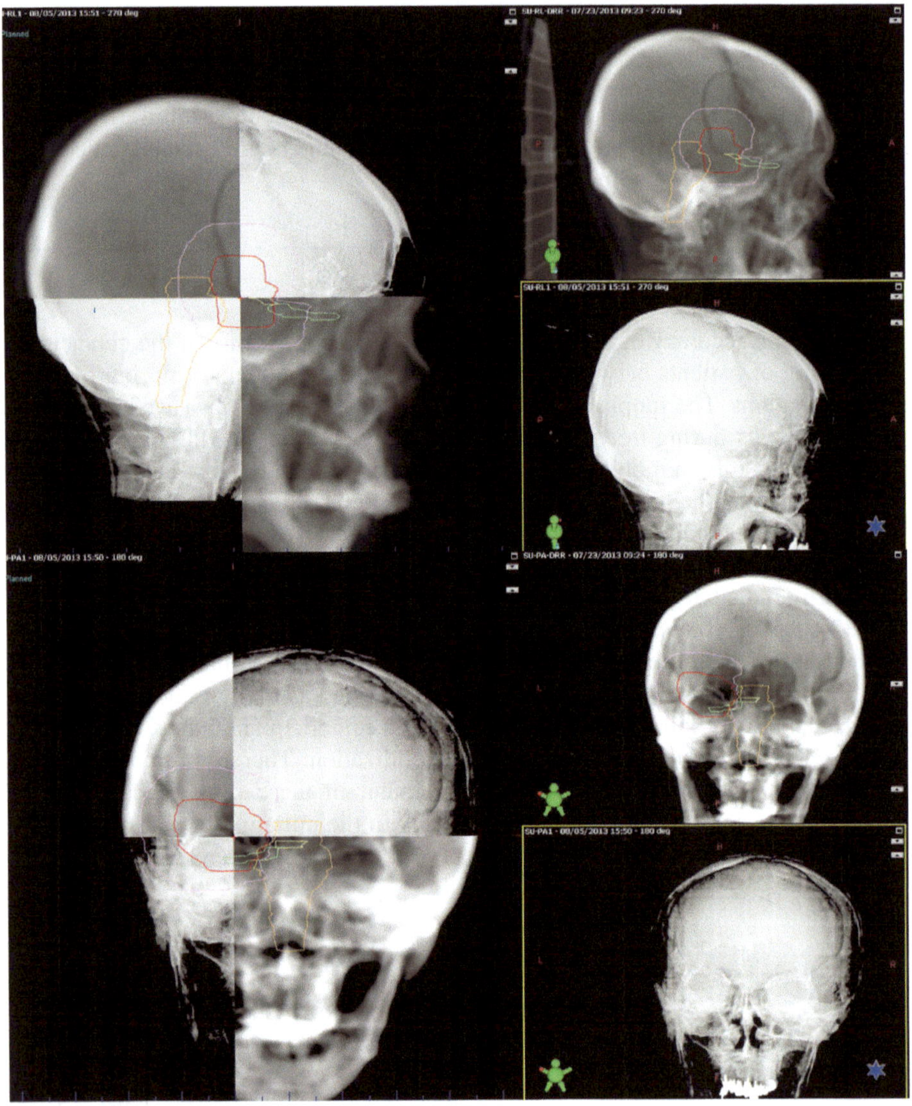

Fig. 7.2 Megavoltage electronic portal images are generated and bony landmarks are used to ensure geometric accuracy before each treatment

initiation of RT at >4 weeks after surgery, probably owing to reoxygenation after surgical procedures [34].

Before each treatment, megavoltage electronic portal images are generated and bony landmarks are used to ensure geometric accuracy. Correction of the position of the head is performed by adjustment of the treatment table in three dimensions prior to treatment (Fig. 7.2).

7.5 Target Delineation

7.5.1 Target Delineation on CT/MRI

After fabrication of the facial mask, thin slice CT is performed for planning. The data from the CT images are also used for absorption correction during the planning of dose delivery. For target delineation, MRIs are also obtained and merged into the planning CT. Target volumes are contoured on these images in line with the International Committee on Radiation Units and Measurements (ICRU) reports 50 and 62 [35, 36]. Gross tumor volume (GTV) is usually defined as the enhanced lesion on T1-weighted MR images and the surgical cavity. In the treatment of patients with GBM, a simultaneous integrated boost technique is usually employed, and layered clinical target volumes (CTVs) are settled to cover both the tumor bulk and microscopic spread of the tumor. Because the dominant pattern of failure is local after standard GBM treatment, higher dose delivery to the regional tumor is required. Therefore, the CTV of the tumor bulk (CTVb) is defined as being equal to the GTV. GBM is a highly infiltrative tumor, and migration of tumor cells into the surrounding brain should also be included in the CTV. Two different approaches exist regarding the definition of the CTV for the infiltrating area (CTVi). The first approach is based on the experiences of treatment failure after conventional RT. The majority of the recurrences have been reported to arise within 15–20 mm of the tissue surrounding the enhanced lesion, and the CTVi may be defined as the expansion of the GTV to encompass the surrounding 15–20 mm zone. This expansion is automatically performed in the planning software without consideration of the anatomical spread of tumor cells. Therefore, excessive volume should be manually removed from the CTVi after automatic expansion (Fig. 7.3). The other approach regarding the definition of the CTVi is based upon the estimated microscopic spread of tumors cells. As described above, the volume of the edema surrounding the enhancing lesion includes microscopically detected migrating tumor cells and is also defined as the CTVi. MRIs of both the T2-weighted images and fluid-attenuated inversion recovery (FLAIR) images are used to visualize edema. Although discordance between CTVs based upon T2-weighted and FLAIR images has been reported [37], FLAIR images are more frequently employed because the CSF is also visualized brightly on T2-weighted images and may impair the visualization of edema. After contouring the CTVs, the planning target volume (PTV) is defined by expanding the CTV with a 3.0–5.0 mm margin, which includes the immobilization and treatment errors.

7.5.2 Target Delineation Based upon Biological Imaging

Although MRIs provide precise geometrical information, they do not directly visualize the location of tumor cells. The contrast-enhanced area is usually diagnosed as tumor bulk, but enhancement is only the result of leakage of contrast medium

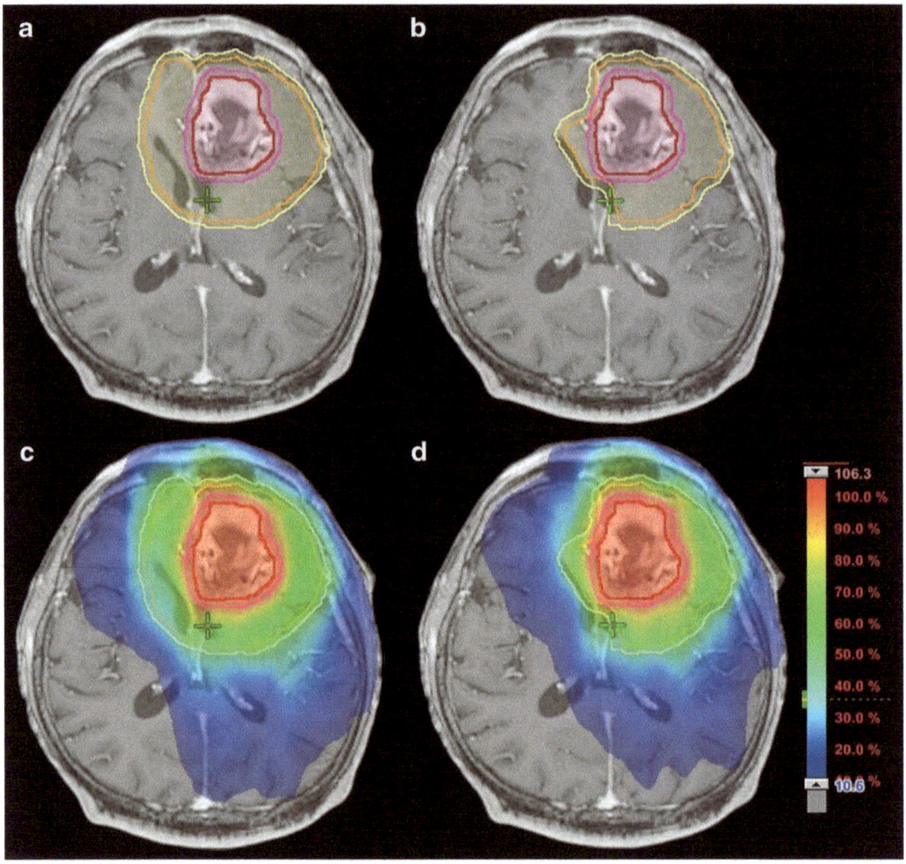

Fig. 7.3 Target delineations and dose deliveries with and without consideration of anatomical spread of tumor cells. The clinical target volume for the infiltrating area (CTVi; *orange line*) was contoured automatically by expanding it by a distance of 20 mm from the gross tumor volume (GTV; *red line*) (**a**). However, this automatic expansion did not consider the anatomical spread of tumor cells. In this case, tumor cells could infiltrate only via the corpus callosum to the contralateral frontal lobe and could not move across the falx. Furthermore, the ventricle wall blocked the way of tumor cells to the caudate head. Therefore, the excessive volumes expanded across the falx and the ventricle was removed from the CTVi (**b**). After contouring the CTVs, PTVb for tumor bulk (*pink line*) and PTVi for the infiltrating tumor (*yellow line*) were defined by expanding the GTV and CTVi with a 3 mm margin. Prescribed doses of 68 Gy for PTVb and 40 Gy for PTVi were delivered by 8 fractions. The different distributions of dose delivery before (**c**) and after removal of excessive volume from the CTVi (**d**) were also demonstrated. This difference indicated the significance of anatomical consideration during delineation of the CTVs to minimize the dose for the surrounding normal brain

through the impaired brain-blood barrier (BBB) of tumor vessels. The high-intensity area on T2-weighted or FLAIR images may include infiltrating tumor cells, but this does not mean that migrating tumor cells exist anywhere in these areas. Furthermore, induced neovascularization as a physiological reaction after surgery also causes contrast enhancement, and changes in the blood circulation after surgery also cause an increase in the intensity of normal brain surrounding the surgical cavity on T2-weighted or FLAIR images. Therefore, CTVs contoured on these images may include the area without the need for dose delivery. To decrease the excessive radiation delivery to the critical structures adjacent to the target, biological imaging is required for the delineation of real targets.

Amino acid (AA) positron emission tomography (PET) has been tried to improve the delineation of the distribution of glioma cells. L-[methyl-^{11}C]methionine (MET) and O-(2-[^{18}F]fluoroethyl)-L-tyrosine (FET) are the most widely used tracers. These tracers are transported across the BBB by means of the membrane transport system, and the accumulation of these tracers reflects the protein synthesis taking place in tumor cells. The uptake of these tracers is not dependent on the disruption of the BBB, and the significantly higher accuracy of AA-PET relative to CT and MRI in defining the distribution of tumor cells has been reported [38–43]. MET-PET and FET-PET provide comparable diagnostic information for the delineation of tumors [44], even though the short half-life of ^{11}C (20 min) limits the use of MET-PET to institutions with an onsite cyclotron. In the majority of the primary cases, the AA tracer accumulates beyond the enhanced lesion and within the high-intensity area on T2-weighted or FLAIR MRIs (Fig. 7.4). Although target delineation based upon these accurate images will contribute to the sparing of neurological functions in patients after IMRT, the major remaining problem regarding the use of PET information is the threshold of uptake. The uptake of AA is semiquantitatively expressed by the ratio of uptake in the lesion to that in the contralateral normal brain (T/N ratio). However, limited data is available regarding the optimal value of the T/N ratio required for drawing the target volume and the optimal radiation dose for the control of these areas.

On the other hand, PET is also used to define the radioresistant area in the tumor. Hypoxia is one of the major causes of radioresistance in GBMs, but heterogeneous distribution of hypoxic cells has also been reported [45, 46]. Although some limitations remain, the significance of [^{18}F]FMISO-PET in the visualization of the hypoxic area in gliomas has been reported [47, 48]. In the treatment of head and neck

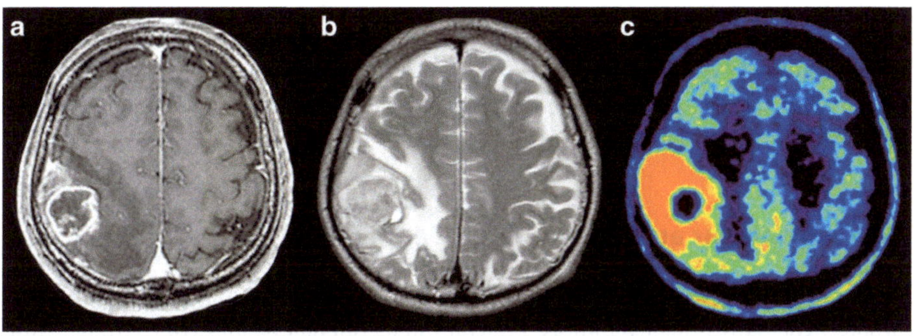

Fig. 7.4 Different visualizations of tumor on enhanced MRI (**a**), FLAIR image (**b**), and methionine-PET (**c**)

cancers, [^{60}Cu]ATSM-PET- and [^{18}F]FMISO-PET-guided radiotherapy planning have been reported [49, 50], and these images may also contribute to the treatment of GBM in the future.

7.5.3 Organ at Risk (OAR)

As the prognosis of patients with GBM is poor, some oncologists may feel little need for the preservation of critical structures from late toxicities [51]. However, temozolomide prolongs the survival of GBM patients, and the 5-year survival rates have improved to 13.8 % for patients with the methylated *O-6-methylguanine-DNA methyltransferase (MGMT)* gene [4]. Therefore, the issue of late toxicity has become more significant.

In treatment planning for brain tumors, OARs include the eye lens, retina, optic nerve, chiasm, cochlea, and brain stem. A study by Emami et al. in 1991 was the first to summarize the clinical experience regarding partial organ tolerance doses for OARs [52]. More recently, QUANTEC (quantitative analysis of normal tissue effects in the clinic) articles have updated/refined these data (Table 7.2) [53–56]. As these data are based upon clinical outcomes after conventional fractionated radiotherapy or radiosurgery involving single fraction treatment, the tolerance doses for OARs have been estimated from these data using the LQ model. However, the reliability of the LQ model and the value of the α/β ratio are still debated in hypofractionated RT (see Chap. 4), and further investigations are required to define the tolerance doses for OARs.

7.6 Dose Delivery

7.6.1 Inverse Planning

After contouring the PTVs and OARs, the prescribed doses and the tolerance doses are inputted into the inverse planning system, and the dose distribution calculated automatically. In cases where the tumor exists adjacent to the critical organs, it may be difficult to deliver the prescribed doses to the PTV homogeneously while keeping the doses to the OARs below the limit of the tolerance doses. In such cases, the preferred dose delivery/avoidance should be decided individually based upon the estimated prognosis and functions of the patient.

7.6.2 Coplanar and Noncoplanar Beam IMRT

Although IMRT has improved the conformity and homogeneity of dose delivery to targets, especially concave tumors, conventional coplanar beam IMRT still has limitations with regard to the sparing of critical organs. In the treatment of tumors located in the posterior or temporal fossa, PTVs and OARs (optic nerve, chiasm, brain stem, and cochlea) may be included in the same axial plane, and the coplanar

Table 7.2 Approximate dose/volume/outcome data for critical organs following conventionally fractionated radiotherapy and stereotactic radiosurgery

Organ	End point	Irradiation	Dose	Rate	
Eye, lens	Cataract	Whole organ	10 Gy	5 %[a]	[51]
			18 Gy	50 %[a]	[51]
Eye, retina	Blindness	Whole organ	45 Gy	5 %[a]	[51]
			65 Gy	50 %[a]	[51]
Optic nerve/chiasm	Optic neuropathy	Whole organ	50 Gy	5 %[a]	[51]
			65 Gy	50 %[a]	[51]
		3D-CRT	Dmax <55 Gy	<3 %[b]	[53]
			Dmax 55–60 Gy	3–7 %[b]	[53]
			Dmax >60 Gy	>7–20 %[b]	[53]
		SRS	Dmax <12 Gy	<10 %[b]	[53]
Cochlea	Hearing loss	3D-CRT	Mean dose ≤45 Gy	<30 %[b]	[54]
		SRS	≤14 Gy	<25 %[b]	[54]
Ear, mid/external	Acute otitis	Whole organ	30 Gy	5 %[a]	[51]
			40 Gy	50 %[a]	[51]
	Chronic otitis	Whole organ	55 Gy	5 %[a]	[51]
			65 Gy	50 %[a]	[51]
Brain stem	Necrosis, infarction	Whole organ	50 Gy	5 %[a]	[51]
			65 Gy	50 %[a]	[51]
	Neuropathy/necrosis	Whole organ	Dmax <54 Gy	<5 %[b]	[55]
		3D-CRT	D1-10 cc ≤59 Gy	<5 %[b]	[55]
			Dmax <64 Gy	<5 %[b]	[55]
		SRS	Dmax <12.5 Gy	<5 %[b]	[55]

[a]Probability within 5 years in "Emami Paper"
[b]QUANTEC data

beam cannot completely avoid critical organs (Fig. 7.5a) [57]. In such cases, noncoplanar beam arrangement may decrease the doses to the OARs (Fig. 7.5b) [58].

7.6.3 Volumetric-Modulated Arc Therapy (VMAT)

IMRT is usually performed using multiple static intensity-modulated beams. However, using static fields, the isodose curves in the range of lower doses extend along the gantry angle [58]. This "tail" of the low-dose field extends more sharply and over a greater distance when the angle or number of fields is limited to decrease the doses to the OARs adjacent to the PTV (Fig. 7.5a). Recently, VMAT, a rotating IMRT, has also become available for the treatment of patients with GBMs. With this method, a more homogeneous distribution can be achieved, not only in the low-dose field but also in the PTV, while maintaining a decreased dose to the OARs (Fig. 7.5c). The range of rotation can be changed and a noncoplanar arc can also be made available to further decrease the doses to the critical organs.

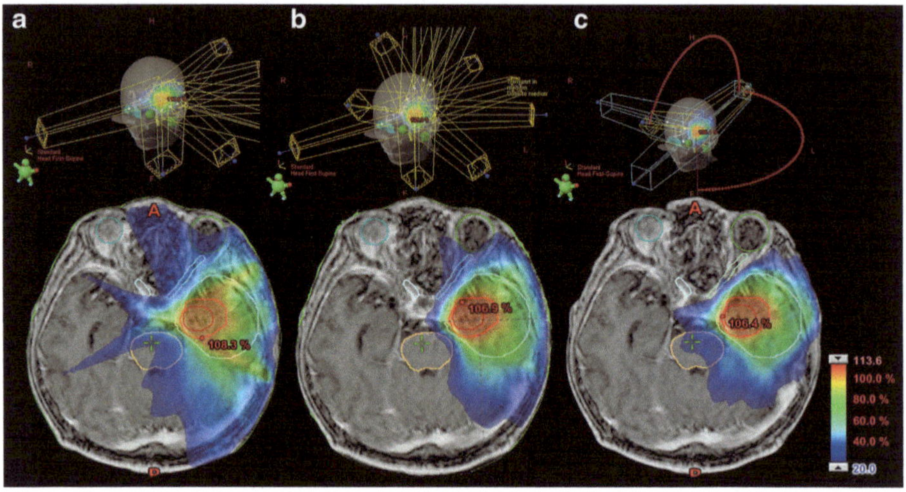

Fig. 7.5 Dose plans for a patient with glioblastoma located in the left temporal fossa. With coplanar planning, the angles of the fields were limited to avoid the organs at risks (OARs) and the isodose curves were stretched along the gantry angles (**a**). Noncoplanar beam arrangement improved the heterogeneity of dose delivery in the PTVs and decreased the dose to the OARs (**b**), even though 11 fields were required for this plan. On the other hand, an excellent distribution of doses was achieved using volumetric-modulated arc therapy (VMAT) with only two arcs of noncoplanar beams (**c**). In addition to the improvement of conformity and homogeneity of dose delivery, beam-on times and monitor units (MU) were decreased using VMAT. Beam-on times in coplanar IMRT, noncoplanar IMRT, and noncoplanar VMAT were 6.9, 8.1, and 3.5 min, respectively, and MU values were 3,430, 4,040, and 2,120, respectively. Prescribed doses were 68 Gy/8 fractions for PTVb (GTV + 5 mm) and 40 Gy/8 fractions for PTVi (GTV + 20 mm)

The other disadvantages of static IMRT are increased beam-on time and monitor units (MU) relative to conventional RT. This increased beam-on time may also cause elevated interleaf leakage during treatment [59, 60], which is significant particularly in pediatric oncology [57, 59, 60]. The beam-on time and MU can also be reduced using VMAT [61]. Reduction of the treatment time may also improve the immobilization error during treatment and facilitates more comfortable treatment especially for pediatric patients [62]. Furthermore, it allows more patients to be treated per day on one linear accelerator.

7.7 IMRT in the Multidisciplinary Treatment of GBM

7.7.1 Chemotherapy

7.7.1.1 Temozolomide (TMZ)

As a recent clinical trial has demonstrated, the combined use of RT and TMZ improves the overall survival of patients with GBM [4], and concurrent and adjuvant TMZ is usually administrated with IMRT. However, it is also well known that the effects of TMZ are dependent upon the status of the *MGMT* gene promoter. Furthermore, methylation of this gene is also correlated with the prolonged survival of patients after RT alone without alkylating agent. Significant effects of these treatments are expected only in patients whose tumor demonstrates the methylated *MGMT* gene promoter [63]. On the other hand, the frequency of methylation of this gene has been reported as only 44 % in a recent meta-analysis [64]; it was found that only half or less of patients benefit from treatment with TMZ. As the reported median survivals of patients with *MGMT*-methylated and *MGMT*-unmethylated tumors are quite different (23.4 months and 12.6 months, respectively), different treatment approaches are required. Although the significance of personalized therapy based upon the methylation status of *MGMT* is still debated, highly conformal IMRT may contribute to better neurological outcome in long-term survivors in *MGMT*-methylated cases; this is achieved by decreasing the dose to the surrounding normal brain, and better tumor control in *MGMT*-unmethylated cases by escalating the doses to the targets.

7.7.1.2 Bevacizumab (BV)

Bevacizumab (BV) is a recombinant humanized monoclonal antibody that targets vascular endothelial growth factor (VEGF). This agent inhibits the growth of new vessels and was initially expected to starve tumors of oxygen and nutrients. However, more recently, the inhibition of new vessel growth has been considered to induce normalization of tumor vasculatures and improved blood flow and oxygenation in tumors [65]. Furthermore, VEGF-targeted therapy also improves vascular function and decreases the tumor interstitial fluid pressure. These improved vascular function and oxygenation are expected to increase the ionizing effects of RT [66].

7.7.1.3 Carmustine Wafer

Carmustine wafers are used for the local delivery of the chemotherapeutic agent to GBM patients after tumor resection. These wafers are implanted in the wall of the surgical cavity and release carmustine continuously. Survival benefits from the use of these wafers have been reported, but enlargement of tumor bed cysts after implantation has also been reported [67, 68]. The enlargement of these cysts continues for several months after surgery, and brain deformity after the planning of RT may decrease the accuracy of treatment, especially in the case of highly conformal RT such as IMRT. Therefore, the indications for IMRT regarding patients implanted with a carmustine wafer should be carefully considered.

7.7.2 Antiepileptic Drugs for Tumor Control

7.7.2.1 Valproic Acid (VPA)

Histone deacetylase (HDAC) leads to the formation of condensed and transcriptionally silenced chromatin by removing the acetyl groups from the core histone. VPA, a common antiepileptic agent (AED), is well known to play a role as a HDAC inhibitor and is expected to sensitize glioma cells to ionizing radiation. The enhanced effects of irradiation on glioma cells in combination with HDAC inhibitors have been demonstrated in in vitro studies [69, 70]; in addition, recent retrospective clinical studies have demonstrated the survival benefit associated with the concurrent use of VPA in combination with RT [71, 72]. However, there still remain some limitations concerning the benefit of the concurrent use of VPA with IMRT; the in vitro studies employed extremely high concentrations of VPA which are clinically unfeasible, and the data from nonrandomized studies may have unsuspected bias. The increased hematologic toxicities associated with this AED are also known, and randomized evaluation is required to confirm the benefit and safety of VPA regarding the survival of patients.

7.7.2.2 Levetiracetam (LEV)

Levetiracetam (LEV) is a newer AED with a favorable safety profile. However, this AED is also known as a potent *MGMT* inhibitor. In vitro, LEV decreases the MGMT protein at concentrations within the clinically therapeutic range for seizure prophylaxis [73]. Although the direct effects of this agent on RT have not been reported, LEV will be frequently selected for the treatment of seizure prophylaxis in glioma-associated epilepsy, with the expectation of an increased antitumor effect regarding TMZ, which is commonly administrated with IMRT.

7.8 Complications

While IMRT contributes to the highly conformal dose delivery and may decrease the risk for radiation injury, we should be aware of its remaining risks because this radiation modality is frequently employed for the escalation of treatment doses with the expectation of increased treatment efficacy.

7.8.1 Acute Toxicity

Acute complications after RT, including headache, insomnia, confusion, partial seizure, nausea, and anorexia, are caused by transient BBB disruption and can be mitigated by steroids. Large radiation fractions are known to pose a risk regarding this pattern of radiation injury [74]. However, this acute reaction is dependent upon the volume of the radiation field, and recent trials of hypofractionated IMRT have reported infrequent acute toxicity [24, 28, 29].

7.8.2 Early-Delayed Toxicity

Transient demyelination as a result of damage to oligodendrocytes caused by irradiation may result in early-delayed complications ($\leq$12 weeks after RT). The clinical symptoms of radiation injury during this phase are characterized by fatigue, somnolence, worsening of preexisting symptoms, and transient cognitive deterioration (the so-called somnolence syndrome). Despite clinical worsening, the radiological change is limited although enlargement of non-enhancing white-matter hyperintensities on T2-weighted or FLAIR MRIs may be observed. These reactions are spontaneously recovered from or can be alleviated by steroids; they also occur infrequently after limited volume IMRT.

7.8.3 Pseudoprogression

The other transient phenomenon at the time the early-delayed reaction is expected is termed "pseudoprogression." This reaction is more frequently observed in patients who received daily TMZ with RT and more often in GBM patients with a methylated *MGMT* gene promoter. Necrosis, edema, and infiltration of inflammatory components are frequently observed in pseudoprogression, and this indicates that the lesion is consistent with it being a subacute radiation reaction and treatment-related necrosis [75]. Although the precise mechanism of pseudoprogression is still unclear, abnormal vessel permeability causes enlargement of the enhanced area with progression of edema on MRIs. These findings mimic tumor progression and it is difficult to distinguish pseudoprogression from tumor recurrence in conventional imaging studies. Magnetic resonance spectroscopy and diffusion-weighted imaging are useful diagnostic tools for distinguishing tumor recurrence from necrosis. The significant sensitivity and specificity of PET using amino acid tracers in the differentiation of viable tumor from necrosis have also been reported [76, 77]. However, we should be aware of the limitations of these imaging studies because they only reflect the dominant feature of the lesions.

As pseudoprogression is induced by local tissue reaction against radiation, high-dose IMRT may increase the risk of this phenomenon. However, the incidence of pseudoprogression after high-dose IMRT has been reported to be comparable to that associated with conventional RT [78].

7.8.4 Late Radiation Necrosis

Radiation necrosis is the most frequently occurring late toxicity after high-dose RT. As the prognosis of patients with GMB is poor, the risk of necrosis after radiation therapy may have been underestimated. However, recent progress in surgical techniques and chemoradiation therapy has prolonged the survival of patients; this has increased the significance of delayed toxicity in the treatment of GBM patients. From our experience of hypofractionated high-dose IMRT, necrosis was diagnosed in more than 30 % of cases, and it required surgical necrotomy in half of these [29].

Radiation-induced BBB disruption causes extravascular infiltration of inflammatory cells through the BBB. These inflammatory cells, such as macrophages and lymphocytes, secrete cytokines and TNF-α, which induce the development of necrosis. BBB disruption also causes increased interstitial fluid pressure, and it worsens the hypoxic condition of the lesions. Furthermore, hypoxia induces

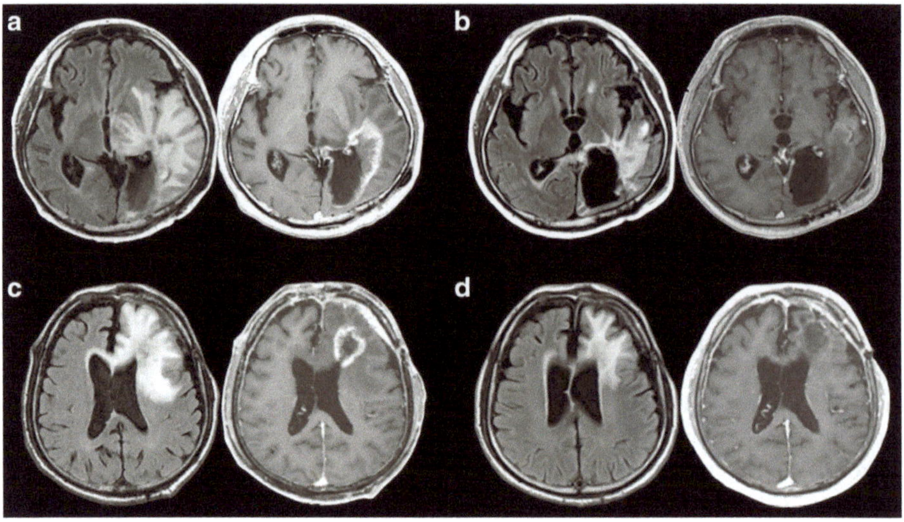

Fig. 7.6 Response of radiation necrosis regarding VEGF-targeted therapy as observed on MRIs. FLAIR images (*left*) and enhanced MRIs of radiation necrosis before (**a**, **c**) and after the administration of bevacizumab (**b**, **d**)

astrocytes to secret VEGF, which increases the permeability of vessels. Therefore, these reactions progress relentlessly and may cause injury even outside of the radiation field.

Corticosteroid is commonly used for the treatment of radiation necrosis. However, long-term usage of this agent risks damaging the patient's health, owing to its chronic side effects. Antiplatelet agents, anticoagulants, vitamin E, and hyperbaric oxygen have been tried for the treatment of this condition, but no clinical trials have been reported which have validated the effectiveness of these treatments.

As VEGF plays a key role in the development of radiation necrosis, blocking VEGF is a logical treatment for suppressing the progression of necrosis. Recently, significant shrinkage of radiation necrosis on images obtained after treatment with BV has been reported (Fig. 7.6) [79, 80]. Further studies are still required to confirm the effectiveness of VEGF-targeted therapy regarding the improvement of patient neurological functions. The establishment of a treatment for radiation necrosis will allow us to challenge GBM with higher-dose IMRT.

7.9 IMRT Targeting of Stem Cells

Neural stem cells (NSCs) reside in the stem cell niche, which is located in the subventricular zone (SVZ) adjacent to the lateral ventricle, and it is hypothesized that NSCs differentiate to cancer stem cells (CSCs), which are also located in the SVZ. However, only half of GBMs show direct contact with the SVZ [81], and conventional PTV delineation by expanding the GTV does not intentionally target the SVZ. This inadequate radiation coverage of the SVZ may be one of the causes of poor control of GBMs. Recently, several studies have reported a positive correlation between increased radiation dose for ipsilateral SVZ and improved control of GBM, indicating the significance of RT targeting of stem cells [82, 83]. Although CSCs are resistant to RT because of the preferential activation of the DNA damage checkpoint and DNA repair response [84], the impairment of CSC niches may sterilize the function of CSCs and decrease the supply of mature GBM cells.

The SVZ also harbors NSCs, and impairment of stem cell niches may also risk functional deterioration of the NSCs. From our experience of hypofractionated high-dose IMRT, the SVZ is vulnerable to irradiation and necrosis has been frequently observed in the SVZ. Necrosis of the SVZ is significantly correlated with prolonged patient survival but is also correlated with the impairment of performance status [29]. Therefore, whether or not we should escalate the radiation dose to the SVZ to control CSCs, or spare this area to protect NSCs, remains controversial. Further investigations are required to establish the treatment targeting of CSCs while maintaining the function of NSCs.

7.10 Summary

IMRT has provided precise delivery of radiation doses. As a result, oncologists have obtained great freedom in the treatment of GBMs. However, this technical freedom should be made use of only with a biological understanding of this tumor. Biological information regarding GBM is still limited, and further accumulation of information is required.

References

1. Walker MD, Alexander E Jr, Hunt WE et al (1978) Evaluation of BCNU and/or radiotherapy in the treatment of anaplastic gliomas. A cooperative clinical trial. J Neurosurg 49:333–343
2. Walker MD, Strike TA, Sheline GE (1979) An analysis of dose-effect relationship in the radiotherapy of malignant gliomas. Int J Radiat Oncol Biol Phys 5:1725–1731
3. Bleehen NM, Stenning SP (1991) A Medical Research Council trial of two radiotherapy doses in the treatment of grades 3 and 4 astrocytoma. The Medical Research Council Brain Tumour Working Party. Br J Cancer 64:769–774
4. Stupp R, Hegi ME, Mason WP et al (2009) Effects of radiotherapy with concomitant and adjuvant temozolomide versus radiotherapy alone on survival in glioblastoma in a randomised phase III study: 5-year analysis of the EORTC-NCIC trial. Lancet Oncol 10:459–466
5. Ruben JD, Dally M, Bailey M et al (2006) Cerebral radiation necrosis: incidence, outcomes, and risk factors with emphasis on radiation parameters and chemotherapy. Int J Radiat Oncol Biol Phys 65:499–508
6. Danova M, Riccardi A, Gaetani P et al (1988) Cell kinetics of human brain tumors: in vivo study with bromodeoxyuridine and flow cytometry. Eur J Cancer Clin Oncol 24:873–880
7. Furneaux CE, Marshall ES, Yeoh K et al (2008) Cell cycle times of short-term cultures of brain cancers as predictors of survival. Br J Cancer 99:1678–1683
8. Budach W, Gioioso D, Taghian A et al (1997) Repopulation capacity during fractionated irradiation of squamous cell carcinomas and glioblastomas in vitro. Int J Radiat Oncol Biol Phys 39:743–750

9. Kelly PJ, Daumas-Duport C, Kispert DB et al (1987) Imaging-based stereotaxic serial biopsies in untreated intracranial glial neoplasms. J Neurosurg 66:865–874
10. Burger PC, Heinz ER, Shibata T et al (1988) Topographic anatomy and CT correlations in the untreated glioblastoma multiforme. J Neurosurg 68:698–704
11. Nagashima G, Suzuki R, Hokaku H et al (1999) Graphic analysis of microscopic tumor cell infiltration, proliferative potential, and vascular endothelial growth factor expression in an autopsy brain with glioblastoma. Surg Neurol 51:292–299
12. Brizel DM, Sibley GS, Prosnitz LR et al (1997) Tumor hypoxia adversely affects the prognosis of carcinoma of the head and neck. Int J Radiat Oncol Biol Phys 38:285–289
13. Brat DJ, Mapstone TB (2003) Malignant glioma physiology: cellular response to hypoxia and its role in tumor progression. Ann Intern Med 138:659–668
14. Jensen RL (2009) Brain tumor hypoxia: tumorigenesis, angiogenesis, imaging, pseudoprogression, and as a therapeutic target. J Neurooncol 92:317–335
15. Brown JM (1979) Evidence for acutely hypoxic cells in mouse tumours, and a possible mechanism of reoxygenation. Br J Radiol 52:650–656
16. Masunaga S, Matsumoto Y, Kashino G et al (2010) Significance of manipulating tumour hypoxia and radiation dose rate in terms of local tumour response and lung metastatic potential, referring to the response of quiescent cell populations. Br J Radiol 83:776–784
17. Tachibana I, Nishimura Y, Shibata T et al (2013) A prospective clinical trial of tumor hypoxia imaging with 18F-fluoromisonidazole positron emission tomography and computed tomography (F-MISO PET/CT) before and during radiation therapy. J Radiat Res 54:1078–1084
18. Beck-Bornholdt HP, Dubben HH, Liertz-Petersen C et al (1997) Hyperfractionation: where do we stand? Radiother Oncol 43:1–21
19. Shibamoto Y, Nishimura Y, Tsutsui K et al (1997) Comparison of accelerated hyperfractionated radiotherapy and conventional radiotherapy for supratentorial malignant glioma. Jpn J Clin Oncol 27:31–36
20. Prados MD, Wara WM, Sneed PK et al (2001) Phase III trial of accelerated hyperfractionation with or without difluromethylornithine (DFMO) versus standard fractionated radiotherapy with or without DFMO for newly diagnosed patients with glioblastoma multiforme. Int J Radiat Oncol Biol Phys 49:71–77
21. Floyd NS, Woo SY, Teh BS et al (2004) Hypofractionated intensity-modulated radiotherapy for primary glioblastoma multiforme. Int J Radiat Oncol Biol Phys 58:721–726
22. Sultanem K, Patrocinio H, Lambert C et al (2004) The use of hypofractionated intensity-modulated irradiation in the treatment of glioblastoma multiforme: preliminary results of a prospective trial. Int J Radiat Oncol Biol Phys 58:247–252
23. Monjazeb AM, Ayala D, Jensen C et al (2012) A Phase I dose escalation study of hypofractionated IMRT field-in-field boost for newly diagnosed glioblastoma multiforme. Int J Radiat Oncol Biol Phys 82:743–748
24. Tsien CI, Brown D, Normolle D et al (2011) Concurrent temozolomide and dose-escalated intensity-modulated radiation therapy in newly diagnosed glioblastoma. Clin Cancer Res 18(1):273–279
25. Panet-Raymond V, Souhami L, Roberge D et al (2009) Accelerated hypofractionated intensity-modulated radiotherapy with concurrent and adjuvant temozolomide for patients with glioblastoma multiforme: a safety and efficacy analysis. Int J Radiat Oncol Biol Phys 73:473–478
26. Morganti AG, Balducci M, Salvati M et al (2010) A phase I dose-escalation study (ISIDE-BT-1) of accelerated IMRT with temozolomide in patients with glioblastoma. Int J Radiat Oncol Biol Phys 77:92–97
27. Chen C, Damek D, Gaspar LE et al (2011) Phase I trial of hypofractionated intensity-modulated radiotherapy with temozolomide chemotherapy for patients with newly diagnosed glioblastoma multiforme. Int J Radiat Oncol Biol Phys 81:1066–1074
28. Reddy K, Damek D, Gaspar LE et al (2012) Phase II trial of hypofractionated IMRT with temozolomide for patients with newly diagnosed glioblastoma multiforme. Int J Radiat Oncol Biol Phys 84:655–660

29. Iuchi T, Hatano K, Kodama T et al (2014) Phase II trial of hypofractionated high-dose intensity-modulated radiation therapy with concurrent and adjuvant temozolomide for newly diagnosed glioblastoma. Int J Radiat Oncol Biol Phys 88:793–800
30. Miyakawa A, Shibamoto Y, Otsuka S et al (2013) Applicability of the linear-quadratic model to single and fractionated radiotherapy schedules: an experimental study. J Radiat Res 55(3):451–454
31. Tsai JS, Engler MJ, Ling MN et al (1999) A non-invasive immobilization system and related quality assurance for dynamic intensity modulated radiation therapy of intracranial and head and neck disease. Int J Radiat Oncol Biol Phys 43:455–467
32. Masi L, Casamassima F, Polli C et al (2008) Cone beam CT image guidance for intracranial stereotactic treatments: comparison with a frame guided set-up. Int J Radiat Oncol Biol Phys 71:926–933
33. Tryggestad E, Christian M, Ford E et al (2011) Inter- and intrafraction patient positioning uncertainties for intracranial radiotherapy: a study of four frameless, thermoplastic mask-based immobilization strategies using daily cone-beam CT. Int J Radiat Oncol Biol Phys 80:281–290
34. Blumenthal DT, Won M, Mehta MP et al (2009) Short delay in initiation of radiotherapy may not affect outcome of patients with glioblastoma: a secondary analysis from the radiation therapy oncology group database. J Clin Oncol 27:733–739
35. International Commission on Radiation Units and Measurements (1993) Prescribing, recording and reporting photon beam therapy. ICRU, Bethesda
36. International Commission on Radiation Units and Measurements (1999) Prescribing, recording and reporting photon beam therapy (supplement to ICRU report 50). ICRU, Bethesda
37. Stall B, Zach L, Ning H et al (2010) Comparison of T2 and FLAIR imaging for target delineation in high grade gliomas. Radiat Oncol 5:5
38. Mosskin M, Ericson K, Hindmarsh T et al (1989) Positron emission tomography compared with magnetic resonance imaging and computed tomography in supratentorial gliomas using multiple stereotactic biopsies as reference. Acta Radiol 30:225–232
39. Goldman S, Levivier M, Pirotte B et al (1997) Regional methionine and glucose uptake in high-grade gliomas: a comparative study on PET-guided stereotactic biopsy. J Nucl Med 38:1459–1462
40. Pirotte B, Goldman S, Massanger N et al (2004) Comparison of 18F- FDG and 11C-methionine for PET-guided stereotactic brain biopsy of gliomas. J Nucl Med 45:1293–1298
41. Kracht LW, Miletic H, Busch S et al (2004) Delineation of brain tumor extent with (11C) L-methionine positron emission tomography: local comparison with stereotactic histopathology. Clin Cancer Res 10:7163–7170
42. Pauleit D, Floeth F, Hamacher K et al (2005) O-(2-[18F]fluoroethyl)- L-tyrosine PET combined with MRI improves the diagnostic assessment of cerebral gliomas. Brain 128:678–687
43. Floeth FW, Pauleit D, Wittsack HJ et al (2005) Multimodal metabolic imaging of cerebral gliomas: Positron emission tomography with [18F]fluoroethyl-L-tyrosine and magnetic resonance spectroscopy. J Neurosurg 102:318–327
44. Grosu AL, Astner ST, Riedel E et al (2011) An interindividual comparison of O-(2-[18F] fluoroethyl)-L-tyrosine (FET)- and L-[methyl-11C]methionine (MET)-PET in patients with brain gliomas and metastases. Int J Radiat Oncol Biol Phys 281:1049–1058
45. Zagzag D, Zhong H, Scalzitti JM et al (2000) Expression of hypoxia-inducible factor 1alpha in brain tumors: association with angiogenesis, invasion, and progression. Cancer 88:2606–2618
46. Brat DJ, Castellano-Sanchez AA, Hunter SB et al (2004) Pseudopalisades in glioblastoma are hypoxic, express extracellular matrix proteases, and are formed by an actively migrating cell population. Cancer Res 64:920–927
47. Szeto MD, Chakraborty G, Hadley J et al (2009) Quantitative metrics of net proliferation and invasion link biological aggressiveness assessed by MRI with hypoxia assessed by FMISO-PET in newly diagnosed glioblastomas. Cancer Res 69:4502–4509

48. Mendichovszky I, Jackson A (2011) Imaging hypoxia in gliomas. Br J Radiol 84:S145–S158
49. Chao KS, Bosch WR, Mutic S et al (2001) A novel approach to overcome hypoxic tumor resistance: Cu-ATSM-guided intensity-modulated radiation therapy. Int J Radiat Oncol Biol Phys 49:1171–1182
50. Dirix P, Vandecaveye V, De Keyzer F et al (2009) Dose painting in radiotherapy for head and neck squamous cell carcinoma: value of repeated functional imaging with (18)F-FDG PET, (18)F-fluoromisonidazole PET, diffusion-weighted MRI, and dynamic contrast-enhanced MRI. J Nucl Med 50:1020–1027
51. Creak AL, Tree A, Saran F (2011) Radiotherapy planning in high-grade gliomas: a survey of current UK practice. Clin Oncol (R Coll Radiol) 23:189–198
52. Emami B, Lyman J, Brown A et al (1991) Tolerance of normal tissue to therapeutic irradiation. Int J Radiat Oncol Biol Phys 21:109–122
53. Marks LB, Yorke ED, Jackson A et al (2010) Use of normal tissue complication probability models in the clinic. Int J Radiat Oncol Biol Phys 76:S10–S19
54. Mayo C, Martel MK, Marks LB et al (2010) Radiation dose-volume effects of optic nerves and chiasm. Int J Radiat Oncol Biol Phys 76:S28–S35
55. Bhandare N, Jackson A, Eisbruch A et al (2010) Radiation therapy and hearing loss. Int J Radiat Oncol Biol Phys 76:S50–S57
56. Mayo C, Yorke E, Merchant TE (2010) Radiation associated brainstem injury. Int J Radiat Oncol Biol Phys 76:S36–S41
57. Breen SL, Kehagioglou P, Usher C et al (2004) A comparison of conventional, conformal and intensity-modulated coplanar radiotherapy plans for posterior fossa treatment. Br J Radiol 77:768–774
58. Panet-Raymond V, Ansbacher W, Zavgorodni S et al (2012) Coplanar versus noncoplanar intensity-modulated radiation therapy (IMRT) and volumetric-modulated arc therapy (VMAT) treatment planning for fronto-temporal high-grade glioma. J Appl Clin Med Phys 13:3826
59. Low DA, Sohn JW, Klein EE et al (2001) Characterization of a commercial multileaf collimator used for intensity modulated radiation therapy. Med Phys 28:752–756
60. Steciw S, Rathee S, Warkentin B (2013) Modulation factors calculated with an EPID-derived MLC fluence model to streamline IMRT/VMAT second checks. J Appl Clin Med Phys 14:4274
61. Shaffer R, Nichol AM, Vollans E et al (2010) A comparison of volumetric modulated arc therapy and conventional intensity-modulated radiotherapy for frontal and temporal high-grade gliomas. Int J Radiat Oncol Biol Phys 76:1177–1184
62. Beltran C, Gray J, Merchant TE (2012) Intensity-modulated arc therapy for pediatric posterior fossa tumors. Int J Radiat Oncol Biol Phys 82(2):e299–e304
63. Rivera AL, Pelloski CE, Gilbert MR et al (2010) MGMT promoter methylation is predictive of response to radiotherapy and prognostic in the absence of adjuvant alkylating chemotherapy for glioblastoma. Neuro Oncol 12:116–121
64. Zhang K, Wang XQ, Zhou B et al (2013) The prognostic value of MGMT promoter methylation in Glioblastoma multiforme: a meta-analysis. Fam Cancer 12:449–458
65. Ellis LM, Hicklin DJ (2008) VEGF-targeted therapy: mechanisms of anti-tumour activity. Nat Rev Cancer 8:579–591
66. Beal K, Abrey LE, Gutin PH (2011) Antiangiogenic agents in the treatment of recurrent or newly diagnosed glioblastoma: analysis of single-agent and combined modality approaches. Radiat Oncol 6:2
67. McGirt MJ, Villavicencio AT, Bulsara KR et al (2002) Management of tumor bed cysts after chemotherapeutic wafer implantation. Report of four cases. J Neurosurg 96:941–945
68. Della Puppa A, Rossetto M, Ciccarino P et al (2010) The first 3 months after BCNU wafers implantation in high-grade glioma patients: clinical and radiological considerations on a clinical series. Acta Neurochir (Wien) 152:1923–1931
69. Chinnaiyan P, Vallabhaneni G, Armstrong E et al (2005) Modulation of radiation response by histone deacetylase inhibition. Int J Radiat Oncol Biol Phys 62:223–229
70. Van Nifterik KA, Van den Berg J, Slotman BJ et al (2012) Valproic acid sensitizes human glioma cells for temozolomide and γ-radiation. J Neurooncol 107:61–67

71. Weller M, Gorlia T, Cairncross JG (2011) Prolonged survival with valproic acid use in the EORTC/NCIC temozolomide trial for glioblastoma. Neurology 77:1156–1164
72. Barker CA, Bishop AJ, Chang M et al (2013) Valproic acid use during radiation therapy for glioblastoma associated with improved survival. Int J Radiat Oncol Biol Phys 86:504–509
73. Bobustuc GC, Baker CH, Limaye A et al (2010) Levetiracetam enhances p53-mediated MGMT inhibition and sensitizes glioblastoma cells to temozolomide. Neuro Oncol 12:917–927
74. Young DF, Posner JB, Chu F et al (1974) Rapid-course radiation therapy of cerebral metastases: results and complications. Cancer 34:1069–1076
75. Brandsma D, Stalpers L, Taal W et al (2008) Clinical features, mechanisms, and management of pseudoprogression in malignant gliomas. Lancet Oncol 9:453–461
76. Terakawa Y, Tsuyuguchi N, Iwai Y et al (2008) Diagnostic accuracy of 11C-methionine PET for differentiation of recurrent brain tumors from radiation necrosis after radiotherapy. J Nucl Med 49:694–699
77. Okamoto S, Shiga T, Hattori N et al (2011) Semiquantitative analysis of C-11 methionine PET may distinguish brain tumor recurrence from radiation necrosis even in small lesions. Ann Nucl Med 25(897):213–220
78. Piroth MD, Pinkawa M, Holy R et al (2012) Integrated boost IMRT with FET-PET-adapted local dose escalation in glioblastomas. Results of a prospective phase II study. Strahlenther Onkol 188:334–339
79. Gonzalez J, Kumar AJ, Conrad CA et al (2007) Effect of bevacizumab on radiation necrosis of the brain. Int J Radiat Oncol Biol Phys 67:323–326
80. Levin VA, Bidaut L, Hou P et al (2011) Randomized double-blind placebo-controlled trial of bevacizumab therapy for radiation necrosis of the central nervous system. Int J Radiat Oncol Biol Phys 79:1487–1495
81. Lim DA, Cha S, Mayo MC et al (2007) Relationship of glioblastoma multiforme to neural stem cell regions predicts invasive and multifocal tumor phenotype. Neuro Oncol 9:424–429
82. Lee P, Eppinga W, Lagerwaard F et al (2013) Evaluation of high ipsilateral subventricular zone radiation therapy dose in glioblastoma: a pooled analysis. Int J Radiat Oncol Biol Phys 86:609–615
83. Chen L, Guerrero-Cazares H, Ye X et al (2013) Increased subventricular zone radiation dose correlates with survival in glioblastoma patients after gross total resection. Int J Radiat Oncol Biol Phys 86:616–622
84. Bao S, Wu Q, McLendon RE et al (2006) Glioma stem cells promote radioresistance by preferential activation of the DNA damage response. Nature 444:756–760

Nasopharyngeal Cancer

8

Moses M. Tam, Nadeem Riaz, and Nancy Y. Lee

Keywords
Nasopharyngeal carcinoma • IMRT • Radiation therapy

8.1 Introduction

Nasopharyngeal carcinoma (NPC) is relatively uncommon in most parts of the world but is endemic to certain regions such as Southern China [1]. NPC is rare in the United States, with an incidence of less than 1/100,000 person-years compared with 27/100,000 person-years in Southern China.

NPC is unique histologically from other head and neck cancers. In the most recent World Health Organization classification in 2005, NPC comprises three main types, namely, keratinizing squamous cell carcinoma (type 1), non-keratinizing carcinoma (type 2), and basaloid squamous cell carcinoma [2]. Non-keratinizing carcinoma (type 2) is subdivided into differentiated (type 2a) and undifferentiated (type 2b). Type 2 is also strongly associated with Epstein-Barr virus (EBV) and is the most common histologic type found in endemic regions.

Intergroup 0099 established concurrent radiation therapy (RT) and chemotherapy as the standard of care for locally advanced NPC [3]. Although surgical resection is often an option for tumors at other head and neck sites, successful resection of NPC is nearly impossible given its location and frequent involvement of the

M.M. Tam, M.D.
New York University School of Medicine, New York, NY, USA

N. Riaz, M.D. • N.Y. Lee, M.D. (✉)
Department of Radiation Oncology, Memorial Sloan-Kettering Cancer Center,
1275 York Avenue, New York, NY 10065, USA
e-mail: leen2@mskcc.org

Y. Nishimura, R. Komaki (eds.), *Intensity-Modulated Radiation Therapy*,
DOI 10.1007/978-4-431-55486-8_8

lateral retropharyngeal lymph nodes. Thus, surgery is mostly limited to radical or selective neck dissections for persistent or recurrent disease after RT.

Toxicity is an issue with the use of conventional RT given the proximity of the nasopharynx to critical normal structures. Intensity-modulated radiation therapy (IMRT) offers advantages over conventional RT by optimizing the delivery of radiation to irregularly shaped volumes so as to spare organs at risk. Further, different doses can be delivered simultaneously to select regions by dose painting or a simultaneous integrated boost. These techniques allow increased sparing of nearby critical normal structures by simultaneously delivering higher radiation doses to gross disease and lower doses to regions suspected of harboring microscopic disease. In the next section, we evaluate the clinical evidence that has established IMRT as the standard of care for definitive RT in NPC.

8.2 Clinical Evidence for Intensity-Modulated Radiation Therapy

8.2.1 Dosimetry

Dosimetrically, IMRT improves coverage of disease while reducing the dose to the numerous critical adjacent structures relative to conventional RT (Fig. 8.1) [4, 5]. Dosimetric comparisons of IMRT versus two-dimensional (2D) RT and three-dimensional (3D) RT plans showed that IMRT led to lower doses to the spinal cord, mandible, temporal lobe, parotid glands, optic chiasm, and brainstem.

8.2.2 Salivary Function and Treatment Compliance

The most common complication of RT for NPC is a decline in salivary function, known as xerostomia, due to a damage of nearby salivary structures. The symptoms of xerostomia can significantly affect a patient's quality of life [6]. The severity of xerostomia depends mostly on the dose and volume of salivary gland within the radiation field. Dosimetric comparisons have revealed that a mean dose of 26 Gy or less to the parotid glands is necessary to preserve salivary function [7, 8].

The main benefit of IMRT over conventional RT for NPC is the ability to spare the parotid glands. Two phase III randomized controlled trials assigned patients to receive either 2D RT or IMRT with parotid-sparing techniques and evaluated outcomes at 1 year after treatment. The first trial found that IMRT was associated with superior quality of life outcomes [9]. The second study also found benefits in observer-rated xerostomia outcomes and preservation of parotid function (measured by parotid flow rate) with the use of IMRT [10], as well as a trend towards improvement in patient-reported xerostomia outcomes. The same study revealed the somewhat surprising finding that xerostomia quality of life scores only correlated weakly

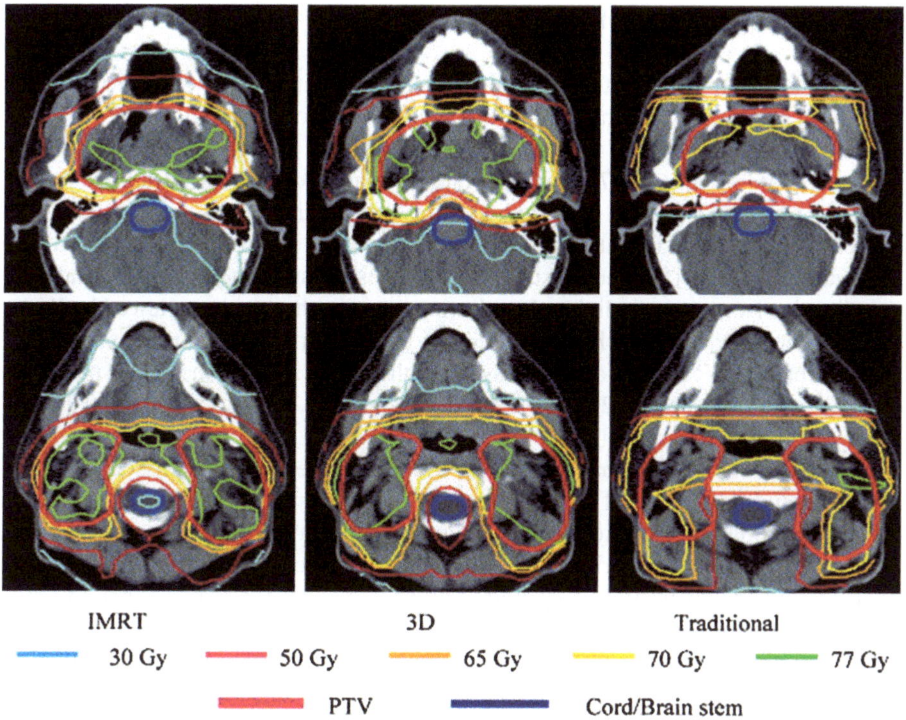

Fig. 8.1 Dosimetric comparison of treatment plans for intensity-modulated radiation therapy (IMRT) vs. 3D-conformal RT vs. traditional RT. Axial dose distributions through the center of the nasopharynx and neck for IMRT (*left*), 3D-conformal RT (*middle*), and traditional treatment plans (*right*). Note the relatively poor coverage of the skull base and medial nodal regions in the traditional plan and the improved dose conformality of the IMRT plan (From: Hunt et al. [4], with permission from Elsevier)

with both salivary flow rates and observer-rated xerostomia outcomes. Therefore, evaluation of both patient-reported and physician-reported outcomes remains important. Regardless, both phase III studies showed improved xerostomia outcomes with the use of IMRT compared with conventional 2D RT.

The lesser toxicity associated with IMRT may also improve treatment compliance or the ability of patients to tolerate the prescribed therapy. A multi-institutional trial of IMRT by the Radiation Therapy Oncology Group, RTOG 0225, showed that 90 % of patients were able to receive the full 70-Gy prescribed dose and that 88 % of the patients with T2b or higher or N+ disease were able to receive the full three cycles of concurrent cisplatin [11]. These findings compare favorably to previous studies that used conventional RT techniques, for example, chemotherapy compliance rates were 63 % in the Intergroup 0099 trial, 71 % in a Singapore randomized trial, and 52 % in the Hong Kong NPC-9901 trial [3, 12, 13].

8.2.3 Disease Control

In addition to improving toxicity outcomes, excellent disease control outcomes have been reported by several institutions. Lee et al. reported findings from an initial series of patients with NPC treated with IMRT, with an incredible 4-year local control rate of 97 %, despite 70 % of patients in that study having locally advanced disease [14]. Kwong et al. reported the first prospective series, with 3-year outcomes of 100 % local control (LC), 92.3 % regional control, and 100 % overall survival (OS) rates [15]. These excellent outcomes are supported by additional published series from many individual institutions, comprehensively reviewed by Wong et al. (Table 8.1) [16]. The RTOG 0225 trial further demonstrated the feasibility of implementing IMRT techniques across the multiple US institutions [11]. That phase II study reported excellent 2-year outcomes of 93 % LC, 89 % local-regional control (LRC), and 80 % OS rates (Table 8.1).

Notably, other factors may contribute to improvements in LRC associated with IMRT, including the use of chemotherapy, better supportive care, and technologic advances in imaging that provide better tumor delineation. Other limitations associated with historical comparisons include changes in the criteria for disease staging over time as well as improved staging with the use of magnetic resonance imaging (MRI) and positron emission tomography (PET) [17]. For example, because MRI is more sensitive than computed tomography (CT) for detecting minimal skull base involvement or intracranial extension, the T category tends to be upstaged when MRI is used rather than CT.

8.3 Techniques

8.3.1 Diagnostic Work-Up for Target Volume Delineation

Disease staging should include both CT and MRI of the head and neck. CT is important for assessing cortical bone involvement; MRI provides superior visualization of skull base involvement and tumor invasion into soft tissue structures compared with CT [18]. Infiltration of disease into the bone marrow is best seen as hypodense regions relative to normal marrow on T1-weighted non-contrast MRI scans. Fusion of the skull base portion of the CT scan with the MRI scan should aid in delineating the gross tumor volume (GTV). MRI also allows retropharyngeal lymph nodes to be distinguished from primary tumor, whereas CT may not.

Enlarged retropharyngeal lymph nodes should be considered a gross disease. Involvement of other lymph node regions is suggested by the presence of central necrosis, extracapsular spread, or nodal diameters of 1 cm or more. PET/CT may help to clarify involvement of borderline lymph nodes. Generally, because NPC has a high likelihood of nodal spread, any nodes suspected of harboring disease should be considered a gross disease.

Table 8.1 Published studies of IMRT for nasopharyngeal cancer

Reference	Stage	No. of patients	Median follow-up time, mo	Time point, y	Local control rate, %	Regional control rate, %	Freedom from DM rate, %	Overall survival rate, %
Lee et al. (2002) (UCSF) [14]	All	67	31	4	97	98	66	88
Kwong et al. (2004) (Hong Kong) [15]	T1 N0–1	33	24	3	100	92	100	100
Kam et al. (2004) (Hong Kong) [22]	All	63	29	3	92	98	79	90
Wolden et al. (2006) (MSKCC) [24]	All	74	35	3	91	93	78	83
Kwong et al. (2006) (Hong Kong) [23]	III–IVB	50	25	2	96	NA	94	92
Lee et al. (2009) (RTOG 0225) [11]	All	68	31	2	93	91	85	80
Lin et al. (2010) (China) [25]	IIB–IVB	370	31	3	95	97	86	89
Lee et al. (2012) (RTOG 0615) [19]	IIB–IVB	42	30	2	NA	NA	91	91

Abbreviations: *DM* distant metastasis, *UCSF* University of California, San Francisco, *RTOG* Radiation Therapy Oncology Group

8.3.2 Simulation and Daily Localization

The patient should be set up for treatment simulation supine, with the neck extended. The immobilization device to be used should include at least the head and neck; if possible, shoulders should also be immobilized to ensure the reproducibility of patient setup from day to day, especially when an extended-field IMRT plan is to be used. A bite block can be placed during treatment simulation and throughout treatment to move the tongue away from the high-dose regions in the nasopharynx.

CT-based treatment simulation should involve 3-mm-thick scan slices with intravenous contrast to help delineate the GTV, particularly the lymph nodes. The isocenter is typically placed immediately above the arytenoids. Image registration and fusion applications with MRI and PET should be used to help delineate target volumes, especially regions of interest that encompass the GTV, skull base, brainstem, and optic chiasm.

8.3.3 Target Volume Delineation and Treatment Planning

Several IMRT dose-fractionation regimens have been used for NPC (Table 8.2). Excellent LRC rates in excess of 90 % have been reported with the use of these regimens.

Several acceptable definitions of target volumes, including the GTV, the clinical target volume (CTV), and planning target volume (PTV), have been used at different institutions, as reviewed by Wong et al. [16]. The RTOG established a guideline for target volume delineation with RTOG 0225, which was successfully implemented in that multi-institutional study [11]. Suggested target volumes for the GTV and high-risk CTV are described in the following sections (Tables 8.3 and 8.4 and Figs. 8.2, 8.3, and 8.4). In a recent RTOG 0615 trial, the lower-than-expected 2-year LRC rate of 84 % was attributed to an increased incidence of major deviations in

Table 8.2 IMRT dose and fractionation schemes

Dose and fractionation	Study institution and reference			
	RTOG [19]	Fujan [20]	SKL [43]	PWH [22]
Gross disease, Gy	69.96	66.0–69.75	68	6,674
Gross disease, Gy/fraction	2.12	2.2–2.25	2.27	2
High-risk region, Gy	59.4	60–60.45	60	60
High-risk region, Gy/fraction	1.8	1.95–2.0	2	1.82
Low-risk region, Gy	50–54.12	54–55.8	50–54	54–60
Low-risk region, Gy/fraction	1.64–2.0	1.8	1.8–2.0	2
Margin around GTV, mm[a]	10	8–13	NA	13

Abbreviations: *RTOG* Radiation Therapy Oncology Group, *SKL* State Key Laboratory of Oncology in Southern China (Guangzhou), *PWH* Prince of Wales Hospital, Hong Kong, *GTV* gross tumor volume, *CTV* clinical target volume, *PTV* planning target volume

[a]Margin is for primary tumor (GTV70), which includes CTV expansion of GTV and PTV expansion

Table 8.3 Definitions of target volumes for gross disease

Target volumes	Definition and description
GTV_{70*} (The subscript 70 denotes the radiation dose delivered)	Primary: All gross diseases on physical examination and imaging (see above regarding the importance of MRI) Neck nodes: All nodes ≥1 cm or those with necrotic center
CTV_{70*}	GTV_{70}+≥5 mm margin; around critical structures like the brainstem, 1 mm margin is acceptable
PTV_{70*}	CTV_{70}+3–5 mm, depending on comfort level of daily patient positioning. Around critical structures like the brainstem, 1 mm margin is acceptable

Table 1.1 from: Lee NY, Le QT, O'Sullivan B, Lu JJ (2003) Chapter 1. Nasopharyngeal carcinoma. *Target Volume Delineation and Field Setup: A Practical Guide for Conformal and Intensity-Modulated Radiation Therapy*, with kind permission from Springer Science+Business Media

*PTV_{70} receives 2.12 Gy/fraction to 70 Gy over 33 fractions. For treatment of nodes that are small (i.e., ~1 cm), the lower dose of 63 Gy (PTV_{63}) can be considered at the discretion of the treating physician

Table 8.4 Definition of target volumes for high-risk subclinical region

Target volumes	Definition and description
$CTV_{59.4*}$	$CTV_{59.4}$ should encompass CTV_{70} with a 5-mm margin and regions at risk for microscopic disease which include Entire nasopharynx Anterior 1/2 or 2/3 of the clivus (entire clivus, if involved) Skull base (ensuring coverage of foramen ovale where V3 resides) Pterygoid fossa Parapharyngeal space Inferior sphenoid sinus (entire sphenoid sinus in T3-T4 disease) Posterior 1/3 of the nasal cavity/maxillary sinuses (ensuring coverage of pterygopalatine fossae where V2 resides) Inferior soft palate Retropharyngeal lymph nodes Retrostyloid space Bilateral nodal levels IB through V** Include cavernous sinus for advanced T3-T4 lesions Importance of reviewing bone window while contouring on CT scan to ensure coverage of skull base foramina
$PTV_{59.4*}$	$CTV_{59.4}$+3–5 mm, depending on the comfort of physician, but around critical structures like the brainstem, 1-mm margin is acceptable

Table 1.2 from: Lee NY, Le QT, O'Sullivan B, Lu JJ (2003) Chapter 1 Nasopharyngeal Carcinoma. *Target Volume Delineation and Field Setup: A Practical Guide for Conformal and Intensity-Modulated Radiation Therapy*. Reproduced with kind permission from Springer Science+Business Media

*High-risk subclinical dose ($PTV_{59.4}$): 1.8 Gy/fraction to 59.4 Gy, for lower-risk subclinical regions *excluding the nasopharynx/skull base regions where they are always considered high risk*, can consider 1.64 Gy/fraction to 54 Gy (PTV_{54}), i.e., N0 neck or low neck (levels IV and VB) at the discretion of the treating physician

**Level IB can be omitted in node-negative disease. At discretion of physician, level 1B may also be spared in low-risk node positive patients (e.g., isolated retropharyngeal nodes or isolated level IV nodes are considered low risk for level 1B involvement). At the same time, treatment of level 1B should be considered in node-negative patients with certain features (e.g., involvement of hard palate or nasal cavity)

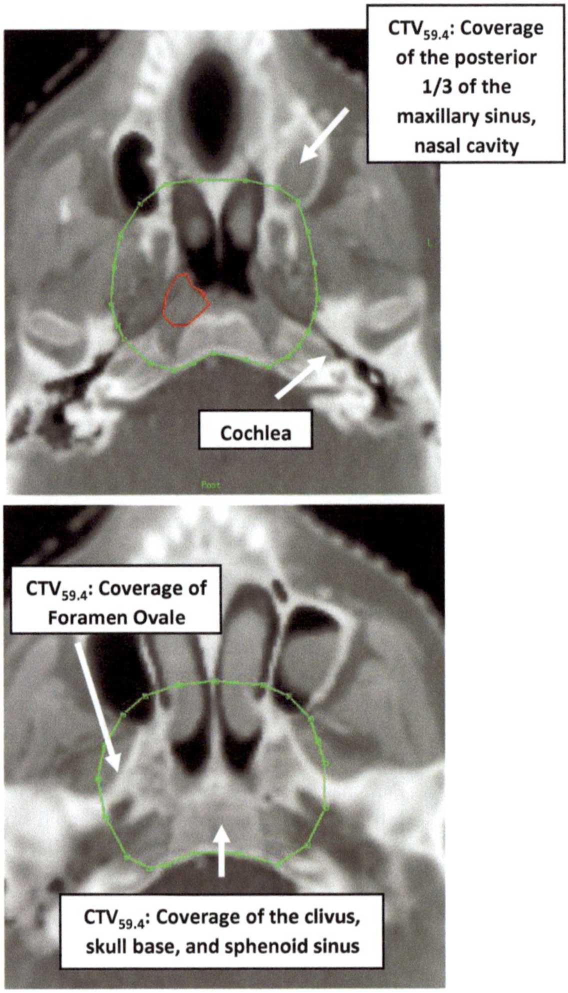

Fig. 8.2 Delineation of target volumes in a case of T1N1 nasopharyngeal carcinoma (NPC). GTV70 (inner contour, *red*) and CTV59.4 (*green*) contours in a patient with T1N1 NPC with coverage of the retropharyngeal and level II nodes (Figure 1.2 from: Lee NY, Le QT, O'Sullivan B, Lu JJ (2003) Chapter 1 Nasopharyngeal Carcinoma. *Target Volume Delineation and Field Setup: A Practical Guide for Conformal and Intensity-Modulated Radiation Therapy*, with kind permission from Springer Science+Business Media)

target volume [19]. Thus, attention must be paid to accurate target delineation to avoid marginal misses when using IMRT.

Reductions in high-risk subclinical volumes with IMRT have also been described. Lin et al. reported a prospective, single-institution study involving 323 patients with NPC; that study reduced the CTV suggested in the RTOG guidelines and resulted in excellent LCR outcomes [20]. One reduction involved the exclusion of upper deep jugular lymph nodes (level IIa above the C1 vertebrae) in the CTV. Treatment volume reductions may be important for reducing toxicity and even secondary primary tumors, the rate of which has been reported to be as high as 1 % among patients with NPC receiving definitive RT [21]. The next section reviews the guidelines used at the authors' institutions and some variations in those guidelines used at other institutions.

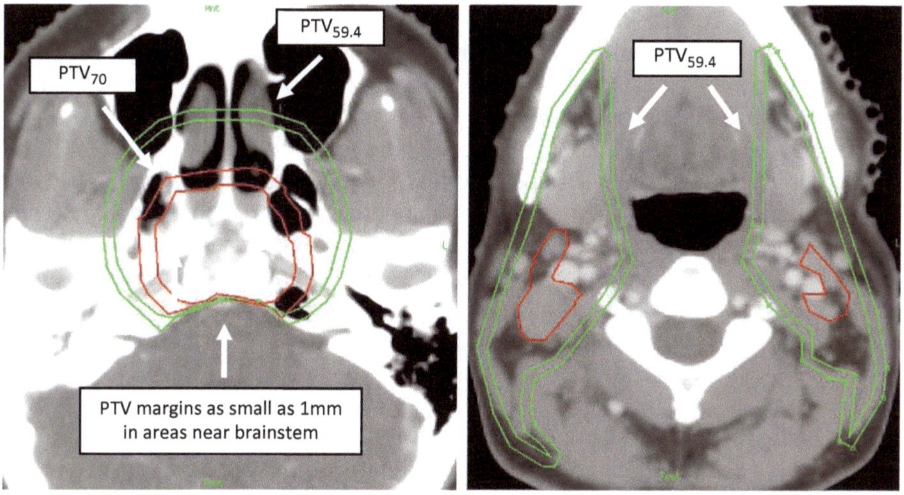

Fig. 8.3 Delineation of target volumes in a case of T3N2 nasopharyngeal carcinoma (NPC). PTV70 (*red*) and PTV59.4 (*green*) in a patient with T3N2 nasopharyngeal carcinoma (Figure 1.4 from: Lee NY, Le QT, O'Sullivan B, Lu JJ (2003) Chapter 1. Nasopharyngeal carcinoma. *Target Volume Delineation and Field Setup: A Practical Guide for Conformal and Intensity-Modulated Radiation Therapy*, with kind permission from Springer Science+Business Media)

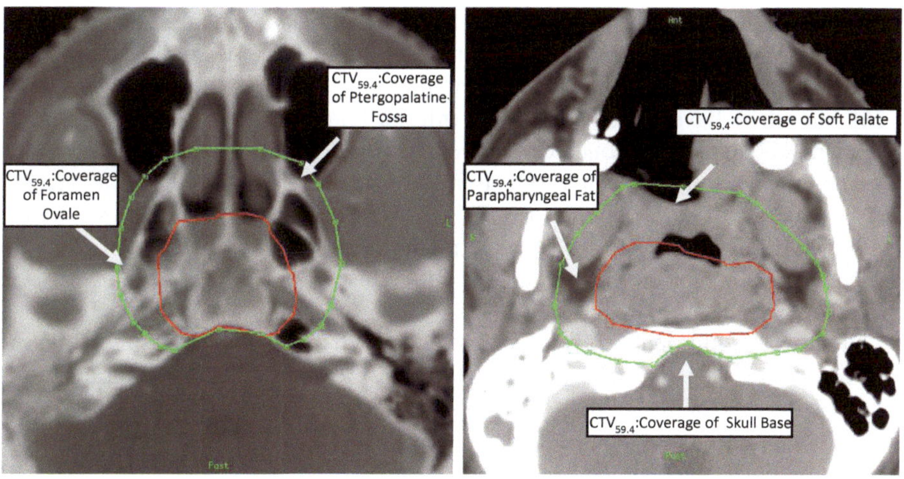

Fig. 8.4 Delineation of target volumes in a case of T3N2 nasopharyngeal carcinoma (NPC) with the use of different CT window settings. GTV70 (*green*) and CTV59.4 (*red*) in bone window (*left*) and soft tissue window (*right*) (Figure 1.5 from: Lee NY, Le QT, O'Sullivan B, Lu JJ (2003) Chapter 1. Nasopharyngeal carcinoma. *Target Volume Delineation and Field Setup: A Practical Guide for Conformal and Intensity-Modulated Radiation Therapy*, with kind permission from Springer Science+Business Media)

8.3.3.1 Gross Tumor Volume

Generally, the GTV is defined as the primary tumor and any involved lymph nodes. Involved lymph nodes are typically defined as any lymph node larger than 1 cm in diameter or those that show avidity on PET scanning.

Expansions around the GTV have included those for both a CTV and a PTV or a single, larger PTV expansion alone. The RTOG studies recommended the use of a CTV70, defined as a 0.5-cm margin with an optional posterior margin reduction of 0.1–0.5 cm (Table 8.3) as well as a PTV70 expansion of 0.5 cm. Variations on these expansions have included a larger CTV expansion of 1 cm [15, 22, 23] or the elimination of a CTV and the use of a larger PTV of 1 cm [12, 24]. The use of the latter method may avoid confusion with the CTVs described below for high-risk and low-risk subclinical regions.

8.3.3.2 High-Risk and Low-Risk Subclinical Regions

The CTV is generally defined as regions at high risk of harboring microscopic disease (Table 8.4). This volume can be treated to a lower dose of 59.4 Gy (CTV59.4), which includes all potential routes of spread for primary and nodal disease. Specifically, CTV59.4 typically covers the clivus, skull base, inferior sphenoid sinus, cavernous sinus, pterygoid fossae, parapharyngeal space, posterior nasal cavity and maxillary sinus, retropharyngeal lymph nodes, and neck levels II through V. The bilateral level IB can be spared in carefully selected patients without compromising LRC [20, 25]. Whether the inferior orbital fissure or the anterior arch of C1 can be spared remains unclear owing to a lack of data [16]. Variations also exist for the inferior border of the retropharyngeal lymph nodes. A consensus guideline published by Gregoire et al. defines the border as the cranial edge of the hyoid bone [26], but others have described it as the inferior border of the hyoid bone [22] and the cranial edge of the second cervical vertebrae [20, 25].

The low anterior neck can also be treated to a lower dose than the GTV because it is at low risk of harboring disease. This low-risk region can be treated separately with a dose of 50.4 Gy in 1.8 Gy per fraction using conventional anteroposterior (AP) or posteroanterior (PA) portals or with a dose of 54 Gy (CTV54) in 1.64 Gy per fraction in a single IMRT plan.

Finally, an additional CTV (CTV63) can be used at the discretion of the treating physician. A lower dose (63 Gy) can be used for a small-volume lymph node disease. Examples of the appropriate application of this intermediate dose would include the presence of small lymph nodes near the mandible or in the lower neck and close to the brachial plexus.

8.3.3.3 Planning Target Volume

The margin for the PTV also varies between institutions [16]. Most institutions have described the PTV as 0.2–0.5 cm beyond the CTV. The use of a PTV margin of 0.3–0.5 cm would be reasonable, as many published studies have shown an efficacy using these limits. Daily image guidance with kV imaging can facilitate margin reduction.

Table 8.5 Normal tissue dose constraints

Critical structures	Constraints
Brainstem	Max <54 Gy or 1 % of PTV cannot exceed 60 Gy
Optic nerves	Max <54 Gy or 1 % of PTV cannot exceed 60 Gy
Optic chiasm	Max <54 Gy or 1 % of PTV cannot exceed 60 Gy
Spinal cord	Max <45 Gy or 1 cm^3 of the PTV cannot exceed 50 Gy
Mandible and TMJ	Max <70 Gy or 1 cm^3 of the PTV cannot exceed 75 Gy
Brachial plexus	Max <66 Gy
Temporal lobes	Max <60 Gy or 1 % of PTV cannot exceed 65 Gy
Other normal structures	Constraints
Oral cavity	Mean <40 Gy
Parotid gland	(a) Mean ≤26 Gy in one gland (b) Or at least 20 cm^3 of the combined volume of both parotid glands will receive <20 Gy (c) Or at least 50 % of one gland will receive <30 Gy
Cochlea	V55 <5 %
Eyes	Mean <35 Gy, Max <50 Gy
Lens	Max <25 Gy
Glottic larynx	Mean <45 Gy
Esophagus, postcricoid pharynx	Mean <45 Gy

8.3.4 Plan Assessment and Dose Constraints

For NPC, the organs at risk include the brainstem, spinal cord, optic nerves, chiasm, parotid glands, pituitary, temporomandibular (TM) joints, middle and inner ears, skin (in the region of the target volumes), oral cavity, mandible, eyes, lens, temporal lobe, brachial plexus, esophagus (including postcricoid pharynx), and glottic larynx (Table 8.5). In cases of advanced disease, we typically prioritize normal structure constraints, specifically the brainstem, spinal cord, and optic chiasm, over full coverage of the tumor. Ideally, at least 95 % of the PTV70 should receive 70 Gy. In addition, the minimum dose to 99 % of the CTV70 should be >65.1 Gy. The maximum dose received by 0.03 cm^3 of the PTV70 should be <80.5 Gy.

For the PTV59.4, 95 % of the volume should receive the prescription dose. The minimum dose to 99 % of the CTV59.4 should be >55.2 Gy. The maximum dose to 0.03 cm^3 of PTV59.4 should be 69.3 Gy.

8.4 Future Directions

Overall, LRC with IMRT is excellent, with rates generally exceeding 90 % in the current era when chemotherapy is included as part of the treatment. Future directions in therapy are now focusing on identifying patients with NPC who are more likely to experience local regional or, more commonly, distant failure after RT. These high-risk patients are likely to benefit from treatment intensification.

Monitoring levels of EBV DNA in plasma samples is one way to stratify patients in terms of risk, as this biomarker is showing great potential in the clinical setting. Many studies, including prospective and phase II studies, have established that pretreatment and posttreatment levels of EBV DNA are reliable indicators of tumor burden, predictors of recurrence and distant failure, and independent prognostic factors in EBV-related NPC [27–32]. Quantification of plasma EBV DNA has also been shown to be useful for monitoring patients with NPC and predicting the outcome of treatment [33]. A recent four-center study sought to harmonize EBV DNA assay methods, to bring us step closer to using EBV in biomarker-driven trials [34]. Indeed, we anticipate that an upcoming phase III study by the RTOG will incorporate plasma EBV DNA levels in treatment stratification.

Several imaging methods are also being used to identify patients with high-risk NPC that is more aggressive and more likely progress despite treatment. On the basis of evidence linking hypoxia with radioresistance [35], Chao et al. tested a PET-based technique to measure hypoxia with a Cu-ATSM [Cu(II)-diacetyl-bis(N(4)-methylthiosemicarbazone] tracer and considered the results promising [36]. Lee et al. also demonstrated the feasibility of using ^{18}F-labeled fluoromisonidazole (^{18}F-FMISO) PET/CT for guiding IMRT so as to allow the dose to radioresistant hypoxic regions to be escalated to 84 Gy (Fig. 8.5) [37]. Findings of

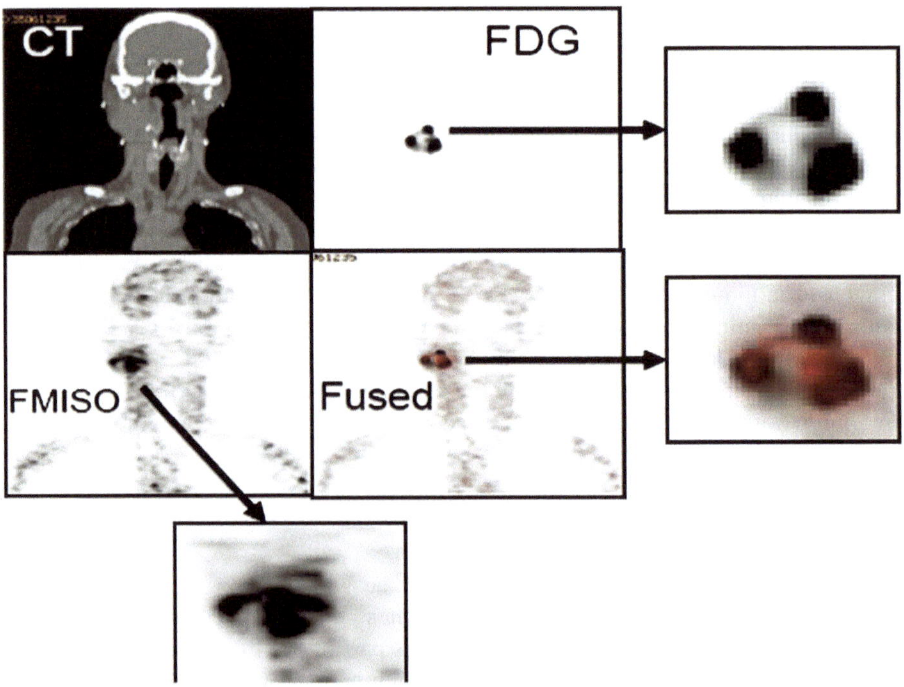

Fig. 8.5 Multimodality image acquisition, processing, and registration for ^{18}F-FMISO PET/CT-guided IMRT. Shown are computed tomography (CT) (*top left*), fluorodeoxyglucose (FDG) (*top right*), ^{18}F-fluoromisonidazole (^{18}F-FMISO) (*bottom left*), and fused FDG-^{18}F-FMISO (*bottom right*) images. Also shown are three enlarged areas from each scan type (From Lee et al. [37], with permission from Elsevier)

an ongoing trial, NCT00606294, are expected to show whether FMISO PET-guided visualization of hypoxia can be used to stratify patients in terms of risk.

Treatment intensification in the form of dose escalation is now an option with IMRT. Previous attempts at dose intensification with conventional RT were limited by toxicity [38, 39]. However, at least two dose-escalation studies have shown that IMRT can allow a safe dose escalation in NPC [22, 23]. In one of those studies, Kwong et al. used a prescribed dose of 76 Gy given in 35 fractions for T3-T4 NPC and found an excellent 2-year LCR rate of 96 % and acceptable toxicity [23]. In another such study, Kam et al. used a boost technique to provide a total dose of 74 Gy and also reported excellent LRC [22]. Although current dose levels have resulted in excellent LRC, dose escalation in selected patients with high-risk NPC may confer further benefits.

The use of adjuvant chemotherapy is another potential form of treatment intensification. Findings from the INT0099 trial indicated that the current standard of care should include adjuvant chemotherapy in addition to concurrent chemoradiation. However, results of a more recent phase III trial found no benefit from the use of adjuvant chemotherapy [40]; in that trial involving 508 patients, the 2-year failure-free survival rate was 86 % in the group with adjuvant chemotherapy and 84 % in the group without adjuvant chemotherapy ($P=0.13$). Additional follow-up is needed, however, as the failure-free survival Kaplan-Meier curves may well separate over time. Moreover, that study was not designed to directly compare this therapy with that of INT0099. Nevertheless, patients with high-risk NPC may be more likely to benefit from adjuvant chemotherapy.

Interest has also been growing in the use of proton therapy in the form of intensity-modulated proton therapy (IMPT). IMPT plans have been shown to provide additional dosimetric advantage over IMRT by improving tumor coverage and reducing the mean dose to organs at risk (Fig. 8.6) [41]. We look forward to identifying potential benefit from protons in the clinical setting. Currently, an ongoing phase II trial at Massachusetts General Hospital is evaluating the potential for reduction in toxicity from the use of proton beam therapy (NCT00592501).

Adaptive RT is also being investigated for its potential to improve clinical outcomes. The rationale for this therapy is that significant anatomic changes during therapy, such as those resulting from loss of body weight or shrinkage in tumor volume (reportedly most severe after the first 2 weeks of treatment [42]), can lead to movement of the organs at risk into the planned radiation field. Conversely, marginal misses may occur if the tumor becomes displaced out of the treatment field, especially given the current efforts to reduce margins and treatment volumes to the greatest possible extent. The potential value of repeated treatment simulations is being considered and has shown some potential [42].

8.5 Conclusions

Clinical outcomes with IMRT have demonstrated clear dosimetric advantages, excellent LRC rates of more than 90 %, and lesser toxicity (specifically by improving salivary function) compared with conventional RT. The use of IMRT with

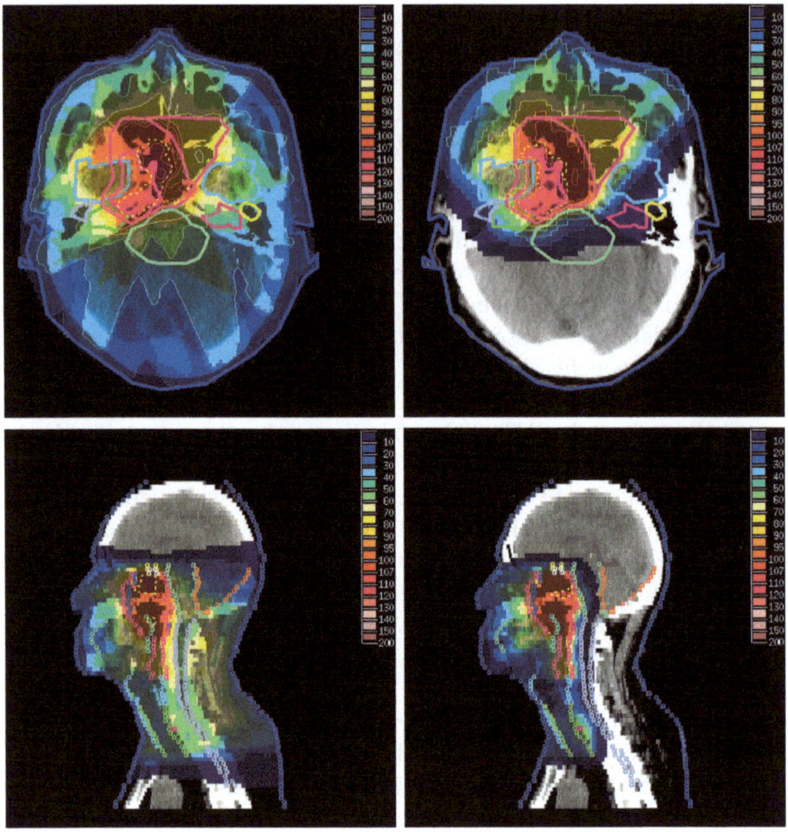

Fig. 8.6 Dosimetric comparison of treatment plans for intensity-modulated radiation therapy (IMRT) vs. intensity-modulated proton therapy (IMPT). Dose distributions are shown for IMRT plans (*left*) and IMPT plans (*right*) for a patient with T4N1N0 nasopharyngeal carcinoma. *Dotted lines* denote 95 % of the prescribed dose to the gross tumor volume. Figure 2 from Taheri-Kadkhoda et al. [41] (License accessible at: http://creativecommons.org/licenses/by/2.0/legalcode)

specific target volume guidelines has been replicated successfully in a multi-institutional setting in the United States.

Further improvements in toxicity after IMRT will rely on either further reductions in margins within treatment volumes or the use of adaptive RT. Proton therapy also shows promise in terms of further sparing critical structures. Regarding approaches to improve disease control, dose escalation with IMRT is now feasible and could be considered for cases of particularly aggressive NPC. The use of imaging parameters and biomarkers, such as EBV DNA levels, also shows promise for risk stratification and consequent treatment intensification for high-risk NPC.

References

1. Yu MC (1990) Diet and nasopharyngeal carcinoma. FEMS Microbiol Immunol 2(4):235–242
2. Thompson LD (2007) Update on nasopharyngeal carcinoma. Head Neck Pathol 1(1):81–86
3. Al-Sarraf M, LeBlanc M, Giri PG, Fu KK, Cooper J, Vuong T, Forastiere AA, Adams G, Sakr WA, Schuller DE, Ensley JF (1998) Chemoradiotherapy versus radiotherapy in patients with advanced nasopharyngeal cancer: phase III randomized Intergroup study 0099. J Clin Oncol 16(4):1310–1317
4. Hunt MA, Zelefsky MJ, Wolden S, Chui CS, LoSasso T, Rosenzweig K, Chong L, Spirou SV, Fromme L, Lumley M, Amols HA, Ling CC, Leibel SA (2001) Treatment planning and delivery of intensity-modulated radiation therapy for primary nasopharynx cancer. Int J Radiat Oncol Biol Phys 49(3):623–632
5. Xia P, Fu KK, Wong GW, Akazawa C, Verhey LJ (2000) Comparison of treatment plans involving intensity-modulated radiotherapy for nasopharyngeal carcinoma. Int J Radiat Oncol Biol Phys 48(2):329–337
6. Chambers MS, Garden AS, Kies MS, Martin JW (2004) Radiation-induced xerostomia in patients with head and neck cancer: pathogenesis, impact on quality of life, and management. Head Neck 26(9):796–807
7. Eisbruch A, Ship JA, Dawson LA, Kim HM, Bradford CR, Terrell JE, Chepeha DB, Teknos TN, Hogikyan ND, Anzai Y, Marsh LH, Ten Haken RK, Wolf GT (2003) Salivary gland sparing and improved target irradiation by conformal and intensity modulated irradiation of head and neck cancer. World J Surg 27(7):832–837
8. Blanco AI, Chao KS, El Naqa I, Franklin GE, Zakarian K, Vicic M, Deasy JO (2005) Dose-volume modeling of salivary function in patients with head-and-neck cancer receiving radiotherapy. Int J Radiat Oncol Biol Phys 62(4):1055–1069. doi:10.1016/j.ijrobp.2004.12.076
9. Pow EH, Kwong DL, McMillan AS, Wong MC, Sham JS, Leung LH, Leung WK (2006) Xerostomia and quality of life after intensity-modulated radiotherapy vs. conventional radiotherapy for early-stage nasopharyngeal carcinoma: initial report on a randomized controlled clinical trial. Int J Radiat Oncol Biol Phys 66(4):981–991
10. Kam MK, Leung SF, Zee B, Chau RM, Suen JJ, Mo F, Lai M, Ho R, Cheung KY, Yu BK, Chiu SK, Choi PH, Teo PM, Kwan WH, Chan AT (2007) Prospective randomized study of intensity-modulated radiotherapy on salivary gland function in early-stage nasopharyngeal carcinoma patients. J Clin Oncol 25(31):4873–4879
11. Lee N, Harris J, Garden AS, Straube W, Glisson B, Xia P, Bosch W, Morrison WH, Quivey J, Thorstad W, Jones C, Ang KK (2009) Intensity-modulated radiation therapy with or without chemotherapy for nasopharyngeal carcinoma: radiation therapy oncology group phase II trial 0225. J Clin Oncol 27(22):3684–3690
12. Lee AW, Lau WH, Tung SY, Chua DT, Chappell R, Xu L, Siu L, Sze WM, Leung TW, Sham JS, Ngan RK, Law SC, Yau TK, Au JS, O'Sullivan B, Pang ES, O SK, Au GK, Lau JT, Hong Kong Nasopharyngeal Cancer Study Group (2005) Preliminary results of a randomized study on therapeutic gain by concurrent chemotherapy for regionally-advanced nasopharyngeal carcinoma: NPC-9901 Trial by the Hong Kong Nasopharyngeal Cancer Study Group. J Clin Oncol 23(28):6966–6975
13. Wee J, Tan EH, Tai BC, Wong HB, Leong SS, Tan T, Chua ET, Yang E, Lee KM, Fong KW, Tan HS, Lee KS, Loong S, Sethi V, Chua EJ, Machin D (2005) Randomized trial of radiotherapy versus concurrent chemoradiotherapy followed by adjuvant chemotherapy in patients with American Joint Committee on Cancer/International Union against cancer stage III and IV nasopharyngeal cancer of the endemic variety. J Clin Oncol 23(27):6730–6738
14. Lee N, Xia P, Quivey JM, Sultanem K, Poon I, Akazawa C, Akazawa P, Weinberg V, Fu KK (2002) Intensity-modulated radiotherapy in the treatment of nasopharyngeal carcinoma: an update of the UCSF experience. Int J Radiat Oncol Biol Phys 53(1):12–22
15. Kwong DL, Pow EH, Sham JS, McMillan AS, Leung LH, Leung WK, Chua DT, Cheng AC, Wu PM, Au GK (2004) Intensity-modulated radiotherapy for early-stage nasopharyngeal

carcinoma: a prospective study on disease control and preservation of salivary function. Cancer 101(7):1584–1593
16. Wang TC, Riaz N, Cheng S, Lu J, Lee N (2012) Intensity-modulated radiation therapy for nasopharyngeal carcinoma: a review. J Radiat Oncol 1(2):129–146
17. Lonneux M, Hamoir M, Reychler H, Maingon P, Duvillard C, Calais G, Bridji B, Digue L, Toubeau M, Gregoire V (2010) Positron emission tomography with [18F]fluorodeoxyglucose improves staging and patient management in patients with head and neck squamous cell carcinoma: a multicenter prospective study. J Clin Oncol 28(7):1190–1195
18. Abdel Khalek Abdel Razek A, King A (2012) MRI and CT of nasopharyngeal carcinoma. AJR Am J Roentgenol 198(1):11–18
19. Lee NY, Zhang Q, Pfister DG, Kim J, Garden AS, Mechalakos J, Hu K, Le QT, Colevas AD, Glisson BS, Chan AT, Ang KK (2012) Addition of bevacizumab to standard chemoradiation for locoregionally advanced nasopharyngeal carcinoma (RTOG 0615): a phase 2 multi-institutional trial. Lancet Oncol 13(2):172–180
20. Lin S, Pan J, Han L, Zhang X, Liao X, Lu JJ (2009) Nasopharyngeal carcinoma treated with reduced-volume intensity-modulated radiation therapy: report on the 3-year outcome of a prospective series. Int J Radiat Oncol Biol Phys 75(4):1071–1078
21. Kong L, Lu JJ, Hu C, Guo X, Wu Y, Zhang Y (2006) The risk of second primary tumors in patients with nasopharyngeal carcinoma after definitive radiotherapy. Cancer 107(6):1287–1293
22. Kam MK, Teo PM, Chau RM, Cheung KY, Choi PH, Kwan WH, Leung SF, Zee B, Chan AT (2004) Treatment of nasopharyngeal carcinoma with intensity-modulated radiotherapy: the Hong Kong experience. Int J Radiat Oncol Biol Phys 60(5):1440–1450
23. Kwong DL, Sham JS, Leung LH, Cheng AC, Ng WM, Kwong PW, Lui WM, Yau CC, Wu PM, Wei W, Au G (2006) Preliminary results of radiation dose escalation for locally advanced nasopharyngeal carcinoma. Int J Radiat Oncol Biol Phys 64(2):374–381
24. Wolden SL, Chen WC, Pfister DG, Kraus DH, Berry SL, Zelefsky MJ (2006) Intensity-modulated radiation therapy (IMRT) for nasopharynx cancer: update of the Memorial Sloan-Kettering experience. Int J Radiat Oncol Biol Phys 64(1):57–62
25. Lin S, Lu JJ, Han L, Chen Q, Pan J (2010) Sequential chemotherapy and intensity-modulated radiation therapy in the management of locoregionally advanced nasopharyngeal carcinoma: experience of 370 consecutive cases. BMC Cancer 10:39
26. Gregoire V, Levendag P, Ang KK, Bernier J, Braaksma M, Budach V, Chao C, Coche E, Cooper JS, Cosnard G, Eisbruch A, El-Sayed S, Emami B, Grau C, Hamoir M, Lee N, Maingon P, Muller K, Reychler H (2003) CT-based delineation of lymph node levels and related CTVs in the node-negative neck: DAHANCA, EORTC, GORTEC, NCIC, RTOG consensus guidelines. Radiother Oncol 69(3):227–236
27. Twu CW, Wang WY, Liang WM, Jan JS, Jiang RS, Chao J, Jin YT, Lin JC (2007) Comparison of the prognostic impact of serum anti-EBV antibody and plasma EBV DNA assays in nasopharyngeal carcinoma. Int J Radiat Oncol Biol Phys 67(1):130–137
28. Leung SF, Chan AT, Zee B, Ma B, Chan LY, Johnson PJ, Lo YM (2003) Pretherapy quantitative measurement of circulating Epstein-Barr virus DNA is predictive of posttherapy distant failure in patients with early-stage nasopharyngeal carcinoma of undifferentiated type. Cancer 98(2):288–291
29. Leung SF, Zee B, Ma BB, Hui EP, Mo F, Lai M, Chan KC, Chan LY, Kwan WH, Lo YM, Chan AT (2006) Plasma Epstein-Barr viral deoxyribonucleic acid quantitation complements tumor-node-metastasis staging prognostication in nasopharyngeal carcinoma. J Clin Oncol 24(34):5414–5418
30. Wang WY, Twu CW, Lin WY, Jiang RS, Liang KL, Chen KW, Wu CT, Shih YT, Lin JC (2011) Plasma Epstein-Barr virus DNA screening followed by (1)(8)F-fluoro-2-deoxy-D-glucose positron emission tomography in detecting posttreatment failures of nasopharyngeal carcinoma. Cancer 117(19):4452–4459

31. Tang LQ, Chen QY, Fan W, Liu H, Zhang L, Guo L, Luo DH, Huang PY, Zhang X, Lin XP, Mo YX, Liu LZ, Mo HY, Li J, Zou RH, Cao Y, Xiang YQ, Qiu F, Sun R, Chen MY, Hua YJ, Lv X, Wang L, Zhao C, Guo X, Cao KJ, Qian CN, Zeng MS, Mai HQ (2013) Prospective study of tailoring whole-body dual-modality [18F]fluorodeoxyglucose positron emission tomography/computed tomography with plasma Epstein-Barr virus DNA for detecting distant metastasis in endemic nasopharyngeal carcinoma at initial staging. J Clin Oncol 31(23):2861–2869
32. Wang WY, Twu CW, Chen HH, Jiang RS, Wu CT, Liang KL, Shih YT, Chen CC, Lin PJ, Liu YC, Lin JC (2013) Long-term survival analysis of nasopharyngeal carcinoma by plasma Epstein-Barr virus DNA levels. Cancer 119(5):963–970
33. Lin JC, Wang WY, Chen KY, Wei YH, Liang WM, Jan JS, Jiang RS (2004) Quantification of plasma Epstein-Barr virus DNA in patients with advanced nasopharyngeal carcinoma. N Engl J Med 350(24):2461–2470
34. Le QT, Zhang Q, Cao H, Cheng AJ, Pinsky BA, Hong RL, Chang JT, Wang CW, Tsao KC, Lo YD, Lee N, Ang KK, Chan AT, Chan KC (2013) An international collaboration to harmonize the quantitative plasma Epstein-Barr virus DNA assay for future biomarker-guided trials in nasopharyngeal carcinoma. Clin Cancer Res 19(8):2208–2215
35. Brizel DM, Sibley GS, Prosnitz LR, Scher RL, Dewhirst MW (1997) Tumor hypoxia adversely affects the prognosis of carcinoma of the head and neck. Int J Radiat Oncol Biol Phys 38(2):285–289
36. Chao KS, Bosch WR, Mutic S, Lewis JS, Dehdashti F, Mintun MA, Dempsey JF, Perez CA, Purdy JA, Welch MJ (2001) A novel approach to overcome hypoxic tumor resistance: Cu-ATSM-guided intensity-modulated radiation therapy. Int J Radiat Oncol Biol Phys 49(4):1171–1182
37. Lee NY, Mechalakos JG, Nehmeh S, Lin Z, Squire OD, Cai S, Chan K, Zanzonico PB, Greco C, Ling CC, Humm JL, Schoder H (2008) Fluorine-18-labeled fluoromisonidazole positron emission and computed tomography-guided intensity-modulated radiotherapy for head and neck cancer: a feasibility study. Int J Radiat Oncol Biol Phys 70(1):2–13
38. Wolden SL, Zelefsky MJ, Hunt MA, Rosenzweig KE, Chong LM, Kraus DH, Pfister DG, Leibel SA (2001) Failure of a 3D conformal boost to improve radiotherapy for nasopharyngeal carcinoma. Int J Radiat Oncol Biol Phys 49(5):1229–1234
39. Jen YM, Lin YS, Su WF, Hsu WL, Hwang JM, Chao HL, Liu DW, Chen CM, Lin HY, Wu CJ, Chang LP, Shueng PW (2002) Dose escalation using twice-daily radiotherapy for nasopharyngeal carcinoma: does heavier dosing result in a happier ending? Int J Radiat Oncol Biol Phys 54(1):14–22
40. Chen L, Hu CS, Chen XZ, Hu GQ, Cheng ZB, Sun Y, Li WX, Chen YY, Xie FY, Liang SB, Chen Y, Xu TT, Li B, Long GX, Wang SY, Zheng BM, Guo Y, Sun Y, Mao YP, Tang LL, Chen YM, Liu MZ, Ma J (2012) Concurrent chemoradiotherapy plus adjuvant chemotherapy versus concurrent chemoradiotherapy alone in patients with locoregionally advanced nasopharyngeal carcinoma: a phase 3 multicentre randomised controlled trial. Lancet Oncol 13(2):163–171
41. Taheri-Kadkhoda Z, Bjork-Eriksson T, Nill S, Wilkens JJ, Oelfke U, Johansson KA, Huber PE, Munter MW (2008) Intensity-modulated radiotherapy of nasopharyngeal carcinoma: a comparative treatment planning study of photons and protons. Radiat Oncol 3:4
42. Bhide SA, Davies M, Burke K, McNair HA, Hansen V, Barbachano Y, El-Hariry IA, Newbold K, Harrington KJ, Nutting CM (2010) Weekly volume and dosimetric changes during chemoradiotherapy with intensity-modulated radiation therapy for head and neck cancer: a prospective observational study. Int J Radiat Oncol Biol Phys 76(5):1360–1368
43. Su SF, Han F, Zhao C, Chen CY, Xiao WW, Li JX, Lu TX (2012) Long-term outcomes of early-stage nasopharyngeal carcinoma patients treated with intensity-modulated radiotherapy alone. Int J Radiat Oncol Biol Phys 82(1):327–333

Oropharyngeal Cancer

9

Takeshi Kodaira, Arisa Shimizu, and Keiichi Takehana

Keywords
Functional preservation • Multidisciplinary therapy • Human papillomavirus

9.1 Introduction

Oropharyngeal cancer (OPC) accounts for almost half of head and neck cancers (HNCs) in the United States and European country [1, 2], while its incidence in Japan was reported to be smaller [3, 4]. Compared to other sites of HNC, patients with OPC have the unique clinical characteristic of a lower incidence of smoking and high rate of human papillomavirus (HPV) infection. Although the number of OPC patients with positive HPV tumors is increasing rapidly, many previous studies have reported its favorable clinical outcome, even for locally advanced clinical stages [5–10]. Thus, the clinical role for radiation therapy (RT) has become more essential in the past decade. The oropharynx plays a crucial role in swallowing and speech; therefore, patients with early stage and moderately advanced OPC are considered as good candidates for definitive RT for functional preservation.

Intensity-modulated radiation therapy (IMRT) was initiated in the mid-1990s [11], and its clinical advantages for OPC have been reported in both prospective multi-institutional trials [12, 13] and clinical encounters at large volume hospitals [14–19] (Table 9.1). The distinct advantage of IMRT for OPC is believed to be the minimization of xerostomia by reducing the dose administered to the parotid gland, and many reports have confirmed this clinical advantages [13, 20–23]. The PARSPORT trial was the first prospective multi-institutional randomized trial to

T. Kodaira (✉) • A. Shimizu • K. Takehana
Department of Radiation Oncology, Aichi Cancer Center Hospital,
1-1 Kanoko-den, Chikusa-ku, Nagoya, Aichi 464-8681, Japan
e-mail: 109103@aichi-cc.jp

Y. Nishimura, R. Komaki (eds.), *Intensity-Modulated Radiation Therapy*,
DOI 10.1007/978-4-431-55486-8_9

Table 9.1 Reported series of definitive IMRT for oropharyngeal cancer patients

	Number	F/U	Age	Tonsil	BT	CS IV	T4	N2b	Dose PTV1	Dose PTV2	Dose PTV3	CRT (%)	SIB	OS	LCR	PFS
Sher	163	36 M	56.1	50	45	75	7	54	70	64	60	100	Yes	86@5Y	86@5Y	
Garden	774	54 M	55	48	46	74	17	58	66–72			54		84@5Y	90@5Y	82@5Y
Setton	442	36.8 M	57	50	46	73	13.8	69.5@N2–3	70	59.4	54	95	Yes	84.9@3Y	LR5.4 %@3Y	
Daly	107	29 M	70 %@60>	44	48	85	29	83@N2-3	66	54	50	98	Yes	83.3@3Y	92@3Y	81@3Y
Huang	71	33 M	55	38	61	76	11	72@N2-3	70	59.4	54	100	Yes	83.3@3Y	90@3Y	81@3Y
Clavel	100	35 M	46	58	39	87	12	87@N2–3	70	59.4	50.4	100	Yes	92.1@3Y	95.1@3Y	85@3Y

Abbreviation
BT bottom of the tongue, *CRT* chemoradiotherapy, *SIB* simultaneous integrated boost, *OS* overall survival, *LCR* local control rate, *PFS* progression-free survival

examine the advantages of IMRT over 3DCRT [13]. Ninety-four patients participated in this trial, of whom 85 % were OPC, and were assigned to receive IMRT or 3DCRT. The rate of xerostomia at 12 months was significantly lower in the IMRT group than in the 3DCRT group (37 % vs. 74 % $p=0.0027$), and the same results were also observed in an objective rate at 24 months and subjective score from a questionnaire on HN35. No significant differences were observed in the locoregional progression-free survival rates between both groups. Thus, IMRT was considered an essential RT technique for HNC, especially in OPC patients.

In this chapter, we have introduced practical guidance and evidence supporting the use of IMRT to treat patients with OPC.

9.2 Epidemiology

The oropharynx comprises the soft palate (SP), uvula, tonsillar fossa (TF) and pillars, glossotonsillar sulci, lateral and posterior pharyngeal wall, vallecula, and base of the tongue (BOT). It is located between the soft palate (superior) and hyoid bone (inferior). The oral cavity is located anterior to the oropharynx. These subsites have been divided into four sections: the anterior wall (AW; BOT, vallecula), lateral wall (LW; TF, pillars, glossotonsillar sulci, lateral pharyngeal wall), superior wall (SW; inferior surface of SP, uvula), and posterior wall (PW; posterior pharyngeal wall). Approximately 60 % of OPC consists of the LW, followed by the AW, SW, and PW.

There is a rich lymphatic network, in which early lymphatic involvement typically develops, with approximately 60 % of diseases being diagnosed as stages III–IVB. The most common histopathology of OPC is squamous cell carcinoma (SCC). Poorly differentiated histopathology is frequently reported in the TF and BOT, while more differentiated histological subtypes have been observed in the other sites. Apart from SCC, histopathological findings have revealed adenocarcinoma, adenoid cystic carcinoma, mucoepidermoid carcinoma, and malignant lymphoma.

A well-known risk factor for OPC is smoking, alcohol consumption, diet, poor oral hygiene, marijuana consumption, and clinical stage. Human papillomavirus (HPV) is becoming more frequently associated with OPC, and incidence has been reported in more than half of HNC patients [1]. HPV is most commonly found in the TF and BOT, with HPV16 being identified as the main population. HPV status has been shown to be a distinct risk factor. Ang et al. proposed a risk model of OPC using recursive-partitioning analysis (RPA) on the basis of four factors [5]: HPV status, pack years of tobacco smoking, tumor stage, and nodal stage. Of 743 patients from the RTOG 0129, who were randomly assigned to receive accelerated fractionation RT with cisplatin (CP) or standard fractionation RT with CP, there were 433 patients (60.1 %) with OPC [24]. HPV status information was obtained from 322 patients in these groups, and they were entered into this analysis. HPV-positive patients ($n=206$) had significantly better overall survival rates (OS, 82.4 % vs. 57.1 %; $p<0.001$) than those of HPV-negative patients. They were further categorized into three groups based on the OS with RPA, and the 3-year OS rates in

Table 9.2 Correlation of HPV status with overall survival in reported series of radiotherapy for oropharyngeal cancer

	Patient number	HPV positive	HPV negative
Ang	223	84 %@3Y	51 %@3Y
Rischin	185	91 %@2Y	74 %@2Y
Shi	111	88 %@3Y	67 %@3Y
Lassen	74	66 %@5Y	28 %@5Y
Fakhry	62	84 %@3Y	50 %@3Y
Nichols	44	89@3Y	65@3Y

low-risk ($n=114$), intermediate-risk ($n=79$), and high-risk groups ($n=73$) were 93 %, 70.8 %, and 46.2 %, respectively. Therefore, HPV status will be regarded as more essential information in future clinical practices [5–10] (Table 9.2).

9.3 Clinical Work-Up

Clinical presentation, patient history, and physical examination are considered essential procedures for a diagnostic work-up. A comprehensive evaluation of the head and neck including the oropharynx, oral cavity, nasopharynx, hypopharynx, and larynx is necessary, and biopsy is a confirmative method used for a diagnostic assessment of the extent of the tumor [25]. Panendoscopy is routinely recommended to detect second primary malignancy [26], and, if possible, narrowband imaging techniques are helpful for finding superficial lesions in both the esophageal and head and neck regions [27]. Early SP tumors sometimes invade superficially with erythroplasia; therefore, careful estimations of the entire mucosa and pathological confirmation should be considered, where necessary.

CT with contrast media is the standard diagnostic imaging technique, and magnetic resonance imaging (MRI) is also recommended to accurately determine both primary tumor extension and normal tissue boundaries with sufficient spatial resolution of the soft tissues. BOT tumors are typically bulky and expand to the vallecula, TF, and mobile tongue; therefore, MRI should be mandatory because of the valuable information obtained regarding the extent of the disease. MRI is also advantageous because it is less sensitive to dental artifacts than CT [28] (Fig. 9.1). Neck lymph nodes are considered to be positive if the smallest diameter measured is greater than 1 cm. Determining the presence or absence of extracapsular invasion is of importance; therefore, thin slice CT or MRI assists in an accurate evaluation. ^{18}Fluorodeoxyglucose (FDG) positron-emission topography (PET) exhibits high accuracy for tumor involvement [29] and can be used to screen for distant metastasis [30]. FDG-PET scans may be recommended especially in the case of advanced disease. FDG-PET scans also have the ability to accurately assess the treatment response of patients treated with RT and/or chemotherapy [31]. The recommended timing of FDG-PET after RT is 2–3 months, because an earlier assessment has been shown to increase the false-positive rate due to acute radiation reactions.

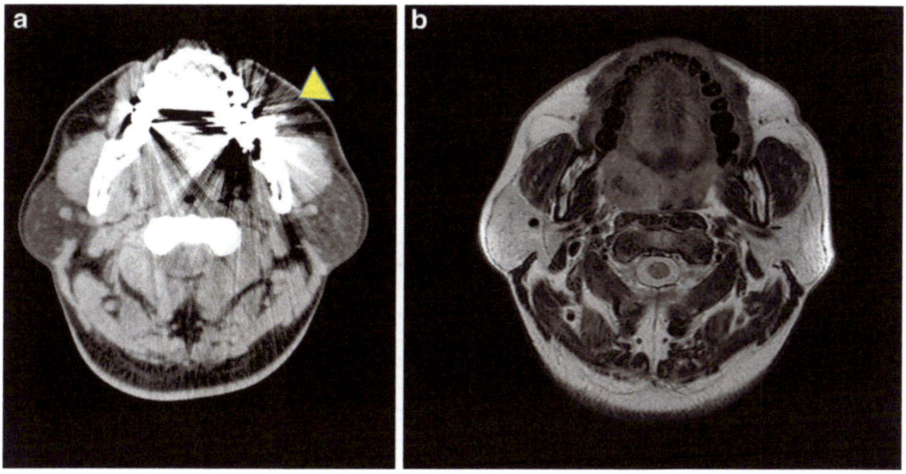

Fig. 9.1 CT image (**a**) and T2-weighted MRI image (**b**) of patients with tonsillar carcinoma (T2N0M0). A dental filling artifact on the CT image (*arrow head*) greatly interfered with the detection of the primary tumor, while it can be clearly detected on the MRI image due to reduced sensitivity to metallic artifacts

Integrated PET/CT imaging has provided interactive information on both anatomical and functional volumes. FDG-PET data can be directly imported to an RT treatment planning workstation, at which researchers reported that integrated PET/CT planning was advantageous for clinical outcomes [32].

Target volume delineation is crucial in the setting of IMRT; therefore, diagnostic information has become more important to refine the quality of treatments used.

A consultation with a dentist regarding management of the disease prior to starting RT is essential, because extraction after RT leads to increase the incidence of critical adverse events, such as osteonecrosis and trismus. Education on how to stop smoking before RT is initiated is also very important because smoking has been shown to decrease the efficacy of RT and increase the risk of both adverse events and secondary cancer.

9.4 Clinical Indication

Since the treatment of OPC is complex, a multidisciplinary team board should ideally determine clinical decisions. The tumor-node-metastasis classification (TNM) stage, diagnostic imaging, pathological findings, patient condition, and social background are essential when determining an adequate treatment modality for OPC (Fig. 9.2).

The mainstay of treatment for OPC is both surgery and RT. External beam RT is typically administered in 70 Gy/35 fractions over 7 weeks with a standard fractionation schedule. A meta-analysis from 15 trials with 6,515 patients revealed that an altered fractionation schedule had significant advantages for local control and OS [33] (level Ia). The majority of cohorts were comprised of OPC patients (47.2 %)

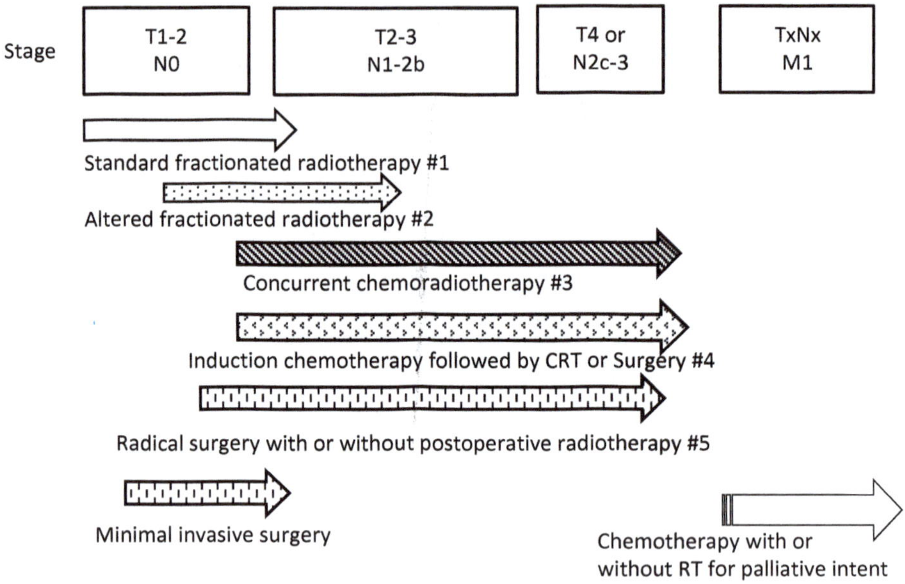

Fig. 9.2 Overview of clinical strategy regarding radiotherapy for patients with oropharyngeal cancer according to clinical stage and practical background
#1 Disease with T2N1 of the tonsil or base of the tongue is candidate for radiotherapy alone
#2 Concurrent use of cytotoxic agents is not supported by evidence
#3 Platinum component (level IA) and cetuximab (level Ib) are considered to be administered
#4 Combination use of taxane, 5-FU, and cisplatin is preferred
#5 Chemoradiotherapy is considered for high-risk features such as positive margin and/or extra-capsular invasion

and stage III patients, which were categorized as an intermediate-risk group, and these groups had slightly better OS in subset analysis. One of the reasons for this patient population is that physicians consider CRT to be a better strategy for patients with advanced nodal disease because of the higher risk of distant metastasis. If altered fractionation is selected, the limitations associated with the combination use of chemotherapy should be considered (which is described later). Brachytherapy [34] and stereotactic radiosurgery [35] are considered better options for minimizing the radiation dose to normal organs such as the parotid gland, mandibula, and pharyngeal constrictor muscle.

A trans-oral approach or endoscopic mucosal resection is considered an adequate option for localized small primary lesions without nodal spread because sufficient outcomes for efficacy and function can be achieved [36]. Postoperative RT should be added if a positive surgical margin is revealed in the pathological specimen; however, this has been associated with a higher risk of adverse events than single modality treatment [37–39]. RT is believed to be more advantageous for both organ and functional preservation than open surgery. IMRT is strongly recommended to minimize the late adverse events associated with RT, especially xerostomia [40, 41].

Patients with moderately advanced disease are also considered as good candidates for definitive RT. Patients with larger tumors (T2–T3) and/or nodal involvement (N1–N2) typically received RT with chemotherapy or molecular-targeted agents. The concurrent use of platinum-based components was distinctly beneficial for OS and disease control [42, 43] (level Ia), and it is considered a standard treatment for locally advanced disease. In the case of concurrent chemoradiotherapy (CCRT), 70 Gy/35 fractions over 7 weeks is the standard RT schedule. The RTOG 0129 trial [24], which enrolled 743 patients with stage III–IV disease, reported that accelerated fractionated (AX) RT with CP did not have any advantage over standard fractionated RT with CP [24]. Therefore, AX combined with CCRT should be used in clinical trial settings.

One randomized trial, RT with cetuximab (CET; anti-epidermal growth factor receptor (EGFR)), showed improvements in OS and local control over those with RT alone [44, 45]. In the Bonner trial, 424 patients with stage III–IV OPC, hypopharynx, and larynx cancer were randomly assigned to the RT with CET arm or the RT arm. The RT with CET arm showed significant improvements in locoregional control ($HR=0.68$, $p=0.005$) and OS ($HR=0.74$, $p=0.03$). Mortality was lower in the RT with CET arm than in the RT arm without an increase in the incidence of adverse events. In this trial, the RT schedule included three types of fractionation schedules without stratification at randomization, such as standard fractionation, hyperfractionation, and the concomitant boost method (AX). Subset analysis revealed better results with the concomitant boost method than the other schedules. We could not obtain the prospective trial results for the direct comparison between CP and CET until the end of 2013. The RTOG 1016, a phase III trial of RT plus CET versus CCRT with CP in HPV-associated OPC, is now ongoing. Results will be obtained from this trial in the near future (RTOG 1016). CET was added to CCRT to assess improvements in the clinical results of CCRT without increasing the toxicity. The RTOG 0522 trial was designed to compare the CCRT with CET arm to the CCRT arm [46]. The findings of this trial failed to show the clear advantages of adding CET to CCRT; however, it revealed the increased incidence of acute adverse events in the combined arm. Based on these findings, CCRT with anti-EGFR should be tested in clinical trials, and care should be taken for its clinical use.

Adequate timing for the addition of RT to chemotherapy is believed to be concurrent administration [43]. When induction and adjuvant setting were compared, CCRT caused the largest benefits in OS and locoregional control associated with increasing acute toxicity. A meta-analysis showed that induction chemotherapy (IC) was moderately beneficial for OS and was more advantageous for the control of distant metastasis [43]. IC with taxane-containing multi-agents showed apparent advantages for OS and disease control than those of CP and 5-FU in both randomized control studies [47, 48] and meta-analysis [49]. According to the results from meta-analysis, the addition of taxane components to CP and 5-FU was distinctly beneficial for both OS (HR 0.79 95 % CI 0.7–0.89) and disease control (HR 0.78 95 % $CI=0.69–0.87$; level Ia). Highly advanced disease (T4 and/or N3 disease) is associated with a higher risk of distant metastasis; therefore, this is a reasonable strategy for these groups. However, intensive IC may lead to low CP compliance (only 50 %) in subsequent CCRT phases [49]. Another clinical utility of IC is as a strategy for organ preservation [50–53]. If a sufficient response is acquired from IC, patients with sufficient responses could take the clinical advantage of selecting RT over up-front surgery. Patients without a sufficient response have to undergo surgery and be subjected to the high-risk features associated with postoperative radiotherapy [37, 38] (see details in Chap. 11).

Since either primary or lymph node bulky lesion may greatly decrease the efficacy of RT, surgical approaches may be necessary for disease control. Functional loss from surgery should also be taken into careful consideration in such cases. A multidisciplinary team board is needed to coordinate both treatment efficacy and morbidity.

Definitive RT (with chemotherapy) is the standard treatment for unresectable disease, such as stage IVB disease. If tumors present with rapid growth and/or a very large volume, the benefits associated with IMRT may be decreased because of both the long preparation time and decreased ability to spare the dose delivered to the normal tissue due to a large CTV. 3D conformal RT may be a more practical application for such cases.

The mainstay of treatment for stage IVC disease is systemic chemotherapy, and RT is chiefly considered as a palliative intent for symptom relief. If patients have significant hazards due to airway obstructions, uncontrollable bleeding, and refractory cancer pain, CCRT still has a sufficient impact on locoregional control for the selected patients with sufficient organ function.

9.5 Target Delineation and Dose Prescription

IMRT can uniquely be administered at variable dose levels to different clinical target volumes (CTVs). The simultaneous integrated boost (SIB) technique is generally used in IMRT for HNC. CTVs for primary lesion and involved nodes are generally delivered to approximately 70 Gy over 6–7 weeks and other CTVs with risk level treated with 54–63 Gy with the same treatment duration. The advantage of the SIB technique is that it is simple and convenient for practical management. The clinical benefits of adaptive replanning have recently been reported in several studies [54–56] (see Chap. 7). IMRT can be administered at steep dose gradients between CTV and normal tissue; therefore, anatomical changes due to tumor shrinkage and/or body weight loss lead to significant differences in the dose delivered. Adaptive replanning may resolve the problem of unexpected dose changes during the treatment course; however, the effort associated with *extra*-planning procedures would increase. Another problem linked to the SIB technique is that CTVs of risk level are treated with fraction sizes smaller than 1.8 Gy, which may decrease the efficacy of its biological radiation effects. The two-step plan, which is similar to conformal RT with the cone-down technique, could extrapolate experiences from standard fractionation schedule [54]. Advances in both the treatment planning apparatus and sophisticated integrated imaging system may warrant exploring more convenient achievements in adaptive replanning.

9.5.1 Gross Tumor Volume (GTV)

The anatomy of the oropharynx is very complex; therefore, integrating all clinical information from physical examinations, such as inspection and palpation, and diagnostic imaging from CT, MRI, fiber-optic images, and PET scans is of importance. These diagnostic images should be taken in identical head positions to those taken during the CT simulation with an immobilization mask. If possible, a metallic marker should be placed at the landmark of the primary tumor to achieve accurate delineation. In addition, the volumetric 3D reconstruction of GTVs in coronal and sagittal views on the treatment planning system is helpful for assessing the extension of the primary lesion. Regarding the lymph node evaluation, if its shortest diameter is equal to or greater than 1 cm, it is defined as lymphatic involvement. In addition, the clinical features of the lymph nodes with peripheral rim enhancement or positive PET finding are also considered to be the feature of lymphatic involvement. Clinical features of borderline findings with either primary lesion or lymph node involvement are regarded as GTV*intermediate-risk*. GTV*intermediate-risk* typically expands with a 1-cm margin surrounding GTVprimary. If patients receive neoadjuvant chemotherapy prior to IMRT, GTV should be carefully determined according to initial diagnostic findings and not those taken after chemotherapy. Insufficient contouring toward microscopic tumor deposits modified by the chemotherapy response may decrease curability by RT, especially in the case of IMRT.

9.5.2 Clinical Target Volume (CTV)

CTV is typically arranged using automated volumetric expansion on a treatment planning system and encompasses GTV*primary*, GTV*node*, and GTVintermediate-risk areas. The CTV*primary* margin is generally defined as 10 mm in three dimensions, minus the anatomical boundaries without invasion. CTV*node* margin typically was defined as 5-mm surrounding GTV*node*, while the extent of the margins should be increased according to adverse features such as extracapsular invasion. Whether normal structures adjacent to GTV are involved or not, both the anatomical location and biological features of the primary tumor should be carefully considered for the delineation of CTV. Clinical encounters and integrated information by radiation oncologists are crucial for deciding the CTV boundary and greatly correlate with performance in IMRT. Thus, expert radiation oncologists must carefully edit *automatically made* CTV. Both CTV*primary* and CTV*node* were generally treated with a dose of approximately 70 Gy. Using SIB technique, doses of 60–63 Gy are commonly delivered to *CTVintermediate-risk.*

CTV for prophylactic nodal areas is commonly defined independently. The lymph node level varies with the T and N stages and the distribution of nodal involvement [57] (Table 9.3). Bilateral levels II to V and retropharyngeal node area will typically be included for locally advanced carcinoma. Level Ib area may also be covered in case of positive nodes in level II and/or oral cavity invasion. If lower neck node involvement is observed, the supraclavicular lymph node would be

Table 9.3 Incidence (%) of lymph node involvement according to subsites and T stage in patients with oropharyngeal cancer

	T stage	N0	N1	N2
Oropharyngeal wall	T1	75	0	25
	T2	70	10	20
	T3	33	23	45
	T4	24	24	52
Soft palate	T1	92	0	8
	T2	64	12	25
	T3	35	26	39
	T4	33	11	56
Tonsillar fossa	T1	30	41	30
	T2	33	14	54
	T3	30	18	52
	T4	11	13	77
Base of the tongue	T1	30	15	55
	T2	29	15	57
	T3	26	23	52
	T4	16	9	76

Ref. Lindberg [57]
The incidence of clinically positive lymph node of oropharyngeal cancer is reported from MD Anderson Cancer Center during two decades (1948–1965)

covered in the prophylactic nodal area. The level V area can be excluded from the prophylactic nodal area in the case of N0 disease. Unilateral nodal CTV may be considered in early and intermediate-risk patients with favorable outcomes, such as those with localized TF lesions with T1–2 N0–1 [58].

9.5.3 Planning Treatment Volume (PTV)

PTVs are defined as CTVs with adequate margin providing sufficient dose coverage and a setup error. The range of the PTV margin may vary depending on the features of the IMRT system used in each institute. Details have been described in Chap. 5.

9.5.4 Organ at Risk (OAR) and Planning Organs at Risk Volume (PRV)

Ten or more OARs are typically prepared for treatment planning. The spinal cord, brainstem, parotid gland, mandibula, inner ear, optic pathway, eye and lens, oral cavity, pharyngeal constrictor muscle, and larynx are commonly used for IMRT planning. If needed, the submandibular gland, thyroid gland, brachial plexus, masticator muscle, and temporomandibular joint can also be used. Special care should be taken for dose sparing for submandibular gland, especially in cases of ipsilateral level II involvement. Dose sparing toward submandibular gland may increase the

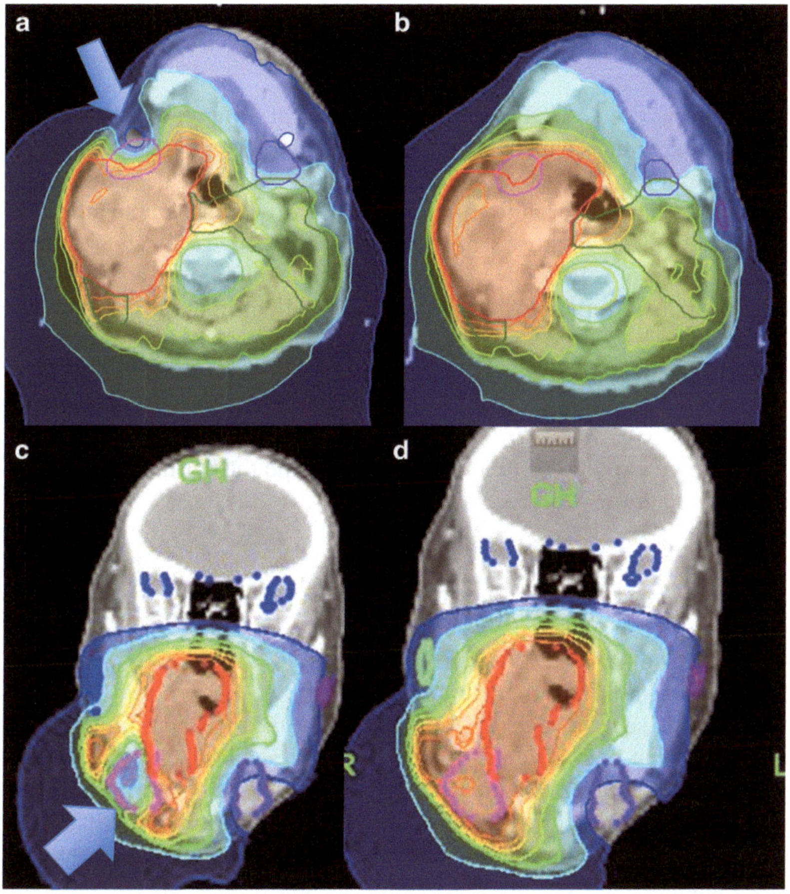

Fig. 9.3 Comparison of IMRT planning with ((**a**) axial view, (**b**) coronal view) or without ((**c**) axial view, (**d**) coronal view) submandibular gland sparing for a tonsillar cancer patient with T4aN2bM0
CTV including the right neck lesion of level II (*red*) is adjacent to the right submandibular gland (*purple*). Submandibular sparing may cause critical uncertainty for the dose distribution of CTV (*arrow*)

risk of an insufficient dose being delivered to neighboring PTV in such cases (Fig. 9.3). As for critical organs such as the spinal cord and brainstem, PRV with expanding 5 mm to the OAR is commonly added (see details in Chap. 5). Dummy objects that improve the conformal dose distribution in IMRT are typically used and depend on the institutional IMRT system including the treatment machine. For example, a dummy object on the back of the neck is commonly used to reduce the dose delivered to the soft tissue in the posterior neck region.

9.6 Case Presentation and Considerations for Treatment Planning

The treatment purpose for OPC should be somewhat different from its clinical features. A high cure rate is expected by either RT or surgery for early disease [12, 59]. Thus, the functional and cosmetic advantages of IMRT may become more important in this population. Carefully limited CTV modifications are an attractive strategy for achieving an improved organ function for low-risk patient treated with definitive RT [58, 60]. IMRT is generally used in combination with intensive chemotherapy for advanced disease [18, 19]. Some patients that received definitive RT needed to undergo salvage surgery in case of residual or locoregional recurrence, while other patients underwent primary surgery and had a consultation for postoperative (chemo-) RT because of unfavorable features, such as a positive margin status or capsular invasion [37, 38]. Strategies for advanced disease need to be documented by the multidisciplinary team board, and the harmonization of surgery, RT, and chemotherapy is a very important issue.

9.6.1 Early Stage Disease, Stages I–II

The National Comprehensive Cancer Network (NCCN) guidelines for OPC with T1–2N0–1 defined the treatment strategy as definitive RT, primary surgery with or without neck dissection, and CCRT (for T2N1 only) [61]. Similar results are expected following both RT and surgery, RT alone is considered the standard therapy [12, 59]. The number of patients with OPC that test positive for HPV has been gradually increasing, and many studies have reported favorable results from definitive RT [5, 9]. These clinical features will impact on treatment decisions in clinical practice. Functional and cosmetic issues are considered to be the clinical advantages of definitive RT over up-front surgery; thus, IMRT is highly recommended, especially for the favorable risk group.

The RTOG 00-22 was conducted to evaluate the efficacy of IMRT without chemotherapy for OPC patients at favorable risk, such as T1–2N0–1M0 [12]. A total of 66 Gy at 2.2 Gy/fraction was delivered to primary and involved node and 54–60 Gy at 1.8–2.0 Gy/fraction to subclinical risk areas. Sixty-nine patients from 14 institutes were registered in this trial. The 2-year estimated locoregional failure-free rate was 9 % at the median 2.8-year follow-up and was 6 % among patients without deviations for underdoses in quality assurance. The 2-year OS and progression-free survival rates were 95.5 % and 82 %, respectively. The rates of xerostomia of grade 2 or more at 6, 12, and 24 months were 55 %, 25 %, and 16 %, respectively. Both excellent local control and few late adverse events could be achieved with IMRT using a slightly accelerated hypofractionated schedule without chemotherapy.

As for CTV, primary lesion with prophylactic bilateral levels II–IV was commonly included in N0 cases [58]. For N1 case a level V and supraclavicular region is recommend to be included. Patients with TF at favorable risk are considered as candidates for unilateral CTV. Al-Mamgani A. et al. performed a retrospective study on 185 patients with OPC who received unilateral neck irradiation using IMRT [60]. A total of 172 (93 %) patients had T1–2 disease and 135 (73 %) had N0–1 disease, respectively. The primary site in 129 (70 %) cases was the TF and the SP in 47 (25 %). A total of 174 patients (94 %) received RT alone, and 80 patients (43 %) subsequently underwent neck dissection. Only 6 failures were reported including 2 contralateral failures. The 5-year local control and OS rates were 91 % and 70 %, respectively. Thirteen patients had grade 2 xerostomia and one had grade 3. Grade 3 dysphagia was not observed. Limited CTV for selected favorable patients is an attractive modality from the viewpoint of reducing the incidence of late adverse events (Fig. 9.4).

9.6.2 Locally Advanced Disease, Stages III–IV

The standard treatment used recently for locally advanced lesions is CCRT in patients hoping to preserve organs and/or these with unresectable lesions. The survival advantage of CCRT over that of the sequential administration of chemotherapy was confirmed in several randomized trials [50] and meta-analysis [43]. RT with CET (Bonner trial) [45] and IC containing taxane with FP (TAX 323 and 324) [47, 49] are also reasonable choices for clinical setting (Fig. 9.5).

CTV should cover primary lesion, involving the lymph nodes, and prophylactic nodal areas from levels II–V and the retropharyngeal node. Prophylactic nodal areas should cover level Ib and the supraclavicular region where needed. Altered fractionation has been shown to increase local control; however, the advantage for OS was not commonly confirmed in most randomized trials [62]. Meta-analysis using individual patient data revealed that OS improved with altered fractionation with the largest advantage in the hyperfractionation arm [62, 63]. On the other hand, altered fractionation was not beneficial for CCRT in the randomized phase III trial at the RTOG 0129 [5]; therefore, standard fractionation is used in CCRT in clinical settings.

The clinical results of IMRT in patients with OPC are summarized in the Table 9.1 [14–19]. In these series, the majority of primary sites were found in the tonsils and BOTs. Excellent local control (>90 %) and favorable OS (83–86 %) were reported in patients in whom over 70 % had stage IV disease. Many institutes used the SIB technique combined with CCRT in the reported series. IMRT led to promising results for patients with advanced OPC in clinical practice.

Multidisciplinary treatment is the mainstay for locally advanced disease, while intensive treatment may sometimes lead to an increase in toxicity [39]. Neck dissection or salvage surgery should be considered if residual tumors exist or locoregional recurrence occurs after definitive RT. Difficulty due to RT damage for surgical procedure is a well-known major problem in clinical practice. Previous studies indicated that RT with CET may be preferable under these conditions. The TRUMERIN

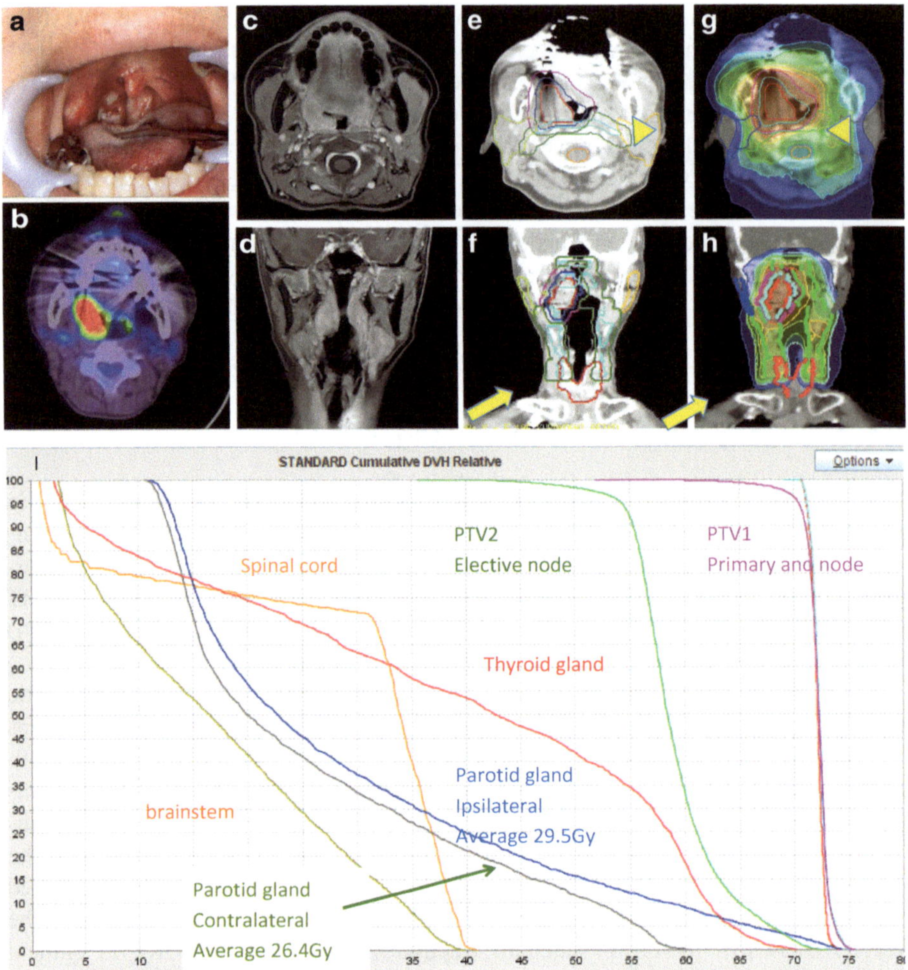

Fig. 9.4 (**a–i**) A 62-year-old female consulted our hospital due to discomfort in the pharynx and was diagnosed with right tonsillar cancer with T2N0M0. A histopathological specimen revealed that the tumor was p-16 positive. An overview photograph of the oropharynx (Fig. 9.4a), FDG-PET scan (Fig. 9.4b), and MRI with contrast media in axial (Fig. 9.4c) and coronal (Fig. 9.4d) images. CTV1 (*blue*) encompassed the primary tumor (*red*) with a 1-cm margin, and PTV1 expanded by 5 mm to surround CTV1 (Fig. 9.4e). CTV2 (*cyan*) included levels II–IV and the retropharyngeal node. Considering a favorable outcome with clinical features, a part of level IV at the middle level of the thyroid gland was spared from CTV2 (*arrow*) to reduce the thyroid dose (Fig. 9.4f). Using the SIB technique, 70 Gy for CTV1 and 54 Gy for CTV2 were delivered in 33 fractions over 6 weeks. On the axial image of dose distribution (Fig. 9.4g), CTV 1 was sufficiently covered at the 95 % dose line (*red*), with dose sparing the lateral part of both parotid glands (*arrow head*). The lower part of the thyroid gland could also keep a lower dose level (Fig. 9.4h). IMRT was performed using helical tomotherapy without a junction to photon portals to the lower neck. The DVH of this case (Fig. 9.4i) showed a conformal dose coverage in PTV1-2, and several OARs (the parotid glands, spinal cord, stem, and thyroid gland) met desirable dose constraints. She had no evidence of disease 4 years after the completion of radiotherapy without any evidence of xerostomia or hypothyroidism

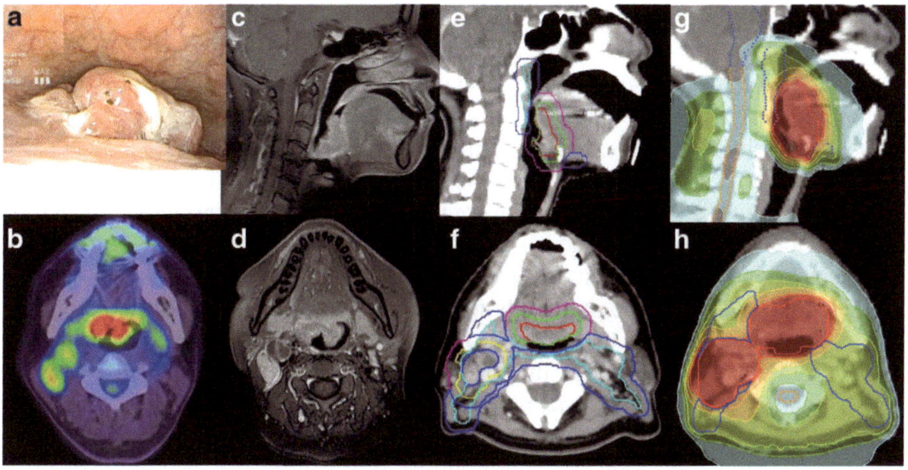

Fig. 9.5 (**a–h**) A 42-year-old man presented with otalgia, discomfort in the throat, and right neck mass. Endoscopy of the pharynx revealed a mass on the bottom of the tongue (Fig. 9.5a), and a biopsy specimen from the tumor revealed poorly differentiated SCC that was positive for HPV. Both FDG-PET scans (Fig. 9.5b) and MRI with contrast media (Fig. 9.5c sagittal view; 9.5d axial view) showed a bulky primary lesion at the bottom of the tongue, and right lymphadenopathy was performed at both levels II and III. He was diagnosed at the bottom of the tongue cancer with T3N2bM0; stage IVA disease (UICC v7). Since the patient requests organ-preserving therapy, two cycles of induction chemotherapy including docetaxel, cisplatin, and 5-FU were administered, which led to a partial response in both the primary lesion and neck node. Definitive radiotherapy was undergone with concurrent administration of cisplatin. CTVprimary (*green*) and CTVnode (*yellow*) encompassed a 0.5–1-cm margin toward GTVprimary (*red*) and GTVnode (*blue*), respectively (Fig. 9.5e sagittal view; 9.5f axial view). CTV1 comprised both CTVprimary and CTVnode, while CTV2 (*cyan*) included levels II–V, the retropharyngeal node, and the supraclavicular region. IMRT planning by helical tomotherapy was prepared using treatment planning system (Fig. 9.5g sagittal view; 9.5h axial view). Using the SIB technique, 70 Gy for CTV1 and 56 Gy for CTV2 were delivered in 35 fractions over 7 weeks. On the axial images for the dose distribution (Fig. 9.5g), both the primary lesion and involved node were covered with the 95 % dose line of prescribed dose (*red*). The mean doses of both parotid glands were 36.1 Gy (*right*) and 23.8 Gy (*light*), respectively. Dose constraints of other OARs (spinal cord, brain stem, mandibula, larynx, and pharyngeal constrictor muscle) met desirable requirement. He recovered well and neither severe dysphagia aspiration nor disease recurrence was observed 4 years after chemoradiotherapy

trial is a randomized phase II study that was conducted to evaluate the efficacy of RT with CET in relation to CCRT in locally advanced patients [64]. A total of 153 patients received 3 cycles of IC containing docetaxel, 5-FU, and CP, and 116 patients who achieved sufficient responses were randomized to receive CCRT ($n=60$) or CET-RT ($n=56$). No significant differences were observed in the OS rates between both groups. Compliance of planned schedule was better in the CET-RT arm than in the CCRT arm and salvage surgery was shown to be feasible in the CET arm. Since intensive multidisciplinary treatment may often increase its treatment toxicity, the application of CET is an attractive strategy to minimize the incidence of adverse events without sacrificing efficacy.

9.7 Toxicity

Acute toxicity is characterized by dermatitis, taste impairment, mucosal reaction with pain, fatigue, and weight loss. Intensive treatment is known to increase the risk of acute toxicity. Smoking cessation is important not only to reduce toxicity but also to retain the efficacy of RT. The prevention and management of mucositis are essential, and systemic oral care hygiene in combination with interventions using analgesics is recommended with the involvement of a multidisciplinary team (nurses, physician, dentist, nutritionist, and pharmacist).

Late toxicities are less common; however, they can be irreversible. Xerostomia is a common late toxicity in the salivary gland. The clinical advantage in parotid sparing with IMRT is supported by the findings of several prospective studies [13, 20, 21, 23] and is considered the standard management for definitive RT [65]. A mean dose <26 Gy has been recommended for the parotid gland in order to preserve the recovery of stimulated salivary excretion [40, 66]. Murdoch-Kinch et al. reported that static salivary excretion can recover, if the dose delivered to the submandibular gland is kept under 39 Gy [67]. Sparing submandibular function is an interesting method [68], but special care should be taken for dose coverage in neighboring CTV especially for level II areas (Fig. 9.3).

An increase in the incidence of aspiration and/or dysphagia has been associated with treatment intensification [39, 69, 70]. Dose constraints for the pharyngeal constrictor muscle and larynx were shown to be risk factors for aspiration and/or dysphagia [71–73]. IMRT is believed to have a distinct benefit of reducing long-term adverse events. Late adverse events associated with RT include carotid stenosis [74, 75] and hypothyroidism [76, 77]. Long-term survivors following RT should be examined thoroughly for these late events in the follow-up.

9.8 Conclusions

IMRT is a promising modality for both early and locally advanced OPC from the viewpoint of disease control and survival with significantly lower toxicities. Definitive RT for OPC is advantageous for functional preservation; however, further improvements are still needed to reduce the incidence of late adverse events. Improvements in the incidence of late adverse events for early disease by customizing CTV according to the disease features could be developed in a sophisticated clinical trial. Regarding locally advanced disease, both an increase in treatment efficacy, especially for high-risk disease, and minimizing adverse events using adaptive radiotherapy will become more important issues. Customized treatment strategies based on risk factors including biological features such as biomarkers are an attractive concept that may maximize the advantages of definitive IMRT for patients with OPC in the future.

Acknowledgments This study was supported in part by a Grant-in-Aid for Cancer Research (H23-009) from the Ministry of Health, Labour and Welfare of Japan.

References

1. O'Rorke MA, Ellison MV, Murray LJ, Moran M, James J, Anderson LA (2012) Human papillomavirus related head and neck cancer survival: a systematic review and meta-analysis. Oral Oncol 48(12):1191–1201. doi:10.1016/j.oraloncology.2012.06.019
2. Tribius S, Hoffmann M (2013) Human papilloma virus infection in head and neck cancer. Deutsches Arzteblatt Int 110(11):184–190, 190e181. doi:10.3238/arztebl.2013.0184
3. Cancer Registry Committee, Cancer JSfHaN (2006) Report of Head and Neck Cancer Registry of Japan Clinical Statistics of Registered Patients, 2002
4. Cancer Registry Committee, Cancer JSfHaN (2007) Report of Head and Neck Cancer Registry of Japan: clinical statistics of registered patients, 2003
5. Ang KK, Harris J, Wheeler R, Weber R, Rosenthal DI, Nguyen-Tan PF, Westra WH, Chung CH, Jordan RC, Lu C, Kim H, Axelrod R, Silverman CC, Redmond KP, Gillison ML (2010) Human papillomavirus and survival of patients with oropharyngeal cancer. N Engl J Med 363(1):24–35. doi:10.1056/NEJMoa0912217
6. Fakhry C, Westra WH, Li S, Cmelak A, Ridge JA, Pinto H, Forastiere A, Gillison ML (2008) Improved survival of patients with human papillomavirus-positive head and neck squamous cell carcinoma in a prospective clinical trial. J Natl Cancer Inst 100(4):261–269. doi:10.1093/jnci/djn011
7. Lassen P, Eriksen JG, Hamilton-Dutoit S, Tramm T, Alsner J, Overgaard J (2009) Effect of HPV-associated p16INK4A expression on response to radiotherapy and survival in squamous cell carcinoma of the head and neck. J Clin Oncol 27(12):1992–1998. doi:10.1200/JCO.2008.20.2853
8. Nichols AC, Faquin WC, Westra WH, Mroz EA, Begum S, Clark JR, Rocco JW (2009) HPV-16 infection predicts treatment outcome in oropharyngeal squamous cell carcinoma. Otolaryngol Head Neck Surg Off J Am Acad Otolaryngol Head Neck Surg 140(2):228–234. doi:10.1016/j.otohns.2008.11.025
9. Rischin D, Young RJ, Fisher R, Fox SB, Le QT, Peters LJ, Solomon B, Choi J, O'Sullivan B, Kenny LM, McArthur GA (2010) Prognostic significance of p16INK4A and human papillomavirus in patients with oropharyngeal cancer treated on TROG 02.02 phase III trial. J Clin Oncol 28(27):4142–4148. doi:10.1200/JCO.2010.29.2904. RTOG 1016. http://clinicaltrials.gov/show/NCT01302834
10. Shi W, Kato H, Perez-Ordonez B, Pintilie M, Huang S, Hui A, O'Sullivan B, Waldron J, Cummings B, Kim J, Ringash J, Dawson LA, Gullane P, Siu L, Gillison M, Liu FF (2009) Comparative prognostic value of HPV16 E6 mRNA compared with in situ hybridization for human oropharyngeal squamous carcinoma. J Clin Oncol 27(36):6213–6221. doi:10.1200/JCO.2009.23.1670
11. Chao KS, Majhail N, Huang CJ, Simpson JR, Perez CA, Haughey B, Spector G (2001) Intensity-modulated radiation therapy reduces late salivary toxicity without compromising tumor control in patients with oropharyngeal carcinoma: a comparison with conventional techniques. Radiother Oncol 61(3):275–280
12. Eisbruch A, Harris J, Garden AS, Chao CK, Straube W, Harari PM, Sanguineti G, Jones CU, Bosch WR, Ang KK (2010) Multi-institutional trial of accelerated hypofractionated intensity-modulated radiation therapy for early-stage oropharyngeal cancer (RTOG 00-22). Int J Radiat Oncol Biol Phys 76(5):1333–1338. doi:10.1016/j.ijrobp.2009.04.011
13. Nutting CM, Morden JP, Harrington KJ, Urbano TG, Bhide SA, Clark C, Miles EA, Miah AB, Newbold K, Tanay M, Adab F, Jefferies SJ, Scrase C, Yap BK, A'Hern RP, Sydenham MA, Emson M, Hall E (2011) Parotid-sparing intensity modulated versus conventional radiotherapy in head and neck cancer (PARSPORT): a phase 3 multicentre randomised controlled trial. Lancet Oncol 12(2):127–136. doi:10.1016/S1470-2045(10)70290-4
14. Clavel S, Nguyen DH, Fortin B, Despres P, Khaouam N, Donath D, Soulieres D, Guertin L, Nguyen-Tan PF (2012) Simultaneous integrated boost using intensity-modulated radiotherapy compared with conventional radiotherapy in patients treated with concurrent carboplatin and

5-fluorouracil for locally advanced oropharyngeal carcinoma. Int J Radiat Oncol Biol Phys 82(2):582–589. doi:10.1016/j.ijrobp.2010.10.061
15. Daly ME, Le QT, Maxim PG, Loo BW Jr, Kaplan MJ, Fischbein NJ, Pinto H, Chang DT (2010) Intensity-modulated radiotherapy in the treatment of oropharyngeal cancer: clinical outcomes and patterns of failure. Int J Radiat Oncol Biol Phys 76(5):1339–1346. doi:10.1016/j.ijrobp.2009.04.006
16. Garden AS, Dong L, Morrison WH, Stugis EM, Glisson BS, Frank SJ, Beadle BM, Gunn GB, Schwartz DL, Kies MS, Weber RS, Ang KK, Rosenthal DI (2013) Patterns of disease recurrence following treatment of oropharyngeal cancer with intensity modulated radiation therapy. Int J Radiat Oncol Biol Phys 85(4):941–947. doi:10.1016/j.ijrobp.2012.08.004
17. Huang K, Xia P, Chuang C, Weinberg V, Glastonbury CM, Eisele DW, Lee NY, Yom SS, Phillips TL, Quivey JM (2008) Intensity-modulated chemoradiation for treatment of stage III and IV oropharyngeal carcinoma: the University of California-San Francisco experience. Cancer 113(3):497–507. doi:10.1002/cncr.23578
18. Setton J, Caria N, Romanyshyn J, Koutcher L, Wolden SL, Zelefsky MJ, Rowan N, Sherman EJ, Fury MG, Pfister DG, Wong RJ, Shah JP, Kraus DH, Shi W, Zhang Z, Schupak KD, Gelblum DY, Rao SD, Lee NY (2012) Intensity-modulated radiotherapy in the treatment of oropharyngeal cancer: an update of the Memorial Sloan-Kettering Cancer Center experience. Int J Radiat Oncol Biol Phys 82(1):291–298. doi:10.1016/j.ijrobp.2010.10.041
19. Sher DJ, Thotakura V, Balboni TA, Norris CM, Haddad RI, Posner MR, Lorch J, Goguen LA, Annino DJ, Tishler RB (2012) Treatment of oropharyngeal squamous cell carcinoma with IMRT: patterns of failure after concurrent chemoradiotherapy and sequential therapy. Ann Oncol Off J Eur Soc Med Oncol/ESMO 23(9):2391–2398. doi:10.1093/annonc/mdr609
20. Gupta T, Agarwal J, Jain S, Phurailatpam R, Kannan S, Ghosh-Laskar S, Murthy V, Budrukkar A, Dinshaw K, Prabhash K, Chaturvedi P, D'Cruz A (2012) Three-dimensional conformal radiotherapy (3D-CRT) versus intensity modulated radiation therapy (IMRT) in squamous cell carcinoma of the head and neck: a randomized controlled trial. Radiother Oncol 104(3):343–348. doi:10.1016/j.radonc.2012.07.001
21. Kam MK, Leung SF, Zee B, Chau RM, Suen JJ, Mo F, Lai M, Ho R, Cheung KY, Yu BK, Chiu SK, Choi PH, Teo PM, Kwan WH, Chan AT (2007) Prospective randomized study of intensity-modulated radiotherapy on salivary gland function in early-stage nasopharyngeal carcinoma patients. J Clin Oncol 25(31):4873–4879. doi:10.1200/JCO.2007.11.5501
22. Lee NY, de Arruda FF, Puri DR, Wolden SL, Narayana A, Mechalakos J, Venkatraman ES, Kraus D, Shaha A, Shah JP, Pfister DG, Zelefsky MJ (2006) A comparison of intensity-modulated radiation therapy and concomitant boost radiotherapy in the setting of concurrent chemotherapy for locally advanced oropharyngeal carcinoma. Int J Radiat Oncol Biol Phys 66(4):966–974. doi:10.1016/j.ijrobp.2006.06.040
23. Pow EH, Kwong DL, McMillan AS, Wong MC, Sham JS, Leung LH, Leung WK (2006) Xerostomia and quality of life after intensity-modulated radiotherapy vs. conventional radiotherapy for early-stage nasopharyngeal carcinoma: initial report on a randomized controlled clinical trial. Int J Radiat Oncol Biol Phys 66(4):981–991. doi:10.1016/j.ijrobp.2006.06.013
24. Ang K, Zhang Q, Wheeler RH, Rosenthal DI, Nguyen-Tan F, Kim H, Lu C, Axelrod RS, Silverman CI, Weber RS (2010a) A phase III trial (RTOG 0129) of two radiation-cisplatin regimens for head and neck carcinomas (HNC): impact of radiation and cisplatin intensity on outcome. J Clin Oncol 28:15s:(suppl; abstr 5507)
25. Head and Neck Tumours; Pharynx (2009) In: Sobin LH, Gospodarowcz MK, Wittekind C (eds) TNM classification of malignant tumors, 7th edn. Wiley-Blackwell, Hoboken, pp 30–38
26. Leon X, Quer M, Diez S, Orus C, Lopez-Pousa A, Burgues J (1999) Second neoplasm in patients with head and neck cancer. Head Neck 21(3):204–210
27. Wanders LK, East JE, Uitentuis SE, Leeflang MM, Dekker E (2013) Diagnostic performance of narrowed spectrum endoscopy, autofluorescence imaging, and confocal laser endomicroscopy for optical diagnosis of colonic polyps: a meta-analysis. Lancet Oncol 14(13):1337–1347. doi:10.1016/S1470-2045(13)70509-6

28. Lell MM, Meyer E, Kuefner MA, May MS, Raupach R, Uder M, Kachelriess M (2012) Normalized metal artifact reduction in head and neck computed tomography. Investig Radiol 47(7):415–421. doi:10.1097/RLI.0b013e3182532f17
29. Yongkui L, Jian L, Wanghan, Jingui L (2013) 18FDG-PET/CT for the detection of regional nodal metastasis in patients with primary head and neck cancer before treatment: a meta-analysis. Surg Oncol 22(2):e11–e16. doi:10.1016/j.suronc.2013.02.002
30. Xu G, Li J, Zuo X, Li C (2012) Comparison of whole body positron emission tomography (PET)/PET-computed tomography and conventional anatomic imaging for detecting distant malignancies in patients with head and neck cancer: a meta-analysis. Laryngoscope 122(9):1974–1978. doi:10.1002/lary.23409
31. Dornfeld K, Hopkins S, Simmons J, Spitz DR, Menda Y, Graham M, Smith R, Funk G, Karnell L, Karnell M, Dornfeld M, Yao M, Buatti J (2008) Posttreatment FDG-PET uptake in the supraglottic and glottic larynx correlates with decreased quality of life after chemoradiotherapy. Int J Radiat Oncol Biol Phys 71(2):386–392. doi:S0360-3016(07)04371-4, [pii] 10.1016/j.ijrobp.2007.09.052
32. Kruser TJ, Bradley KA, Bentzen SM, Anderson BM, Gondi V, Khuntia D, Perlman SB, Tome WA, Chappell RJ, Walker WL, Mehta MP (2009) The impact of hybrid PET-CT scan on overall oncologic management, with a focus on radiotherapy planning: a prospective, blinded study. Technol Cancer Res Treat 8(2):149–158
33. Bourhis J, Overgaard J, Audry H, Ang KK, Saunders M, Bernier J, Horiot JC, Le Maitre A, Pajak TF, Poulsen MG, O'Sullivan B, Dobrowsky W, Hliniak A, Skladowski K, Hay JH, Pinto LH, Fallai C, Fu KK, Sylvester R, Pignon JP (2006) Hyperfractionated or accelerated radiotherapy in head and neck cancer: a meta-analysis. Lancet 368(9538):843–854
34. Eisbruch A, Levendag PC, Feng FY, Teguh D, Lyden T, Schmitz PI, Haxer M, Noever I, Chepeha DB, Heijmen BJ (2007) Can IMRT or brachytherapy reduce dysphagia associated with chemoradiotherapy of head and neck cancer? The Michigan and Rotterdam experiences. Int J Radiat Oncol Biol Phys 69(2 Suppl):S40–S42. doi:S0360-3016(07)00952-2, [pii] 10.1016/j.ijrobp.2007.04.083
35. Nijdam W, Levendag P, Fuller D, Schulz R, Prevost JB, Noever I, Uyl-de Groot C (2007) Robotic radiosurgery vs. brachytherapy as a boost to intensity modulated radiotherapy for tonsillar fossa and soft palate tumors: the clinical and economic impact of an emerging technology. Technol Cancer Res Treat 6(6):611–620. doi:d=3029&c=4245&p=16281&do=detail [pii]
36. Nichols AC, Yoo J, Hammond JA, Fung K, Winquist E, Read N, Venkatesan V, MacNeil SD, Ernst DS, Kuruvilla S, Chen J, Corsten M, Odell M, Eapen L, Theurer J, Doyle PC, Wehrli B, Kwan K, Palma DA (2013) Early-stage squamous cell carcinoma of the oropharynx: radiotherapy vs. trans-oral robotic surgery (ORATOR) – study protocol for a randomized phase II trial. BMC Cancer 13:133. doi:10.1186/1471-2407-13-133
37. Bernier J, Domenge C, Ozsahin M, Matuszewska K, Lefebvre JL, Greiner RH, Giralt J, Maingon P, Rolland F, Bolla M, Cognetti F, Bourhis J, Kirkpatrick A, van Glabbeke M (2004) Postoperative irradiation with or without concomitant chemotherapy for locally advanced head and neck cancer. N Engl J Med 350(19):1945–1952
38. Cooper JS, Pajak TF, Forastiere AA, Jacobs J, Campbell BH, Saxman SB, Kish JA, Kim HE, Cmelak AJ, Rotman M, Machtay M, Ensley JF, Chao KS, Schultz CJ, Lee N, Fu KK (2004) Postoperative concurrent radiotherapy and chemotherapy for high-risk squamous-cell carcinoma of the head and neck. N Engl J Med 350(19):1937–1944
39. Machtay M, Moughan J, Trotti A, Garden AS, Weber RS, Cooper JS, Forastiere A, Ang KK (2008) Factors associated with severe late toxicity after concurrent chemoradiation for locally advanced head and neck cancer: an RTOG analysis. J Clin Oncol 26(21):3582–3589. doi:JCO.2007.14.8841, [pii] 10.1200/JCO.2007.14.8841
40. O'Daniel JC, Garden AS, Schwartz DL, Wang H, Ang KK, Ahamad A, Rosenthal DI, Morrison WH, Asper JA, Zhang L, Tung SM, Mohan R, Dong L (2007) Parotid gland dose in

intensity-modulated radiotherapy for head and neck cancer: is what you plan what you get? Int J Radiat Oncol Biol Phys 69(4):1290–1296. doi:10.1016/j.ijrobp.2007.07.2345
41. Wang ZH, Yan C, Zhang ZY, Zhang CP, Hu HS, Tu WY, Kirwan J, Mendenhall WM (2011) Impact of salivary gland dosimetry on post-IMRT recovery of saliva output and xerostomia grade for head-and-neck cancer patients treated with or without contralateral submandibular gland sparing: a longitudinal study. Int J Radiat Oncol Biol Phys 81(5):1479–1487. doi:10.1016/j.ijrobp.2010.07.1990
42. Pignon JP, Bourhis J, Domenge C, Designe L (2000) Chemotherapy added to locoregional treatment for head and neck squamous-cell carcinoma: three meta-analyses of updated individual data. MACH-NC Collaborative Group. Meta-Analysis of Chemotherapy on Head and Neck Cancer. Lancet 355(9208):949–955
43. Pignon JP, le Maitre A, Maillard E, Bourhis J (2009) Meta-analysis of chemotherapy in head and neck cancer (MACH-NC): an update on 93 randomised trials and 17,346 patients. Radiother Oncol 92(1):4–14. doi:10.1016/j.radonc.2009.04.014
44. Bonner JA, Harari PM, Giralt J, Azarnia N, Shin DM, Cohen RB, Jones CU, Sur R, Raben D, Jassem J, Ove R, Kies MS, Baselga J, Youssoufian H, Amellal N, Rowinsky EK, Ang KK (2006) Radiotherapy plus cetuximab for squamous-cell carcinoma of the head and neck. N Engl J Med 354(6):567–578
45. Bonner JA, Harari PM, Giralt J, Cohen RB, Jones CU, Sur RK, Raben D, Baselga J, Spencer SA, Zhu J, Youssoufian H, Rowinsky EK, Ang KK (2010) Radiotherapy plus cetuximab for locoregionally advanced head and neck cancer: 5-year survival data from a phase 3 randomised trial, and relation between cetuximab-induced rash and survival. Lancet Oncol 11(1):21–28. doi:10.1016/S1470-2045(09)70311-0
46. Ang KK, Zhang QE, Rosenthal DI, Nguyen-Tan P, Sherman EJ, Weber RS, Galvin JM, Schwartz DL, El-Naggar AK, Gillison ML, Jordan R, List MA, Konski AA, Thorstad WL, Trotti A, Beitler JJ, Garden AS, Spanos WJ, Yom SS, Axelrod RS (2011) A randomized phase III trial (RTOG 0522) of concurrent accelerated radiation plus cisplatin with or without cetuximab for stage III-IV head and neck squamous cell carcinomas (HNC). J Clin Oncol 29:(suppl; abstr 5500)
47. Posner MR, Hershock DM, Blajman CR, Mickiewicz E, Winquist E, Gorbounova V, Tjulandin S, Shin DM, Cullen K, Ervin TJ, Murphy BA, Raez LE, Cohen RB, Spaulding M, Tishler RB, Roth B, Viroglio Rdel C, Venkatesan V, Romanov I, Agarwala S, Harter KW, Dugan M, Cmelak A, Markoe AM, Read PW, Steinbrenner L, Colevas AD, Norris CM Jr, Haddad RI (2007) Cisplatin and fluorouracil alone or with docetaxel in head and neck cancer. N Engl J Med 357(17):1705–1715
48. Vermorken JB, Remenar E, van Herpen C, Gorlia T, Mesia R, Degardin M, Stewart JS, Jelic S, Betka J, Preiss JH, van den Weyngaert D, Awada A, Cupissol D, Kienzer HR, Rey A, Desaunois I, Bernier J, Lefebvre JL (2007) Cisplatin, fluorouracil, and docetaxel in unresectable head and neck cancer. N Engl J Med 357(17):1695–1704
49. Blanchard P, Bourhis J, Lacas B, Posner MR, Vermorken JB, Hernandez JJ, Bourredjem A, Calais G, Paccagnella A, Hitt R, Pignon JP (2013) Taxane-cisplatin-fluorouracil as induction chemotherapy in locally advanced head and neck cancers: an individual patient data meta-analysis of the meta-analysis of chemotherapy in head and neck cancer group. J Clin Oncol 31(23):2854–2860. doi:10.1200/JCO.2012.47.7802
50. Forastiere AA, Goepfert H, Maor M, Pajak TF, Weber R, Morrison W, Glisson B, Trotti A, Ridge JA, Chao C, Peters G, Lee DJ, Leaf A, Ensley J, Cooper J (2003) Concurrent chemotherapy and radiotherapy for organ preservation in advanced laryngeal cancer. N Engl J Med 349(22):2091–2098
51. The Department of Veterans Affairs Laryngeal Cancer Study Group (1991) Induction chemotherapy plus radiation compared with surgery plus radiation in patients with advanced laryngeal cancer. N Engl J Med 324(24):1685–1690
52. Lefebvre JL, Chevalier D, Luboinski B, Kirkpatrick A, Collette L, Sahmoud T (1996) Larynx preservation in pyriform sinus cancer: preliminary results of a European Organization for

Research and Treatment of Cancer phase III trial. EORTC Head and Neck Cancer Cooperative Group. J Natl Cancer Inst 88(13):890–899
53. Machtay M, Rosenthal DI, Hershock D, Jones H, Williamson S, Greenberg MJ, Weinstein GS, Aviles VM, Chalian AA, Weber RS (2002) Organ preservation therapy using induction plus concurrent chemoradiation for advanced resectable oropharyngeal carcinoma: a University of Pennsylvania Phase II trial. J Clin Oncol 20(19):3964–3971
54. Nishimura Y, Shibata T, Nakamatsu K, Kanamori S, Koike R, Okubo M, Nishikawa T, Tachibana I, Tamura M, Okumura M (2010) A two-step intensity-modulated radiation therapy method for nasopharyngeal cancer: the Kinki University experience. Jpn J Clin Oncol 40(2):130–138. doi:10.1093/jjco/hyp136
55. Schwartz DL, Garden AS, Shah SJ, Chronowski G, Sejpal S, Rosenthal DI, Chen Y, Zhang Y, Zhang L, Wong PF, Garcia JA, Kian Ang K, Dong L (2013) Adaptive radiotherapy for head and neck cancer–dosimetric results from a prospective clinical trial. Radiother Oncol 106(1):80–84. doi:10.1016/j.radonc.2012.10.010
56. Yang H, Hu W, Wang W, Chen P, Ding W, Luo W (2013) Replanning during intensity modulated radiation therapy improved quality of life in patients with nasopharyngeal carcinoma. Int J Radiat Oncol Biol Phys 85(1):e47–e54. doi:10.1016/j.ijrobp.2012.09.033
57. Lindberg R (1972) Distribution of cervical lymph node metastases from squamous cell carcinoma of the upper respiratory and digestive tracts. Cancer 29(6):1446–1449
58. Yeung AR, Garg MK, Lawson J, McDonald MW, Quon H, Ridge JA, Saba N, Salama JK, Smith RV, Yom SS, Beitler JJ (2012) ACR appropriateness criteria(R) ipsilateral radiation for squamous cell carcinoma of the tonsil. Head Neck 34(5):613–616. doi:10.1002/hed.21993
59. Parsons JT, Mendenhall WM, Stringer SP, Amdur RJ, Hinerman RW, Villaret DB, Moore-Higgs GJ, Greene BD, Speer TW, Cassisi NJ, Million RR (2002) Squamous cell carcinoma of the oropharynx: surgery, radiation therapy, or both. Cancer 94(11):2967–2980. doi:10.1002/cncr.10567
60. Al-Mamgani A, van Rooij P, Fransen D, Levendag P (2013) Unilateral neck irradiation for well-lateralized oropharyngeal cancer. Radiother Oncol 106(1):69–73. doi:10.1016/j.radonc.2012.12.006
61. Pfister DG, Ang KK, Brizel DM, Burtness BA, Busse PM, Caudell JJ, Cmelak AJ, Colevas AD, Dunphy F, Eisele DW, Gilvert J, Gillison ML, Haddad RI, Haughey BH, Hicks WL, Hitchcock YJ, Kies MS, Lydiatt WM, Maghami E, Martins R, McCaffrey T, Mittal BB, Pinto HA, Ridge JA, Samant S, Schuller DE, Shah JP, Spencer S, Weber RS, Wolf GT, Worden F, Yom SS, Gallagher L, Hughes M, McMillian N (2013) NCCN clinical practice guidelines in oncology; head and neck cancers: cancer of the oropharynx
62. Brizel DM, Albers ME, Fisher SR, Scher RL, Richtsmeier WJ, Hars V, George SL, Huang AT, Prosnitz LR (1998) Hyperfractionated irradiation with or without concurrent chemotherapy for locally advanced head and neck cancer. N Engl J Med 338(25):1798–1804
63. Bourhis J, Sire C, Graff P, Gregoire V, Maingon P, Calais G, Gery B, Martin L, Alfonsi M, Desprez P, Pignon T, Bardet E, Rives M, Geoffrois L, Daly-Schveitzer N, Sen S, Tuchais C, Dupuis O, Guerif S, Lapeyre M, Favrel V, Hamoir M, Lusinchi A, Temam S, Pinna A, Tao YG, Blanchard P, Auperin A (2012) Concomitant chemoradiotherapy versus acceleration of radiotherapy with or without concomitant chemotherapy in locally advanced head and neck carcinoma (GORTEC 99-02): an open-label phase 3 randomised trial. Lancet Oncol 13(2):145–153. doi:10.1016/S1470-2045(11)70346-1
64. Lefebvre JL, Pointreau Y, Rolland F, Alfonsi M, Baudoux A, Sire C, de Raucourt D, Malard O, Degardin M, Tuchais C, Blot E, Rives M, Reyt E, Tourani JM, Geoffrois L, Peyrade F, Guichard F, Chevalier D, Babin E, Lang P, Janot F, Calais G, Garaud P, Bardet E (2013) Induction chemotherapy followed by either chemoradiotherapy or bioradiotherapy for larynx preservation: the TREMPLIN randomized phase II study. J Clin Oncol 31(7):853–859. doi:10.1200/JCO.2012.42.3988
65. Lee TF, Fang FM (2013) Quantitative analysis of normal tissue effects in the clinic (QUANTEC) guideline validation using quality of life questionnaire datasets for parotid gland constraints to

avoid causing xerostomia during head-and-neck radiotherapy. Radiother Oncol 106(3):352–358. doi:10.1016/j.radonc.2012.11.013
66. Eisbruch A, Ten Haken RK, Kim HM, Marsh LH, Ship JA (1999) Dose, volume, and function relationships in parotid salivary glands following conformal and intensity-modulated irradiation of head and neck cancer. Int J Radiat Oncol Biol Phys 45(3):577–587
67. Murdoch-Kinch CA, Kim HM, Vineberg KA, Ship JA, Eisbruch A (2008) Dose-effect relationships for the submandibular salivary glands and implications for their sparing by intensity modulated radiotherapy. Int J Radiat Oncol Biol Phys 72(2):373–382. doi:10.1016/j.ijrobp.2007.12.033
68. Saarilahti K, Kouri M, Collan J, Kangasmaki A, Atula T, Joensuu H, Tenhunen M (2006) Sparing of the submandibular glands by intensity modulated radiotherapy in the treatment of head and neck cancer. Radiother Oncol 78(3):270–275. doi:10.1016/j.radonc.2006.02.017
69. Caudell JJ, Schaner PE, Meredith RF, Locher JL, Nabell LM, Carroll WR, Magnuson JS, Spencer SA, Bonner JA (2009) Factors associated with long-term dysphagia after definitive radiotherapy for locally advanced head-and-neck cancer. Int J Radiat Oncol Biol Phys 73(2):410–415
70. van der Molen L, Heemsbergen WD, de Jong R, van Rossum MA, Smeele LE, Rasch CR, Hilgers FJ (2013) Dysphagia and trismus after concomitant chemo-Intensity-Modulated Radiation Therapy (chemo-IMRT) in advanced head and neck cancer; dose-effect relationships for swallowing and mastication structures. Radiother Oncol 106(3):364–369. doi:10.1016/j.radonc.2013.03.005
71. Caglar HB, Tishler RB, Othus M, Burke E, Li Y, Goguen L, Wirth LJ, Haddad RI, Norris CM, Court LE, Aninno DJ, Posner MR, Allen AM (2008) Dose to larynx predicts for swallowing complications after intensity-modulated radiotherapy. Int J Radiat Oncol Biol Phys 72(4):1110–1118
72. Feng FY, Kim HM, Lyden TH, Haxer MJ, Feng M, Worden FP, Chepeha DB, Eisbruch A (2007) Intensity-modulated radiotherapy of head and neck cancer aiming to reduce dysphagia: early dose-effect relationships for the swallowing structures. Int J Radiat Oncol Biol Phys 68(5):1289–1298. doi:10.1016/j.ijrobp.2007.02.049
73. Levendag PC, Teguh DN, Voet P, van der Est H, Noever I, de Kruijf WJ, Kolkman-Deurloo IK, Prevost JB, Poll J, Schmitz PI, Heijmen BJ (2007) Dysphagia disorders in patients with cancer of the oropharynx are significantly affected by the radiation therapy dose to the superior and middle constrictor muscle: a dose-effect relationship. Radiother Oncol 85(1):64–73. doi:S0167-8140(07)00345-3, [pii] 10.1016/j.radonc.2007.07.009
74. Gujral DM, Chahal N, Senior R, Harrington KJ, Nutting CM (2013) Radiation-induced carotid artery atherosclerosis. Radiother Oncol 110(1):31–38. doi:10.1016/j.radonc.2013.08.009
75. Smith GL, Smith BD, Buchholz TA, Giordano SH, Garden AS, Woodward WA, Krumholz HM, Weber RS, Ang KK, Rosenthal DI (2008) Cerebrovascular disease risk in older head and neck cancer patients after radiotherapy. J Clin Oncol 26(31):5119–5125. doi:10.1200/JCO.2008.16.6546
76. Boomsma MJ, Bijl HP, Langendijk JA (2011) Radiation-induced hypothyroidism in head and neck cancer patients: a systematic review. Radiother Oncol 99(1):1–5. doi:10.1016/j.radonc.2011.03.002
77. Vogelius IR, Bentzen SM, Maraldo MV, Petersen PM, Specht L (2011) Risk factors for radiation-induced hypothyroidism: a literature-based meta-analysis. Cancer 117(23):5250–5260. doi:10.1002/cncr.26186

Postoperative Intensity-Modulated Radiation Therapy for Head and Neck Cancers: A Case-Based Review

10

G. Brandon Gunn and Adam S. Garden

Keywords
Postoperative • Intensity-modulated radiation therapy • Head and neck • Target volumes

10.1 Introduction

Certain subsets of primary neoplasms of the head and neck, influenced by the histologic type, anatomic site/location, and tumor extent, are preferably managed with a primary surgical approach. Depending on the clinical, surgical, and pathology findings, postoperative radiation therapy (or chemoradiation) is frequently implemented to improve the local and/or regional control and possibly survival. Since postoperative radiation therapy targets of the head and neck region are often in close proximity to nearby critical and avoidance structures, intensity-modulated radiation therapy (IMRT) is commonly preferred primarily due to (1) the ability to conform the high-dose regions around geometrically complex targets, (2) the ability to build steep dose gradients between targets and nearby critical structures, and (3) the ability to generate relative sparing of various surrounding nontarget normal structures from clinically significant doses without compromise of desired target coverage.

Over the past decade or so, IMRT has become widely implemented into clinical practice for head and neck cancers in the United States [1] and is now the standard technique of postoperative radiation therapy planning and delivery in ongoing Radiation Therapy Oncology Group clinical trials in head and neck cancers (www.rtog.org). While results from recent randomized trials have yielded positive results

G.B. Gunn, M.D. (✉) • A.S. Garden, M.D.
Department of Radiation Oncology, Unit 97, The University of Texas MD Anderson Cancer Center, 1515 Holcombe Boulevard, Houston, TX 77030, USA
e-mail: gbgunn@mdanderson.org

Y. Nishimura, R. Komaki (eds.), *Intensity-Modulated Radiation Therapy*,
DOI 10.1007/978-4-431-55486-8_10

in terms of disease control, functional outcomes, and quality of life for patients treated with definitive IMRT versus traditional techniques for nasopharyngeal [2, 3] and oropharyngeal [4] cancers, the level of evidence supporting the use of IMRT in the postoperative setting for head and neck cancers is generally limited to dosimetric studies, strong clinical rationale coupled with extrapolation from studies performed in the definitive setting, and institutional reports. Our goals here are to (1) provide an expanded case-based illustration of our approach to postoperative IMRT target volume delineation and treatment planning for various disease sites within the head and neck and (2) highlight some the literature, including reports regarding patterns of failure, which helped formulate our current approach.

10.2 Postoperative IMRT Technique and Target Volumes

Postoperative radiation therapy for head and neck cancers has generally been recommended for the presence of adverse surgical and pathology findings at the primary site and/or the regional lymph nodes, such as compromised or positive surgical margin, neural or perineural invasion, lymphovascular space invasion, locally advanced primary tumors, bone or cartilage invasion, lymph node involvement, and presence of nodal extracapsular spread [5]. Concurrent chemotherapy is recommended for medically fit patients with high-risk pathologic features, defined as positive surgical margins or nodal extracapsular spread with category level 1 evidence for squamous carcinomas of the oral cavity, oropharynx, larynx, and hypopharynx [6]. The histologic subtype, grade, or differentiation of the primary tumor is also an important factor in the postoperative radiation therapy decision making process for neoplasms of salivary gland origin or originating in the paranasal sinuses. The relative indications for postoperative radiation therapy for thyroid cancer are discussed separately below.

IMRT has the ability to treat the entirety of head and neck target volumes (i.e., whole-field IMRT), or it can be used to treat only the upper portions while using a more traditional anterior supraclavicular field to treat the lower neck, which is matched to the upper IMRT fields. This split-field IMRT approach affords the maximal laryngeal and esophageal inlet sparing and is our preferred postoperative IMRT technique for resected primaries of the more proximal upper aerodigestive tract, such as the oral cavity, naso- and oropharynx, and nasal cavity and paranasal sinuses. However, postoperative whole-field IMRT technique is generally used when the larynx is a component of the radiation target or when larynx shielding is a nonissue, such as following laryngectomy or laryngopharyngectomy for larynx or hypopharynx cancer [7].

Much emphasis has been placed on how to best translate treatment concepts including field and block design used with traditional conventional techniques (including the classic 3-field head and neck technique) to clinical target volume delineation used with IMRT [8]. Consensus groups sought to bridge this knowledge gap with proposals of target delineation guidelines, such as the CT atlas-based delineation guidelines for nodal clinical target volumes in the node-negative neck (available online: http://www.rtog.org/atlases/hnatlas/main.html). Likewise, guidelines regarding target volume delineation in the node-positive neck and postoperative neck have also been proposed [9]. Compared to clinical target volume delineation of intact regions (i.e., unoperated or unviolated neck), postoperative clinical target volumes are generally broader, in that they cross adjacent disturbed anatomic boundaries, are inclusive of surgical tracts and suture lines, and come nearer to the patient surface.

When treating with postoperative IMRT, we have preferred a simultaneous integrated boost approach over sequential plans, typically treating multiple targets/dose levels in 30–33 daily fractions using a single IMRT plan. Postoperative IMRT high-risk clinical target volumes (HR-CTV) include the excised primary tumor bed with margin (including reconstruction sites such as attachments of any rotational or free tissue flaps and mucosal anastomosis) and the positive lymph node bed or involved nodal level, all of which are determined by thorough integration of initial preoperative clinical and radiographic findings, operative findings, surgeon impression, and surgical pathology features. Our postoperative intermediate-risk CTVs (IR-CTVs) are inclusive of the HR-CTVs with a surrounding customized margin incorporating the operative bed, which generally consists of the operative sites, surgical scars, operated soft and bony tissues, and violated but otherwise pathologically negative neck. The "elective" or standard-risk CTVs (SR-CTVs) are those unoperated (unviolated) regions felt to be at substantial (~ ≥5 %) risk of harboring microscopic disease, such as undissected cervical lymph node regions at risk, undissected adjacent soft tissues at risk, and portions of undissected proximal nerve tracts at risk in selected scenarios. Generally, our HR-CTV, IR-CTV, and SR-CTV are treated to 60, 57, and 54 Gy, respectively, in 30 daily fractions. A 3–6 Gy boost is considered for areas of positive microscopic surgical margin, often integrated into the a 30 fraction plan, though infrequently this boost will be done sequentially with additional fractions if the boost is adjacent to a critical structure to avoid fraction sizes of greater than 2 Gy. Areas of residual or unresected gross disease are typically treated to 70 Gy in 33–35 fractions.

10.3 Oral Cavity

Our discussion here pertains to squamous carcinomas of the oral tongue, floor or mouth, lip, gingiva, retromolar trigone, and buccal mucosa. The presence of clinical regional lymph node involvement or perceived substantial risk for occult microscopic lymph node involvement typically dictates the extent of surgical and/or radiation therapy neck management. Central oral cavity subsites, such as the oral tongue, floor of the mouth, and lip, are rich in crossover lymphatic drainage pathways, and as such, our radiation target volumes typically include both sides of the neck for cancers of these subsites. However, for well-lateralized primaries of the buccal mucosa, gingiva, and retromolar trigone, ipsilateral neck radiotherapy is often considered [10]. See case examples 1–3 (Figs. 10.1, 10.2 and 10.3) below. We recognize that there is great controversy regarding ipsilateral versus bilateral therapy for lateralized oral tongue cancers. With rare exception, we have been proponents of bilateral therapy, as these tumors are very aggressive, and salvage even in unirradiated tissues is challenging.

10.4 Oropharynx

The oropharynx consists of the tonsils, base of tongue, soft palate, and oropharyngeal walls. Tumors in these sites are often treated with definitive (chemo)radiation, but on occasion will be operated first. Surgery, with the increasing popularity of minimally invasive intraoral approaches, is being performed more frequently. Conceptually, postoperative irradiation for tumors of these subsites within the oropharynx is similar to the treatment of oral cavity cancers. The majority of subsites can potentially drain to both sides of the neck, so targets will include the primary tumor and draining lymphatics. The principal difference between the oropharynx and oral cavity is that oropharyngeal tumors can metastasize to retropharyngeal nodes in addition to cervical nodes, so the lateral retropharyngeal nodal beds are included in the target volumes [11]. While the vast majority of tumors will require bilateral irradiation, well-lateralized tonsillar tumors that are confined clinically and pathologically to the tonsillar bed can be treated just to the primary tumor bed and ipsilateral neck (provided there is no contralateral disease) [12].

10.5 Larynx and Hypopharynx

Primary surgical management (e.g., laryngectomy and neck dissections) followed by postoperative radiation therapy (with concurrent chemotherapy for high-risk features) is the current standard approach for patients with squamous carcinomas with full-thickness cartilage destruction at presentation or poor candidates for functional laryngeal preservation approaches, whether due to primary tumor features, patient factors, or poor baseline or predicted poor laryngeal function. See case example 4 (Fig. 10.4).

10.6 Nasal Cavity and Paranasal Sinus

Regarding postoperative IMRT for tumors of this region, Hoppe et al. have reported good dosimetric results in terms of target coverage and normal structure sparing with early encouraging disease control and toxicity results. A number of nearby critical structures must be considered during treatment planning, particularly optic and central nervous system structures [13]. Given our observed nodal relapse rates for patients treated in previous eras without routine elective lymph node irradiation, for clinically node-negative patients, we currently treat upper cervical lymph node regions at risk with squamous or poorly differentiated histologies [14] and olfactory neuroblastoma [15]. See case examples 5 and 6 (Figs. 10.5 and 10.6).

10.7 Major and Minor Salivary Glands

Primary surgical resection is the preferred treatment approach for the majority of minor and major salivary gland neoplasms managed at our institution. Beyond the aforementioned traditional indications for postoperative radiation therapy, the diversity of histologic types that arise from these regions and associated unique patterns of spread also guide indications for postoperative radiation therapy and target volume selection. For example, adenoid cystic carcinoma has a propensity for perineural and neural spread, and IMRT allows for selective targeting of nerve pathways at risk to the skull base or in some cases beyond [16]. Lymph node involvement in adenoid cystic carcinoma is quite rare, and as such the neck is rarely included as an elective radiation target volume (with the exception of high grade or solid type), but for higher-grade malignancies (i.e., high-grade mucoepidermoid carcinoma, carcinoma ex pleomorphic adenoma, and salivary duct carcinoma), the risk of regional lymph node occult microscopic involvement justifies elective management [17]. See case examples 7 and 8 (Figs. 10.7 and 10.8).

10.8 Thyroid

External beam radiation can be considered in the postoperative/adjuvant setting for patients at highest risk for local regional recurrence. For differentiated thyroid cancers, relative indications for postoperative radiation include incomplete resection of nonfunctioning (noniodine-avid) cancers; direct invasion of central structures, such as the trachea, cricoid, esophagus, strap muscles, and deep neck musculature; extracapsular nodal extension; recurrent disease; or mediastinal structure invasion or extensive or mediastinal nodal disease. For medullary thyroid carcinoma, in addition the aforementioned relative indications, postoperative radiation is also considered for persistently elevated calcitonin after complete surgery and no evident distant metastatic disease. For poorly differentiated and undifferentiated thyroid cancers, combinations of surgery and postoperative radiation are typically preferred if feasible (American Thyroid Association Guidelines at http://www.thyroid.org/thyroid-guidelines/). Given the difficulty in delivering desired doses to the low neck, namely, due to the complex and mostly concave geometry and close proximity of the spinal cord and lungs, IMRT technique is favored over conventional methods [18]. See case example 9 (Fig. 10.9).

10.9 Conclusions

As illustrated in the presented case examples, the inherent advantages of IMRT in the postoperative setting for head and neck cancers are due to dose conformality, which is important for this region rich in critical and avoidance structures. However, the selection and delineation of these postoperative IMRT targets requires careful integration of the presurgical clinical and radiographic extent of disease, intraoperative and surgical pathology findings, and understanding of the case-specific likely patterns of local and regional spread.

Conflicts of Interest The authors have none to disclose.

Case Examples

Case 1 (Fig. 10.1)

A 63-year-old former smoker presented with a 3.5 cm painful mass of the left lateral border of the oral tongue (photograph). Biopsy of this mass showed moderately differentiated, invasive squamous carcinoma. Contrast-enhanced diagnostic CT of the neck (with puffed-cheek technique) demonstrated the primary mass (outlined in *green*) and revealed a radiographically positive lymph node in left level IIa due to peripheral enhancement and central hypodensity (*arrow*). There were no metastases on chest X-ray. Clinical stage was T2N1M0. He underwent primary surgical resection, including left partial glossectomy, left modified radical neck dissection (levels Ia through IV), and radial forearm free flap reconstruction of the left lateral tongue defect. Final surgical pathology revealed a 3.5 cm primary tumor, closest surgical margin was 2.0 cm, and there was no perineural or lymphovascular invasion. Two of 6 and 2 of four lymph nodes were positive for carcinoma in left neck levels Ib and IIa, respectively, each with extracapsular spread. The remainder of the left neck lymph nodes (levels IIb, III, and IV) were negative for carcinoma. Final pathologic stage was T2N2b. He was treated with postoperative chemoradiation (concurrent cis-platinum) using IMRT. Representative axial, sagittal, and coronal views from planning CT are shown. The primary tumor bed, free flap, and its attachments to the native tongue and positive neck levels (I and IIa) plus ~ 1 cm were delineated as high-risk CTV (CTV 60 Gy). The remainder of the oral tongue and operative bed of the left neck were treated to 57 Gy (intermediate risk), and the contralateral neck levels I–IV were treated electively to 54 Gy (standard risk). This was accomplished in a single integrated IMRT plan in 30 fractions. Representative isodose distributions are shown in multiple axial planes.

Case 1 Key Technical Points

1. At the time of simulation, the left neck suture line/surgical scar was marked with CT-compatible radiopaque wire, and 3 mm bolus was placed over the scar to ensure adequate surface dose to the scar. The left neck CTV contours were inclusive of the operative bed and surgical tracts and were taken out to the skin surface at the level of the scar.
2. A customized mouth-opening and tongue-depressing/tongue-immobilizing intraoral stent was used in this case. As demonstrated in the sagittal view, this provided physical separation of the tongue from the nontarget upper lip and hard palate.

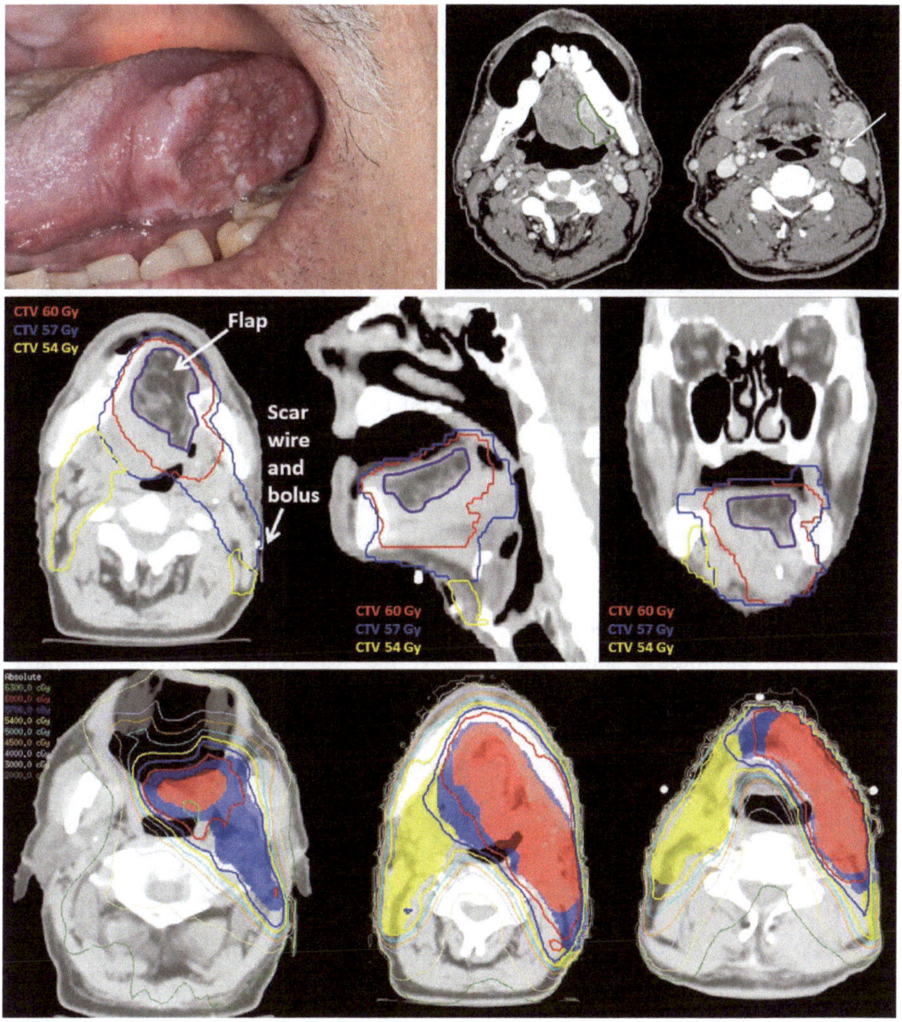

Fig. 10.1

3. In cancers of the oral tongue, in addition to targeting of the primary site tumor bed, we typically treat the remainder of the involved organ (entirety of the oral tongue) to operative bed or elective doses.
4. As demonstrated in the sagittal and coronal views, we ensured the CTVs were carried through the adjacent floor of the mouth, root of tongue, and submental region to the level of the hyoid.
5. The contralateral undissected (*right*) neck was treated electively to 54 Gy. The CTV for this unviolated/unoperated neck differed from that of the operated left neck, in that the right neck lymph node levels were delineated restricted by their

anatomic boundaries (e.g., deep surface of the sternocleidomastoid and platysma), while in the left neck, the entirety of the sternocleidomastoid muscle was targeted as part of the operative bed target volume.

Case 2 (Fig. 10.2)

A 65-year-old former smoker presented with a 4.0 cm painful expansile mass of the left lower alveolus (photograph). Biopsy of this mass showed poorly differentiated, invasive squamous carcinoma. Contrast-enhanced diagnostic CT of the neck (with puffed-cheek technique) demonstrated the primary mass interdigitated with multiple teeth and bone erosion (outlined in *green*). There were lymph node metastases seen on CT. There were no metastases seen on chest X-ray. Clinical stage was T4aN0M0. He underwent primary surgical resection, including left segmental mandibulectomy, left modified radical neck dissection (levels Ia through IV), and osseocutaneous fibular free flap reconstruction. Final surgical pathology revealed a 4.2 cm primary tumor with bone involvement, closest surgical margin was 1.0 cm, and there was perineural and lymphovascular invasion. One of 4 lymph nodes was positive for carcinoma in left neck level Ib, with extracapsular spread. The remainder of the left neck lymph nodes (levels IIa/b, III, and IV) were negative for carcinoma. Final pathologic stage was T4aN1. He was treated with postoperative chemoradiation (concurrent carboplatinum rather than cis-platinum due to profound

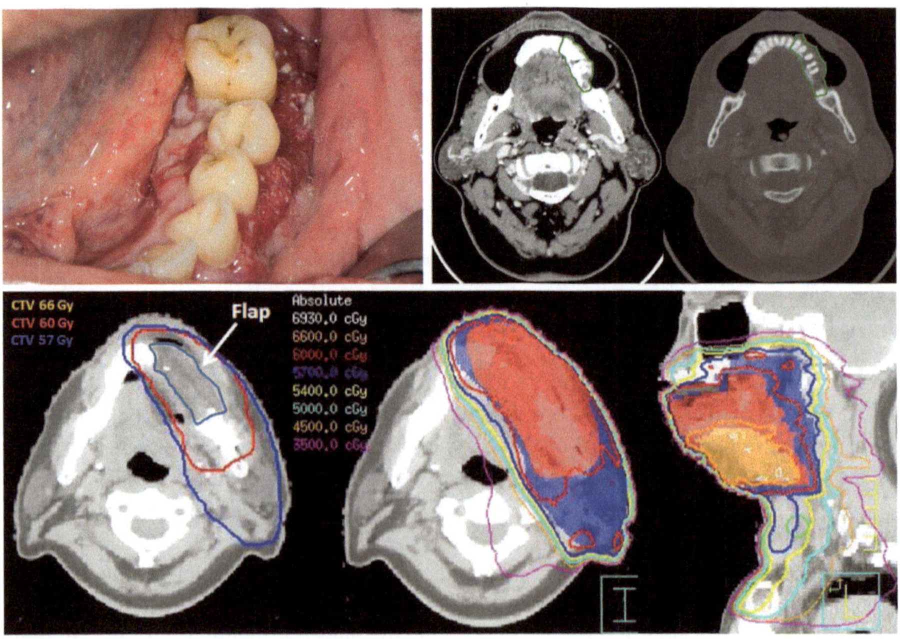

Fig. 10.2

baseline hypoacusis) using IMRT targeting the tumor bed, operative bed, and ipsilateral neck. A representative axial planning CT image with CTVs is shown. The primary tumor bed, free flap and its attachments, and positive neck level (Ib) plus margin were delineated as high-risk CTV (CTV 60 Gy). The remainder of the operative bed of the left neck and segment of the inferior alveolar nerve were treated to 57 Gy (intermediate risk). The left level Ib (area of ECE) was boosted to 66 Gy. This was accomplished in a single integrated IMRT plan in 30 fractions. Representative isodose distributions are shown in axial and sagittal plane.

Case 2 Key Technical Points

1. A customized mouth-opening and tongue-deviating intraoral stent was used in this case in order to displace the tip of the oral tongue, upper lip, and upper gingiva from the high-dose region.
2. The primary tumor bed, flap and its attachments, and bony cuts of the mandible were included in the CTV 60 Gy.
3. Given microscopic perineural invasion, the inferior alveolar nerve was targeted; the most proximal aspect of nerve coverage was to the interspace of the medial and lateral pterygoid muscles.
4. Given the surgical exposure associated with mandibulectomy and neck dissection, the tail of the left parotid was included in the operative bed target (CTV 57 Gy), as recurrences near a spared parotid in a pathologically involved neck have been described (10).

Case 3 (Fig. 10.3)

A 63-year-old former smoker presented with a painful 1.7 cm mass centered in the right floor of the mouth, which was crossing midline (see photograph). Contrast-enhanced diagnostic CT of the neck demonstrated the primary mass (outlined in *green*) close to but not invading an adjacent mandibular torus. There were no lymph node metastases seen on CT. There were no metastases seen on chest X-ray. Clinical stage was T1N0M0. He underwent resection of the anterior floor of the mouth and bilateral neck dissection (levels II–IV) with radial forearm free flap reconstruction of the oral defect. Final pathology showed a 1.2 cm moderately differentiated tumor, 0.6 cm depth of invasion, no perineural or lymphovascular invasion, and free margins. There was a single lymph node involved in right level IB (without extracapsular extension) and a focus of metastatic carcinoma in the right submandibular gland, suspected to be direct spread down Wharton's duct. He was treated with postoperative IMRT. Representative axial, sagittal, and coronal views from planning CT with CTVs are shown. The primary tumor bed, floor of the mouth, free flap and its attachments, and submandibular triangle plus margin were delineated as high-risk CTV (CTV 60 Gy). The remainder of the uninvolved operated bilateral neck above the junction was treated to 57 Gy (intermediate risk). Representative axial, sagittal, and coronal views from planning CT are shown with CTVs in color wash with accompanying dose distributions.

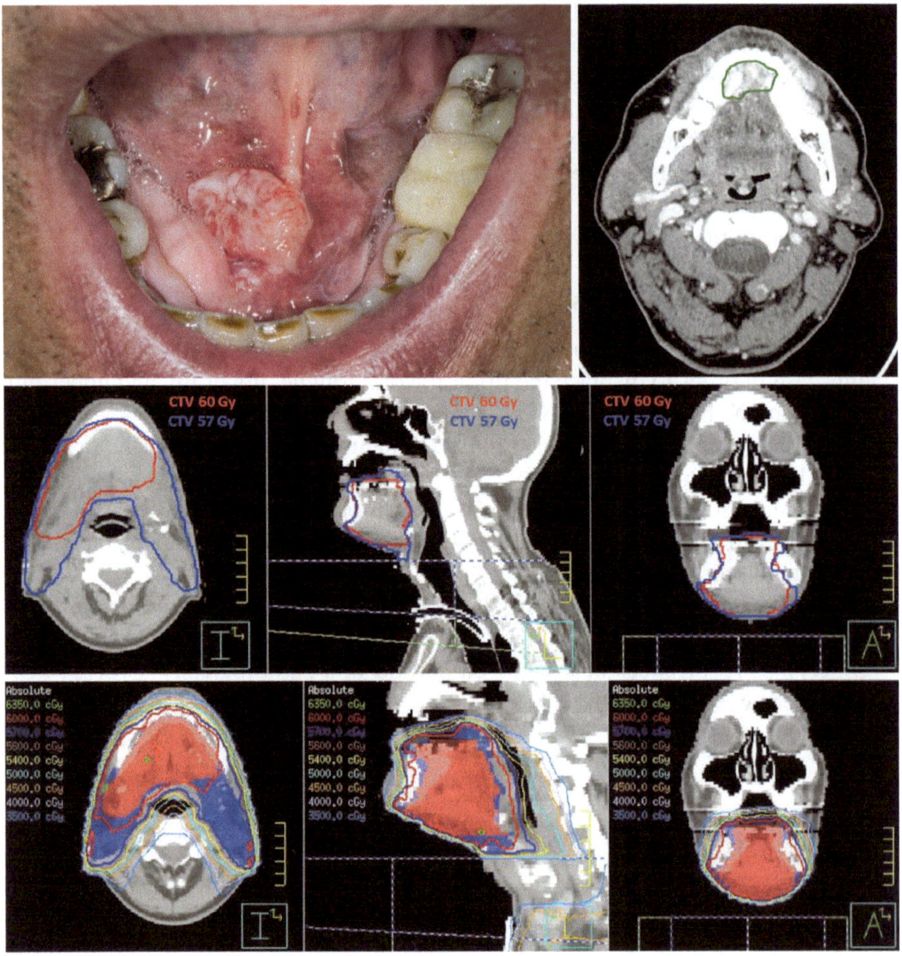

Fig. 10.3

Case 3 Key Technical Points

1. A customized mouth-opening and tongue-elevating intraoral stent was used in this case in order to displace the mobile tongue, upper lip, and upper gingiva from the floor of the mouth.
2. For maximal, inferior constrictor muscle, larynx, and esophageal inlet sparing, a split-field IMRT technique utilizing half-beam block was used with the isocenter placed just above the arytenoids. The lower neck and supraclavicular fossae were treated using a low anterior neck field with a larynx block, which was extended to a full midline block after 40 Gy. Total dose to the bilateral midneck below the junction was 56 Gy in 28 fractions. A patched low-energy electron field was used to treat the central aspect of the neck surgical scar that fell under the larynx and full midline block.

Case 4 (Fig. 10.4)

As 58-year-old smoker presented with hoarseness of voice and globus sensation. Biopsy of an anterior glottic mass revealed invasive squamous carcinoma. Contrast-enhanced CT neck demonstrated an enhancing mass in the anterior larynx, with full-thickness cartilage erosion with invasion of the anterior neck soft tissues (outlined in green). Bilateral cervical lymphadenopathy was seen (white *arrows*). There were no metastases on CXR. Final clinical stage was T4aN2cM0. He underwent total laryngectomy, with total thyroidectomy, and bilateral modified radical neck

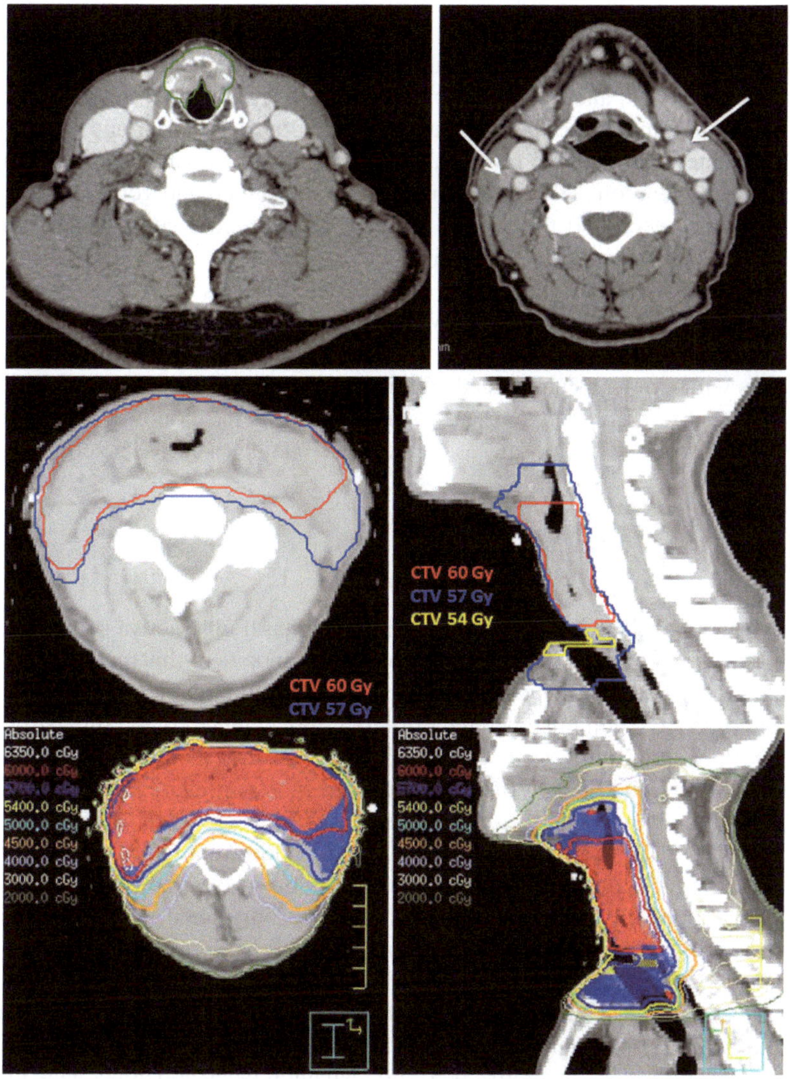

Fig. 10.4

dissection and bilateral paratracheal dissections. Final pathology showed cartilage destruction and anterior soft tissue invasion, and surgical margins were free. There was multilevel adenopathy in the right neck with multiple positive nodes in levels II and III and a single involved node in left level II and with extracapsular spread of all positive nodes. He was treated with postoperative IMRT with concurrent cisplatinum. The tumor bed, larynx bed, reconstructed neopharynx and anastomosis, and bilateral positive nodal levels were treated to 60 Gy in 30 fractions. The remainder of the operative bed was treated to 57 Gy. Axial and sagittal planning CT images with CTVs and corresponding dose distributions are shown.

Case 4 Key Technical Points

1. At the time of simulation, the surgical scar was marked with CT-compatible radiopaque wire, and 3 mm bolus was placed over the scar both during simulation and treatment to ensure adequate surface dose to the scar and surgical tract.
2. Given anterior soft tissue invasion and multilevel adenopathy with extracapsular extension, generous HR-CTV delineation was pursued and generously included the anterior neck soft tissues, the entirety of the central compartment, positive nodal levels, and suprastomal region.
3. The tracheostoma was included in the lower-dose CTVs in this case. We consider including the stoma in the IR-CTV if there was extensive subglottic disease, narrow subglottic surgical margin, or low neck soft tissue invasion such as low neck nodes with extracapsular spread and in the HR-CTV if the location of the stoma is adjacent to preoperative disease. The risk of this location harboring disease is balanced with the risk of microstomia, which can be a severe late effect.

Case 5 (Fig. 10.5)

A 68-year-old smoker presented with right-sided nasal obstructive symptoms and a right maxillary sinus mass. Transnasal endoscopic biopsy of this mass showed moderately differentiated invasive squamous carcinoma. Coronal views of an MRI of the face and a PET/CT demonstrated the primary tumor centered in the suprastructure of the maxillary sinus with extension to the lateral aspect of the nasal cavity as well as invasion through the floor of the orbit (outlined in *green*; *white arrow*). There were no lymph node metastases seen on MRI or PET/CT. There were no distant metastases on PET/CT. Clinical stage was T3N0M0. He underwent primary surgical resection, including subtotal maxillectomy and resection of the floor or orbit via lateral rhinotomy approach. The floor of the orbit was reconstructed with a titanium plate and the maxilla with an anterolateral thigh free flap. Final surgical pathology revealed a 4 cm primary tumor involving bone and invasion of the infraorbital nerve. The infraorbital nerve was dissected back to the foramen rotundum. Final surgical margins on the nerve and maxilla were clear. He was treated with postoperative IMRT, targeting the tumor bed, operative bed, and ipsilateral facial and upper neck lymphatics (levels IB–II). Representative axial and coronal planning CT images with CTVs in color wash and associated isodose lines are shown. The primary tumor bed, flap, reconstruction plate, floor of the orbit, right nasal cavity, and bony cuts of the maxilla

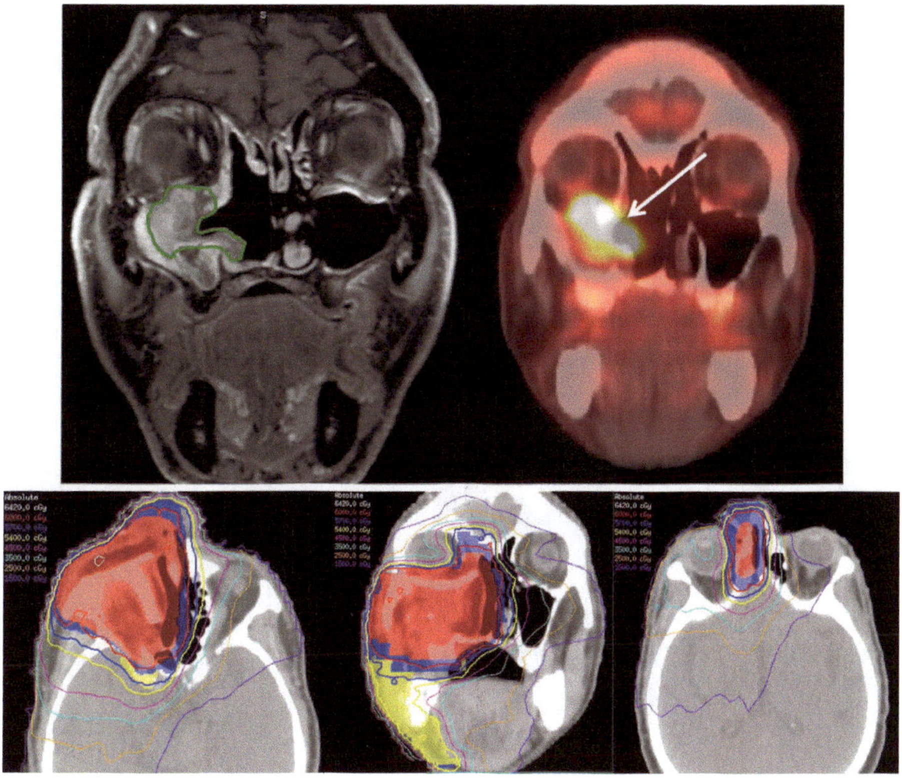

Fig. 10.5

were included in the CTV 60 Gy. The remainder of the operative bed was treated to 57 Gy. Right levels Ib and II were treated electively to 54 Gy. Treatment was accomplished in a single integrated IMRT plan in 30 fractions.

Case 5 Key Technical Points

1. A customized mouth-opening and tongue-depressing intraoral stent was used in this case in order to displace the lower oral cavity from the high-dose region.
2. Given gross neural spread along the infraorbital nerve, the most proximal aspect of nerve coverage was extended intracranially to include elective coverage of the trigeminal nerve ganglion (cavernous sinus).
3. IMRT facilitated excellent coverage of the high right nasal cavity and adjacent ethmoid sinuses and sparing of left optic structures.

Case 6 (Fig. 10.6)

A 65-year-old man presented with persistent sinus congestion. A right-sided naso-ethmoid tumor was discovered and biopsy showed olfactory neuroblastoma. Sagittal T1 MRI with gadolinium and a coronal contrast-enhanced CT showed this tumor to

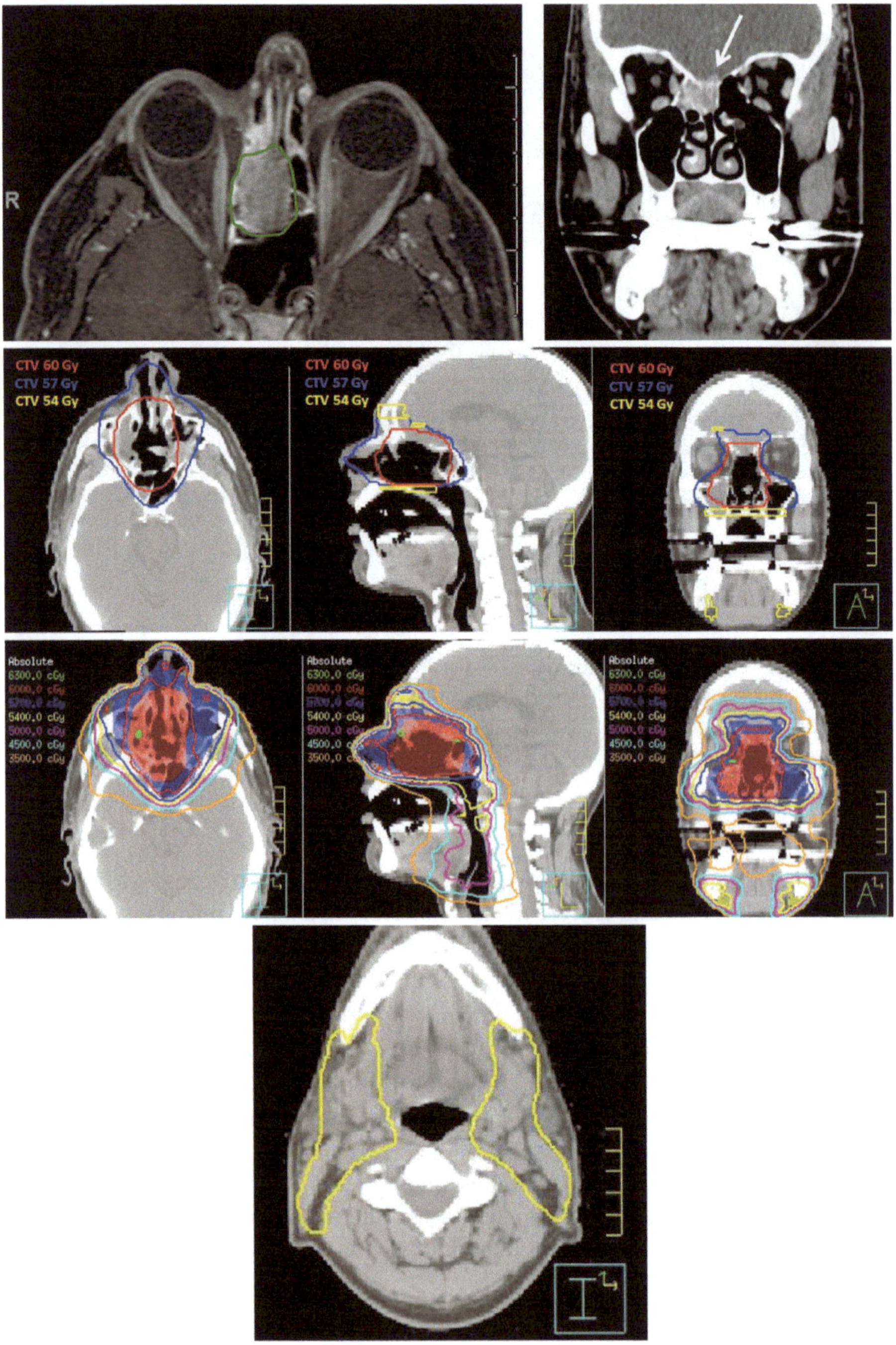
R
CTV 60 Gy
CTV 57 Gy
CTV 54 Gy
CTV 60 Gy
CTV 57 Gy
CTV 54 Gy
CTV 60 Gy
CTV 57 Gy
CTV 54 Gy
Absolute
6300.0 cGy
6000.0 cGy
5400.0 cGy
5000.0 cGy
4500.0 cGy
3500.0 cGy
Absolute
6300.0 cGy
6000.0 cGy
5400.0 cGy
Absolute
6300.0 cGy
6000.0 cGy
5400.0 cGy
5000.0 cGy
4500.0 cGy
3500.0 cGy

Fig. 10.6

be 2.8 cm, widening the right posterior ethmoid sinus, crossing midline, and eroding the anterior skull base with dural involvement (outlined in *green*; *white arrow*). There were no lymph nodes or distant metastases, representing Kadish C disease. He underwent endonasal endoscopic ethmoidectomy with resection of tumor, cribriform plate, and adjacent frontal lobe dura. He was treated with postoperative IMRT, 60 Gy to the positive tumor bed and adjacent skull base and dura, 57 Gy to the operative bed and adjacent sinuses at risk, and 54 Gy electively to the draining lymphatics at risk (bilateral levels IB and II), all in a single integrated IMRT plan in 30 fractions. Axial, sagittal, and coronal planning CT images with CTVs and corresponding dose distributions are shown.

Case 6 Key Technical Points

1. A customized mouth-opening and tongue-depressing intraoral stent was used in this case in order to displace the lower oral cavity from the high-dose region.
2. IMRT utilizing multiple noncoplanar beam angles facilitated excellent coverage of the high nasoethmoid region, skull base, and dura of the anterior cranial fossa while generating a steep dose gradient near the adjacent critical structures of the central nervous system. The maximum dose to each optic nerve and optic chiasm was <54 Gy and the corneas were both <30 Gy. Mean dose to each parotid was <24 Gy.

Case 7 (Fig. 10.7)

A 50-year-old man developed an asymptomatic left preauricular lump. FNA suggested pleomorphic adenoma. He underwent excisional biopsy revealing adenoid cystic carcinoma predominately cribriform type but with a focal solid component. There was intraoperative concern for residual disease near the mastoid tip and tumor adherence to the main trunk of the facial nerve. Post-biopsy diagnostic CT showed enhancing soft tissue just inferior and lateral to the stylomastoid foramen (*white arrow*). There were no lymph nodes or distant metastases on imaging. There were no cranial neuropathies. Clinical stage was TX(1)N0M0. He then underwent surgical resection consisting of lateral temporal bone resection with decompression of the facial nerve, superficial parotidectomy, and left supraomohyoid neck dissection. The tumor was completely resected and the facial nerve was dissected and preserved in its entirety. Final pathology showed a small focus of residual carcinoma and perineural and lymphovascular invasion. All lymph nodes were negative and surgical margins were free. He was then treated with postoperative IMRT, 60 Gy to the tumor bed (including the superficial parotid bed and stylomastoid foramen) and 57 Gy to the operative bed, which included the lateral temporal bone surgical site and facial nerve pathway in the descending facial canal and left upper neck. Axial, sagittal, and coronal planning CT images with CTVs and corresponding dose distributions are shown.

Case 7 Key Technical Points

1. A customized tongue-deviating intraoral stent was used in this case in order to displace the mobile tongue from the high- and intermediate-dose gradient. As

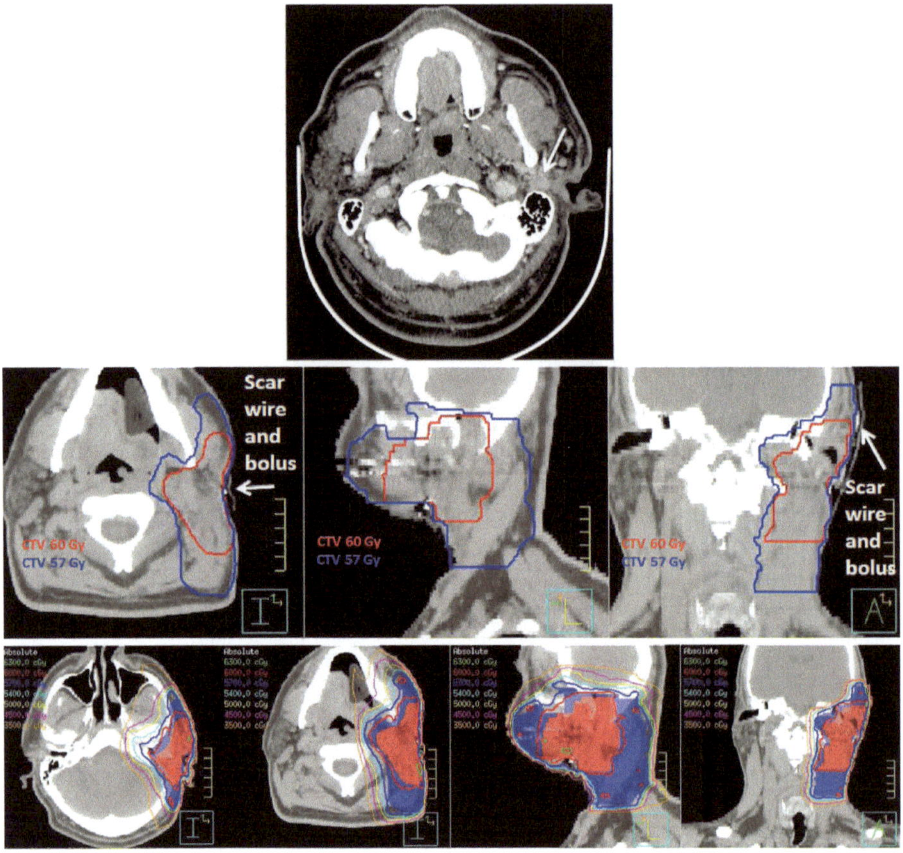

Fig. 10.7

seen in the axial image, the tongue was able to be displaced from the 35 Gy isodose line, which largely fell in the stent itself rather than the tongue.

2. At the time of simulation, the surgical scar was marked with CT-compatible radiopaque wire, and 3 mm bolus was placed over the scar to ensure adequate surface dose to the scar and surgical tract.
3. To ensure adequate coverage of the entire parotid bed, the anterior aspect of CTV 57 was extended to the level of the anterior edge of the masseter muscle and the deep border was taken down to the pharyngeal wall to ensure coverage of the deep lobe of parotid.
4. IMRT facilitated excellent coverage of the operative bed of the left temporal bone while minimizing dose to the underlying brain (left temporal lobe) as demonstrated in the coronal view.

Case 8 (Fig. 10.8)

A 67-year-old presented with a painless pea-sized lump of the right side of the hard palate (photograph). Biopsy of this mass showed adenoid cystic carcinoma, tubular

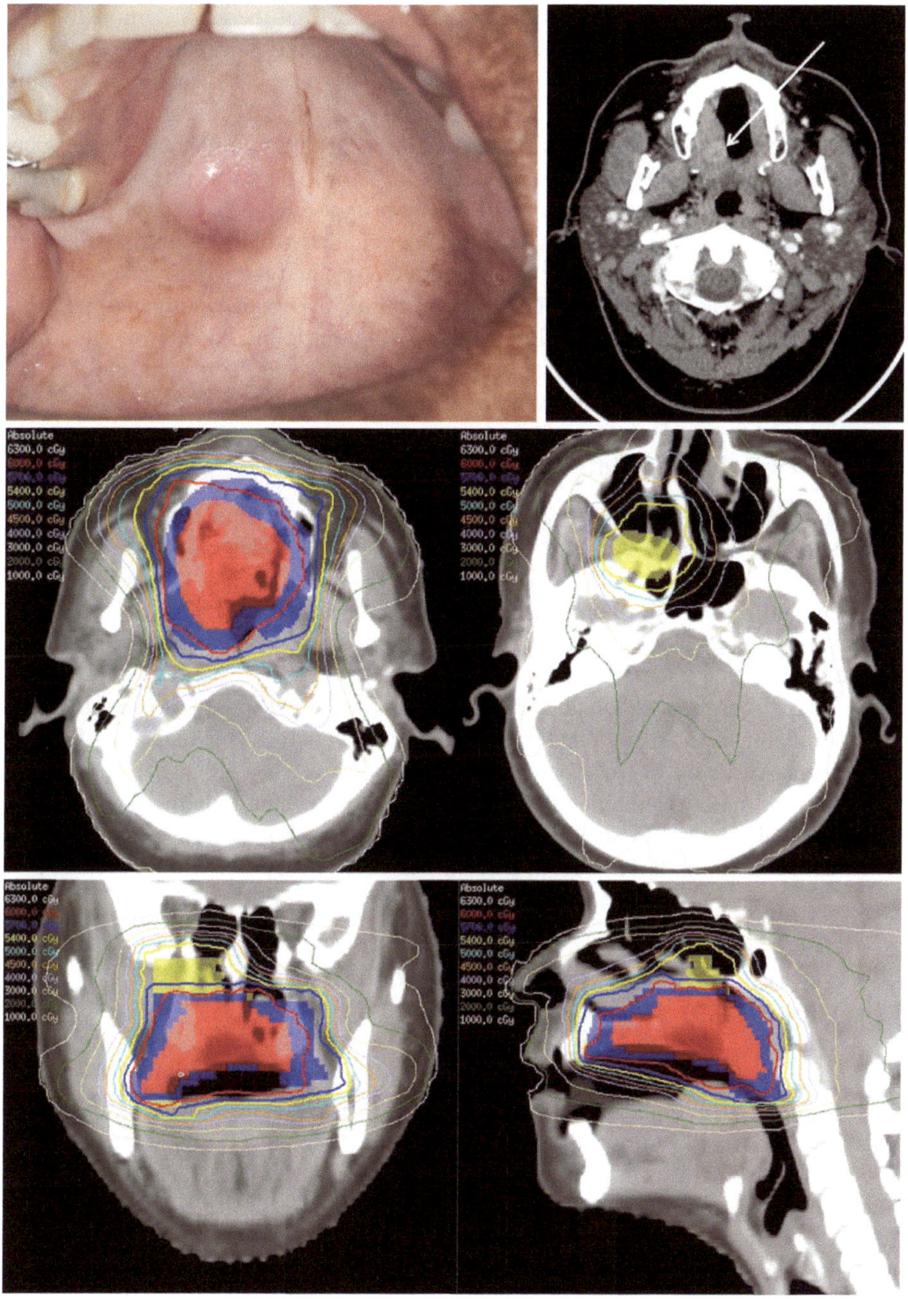
Absolute
6300.0 cGy
5400.0 cGy
5000.0 cGy
4500.0 cGy
4000.0 cGy
3000.0 cGy
1000.0 cGy

Fig. 10.8

and cribriform type. Contrast-enhanced diagnostic CT of the neck demonstrated a 1.5 cm primary mass (*white arrow*) at the hard-soft palate junction. There were no metastases on chest X-ray. She underwent primary surgical resection, including right infrastructure maxillectomy and placement of a surgical obturator in the palate defect. Final surgical pathology revealed a 1.5 cm primary tumor, closest surgical margin was 1.0 cm, and there was perineural invasion. There was no bone invasion. She was treated with postoperative IMRT. The primary tumor bed and margin were delineated as high-risk CTV (CTV 60 Gy). The remainder of the adjacent palate and nasal floor were treated to 57 Gy (intermediate risk), and the right greater palatine nerve pathway was treated to 54 Gy at the proximal aspect of its targeted pathway. This was accomplished in a single integrated volumetric modulated arc plan in 30 fractions. Representative axial, sagittal, and coronal views from planning CT are shown with CTVs in color wash with accompanying dose distributions.

Case 8 Key Technical Points

1. The patient's wax obturator was left in during simulation and treatment delivery to occupy the palate defect to help ensure adequate surface dose of the tumor bed.
2. A customized mouth-opening and tongue-depressing intraoral stent was used in this case in order to displace the lower oral cavity from the high-dose region of the palate.
3. Given perineural invasion and propensity for perineural spread of adenoid cystic carcinoma, the most proximal aspect of nerve coverage was extended to the skull base and inclusive of the pterygopalatine fossa.
4. IMRT facilitated excellent coverage of the palate and neural pathways at risk and sparing of bilateral major salivary glands and central nervous system structures (mean dose to the left and right parotid was each <6 Gy).

Case 9 (Fig. 10.9)

An 80-year-old female presented with hoarseness of voice and right true vocal cord paresis, and a 5 cm right thyroid mass was discovered. Biopsy of the right thyroid mass showed papillary thyroid carcinoma. Contrast-enhanced diagnostic CT neck showed a necrotic mass in the right lobe of the thyroid with tracheal and esophageal invasion, involvement of the common party wall, and a positive right level II/III lymph node (*white arrow*). There were no distant metastases seen imaging. She underwent total thyroidectomy, bilateral paratracheal and superior mediastinal lymph node dissections, right neck dissection (levels II–V), partial esophageal muscularis resection, partial tracheal resection, and partial cricoid resection. Final pathology showed 5.0 cm papillary thyroid carcinoma with foci of squamoid and poorly differentiated components, cartilage invasion, extrathyroidal extension, 6 positive paratracheal lymph nodes with extracapsular extension, and a single

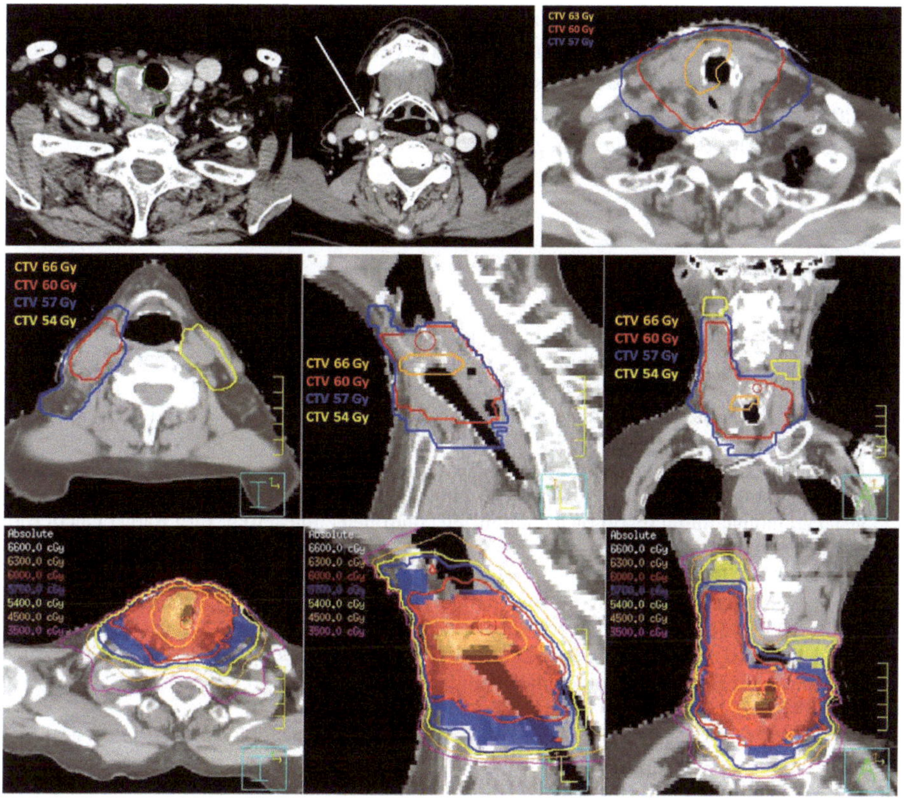

Fig. 10.9

positive lymph node in the right neck specimen. The inferior tracheal margin was clear but the superior tracheal margin was focally close (<1 mm) from the soft tissue edge. She was then treated with postoperative IMRT. The primary tumor bed, central compartment, and positive nodal levels were delineated as high-risk CTV (CTV 60 Gy). The remainder of the operative bed was in CTV 57 Gy (intermediate risk), and the undissected elective left midneck was included in CTV 54 Gy. The right aspect of the cricotracheal anastomosis was selected for an integrated boost to 63 Gy. This was accomplished in a 30 fraction plan. Representative axial, sagittal, and coronal views from planning CT are shown with CTVs in color wash with accompanying dose distributions.

Case 9 Key Technical Points

1. High-risk CTVs in this case included the entirety of the central compartment with generous coverage of the bilateral tracheoesophageal grooves, common party wall, laryngeal inlet on the right, cricoid cartilage, resected tracheal bed, resected esophageal muscularis bed, thyroid bed, anterior soft tissue of the low neck, paratracheal nodal bed, and positive nodal bed of the right neck.

2. Regarding radiation target volumes in the lateral neck, levels II–V were targeted in the node-positive neck (*right* side) and levels III–IV were targeted in the node-negative neck (*left* side). This allowed for sparing of the left major salivary glands. Mean dose to the left submandibular gland and that of left parotid gland were 12 and 6 Gy, respectively.

References

1. Beadle BM, Liao K-P, Elting LS et al (2014) Improved survival using intensity-modulated radiation therapy in head and neck cancers: a SEER-Medicare analysis. Cancer 120:702–710
2. Kam MKM, Leung S-F, Zee B et al (2007) Prospective randomized study of intensity-modulated radiotherapy on salivary gland function in early-stage nasopharyngeal carcinoma patients. J Clin Oncol Off J Am Soc Clin Oncol 25:4873–4879
3. Pow EHN, Kwong DLW, McMillan AS et al (2006) Xerostomia and quality of life after intensity-modulated radiotherapy vs. conventional radiotherapy for early-stage nasopharyngeal carcinoma: initial report on a randomized controlled clinical trial. Int J Radiat Oncol Biol Phys 66:981–991
4. Nutting CM, Morden JP, Harrington KJ et al (2011) Parotid-sparing intensity modulated versus conventional radiotherapy in head and neck cancer (PARSPORT): a phase 3 multicentre randomised controlled trial. Lancet Oncol 12:127–136
5. Ang KK, Trotti A, Brown BW et al (2001) Randomized trial addressing risk features and time factors of surgery plus radiotherapy in advanced head-and-neck cancer. Int J Radiat Oncol 51:571–578
6. Bernier J, Cooper JS, Pajak TF et al (2005) Defining risk levels in locally advanced head and neck cancers: a comparative analysis of concurrent postoperative radiation plus chemotherapy trials of the EORTC (#22931) and RTOG (# 9501). Head Neck 27:843–850
7. Dabaja B, Salehpour MR, Rosen I et al (2005) Intensity-modulated radiation therapy (IMRT) of cancers of the head and neck: comparison of split-field and whole-field techniques. Int J Radiat Oncol Biol Phys 63:1000–1005
8. David MB, Eisbruch A (2007) Delineating neck targets for intensity- modulated radiation therapy of head and neck cancer. What we learned from marginal recurrences? Front Radiat Ther Oncol 40:193–207
9. Grégoire V, Eisbruch A, Hamoir M et al (2006) Proposal for the delineation of the nodal CTV in the node-positive and the post-operative neck. Radiother Oncol J Eur Soc Ther Radiol Oncol 79:15–20
10. Chan AK, Huang SH, Le LW et al (2013) Postoperative intensity-modulated radiotherapy following surgery for oral cavity squamous cell carcinoma: patterns of failure. Oral Oncol 49:255–260
11. Gunn GB, Debnam JM, Fuller CD et al (2013) The impact of radiographic retropharyngeal adenopathy in oropharyngeal cancer. Cancer 119(17):3162–3169
12. Chronowski GM, Garden AS, Morrison WH et al (2012) Unilateral radiotherapy for the treatment of tonsil cancer. Int J Radiat Oncol Biol Phys 83:204–209
13. Hoppe BS, Wolden SL, Zelefsky MJ et al (2008) Postoperative intensity-modulated radiation therapy for cancers of the paranasal sinuses, nasal cavity, and lacrimal glands: technique, early outcomes, and toxicity. Head Neck 30:925–932
14. Bristol IJ, Ahamad A, Garden AS et al (2007) Postoperative radiotherapy for maxillary sinus cancer: long-term outcomes and toxicities of treatment. Int J Radiat Oncol Biol Phys 68:719–730

15. Rosenthal DI, Barker JL Jr, El-Naggar AK et al (2004) Sinonasal malignancies with neuroendocrine differentiation: patterns of failure according to histologic phenotype. Cancer 101:2567–2573
16. Garden AS, Weber RS, Morrison WH et al (1995) The influence of positive margins and nerve invasion in adenoid cystic carcinoma of the head and neck treated with surgery and radiation. Int J Radiat Oncol Biol Phys 32:619–626
17. Garden AS, el-Naggar AK, Morrison WH et al (1997) Postoperative radiotherapy for malignant tumors of the parotid gland. Int J Radiat Oncol Biol Phys 37:79–85
18. Bhatia A, Rao A, Ang K-K et al (2010) Anaplastic thyroid cancer: clinical outcomes with conformal radiotherapy. Head Neck 32:829–836

Sequelae of Therapy of Head and Neck Cancer: Their Prevention and Therapy

11

Amrut S. Kadam and Avraham Eisbruch

Keywords
Xerostomia • Parotid gland • Submandibular • Sublingual glands • Minor salivary glands • Dysphagia • Hearing loss • Larynx • Neural structures • Spinal Cord • Reirradiation of the spinal cord • Radiation-induced brachial plexopathy (RIBP) • Radiation-induced optic neuropathy (RION) • Osteoradionecrosis of the mandible • Hyperbaric oxygen therapy • Dental caries

11.1 Introduction

Radiotherapy (RT) has an integral role in the treatment of head and neck cancer (HNC) with a goal for organ preservation. Intensification of RT for locally advanced HNC along with systemic chemotherapy, and more recently targeted therapy, has led to significantly improved locoregional control and survival compared with conventional RT. However, these improvements are accompanied with increased toxicity [1, 2].

IMRT allows highly conformal dose distributions to target volumes. Appropriate selection and accurate delineation of the target volumes and avoidance organs gain

---

This chapter emphasizes the role of IMRT in reducing complications of irradiation of head and neck cancer.

A.S. Kadam, M.D.
Department of Radiotherapy, Victoria Hospital, Bangalore Medical College and Research Institute, Fort, K R road, Bangalore 560002, Karnataka, India
e-mail: raysoflife@gmail.com

A. Eisbruch, M.D. (✉)
Department of Radiation Oncology, University of Michigan Hospital, 1500 E Medical Center Drive, UHB2C490, Ann Arbor, MI 48109, USA
e-mail: Eisbruch@umich.edu

Y. Nishimura, R. Komaki (eds.), *Intensity-Modulated Radiation Therapy*,
DOI 10.1007/978-4-431-55486-8_11

critical importance [3]. In HNC, intensity-modulated radiotherapy is especially attractive due to its unique ability to treat concave target shapes, the close vicinity of the targets, and many dose-limiting and noninvolved organs at risk (OARs), and its accuracy is enhanced by the lack of breathing-related motion in these tumors.

The major OARs in HNC and IMRT role in its prevention of radiotherapy-related complications in the HNC are discussed below.

11.2 Xerostomia

Permanent xerostomia is the most prevalent late consequence of irradiation (RT) of HNC and a major cause of reduced quality of life (QOL). In addition to patient perception of dryness, diminished salivary output has other effects, including making mastication and deglutition difficult, which may contribute to nutritional deficiencies, predisposing the patient to mucosal fissures and ulcerations, and changing the composition of oral flora, promoting dental caries and contributing to osteoradionecrosis (ORN, discussed later; see under Sect. 11.6.1). The prevalence of xerostomia relates to the extreme radiosensitivity of the salivary glands [4]. Radiation to the salivary glands alters the volume, consistency, and pH of secreted saliva [5].

The severity of the damage to the salivary glands is dependent both on the total radiation dose and on the volume of irradiated salivary glands (major and minor salivary glands) [6].

11.2.1 Parotid Gland

Especially for HNC squamous cell carcinoma, the necessity of treating bilateral level II lymph nodes makes it difficult to spare the parotid glands using laterally opposed RT techniques. However, with 3D CRT or IMRT, it is possible to partly spare at least one parotid gland in selected patients. A high dose is delivered to only a small part of the parotid gland that is located closest to the target volumes, typically the tail of the parotid gland, while the rest of the parotid receives a low dose or no dose at all [4, 7].

The parotid glands under stimulated status produce 60–25 % of saliva, 20–30 % by the submandibular glands (SMGs), and 2–5 % by the sublingual glands. However, in the nonstimulated state, the SMGs contribute up to 90 % of the salivary output [8]. The serous part of the saliva is secreted by the parotid glands, whereas mucins in saliva, which chiefly contribute to the patient's subjective sense of moisture, are contributed by the SMGs [6]. It becomes imperative to protect functions of the SMGs as well as the parotid glands.

Over the years, literature has demonstrated the ability of IMRT to deliver dose distributions that allow partial preservation of parotid function, assessed by either salivary flow measurements or salivary gland scintigraphy (Table 11.1).

Table 11.1 Prospective trials on parotid-sparing assessments treated with IMRT

Year	Author	*n*	Stage	Mean dose (Gy) constraint	End point assessments	
					Objective	Subjective
2013	Chen et al. [9], Blanco et al. [10]	31	I–IV	<25.8	SGS	NS
2010	Eisbruch et al. [11]	69	I–II	<26	SF	XQ
2007	Scrimger et al. [12]	47	I–IV	≤26	SF	XQ
2005	Saarilahti et al. [13]	17	II–IV	≤25.5	SF	NS
2004	Parliament et al. [14]	23	I–IV	≤26	SF	XQ
2004	Münter et al. [15]	18	I–IV	≤26	SGS	NS
2001	Eisbruch et al. [16]	84	I–IV	≤26	SF	XQ
2001	Chao et al. [17]	41	II–IV	≤32	SF	XQ

SF salivary flow, *XQ* xerostomia questionnaire, *NS* not stated, *SGS* salivary gland scintigraphy

A number of prospective clinical trials have demonstrated that parotid-sparing IMRT reduces long-term xerostomia without jeopardizing local-regional control for nasopharyngeal cancer (NPC) compared with conventional RT [18–20] and preserved salivary flow in oropharyngeal cancer [10, 21].

Practice guidelines on parotid-sparing IMRT for HNC have been formulated with use of data on locoregional failure after IMRT. Based on the location of tumor, the parotid glands can be safely spared. For N0 at least one, but usually both, parotid glands; for N1 and 2 a, b disease, sparing of the contralateral parotid gland does not result in increased marginal failures [22, 23], and lower priority is given to the ipsilateral parotid if lymph nodes at level II are involved [5, 24, 25].

In addition, detailed delineation have been given about the cranial border of level II, as it has clear pertinence with the possibility of sparing the parotid [26]. For N0 disease, the upper boundary of level II is placed at the caudal edge of the lateral process of the first vertebra [27]. For N+, level II on the involved neck side is extended to the skull base and includes the retrostyloid space [28].

11.2.1.1 Dose Response

The prevalence and extent of dry mouth can be greatly reduced over time by reducing the mean dose to at least one parotid gland as salivary function can be partially preserved which improves gradually over time.

Among the variety of salivary endpoints, subjective xerostomia and objective stimulated/unstimulated salivary flow most commonly used have been correlated with the dosimetric dose–volume parameters. The mean parotid gland dose [12, 29–31] in particular has been correlated with whole mouth or individual gland salivary production.

Definition of dose–volume–response relationships for the parotid glands has been well established from the data regarding correlation of residual salivary function with radiation dose. Table 11.2 summarizes the reported dose–volume predictors for salivary flow and salivary function recovery. Minimal gland function reduction occurs at <10–15 Gy mean dose. Gland function reduction gradually increases at radiation doses of 20–40 Gy, with a strong reduction (usually by >75 %) at >40 Gy [12, 17].

Table 11.2 Dose–volume predictors for salivary flow

Author (year)	*N*	Tumor dose (Gy)	Follow-up (months)	End point stimulated	Dose–volume parameters (mean)
Little et al. (2012) [32]	78	50–25	24	Saliva flow	25.4 Gy
Li et al. (2007) [30]	142	60–25	24	Saliva flow	<25–30 Gy
Blanco et al. (2005) [10]	55	50–21	12	SOMA	<25.8 Gy#
Eisbruch (1999) [33]	88	58–22	12	Saliva Flow	≤25–26 Gy*

RT 66–70 Gy to primary tumor and pathologic nodes, 50–70 Gy to tumor bed if postoperatively, 46–50 Gy to elective nodes. All 1.8–2.0 Gy/fraction, 1.5–1.8-Gy/fractions in low-risk target volumes: SOMA = Grade 4 xerostomia using subjective, objective, management, analytic (SOMA) method; #Mean dose to single parotid gland to reduce stimulated salivary flow from that gland to <25 % of pre-RT saliva:* 26 Gy at 1, 3, and 6 months, 25 Gy at 12 months

Eisbruch et al. [33] studied the dose–volume effect relationships for the parotid glands, and they analyzed the correlations between fractional gland volumes receiving various doses and the mean doses and concluded that they were highly correlated; therefore, they concluded that mean dose was an adequate metric. Also a medial shift of the parotid glands during therapy in some patients may increase their mean doses compared with the treatment plans [31, 34].

11.2.1.2 Dose Recommendation

Definition of dose–volume–response relationships for the parotid glands has been well established from the data regarding correlation of residual salivary function with radiation dose. The consensus has been reached that xerostomia can be substantially reduced by limiting the mean parotid gland dose to <26–30 Gy as a planning criterion [35]. Xerostomia risk is reduced with sparing of at least one parotid gland or even one submandibular gland [13]. Severe xerostomia (long-term salivary function <25 % of baseline) can usually be avoided if at least one parotid gland has been spared to a mean dose of <20 Gy or if both glands have been spared to a mean dose of < 25 Gy [36]. At present, the mean noninvolved parotid mean dose is set to be ≤ 26 Gy in the Department of Radiation Oncology, University of Michigan.

11.2.2 Submandibular/Sublingual Glands

These glands lie anterior to the level II lymph node targets in the neck. In many advanced HNC treating bilateral neck disease, it is harder to spare a substantial amount of these glands, especially in cases of bilateral lymphadenopathy, resulting in no measurable salivary output from the majority of these glands after radiation.

Direct comparisons of parotid vs. submandibular gland (SMG) radiation sensitivity studies show a lesser sensitivity of the SMGs compared with the parotid glands [37–41]. Also, lower sensitivity of mucinous compared with serous cells is reported [42, 43]. These findings are confirmed further with the common symptoms of thick and sticky saliva during and shortly after the completion of RT, related to the faster decline in the watery content of the saliva produced by the serous parotid glands, compared with the decline of the mucinous component produced predominantly by the SMGs and the minor salivary glands

11.2.2.1 Dose–Volume Effects

While sparing SMGs, care must be taken so that it does not spare the tumor as it lies in close proximity to the base of tongue, tonsil, and level IIa lymph nodes, which require the full-prescribed radiation dose if gross and/or microscopic disease is in the abovementioned area. At present, available evidence has emerged regarding the efficacy and safety of SMGs-sparing IMRT.

Various studies support a lesser sensitivity of the SMGs compared with the parotid glands, and lower sensitivity of mucinous compared with serous cells is also reported [36, 39–43]. The common symptom of thick and sticky saliva was reported during and shortly after the completion of RT, which was related to the faster decline in the watery content of the saliva produced by the serous parotid glands, compared with the decline of the mucinous component produced predominantly by the SMGs and the minor salivary glands. A study of dose–effect relationships for the submandibular glands showed that their salivary output increased as mean dose was reduced from 40 to 30 Gy and then plateaued. No output was observed in glands receiving mean >40 Gy [44]. The treatment policy at the University of Michigan is to try and reduce the mean dose to 30 Gy depending on the need to treat level II (esp. the jugulodigastric nodes) which lay immediately posterior to the SMG.

11.2.3 Minor Salivary Glands

The minor salivary glands are dispersed throughout the oral cavity. It is well documented that they produce up to 70 % of the total mucins secreted by the salivary glands [45]. Hence, if the dose to the oral cavity is minimized, it might contribute to patient-reported xerostomia and also additional benefits like preventing mucositis and loss of taste [46]. Oral cavity should be contoured and delineated as an OAR and dose constraint given in designing IMRT plan whenever possible. In the Department of Radiation Oncology, University of Michigan, the mean dose for the noninvolved oral cavity is set to be ≤30 Gy with very low priority.

11.2.4 Salivary Glands Dose Recommendations [34]

The improvement in objective parotid function as measured by salivary flow is not always accompanied with improved patient-reported xerostomia [28, 31, 36]. Kam et al. indicated that the observer-based grades underestimated the severity of xerostomia compared with the patient self-reported scores [20]. Hence, not only the objective parotid function but also patient's subjective scores should be the main end points in evaluating xerostomia. As xerostomia is mainly an issue of QOL, patient-reported symptoms are more suggestive of its true severity.

One of the strategies to eliminate xerostomia is to spare at least one parotid gland and to spare at least one submandibular gland to reduce xerostomia risk and increase stimulated and unstimulated salivary function. For complex partial volume RT patterns (IMRT), the mean dose to each parotid gland should be kept as low as possible, consistent with the desired clinical target volume coverage. Severe xerostomia (long-term salivary function <25 % of baseline) can usually be avoided if at least one parotid gland has been spared to a mean dose of less than 20 Gy or if both glands have been spared to a mean dose of less than 25 Gy. A lower mean dose to the parotid gland usually results in better function, even for relatively low mean doses (<10 Gy). Similarly, the mean dose to the parotid gland should still be minimized, consistent with adequate target coverage, even if one or both cannot be kept to a threshold of <20 or <25 Gy. Published variations in response among different patient cohorts were probably related to the lack of an accurate model that correctly includes the effects of multiple salivary glands and intra-gland sensitivity variations. When it can be deemed oncologically safe, submandibular gland sparing to modest mean doses (<35 Gy) might reduce xerostomia symptoms.

11.3 Dysphagia

Radiotherapy for HNC inevitably results in significant dose delivery to some of the critical structures necessary for normal deglutition (such as the tongue, soft palate, pharyngeal and laryngeal muscles) which leads to unavoidable mucositis and swallowing difficulty (dysphagia) [47–49]. These difficulties have become major issues after the wide adoption of concurrent chemotherapy–radiotherapy in the past decade.

IMRT use in head and neck malignancies increased from 1.3 to 46.1 % between 2000 and 2005 [50]. The rationale of using IMRT technique as a strategy to avoid dysphagia is based on the established relationship between functional status of the swallowing-related structures and irradiation dose distribution in these structures and on the ability of the IMRT to shape the high-dose volume in accord with the 3-dimensional outline of the target(s) [3].

Swallowing and mastication involve several nerves, muscles, and connective tissue structures. The inferior, middle, and superior constrictors, innervated by the vagal nerve are the three most important muscles [50, 51]. The mastication structures involved are the pterygoid, masseter, and temporalis muscles, and the mandibular condyle [45, 51–56]. Restricted and/or painful mouth opening affect normal chewing and eating and impair speech and oral hygiene [57, 58].

Normal tissue changes like edema, neuropathy, and fibrosis may impair the swallowing function. Acute toxicities like mucositis and edema commonly disrupt normal swallowing during treatment, but improve substantially in the months following radiotherapy or chemoradiotherapy in a majority of patients. However, neuropathy and fibrosis of the oral, laryngeal, and pharyngeal musculature may develop or persist long after the completion of treatment. These late effects ultimately impair the range of motion of key swallowing structures and have been implicated as the primary mechanisms of long-term dysphagia in HNC survivors [1]. Dysphagia may lead to (silent) aspiration, laryngeal penetration, and excess residue after the swallow and/or reflux [44]. Residue after the swallow is a common source of post-swallow aspiration.

Eisbruch et al. [59] were the first to report that radiation damage to the pharyngeal constrictors and the glottic/supraglottic larynx were implicated in post-RT dysphagia. They suggested that reducing the dose to DARS may lead to improved outcomes. Studies have found significant correlation with dysphagia/aspiration and various dose–volume parameters for the pharyngeal constrictor muscles (superior, medial, and inferior group), esophageal inlet, and glottic and supraglottic larynx [45, 60–62] (Table 11.3).

11.3.1 Assessment

Dysphagia has been evaluated by both objective and subjective methods. Many researchers in clinical trials have analyzed the relationship between irradiated structures and dysphagia; the findings of published studies are nearly consistent regarding the crucial structures associated with swallowing dysfunctions. Few of them are mentioned briefly here.

Roe et al. [53] had done a systematic review of the literature on swallowing outcomes after IMRT (1998–2009), identifying 16 papers regarding methodologic quality and method of swallowing assessment. They conclude that if radiation dose to certain structures is limited, a favorable swallowing outcome may be possible. It is evident that it is impossible to compare results across studies due to heterogeneity in the patient population, use of a range of outcome measures that have not been shown to correlate with each other, and limited use of instrumental assessment (i.e.,

Table 11.3 Studies assessing correlation of dose and DARS

Author	Year	Dosimetric structure correlated
Eisbruch et al. [59]	2004	PCMs (V50) and the glottic and supraglottic larynx (V50)
Feng et al. [63]	2007	PCMs (mean dose, V50, V60, V65) and larynx (mean dose, V50)
Levendag et al. [54]	2007	Superior and middle PCMs (mean dose)
Jensen et al. [64]	2007	Supraglottic larynx (mean dose, median dose, V60, V65)
Caglar et al. [47]	2008	Inferior PCMs and Larynx (both mean dose, V50, D60)
Caudell et al. [48]	2009	Inferior PCMs (V60, V65) and larynx (mean dose, V55, V60, V65, V70)
Dirix et al. [49]	2009	Middle PCMs (mean dose, V50) and supraglottic larynx (mean dose)

PCMs Pharyngeal constrictor muscles, *V50* volume receiving ≥50 Gy, *V60* volume receiving ≥60 Gy, *V65* volume receiving ≥65 Gy, *D60* minimum dose received by 60 % of a structure, *V70* volume receiving ≥70 Gy fiberoptic endoscopic evaluation of swallowing [FEES] and VFS). Also, the methods used to delineate and reduce dose to swallowing organs at risk varied.

Eisbruch et al. [59] recognized that muscular components of the swallowing apparatus, critical to the development of dysphagia in irradiated patients, can be spared by IMRT.

A series of trials have been done to establish whether dose reduction to DARS can improve swallowing outcomes for HNC treated by IMRT. The consistent finding that increased radiation dose to a larger volume of the pharyngeal constrictors resulted in higher levels of dysphagia was seen.

Studies that focused on radiation dose reduction and/or structure avoidance, unfortunately, cannot easily be compared, because of their heterogeneity in tumor sites, treatment protocols, and their overall retrospective nature [51, 64]. It is found that there is significant correlation with dysphagia/aspiration and various dose–volume parameters for the pharyngeal constrictor muscles (superior, medial, and inferior group), esophageal inlet, and glottic and supraglottic larynx [60–62, 65].

At the University of Michigan, Feng et al. [66] demonstrated in a group of 73 patients with oropharyngeal cancer that sparing these structures using IMRT is feasible with high LRC rates and very low treatment-related dysphagia. During delineation of the neck nodes, including only the lateral (lying medial to the carotid arteries) retropharyngeal (RP) nodes which are at risk in HNC, Feng et al. [66] could spare the parts of the pharyngeal constrictors medial to the RP nodes, resulting in mild or no dysphagia in almost all patients.

11.3.2 Dose Effect

The use of high-intensity treatments, especially chemoirradiation, has resulted in considerable rates of swallowing dysfunction, both acute (15–23 %) and long term (3–21 %) [65, 67–72].

The dose delivered to the pharyngeal constrictor muscles plays a crucial role in the development of severe late dysphagia/aspiration [45, 47–49, 55–58, 61–73]. With the technical ability of IMRT, this knowledge provides us with both the rationale and the means to reduce the dose to these structures. To obtain this, we can include these structures in the IMRT optimization process; however, the central location of the swallowing structures and the close proximity between tumor and crucial structures make this often an arduous task with only limited dosimetric gain. Multivariate analysis has identified that the bilateral lymph node irradiation was an important independent predictor for swallowing dysfunction 6 months after treatment [74]. The necessity of elective lymph node irradiation in the eradication of subclinical disease has long been established in HNC [75].

Snadra et al. [60] in their analysis demonstrated that dose de-escalation to the elective lymph nodes significantly reduces the volume of the swallowing apparatus irradiated up to a high dose without compromising target coverage and dose homogeneity. This clinically resulted into significantly less grade 3 dysphagia in the de-escalated arm 3 months after treatment with similar LRC and DFS rates. A combination of mucosal swelling and fibrosis of the swallowing muscles causes late dysphagia after radiation [76].

Various authors have confirmed the steep dose–response relationships between dose to different parts of swallowing apparatus and dysphagia in the short to medium term.

Levendag et al. [54] found a 19 % increase in the probability of late dysphagia grade 3/4 (>3 months after completion of the therapy) with every additional 10 Gy after a dose of 55 Gy in superior constrictor muscles

Eisbruch et al. [59] correlated doses with various outcome measures (objective and subjective outcomes) and noted varying correlation of the doses with each outcome measure. It is likely that mean pharyngeal constrictor doses above 45–60 Gy are associated with worse dysphagia.

Caudell et al. [48] have reported a 7–11 % increase in risk for gastrostomy dependence or aspiration with every 1-Gy increase in a mean dose to the larynx or inferior constrictor.

Van der Laan et al. [77] compared in their planning study 30 standard IMRT treatment plans with swallowing-sparing IMRT plans that aimed to reduce the dose to organs at risk for swallowing dysfunction in the same patients. The dose characteristics of the target volumes and normal structures were comparable. After adequate coverage of target volumes and dose to critical structures within acceptable limits were achieved, the mean doses to the various swallowing-related structures were reduced, depending on N classification and primary tumor location. In addition, the observed dose reductions were reflected in reduced estimates of the NTCP (normal tissue complication probability) values for both physician-rated [RTOG grade 3/4, for 9 %] and patient-rated measures for swallowing dysfunction (moderate to severe complaints: for solid food 7.9 %, for soft food 2.4 %, for liquid food 1.4 %, for choking when swallowing 0.9 %).

Lisette van der Molen et al. [78] reported dose–effect relationships between the radiation doses to the critical swallowing and mastication structures and dysphagia and trismus end points. They summarized that objective dysphagia (PAS) correlated significantly to the inferior constrictor (IC) and subjective patient-reported problems with swallowing solids at 10 weeks posttreatment correlated with the radiation dose to the IC and masseter muscle and at 1-year posttreatment to the masseter muscle. Significant associations were found with the radiation doses to the masseter and pterygoid muscles at 10 weeks for trismus. The radiation doses to the masseter, pterygoid, and temporalis muscles and the mandibular condyle at 1 year significantly correlated between patient-perceived limited mouth opening and at 10 weeks posttreatment with only masseter muscle. He concluded that both objective and subjective measurements are valuable for finding dose relationships.

In Feng's study, significant correlations were observed between aspirations and the mean doses to the PC and GSL, as well as the partial volumes of these structures receiving 50–65 Gy [45]. Both the mean dose to the pharyngeal constrictor muscles and the larynx and the volume of structures receiving 50–60 Gy have been shown to remarkably correlate with the prevalence of dysphagia [33, 49, 55, 60, 66, 68, 73, 76, 78]. These findings imply that limiting the dose to the crucial swallowing structures might decrease both the incidence and severity of radiation-induced dysphagia.

At present, the routine IMRT practice for HNC at the University of Michigan is to keep the mean dose to the noninvolved PC and GSL ≤50 Gy. However, avoiding underdosing to the targets in the vicinity remains the highest priority. In cases of oropharyngeal cancer, the lower neck is either treated with split-field technique in which the glottis larynx and upper esophagus are shielded. Alternatively, whole-neck IMRT is performed while reducing mean doses to the larynx, inferior constrictors, and upper esophagus toward 20 Gy, a dose that is similar to that achieved with split-field IMRT using laryngeal block.

11.4 Hearing Loss

Hearing loss is a common but frequently ignored late complication after RT for HNC. In general, radiation-induced hearing loss includes conductive hearing loss due to damage to the outer and middle ear and sensorineural hearing loss (SNHL) caused by damage to the cochlea and/or the auditory nerve [79], which may result in long-lasting compromise of the quality of life [80]. Good hearing plays an important role in maintaining relationships. The frequent request of others to repeat what has been said might lead to misunderstanding and disruption of relationships and social isolation. It is known that hearing loss may result in serious depression, vertigo, cognitive impairment, and reduction in functional status [79].

Current data from the literature have shown that total radiation dose, cisplatin-based chemotherapy, age, male sex, and hearing deficit before RT are associated with the risk of hearing loss [79].

Patients treated with head and neck IMRT based on audiometric evaluation studies have reported about 0–63 % [79] of hearing loss. Chemoirradiation with cisplatin-based chemotherapy may increase the degree of hearing loss. By decreasing the dose to the cochlea, the incidence of hearing loss may significantly improve.

Cochlea is small in size and lies adjacent to the inner ear; hence, a small deviation of contouring will have a profound effect on hearing loss and defeat the purpose of IMRT. Pacholke et al. [81] established the first guidelines for contouring the middle ear and the two major components of the inner ear (the vestibular apparatus and cochlea). These guidelines have been of practical help to radiation oncologists in the process of radiotherapy planning.

Several studies have attempted to relate mean or median cochlear dose to persistent hearing loss [63, 82, 83] (Table 11.4).

Oh et al. [74] observed a reduction in the radiation dose to the inner ear from 69.6 ± 11.8 to 63.4 ± 9.1 Gy and its resultant reduction in the incidence of SNHL from 68.2 % (15/22) to 0 % (0/8), even though no attempt was made to limit the radiation dose to the inner ear structure on the radiation planning.

Pan et al. [89] did a prospective study to determine the relationship between the RT dose to the inner ear and long-term hearing loss in HNC patients treated with RT. In each patient, hearing in the side that received a high dose to the cochlea was

Table 11.4 Studies related to radiation doses and hearing loss

Author	Year	*N*	Rx	Threshold dose (Gy)	Clinically cochlear damage studied
Grau et al. [84]	1991	22	RT	50	SNHL at high frequencies (2–4 kHz)
Anteunis et al. [85]	1994	18	RT	50	Conductive and/or SNHL
Chen et al. [86]	1999	21	RT	60	SNHL
Honore et al. [87]	2002	20	RT	15	Hearing impairment 0.3 dB/Gy and 15 % risk of SNHL with 15 Gy
Johannesen et al. [88]	2002	33	RT or CRT	54	>4,000 Hz
Oh et al. [74]	2004	24	CRT	63.4±9.1	High frequency
Merchant et al. [75]	2004	72	RT or CRT	32	Low and intermediate frequency (<32 Gy, shunt only); high frequency (>32 Gy, with or without shunt)
Pan et al. [89]	2005	31	RT	45	≥2,000 Hz
Herrmann et al. [90]	2006	32	RT	20–25	ED50 was in range of 20–25 Gy
Hitchcock et al. [91]	2009	62	RT CRT	40 for RT 10 for CRT	SNHL High-frequency SNHL

CRT chemoradiotherapy, *RT* radiation therapy, *SNHL* sensorineural hearing loss, *ED50* dose at which 50 % incidence is expected

compared longitudinally with hearing on the contralateral side which received substantially lesser dose. This unique analysis potentially reduced bias and effects of aging and systemic therapy on the dose–response relationships. They reported that an increase in the mean dose to the inner ear was associated with increased hearing loss at high frequency (>2,000 Hz) and also clinically apparent hearing loss started at a threshold dose of 45 Gy.

Honore et al. [87] retrospectively estimated mean cochlear doses in 20 patients with HNC (1.8–2.3 Gy/fraction) and observed ΔBCT 7–79 months post-RT. A dose–response relationship was observed only for 4 kHz where BCT > 15 dB.

Chen et al. [86] retrospectively studied 22 patients treated with RT for NPC (1.6–2.3 Gy/fraction and concurrent/adjuvant chemotherapy) and studied ΔBCT 12–79 months post-RT. A significant increase in hearing loss (BCT of 20 dB at one frequency or 10 dB at two consecutive frequencies) was observed for all frequencies (0.5–2 kHz) when the mean dose received by the cochlea was >48 Gy.

Van de Putten [92] retrospectively evaluated ΔBCT 2–7 years after RT in 21 patients with unilateral parotid tumors (fraction sizes 1.8–2.0 Gy). Using the contralateral ear as a control, SNlll (BCT >15 dB difference in three frequencies between 0.25 and 22 kHz) was seen when mean doses received by the cochlea were >50 Gy.

Grau et al. [84] demonstrated that the probability and severity of the loss were correlated with dose and the sound frequency when cochlea received doses of 50–70 Gy leads to hearing loss within 18 months. Also in ¾ of the patients of NPC cases analyzed retrospectively showed that the cochleae lie within the PTV and that their position greatly affects the dose they receive.

At MSKCC, the dose limit to the cochlea for NPC patients is 60 Gy wherever possible given the target constraints with an additional constraint requiring ≥99 % of the GTV to receive ≥70 Gy is used to prevent underdosing the tumor. By IMRT dose painting and the simultaneous boost technique, lower cochlear maximum doses of 30 Gy without compromising target coverage can be achieved.

In a retrospective study of 26 patients with medulloblastoma treated by either conventional RT or IMRT, IMRT delivered 68 % of the radiation dose to the auditory apparatus (mean dose: 36.7 vs. 54.2 Gy), without compromising the target dose [82]. Grade 3 or 4 hearing loss was also reduced at 13 % vs. 64 % ($p<0.014$) [93]. Similarly, hearing loss in NPC treated with IMRT is relatively uncommon and less severe.

Sultanem et al. [83] studied IMRT with concurrent and adjuvant cisplatin-based chemotherapy in 32 NPC. From dose–volume histograms (DVHs) analyzed, the average dose to 50 % of the right and left ear volume was 52.0 Gy (range 34.0–71.7 Gy) and 52.2 Gy (range 34.6–64.2 Gy), respectively. At a median follow-up of 21.8 months, only 5 cases developed Grade 3 hearing loss.

Wolden et al. [94] studied IMRT using accelerated fractionation along with concurrent and adjuvant platinum-based chemotherapy in 74 NPC patients (69/74 received CT). No patient had Grade 4 hearing loss at the end of 6 months, and only 15 % had Grade 3 SNHL based on audiograms. The average mean dose received to cochlea was 55.6 Gy.

Due to the large variance in literature reported incidence of hearing loss, future toxicity studies are still needed, such as multi-institutional or larger single prospective trials utilizing both pre- and posttreatment hearing tests, in order to determine absolute hearing loss as a function of frequency and the absolute radiation dose received by each cochlea. Moreover, due to the wide use of combined chemoradiotherapy, the response of SNHL to chemoradiation also needs to be established in prospective trials as a function of both cisplatin and radiation doses, as well as chemoregimen (neoadjuvant, concurrent, or adjuvant).

At the University of Michigan, the dose constraint for cochlea is 40 Gy if the target is close to the cochlea, as may be the case in advanced nasopharyngeal cancer. If the target is not close, we prefer to reduce the dose as much as possible to <30 Gy.

11.5 Larynx

The primary goals of larynx preservation are to assure speech and swallowing function. Radiotherapy may cause progressive edema, lymphatic disruption, and associated fibrosis, which can lead to long-term problems with phonation and swallowing [95].

Locally advanced laryngeal cancer frequently causes voice and swallowing dysfunction, including aspiration that might not improve even if the cancer has been eradicated. These are the main reasons patients presenting with marked laryngeal dysfunction are advised to undergo laryngectomy rather than a trial of chemo-RT. The addition of concurrent chemotherapy to high-dose, extensive-field RT worsens substantially the risk of laryngeal edema and dysfunction. RT without chemotherapy, delivered to small fields for stage T1 glottic larynx cancer, usually results in excellent voice quality [96].

Vocal function is assessed objectively using instruments like videostroboscopy for direct visualization to assess supraglottic activity, vocal fold edge, amplitude, mucosal wave, phase symmetry, and glottis closure [97], aerodynamic measurements of phonation time [98], or human observation [99] as subjective assessments can be made with validated patient-focused questionnaires to assess various combinations of voice, eating, speech, and social function.

Flexible fiberoptic examination is used to assess edema. As per RTOG scale [100], Grade 1 edema would correspond to "minimal" thickening of the epiglottis, aryepiglottic folds, arytenoids, and false cords. Grade 2 is a more diffuse and evident edema, although still without significant or symptomatic airway obstruction.

Due to the small size and close proximity of these vocal structures, high-resolution, contrast-enhanced computed tomography facilitates accurate substructure definition. The identification of the most important anatomic sites, whose dose–volume parameters would primarily affect vocal function, is of paramount importance. Various authors suggest different structures for planning purposes; Sanguineti et al. [99] considered the larynx from the tip of the epiglottis superiorly to the bottom of the cricoid inferiorly; the external cartilage framework was excluded from the laryngeal volume. Dornfeld et al. [100] considered the dose points in various structures (e.g., base of tongue, epiglottis, lateral pharyngeal walls, pre-epiglottic space, aryepiglottic folds, false vocal cords, and upper esophageal sphincter) to be related to vocal injury.

The exact correlation between voice abnormalities and the degree of laryngeal edema has not been assessed. Pre-RT voice abnormalities have not been considered in most studies and hence might have overestimated the degree of RT-related damage. Many studies have shown a good voice outcome after RT for Stage T1 laryngeal cancer (typically 60–66 Gy without chemotherapy). In the locally advanced setting, less information is available regarding voice quality after treatment.

Dornfeld et al. [100] found a strong correlation between speech and doses delivered to the aryepiglottic folds, pre-epiglottic space, false vocal cords, and lateral pharyngeal walls at the level of the false vocal cords. In particular, they noted a steep decrease in function after 66 Gy to these structures. Their study was limited by not having full three-dimensional dose metrics.

Fung et al. [96] evaluated the subjective and objective parameters of vocal function. Changes in voice were related to doses to the larynx and pharynx and oral cavity. They suggest that saliva, pharyngeal lubrication, and soft tissue/structural changes within the surrounding musculature play an important role in voice function.

Sanguineti et al. [99] found that neck stage, nodal diameter, mean laryngeal dose, and percentage of laryngeal volume receiving ≥30–70 Gy were all significantly associated with edema Grade 2 or greater on univariate analysis. On multivariate analysis, the mean laryngeal dose or percentage of volume receiving ≥50 Gy and neck stage were the only independent predictors.

Rancati et al. [101] studied two normal tissue complication probability models (the Lyman–Kutcher–Burman model and the logit model) with the dose–volume histogram reduced to the equivalent uniform dose (EUD). A significant volume effect was found for edema, consistent with a prevalent parallel architecture of the larynx for this endpoint. These findings suggested an EUD of <30–35 Gy to reduce the risk of Grade 2–3 edema.

Longitudinal studies consisting of objective scoring of laryngeal edema, voice quality, and patient-reported measures are necessary to assess the intercorrelations among these measures. Such studies should include pretherapy assessments to account for tumor-related voice abnormalities and should concentrate on patients receiving concurrent chemo-RT who are at the greatest risk of laryngeal toxicity.

The surrounding tissues might be indirectly affected by a reduction in salivary function or directly by effects on the intrinsic musculature and soft tissue. From the published data, it seems reasonable to suggest limiting the mean noninvolved larynx dose to 40–45 Gy and limiting the maximal dose to <63–66 Gy, if possible, according to the tumor extent. To minimize the risks of laryngeal edema, it is recommended that the percentage of larynx volume receiving ≥50 Gy be < 27 % and the mean laryngeal dose ≤ 44 Gy [102].

At the University of Michigan, 91 patients with stage III/IV oropharyngeal cancer were treated on two consecutive prospective studies of definitive chemoradiation using whole-field IMRT from 2003 to 2011. Patient-reported voice and speech quality were longitudinally assessed from pretreatment through 24 months using the Communication Domain of the Head and Neck Quality of Life (HNQOL-C) instrument and the speech question of the University of Washington QOL (UWQOL-S) instrument, respectively. Factors associated with patient-reported voice quality worsening from baseline and speech impairment were assessed. The results demonstrated that patient-reported voice quality decreased maximally at 1 month, with 68 and 41 % of patients reporting worse HNQOL-C and UWQOL-S scores compared to pretreatment, and improved thereafter, recovering to baseline by 12–18 months on average. In contrast, observer-rated larynx toxicity was rare (7 % at 3 months; 5 % at 6 months). Among patients with mean glottic larynx (GL) dose ≤20 Gy, >20–20 Gy, >30–20 Gy, >40–20 Gy, and >50 Gy, 10, 32, 25, 30, and 63 % reported worse voice quality at 12 months compared to pretreatment ($p=0.011$). Results for speech impairment were similar. GL dose, N-stage, neck dissection, oral cavity dose, and time since chemo-IMRT were univariately associated with either voice worsening or speech impairment. On multivariate analysis, mean GL dose remained independently predictive for both voice quality worsening (8.1 % per Gy) and speech impairment (4.3 % per Gy). We concluded that voice quality worsening and speech impairment after chemo-IMRT were frequently reported by patients, under-recognized by clinicians, and independently associated with GL dose. These findings support limiting mean GL dose to <20 Gy during whole-neck IMRT when the larynx is not a target [103].

11.6 Neural Structures

11.6.1 Spinal Cord

The Spinal cord extends from the base of the skull through the top of the lumbar spine; individual nerves continue down to the spinal canal to the level of the pelvis. The spinal cord consists of gray and white matter similar to the brain. However, the gray matter is internalized with a butterfly arrangement in transverse section consisting of dorsal and anterior columns. The white matter is the external coat composed of nerve fibers, an inconspicuous vasculature, and glial cells. The concentric myelin sheath that surrounds the axon is formed by the cytoplasmic processes of oligodendrocytes. The reflex motor actions following stimulus enter via the afferent posterior sensory arc and leave delivered by the efferent anterior synapse or arc. The blood supply is via the anterior spinal artery and two posterior spinal arteries, with the longest branches occurring in the thoracic cord. Unique to the spinal cord is interruption or transection of any segment defunctionalizes the caudal extension. One of the most dreaded radiation injuries is a transverse myelopathy that leads to paraparesis or paralysis of the limbs with loss of bladder and rectal sphincter control.

Portions of the spinal cord are often included in radiotherapy (RT) fields for treatment of malignancies involving the neck. RT-induced spinal cord injury (i.e., myelopathy) can be severe, resulting in pain, paresthesias, sensory deficits, paralysis, Brown-Sequard syndrome, and bowel/bladder incontinence [104].

Radiation-induced transverse myelitis is an irreversible process with no effective treatment [105]. The damage that occurs after excessive doses of radiation does not manifest itself until late, 6–24 months after treatment, because the tissues that are damaged are slowly proliferating [105, 106]. Symptoms, usually paresthesias, appear from 6 months to 2 years after completion of radiation therapy and may progress to total paralysis [107].

Lhermitte's sign (LS) is an electric shock-like sensation in the spine and extremities exacerbated by neck flexion. It is caused by reversible demyelination of ascending sensory neurons due to inhibition of oligodendrocyte proliferation after radiotherapy (RT) of the cervical or thoracic spine [108–110]. The denuded axons become sensitive to irritation from neck flexion, causing the characteristic shock sensations. Once oligodendrocytes recover and myelin synthesis is resumed, symptoms subside. Although LS is not usually associated with a progression to chronic progressive irreversible myelitis, delayed radiation myelopathy causing paralysis may be preceded by LS [111].

There are four clinical syndromes of radiation myelopathy as described by Reagan et al. [105]. The first is acute transient radiation myelopathy, distinguished by the presence of Lhermitte's sign. It is the most common myelopathy and is associated with no other abnormalities on neurologic examination. The second syndrome is an acutely developing paraplegia or quadriplegia, presumably secondary to an infarction of the spinal cord because of radiation damage to the blood vessels. The third syndrome is manifested by signs of lower motor-neuron disease in the upper or lower extremities, presumably the result of selective anterior horn-cell damage. These latter two syndromes are exceedingly rare; the final syndrome is chronic progressive radiation myelopathy, the only syndrome for which pathologic findings have been described. This is the syndrome that most concerns radiation oncologists, for it is progressive and permanent and often leads to the development of a fatal complication such as infection or pulmonary embolus [104].

The incidence of permanent injury to the spinal cord as a complication of radiation therapy generally correlates positively with total radiation dosage. The pathogenesis of radiation injury is believed to be primarily from vascular/endothelial damage, glial cell injury, or both [112, 113]. The clinical endpoint in most studies is paralysis, with the spinal cord showing nonspecific white matter necrosis. Several animal study reports suggest regional differences in radiosensitivity across the spinal cord [112, 114].

Phillips and Buschke [113] concluded that the number of fractions was the most important factor in the development of late radiation damage to the spinal cord. They reported on the incidence of myelitis after treatment to portions of the cervical or thoracic spine correlating dosages with number of fractions. They found that a line connecting 6,000 cGy in 35 fractions in 7 weeks (171.5 *cGy* per fraction) or 5,000 cGy in 25 fractions in 5 weeks (200 cGy per fraction) excluded all their myelitis cases. Van der Kogel [114] also concluded that the spinal cord tolerance depended more on the number of fractions and less on overall treatment time.

Abbatucci [115] concluded that 5,000 cGy in 25 fractions could be tolerated safely if not more than 3–5 vertebral segments were involved. Kim and Fayos [116] also concluded that short lengths of spinal cord could safely tolerate 6,000 cGy in fractions of 180–200 cGy. They also suggested that fraction size was one of the most important factors in determining myelopathy.

McCunniff and Liang [117] observed that when doses higher than 6,000 cGy to the cervical spinal cord with fractions ranging from 133 to 180 cGy, only 1 developed spinal cord damage. They concluded that radiation injuries correlated not only with the total dose but also with fraction size.

Marcus and Million [118] reported that the risk of myelitis was 0/124 for a dose range 3,000–3,999 cGy, 0/442 for 4,000–4,499 cGy, 2/471 for 4,500–4,999, and 0/75 for 5,000 cGy or greater. They concluded that a risk of less than 0.5 % is often worth taking if it is necessary to treat a tumor near the spinal cord to a dose near 5,000 cGy and that a total dose of 5,500 given in fractions of less than 200 cGy was associated with a very low risk of permanent neurologic damage.

With conventional fractionation of 2 Gy per day, a total dose of 50 Gy, 60 Gy, and 69 Gy is associated with a 0.2, 6, and 50 % rate of myelopathy [119] (Table 11.5).

11.6.2 Reirradiation of the Spinal Cord

In evaluating reirradiation of the spinal cord, one must not only consider the dose regimen for each course and the volume and region (re)irradiated but also the time interval between the courses of RT [122]. For purposes of comparing different

Table 11.5 Studies related to myelopathy

Author	Total dose	Fraction	2 Gy dose equivalent[a]	n	Myelopathy	Myelopathy probability[b]
Marcus and Million [118]	47.5	1.9	45	211	0	0.000
	52.5	1.9	49.8	22	0	0.000
	60	2	60	19	2	0.118
Abbatucci et al. [115]	54	3	72.8	15	7	0.622
Atkins and Tretter [120]	19	9.5	68.6	13	4	0.437
McCunniff and Liang [117]	60	2	60	12	1	0.090
	65	1.63	56.6	24	0	0.000
Jeremic et al. [121]	65	1.63	56.6	19	0	0.000

[a] $\alpha/\beta = 0.87$ Gy
[b] Schultheiss et al. [110]

regimens (published reports) involving reirradiation of the spinal cord using both conventional, full-circumference external beam RT and SBRT [119], an α/β of 3 Gy was used to calculate the biologically equivalent dose in Gy3 and both α/β values of 1 and 3 Gy were employed to calculate the 2-Gy per fraction equivalent dose. In all of these studies, the median interval between courses was at least 6 months, and only a small number of cases were treated at intervals less than 6 months. Note that few cases of myelopathy are reported despite large cumulative doses, with essentially no cases of myelopathy observed for cumulative doses 60 Gy in 2-Gy equivalent doses. These data are consistent with the observations of post-RT repair observed in the animal models. For reirradiation at 2 Gy per day after prior conventionally fractionated treatment, cord tolerance appears to increase at least 25 % 6 months after the initial course of RT based on animal and human studies [119].

Ang [122] treated the thoracic and cervical spines of Rhesus monkeys to 44 Gy and then reirradiated these animals with an additional 57 Gy at 1–2 years or 66 Gy at 2–3 years, yielding aggregate doses of 101 and 110 Gy, respectively. The study end point was lower extremity weakness or balance disturbances at 2.5 years after reirradiation. Of 45 animals evaluated at the end of the observation period, 4 developed end point symptoms.

A reirradiation tolerance model developed by combining these data with those of a prior study of single-dose tolerance in the same animal model [123] resulted in an estimated recovery of 34 Gy (76 %), 38 Gy (85 %), and 45 Gy (101 %) at 1, 2, and 3 years, respectively. Under conservative assumptions, an estimated overall recovery of 26 Gy (61 %) was calculated.

Treatment with IMRT is characterized by much smaller spinal cord volumes irradiated to high doses, and, if a single IMRT plan is used, the dose/fraction delivered to the SC is usually much lower than the fraction dose around 2 Gy delivered to the target. Adopting maximum allowable spinal cord doses that were used in the 2D and 3D eras (e.g., maximal dose of 50 Gy over a treatment course of 7 weeks) is therefore more conservative than in previous eras, where full fraction doses of

1.8–2.0 Gy were typically delivered homogeneously to relatively large spinal cord segments. However, clinical data showing that escalation of the permissible cord dose is safe are lacking, therefore standard recommendations limit maximal cord dose to 45–50 Gy. If tumors extend very near the spinal cord (uncommon in most HN cancers), some relaxation of this rule is likely quite safe. Rather than using absolute maximum derived from the DVH, which is often received by a single voxel, it is preferable to limit maximal dose to a small volume such as 1 %, to avoid being overly conservative.

11.6.3 Radiation-Induced Brachial Plexopathy (RIBP)

The brachial plexus is a network of nerve bundles that originate from the cervical and upper thoracic spinal cord and is intimately responsible for the cutaneous and muscular innervation of the chest, shoulder, and upper extremity. It begins at the ventral rami of nerve roots at the 5th cervical vertebrae and continues inferiorly to include the nerve roots exiting the neural foramen of the first thoracic vertebrae. It then passes inferolaterally between the anterior and middle scalene muscles to innervate the cutaneous skin of the upper extremity and numerous muscles including the latissimus dorsi, pectoralis major and minor, levator scapulae, deltoid, and biceps brachii.

Though damage to this complex can occur as a result of trauma, tumors, or inflammation, it is believed to be susceptible to injury from therapeutic radiation. Neuropathies attributed to radiation therapy including upper extremity paresthesia, pain, weakness, and motor dysfunction were described as early as 1966 among patients irradiated to the neck and supraclavicular region [124]. The brachial plexus (BP) is adjacent to metastatic lymph nodes and high-risk nodal volume in the neck and supraclavicular area, and it is inevitably covered in the radiation portals using conventional techniques and irradiated to 54–70 Gy. This prompted the Radiation Therapy Oncology Group (RTOG) to include the BP as an OAR in many recent protocols. RIBP is a relatively common late complication after postoperative RT for breast cancer: The reported prevalence is 1–6 % at 5 years after 45–54 Gy irradiation [123, 125, 126]. For patients undergoing radiation therapy for cancers of the head and neck, which typically require significantly higher doses to these areas, there are very few published data on these side effects.

RIBP is a potentially painful and debilitating complication of radiotherapy, characterized by sensory changes and motor deficits [127]. There may be a gradual evolution of symptoms or a more rapid progression with time, which may on occasion culminate in complete loss of function of the affected arm [123, 125]. The median time to onset of symptoms ranging from 6.5 months to 4 years after the completion of radiotherapy and ranges of 1.4 months to 26 years [128, 129].

Postmortem examination in clinically symptomatic patients have revealed such changes as fibrosis surrounding the nerves of the brachial plexus, fibrous thickening of the nerve sheath, demyelination, and fibrous replacement of individual nerve fibers [124]. Notably, these changes are limited to the irradiated region, with extensive demyelination and atrophy seen distal to the radiation field.

The natural course of radiation injury to the brachial plexus is uncertain and has been reported to range widely. Although the development of radiation-induced nerve injury is believed to be a slow process, with a latency period of 1–4 years, some investigators have reported a much more rapid onset of symptoms [130]. What is clear, however, is that peripheral neuropathies increase with prolonged follow-up. In the largest published series, Powell et al. [123] reported that the actuarial incidence of radiation-induced brachial plexopathy was approximately 5 % at 5.5 years among 449 patients treated with postoperative radiation therapy for breast cancer. The investigators did identify a significant difference among patients irradiated using 3 Gy per fraction compared with 1.8 Gy. Notably, the calculated dose to the brachial plexus was 54 Gy using conventional fractionation, which represented a dose approximately 20 % lower than Chen et al. [131]. Platteaux et al. [132] found no RIBP when he retrospectively analyzed 43 cases of HNC treated with IMRT, with a median follow-up of 24 months. The max dose of 64.2 Gy and mean dose of 44.1 Gy were documented.

Chen et al. [131] was the first to report on nerve injuries after high-dose radiation therapy for head and neck cancer. He suggested that a significant proportion of patients experience symptoms, consistent with prior reports from the breast cancer literature; the most common symptoms observed in the studies were paresthesia and dysesthesia of the upper extremities. The use of surgical neck dissection and concurrent chemotherapy each appeared to increase the risk for developing neuropathies; it was notable that 9 and 8 % of patients who never underwent neck dissection and were treated without chemotherapy, respectively, developed symptoms. In the multivariate analysis, when the brachial plexus was irradiated to doses >70 Gy, a significantly higher proportion of patients developed neuropathic symptoms, suggesting a threshold effect. This study was retrospective, relying on patient complaints such as shoulder weakness or pain, and no physical examination or electromyographs were used. The fact that neck surgery increased the risk in this series suggests that at least some of the cases were related to accessory nerve damage from surgery rather than brachial plexopathy. Other publications showed no brachial plexopathy in patients with HNC whose BP received high doses [132, 133]. It is likely that the report by Chen et al. overestimated substantially the risk of BP damage by RT.

Emami et al. [134] had suggested that the TD 5/5 to the entire brachial plexus was 60 Gy. More recently, several studies with over 20 years of follow-up have suggested that the incidence of brachial plexopathy continues to rise after 5 years and may not be apparent for up to 20 years after radiotherapy [126, 135]. The brachial plexus appears to be especially sensitive to fractionation schedules, with the risk of injury much higher for larger fractions despite equivalent BED [123]. With standard fractionation, the risk of clinically apparent nerve damage seems to be <5 %, after 5 years of completing radiotherapy, when the brachial plexus is limited to 60 Gy. At the University of Michigan, adequate treatment of gross disease in the low neck takes precedence over BP doses, and patients with complaints suggesting possible brachial plexopathy have been referred for neurological consultation and electromyographs. Thus far, we have not observed any proven case of brachial plexopathy.

11.6.4 Radiation-Induced Optic Neuropathy (RION)

Developmentally, the optic nerve is part of the brain; its fibers are surrounded by glial, not Schwann cell sheaths [136]. The optic nerve is comprised of over 1 million retinal ganglion cell axons. Its intraocular portion is not myelinated and derives its blood supply from the retinal arterioles and branches of the posterior ciliary artery. The myelinated intraorbital, intracanalicular, and intracranial portions of the optic nerve are supplied by branches of the central retinal artery and choroidal vasculature, the ophthalmic artery, and branches of the internal carotid and ophthalmic arteries.

The therapeutic dose levels for tumors in the central nervous system and head and neck area are often constrained by the radiation tolerance of the optic apparatus. Visual impairment from radiation-induced optic neuropathy (RION) is uncommon [137, 138]. It usually results in rapid painless visual loss [139]. It usually presents with painless rapid visual loss. Vasculature injury has been suggested as a significant contributor to RION [140, 141].

Two types of optic neuropathy [142, 143] (anterior ischemic optic neuropathy and retrobulbar ischemic optic neuropathy) may be seen after irradiation. Both types are believed to be caused by vascular occlusive disease, with interruption of blood supply to either the nerve head or the retrobulbar portion of the nerve, respectively. Patients with preexisting small vessel occlusive disease are probably at increased risk. Ophthalmoscopic findings in anterior ischemic optic neuropathy include disk pallor and edema with splinter hemorrhages on or adjacent to the disk; the clinical picture may be similar to that seen in patients with papilledema. The optic disks are initially normal in most patients with retrobulbar optic neuropathy. In both types of injury, optic atrophy typically develops within 6–8 weeks. Visual loss secondary to optic neuropathy is initially monocular in most cases. In some patients, the second optic nerve is affected, usually weeks or months, but occasionally several years, after the first eye was affected [142, 143]. Some patients are not aware of visual loss in the first eye until the second eye becomes affected or until the patient happens to cover or close the normal eye and note vision loss in the other eye. Some patients present with visual field deficits, which are usually nerve fiber bundle defects, particularly altimdinal defects that are highly suggestive of optic nerve ischemia. Other observed deficits include cecocentral scotomas, temporal field loss, and generalized constricted fields [143].

Optic nerve injury typically results in monocular visual loss, except if it occurs very close to the optic chiasm, where fibers looping up from the contralateral medial eye/retina can be affected. Injury to the entire chiasm can cause bilateral vision loss. Temporary injury limited to the inferior central optic chiasm from pituitary adenoma results in bilateral upper outer quadrant visual field impairment. The loss of a proximal optic tract causes loss of the same half of the visual field in each eye. Because the optic tracts spread out on their way toward the occipital cortex, injuries along the way typically result in small visual field cuts. Uncertainties exist in scoring the toxicity. Acuity problems can result from cataracts, dry eye, or radiation retinopathy (usually distinguishable by examination). Vascular insufficiency to the retina, optic nerves, tracts, or occipital lobes can also cause visual impairments, particularly visual field deficits. Because patients often undergo RT to many of these areas concurrently, it can be challenging to know how to accurately ascribe the clinical events. Lesions anterior to the chiasm will affect the ipsilateral eye, lesions of the chiasm will affect the bilateral temporal visual fields, and lesions posterior to the chiasm will affect visual fields in both eyes.

Based on the QUANTEC [144] review, a whole organ dose of 50 Gy is associated with <1 % risk of blindness. In fact, blindness was quite rare until a dose of ≥55 Gy. Between 55 and 60 Gy, the risk of blindness is approximately 3–7 %. At doses >60 Gy, the risk of RION greatly increases. Goldsmith et al. [145] reported on 49 patients whose optic nerves were incidentally irradiated to doses of 45–59.4 Gy in 28–45 fractions (mean, 53.6 Gy in 30 fractions) postoperatively after resection of meningiomas; optic neuropathy developed in one patient (54 Gy in 30 fractions over 43 days). Optic neuropathy did not develop in any of 106 nerves in the current study that received<59 Gy. In a review in 2008, Danesh–Meyer [138] hypothesized that RION likely results from a multifactorial process, whereby free radicals lead to damage both of the vascular endothelia and the neuroglial cell progenitors, which leads to the clinical vision loss.

Treatment of radiotherapy (RT)-associated visual loss is limited. A successful treatment with steroids, pentoxifylline, warfarin, and vitamin E is reported by Weintraub et al. [146].

11.6.4.1 Dose Effect

The incidence of RION was unusual for aDmax<55 Gy, particularly for fraction sizes <2 Gy. The risk increases (3–7 %) in the region of 55–60 Gy and becomes more substantial (>7–20 %) for doses >60 Gy (1.8–2.0 Gy/Fr). The patients with RION treated in the 55–60 Gy range were typically treated to doses in the very high end of that range (i.e., 59 Gy). One exception to this range was for pituitary tumors, in which investigators used a constrained Dmax of<46 Gy for 1.8 Gy/fraction [144].

11.7 Osteoradionecrosis of the Mandible

Osteoradionecrosis (ORN) of the mandible is one of the known late complications after radiation therapy for HNC [147, 148]. In general, bones are radioresistant and will sustain any overt damage as long as the Overlying soft tissue remains intact and the bone is not subjected to excessive stress or trauma. A number of risk factors are associated with ORN: treatment type, total radiation dose, and associated trauma such as teeth extraction before or after radiation therapy in addition to age, general health, dentition status, oral hygiene, proximity of the tumor to the mandible, or its invasion [147].

Mandible sparing is not easily achieved, given the location of the tumors, and should not be attempted when the mandible is an "at-risk" structure [149]. ORN is most prevalent following radiation treatment of tumors of the tongue, floor of mouth, retromolar trigone, and tonsil [150]. This relates to the mandible being in the high-dose region for these tumor types.

ORN is characterized by a nonhealing area of exposed bone of at least 6 months duration in a patient who has been treated with radiation therapy for cancer. ORN is associated with pain and morbidity and, in advanced stages, typically requires surgical resection and reconstruction for management. It is a progressive disease in which radiation injuries involve soft tissue, cartilage, and bone, resulting in gradual devitalization of these tissues [151]. The associated morbidity of this condition and its subsequent treatment ranges from close observation to radical mandibulectomy [152].

In recent years, there has been a reduction in the incidence of osteoradionecrosis, from 11.8 % before 1968 to 5.4 % from 1968 to 1992, and it decreased again after 1997 to approximately 3%. [153]. The main factors contributing to the reduction in the osteoradionecrosis rates are more conformal dose distributions, which spare parts of the mandible that might have received a higher dose using conventional radiation therapy, and also better prophylactic and ongoing dental care [153].

The risk of ORN increases with radiation dosages above 6,000 cGy, previous cancer resection, advanced dental disease status, and postradiation dental extractions [152, 154, 155]. The risks of developing ORN of the mandible with older radiation techniques are in the order of 5–15 % [149] compared to 6 % by newer techniques [156]. Spontaneous ORN is dose dependent (60 Gy) [157] and relates to the volume of mandible within the treatment field [158]. However, trauma-related ORN can occur at lower doses with no obvious threshold dose. Reuther et al. [150] demonstrated a significant correlation between radiation dose and extent of ORN. Most ORN lesions measuring 2 cm or more occurred at doses of 60 Gy or higher. At this dose level, there was also a preponderance of smaller ORN lesions. Below 60 Gy, most lesions were <2 cm in size [159].

Reuther et al. studied in a cohort of 68 patients with ORN; 34 (50 %) had a tooth extraction closely associated with the onset of ORN [150]. However, not all patients who develop ORN have identifiable specific causes and not all risk factors (e.g., severely compromised dentition) predictably cause ORN [160–163]. These studies collectively demonstrated that the weighted prevalence with IMRT was lower than with conventional RT (5.2 % vs. 7.3 %, respectively)

At the University of Michigan, total 176 patients were analyzed with a minimal follow-up of 6 months. Of these, 31 (17 %) had undergone teeth extractions before RT and 13 (7 %) after RT. Of the 176 patients, 75 % and 50 % had received >65 Gy and >70 Gy to >1 % of the mandibular volume, respectively. At a median follow-up of 34 months, no cases of ORN had developed (95 % confidence interval, 0–2 %). They have followed a strict prophylactic dental care policy and IMRT with no case of clinical ORN.

Gomez et al. [155] further provided evidence that the mechanisms for radiation-induced dental caries and dental extractions differ, with the incidence of dental caries being more related to the dose to the salivary glands and dental extractions being a consequence of radiation directly to the mandible.

I t is believed that reductions in dose to the salivary glands and mandible are likely to translate into reduced incidence of xerostomia and osteoradionecrosis for patients with HNC [162].

11.7.1 Prevention

To minimize all of these risks, a strict prophylactic dental policy and adherence to dose constraints of the mandible and salivary glands as much as feasible so as not to compromise target coverage, as well as routine and aggressive dental care, before, during, and after radiation treatment, while keeping doses to the salivary glands and mandible as low as possible are needed [155].

The best treatment of ORN is prevention. Careful oral examination prior to radiation, treatment of dental caries, extraction when necessary, fluoride trays after treatment, and good oral hygiene all contribute to reducing the incidence of ORN after radiation treatment [164].

11.7.2 Hyperbaric Oxygen Therapy

Head and neck radiation also affects a variety of critical surrounding normal tissues, which can become hypocellular, hypovascular, and hypoxic, frequently eluded to as "3 H tissue."

Tissue hypoxia was accepted as the primary cause of ORN, and this led to the use of hyperbaric oxygen (HBO) for both treatment and prevention of the complications of RT in the head and neck. Over the past decades, a number of theories about the pathogenesis of ORN have been proposed, with consequent implications for its treatment. Recently a new theory proposes that damage to bone is caused by radiation-induced fibrosis, cells in the bone are damaged as a result of acute inflammation and free radicals, and there is chronic activation of fibroblasts by a series of growth factors [164].

Hyperbaric oxygen therapy (HBOT) involves inhalation of 100 % oxygen under a pressure of greater than one atmosphere absolute (ATA). The high pressure increases the amount of oxygen being circulated within the body, which is believed to have various effects including the promotion of vascularization (growth of blood vessels) [165]. Patients inhale pressurized oxygen either through a hood or mask in a multiplace hyperbaric chamber or through an oxygen-filled monoplace chamber in which the patient lies. The Marx protocol is the most widely used protocol for HBOT in the prevention and treatment of osteoradionecrosis developed by Marx et al. [166].

HBOT has been used to assist in the repair of radiation-induced damage [167]. In theory, HBO may stimulate monocytes and fibroblasts function and collagen synthesis [168, 169] and may increase vascular density [170]. The effects of hyperbaric oxygen can be short term due to enhanced oxygen delivery, reduction of edema, and phagocytosis activation, and anti-inflammatory effects are long term due to neovascularization, osteoneogenesis, and stimulation of collagen formation by fibroblasts [171].

HBO increases vascular density and oxygenation in radiation-damaged tissue [172]. It improves tissue oxygen gradients and angiogenesis and enhances leukocyte bactericidal activity. Oxygen tension is increased, enabling fibroblast proliferation, collagen formation, and angiogenesis at the wound edges, further improving oxygenation and re-epithelialization [173].

Marx et al. [166] showed that up-front hyperbaric oxygen reduced the incidence of osteoradionecrosis to 5.4 % compared with the antibiotic group of 29.9 % in a high-risk population who required tooth removal from irradiated mandibles. He recommends that HBO be considered as prophylactic therapy for all patients who require postradiation dental care involving significant trauma to bone.

ORN appears to be highly responsive to HBOT (81 %) [174]. Results show that among H&N subjects who showed favorable response of bone or nonbone symptoms to HBOT (21 of 28), a higher percentage improved after HBOT alone (62 %) compared to the combined treatment of HBOT and surgery (38 %).

Uncontrolled studies showed recovery rates from ORN of 15–45 % with HBO alone and 20–90 % with HBO combined to surgery [166, 167]. Studies have shown that HBOT effectively treats irradiated soft tissue necrosis [175, 176] and has also been used empirically to treat mandibular ORN. Mainous et al. [177] has also reported improved mandibular healing with HBO after radiotherapy for head neck.

Limited evidence suggests HBOT may be effective in promoting the healing of ORN of the mandible following surgical interventions. HBOT alone is not effective in treating ORN of the mandible compared to placebo. There is a lack of evidence regarding the effect of HBOT on tooth implants and associated ORN.

11.7.2.1 Dental Caries

Postradiation, there is a profound shift in the oral microflora to a predominance of acidogenic microbes, primarily *Streptococcus mutans* and lactobacilli, coincident with a decrease in salivary flow and an increase in caries risk [178–180]. Dental caries may occur as early as 3 months after RT. It typically involves the cervical portions of the teeth; however, caries may affect any tooth surface, including those typically resistant to dental caries such as the incisal edges of the mandibular incisors [181, 182].

11.7.2.2 Prevention of Dental Caries

A strict daily oral hygiene regimen should be followed. It includes fluoride and meticulous plaque removal to prevent the development of caries [179, 181] and chlorhexidine gel/mouth wash to clinically reduce caries risk by lowering counts of mutants streptococci and lactobacilli in patients undergoing RT [183, 184]. Alcohol-free formulations should be selected to reduce discomfort in patients with dry mouth. Caries lesions should be restored before RT to prevent progression of disease and reduce microbial load.

The patient will be more comfortable during treatment if the oral mucosa is intact. Diet counseling regarding cariogenic food and its effect on dentition needs to be done prior to starting radiation. Vissink et al. [185] concluded that a lifelong commitment to improved oral hygiene and home care should include meticulous oral hygiene and frequent self-applications of fluoride, either neutral NaF 1 % gel applied at least every other day [182, 186] in custom-made fluoride carriers or NaF 3 % toothpaste twice per day [181]. The daily use of 4 % stannous fluoride also is effective [187, 188]. Presently, there is inadequate evidence to support one type of fluoride product over another for patients undergoing RT; the frequency of application appears to be more important. Because hyposalivation is irreversible in most head and neck irradiation patients, especially those treated with standard therapy, the application of fluoride must be continued indefinitely; otherwise, caries will develop within months [186, 189–192]. On the other hand, the experience at the University of Michigan suggests that sparing saliva by IMRT results in a substantial reduction of teeth decay rates. We continue to recommend the use of prophylactic therapy, including fluoride, even in patients with mild xerostomia after IMRT.

References

1. Nuyts S, Dirix P, Clement PM et al (2009) Impact of adding concomitant chemotherapy to hyperfractionated accelerated radiotherapy for advanced head and neck squamous cell carcinoma. Int J Radiat Oncol Biol Phys 73:1088–1095
2. Garden AS, Harris J, Trotti A et al (2008) Long-term results of concomitant boost radiation plus concurrent cisplatin for advanced head and neck carcinomas: a phase II trial of the radiation therapy oncology group (RTOG 99–14). Int J Radiat Oncol Biol Phys 71(5):1351–1355
3. Harari PM (2008) Beware the swing and a miss: baseball precautions for conformal radiotherapy. Int J Radiat Oncol Biol Phys 70:657
4. Eisbruch A et al (2007) Editorial: reducing xerostomia by IMRT: what may, and may not, be achieved. J Clin Oncol 25(31):4863–4864
5. Dirix P, Nuyts S, Van den Bogaert W (2006) Radiation-induced xerostomia in patients with head and neck cancer: a literature review. Cancer 107:2525–2534
6. Dirix P, Nuyts S (2010) Evidence-based organ-sparing radiotherapy in head and neck cancer. Lancet Oncol 11(1):85–91
7. Eisbruch A, Ship JA, Martel MK et al (1996) Parotid gland sparing in patients undergoing bilateral head and neck irradiation: techniques and early results. Int J Radiat Oncol Biol Phys 36:469–480
8. Eisbruch A, Rhodus N, Rosenthal D et al (2003) How should we measure and report radiotherapy-induced xerostomia? Semin Radiat Oncol 13:226–234
9. Chen W-C, Lai CH et al (2013) Scintigraphic assessment of salivary function after intensity-modulated radiotherapy for head and neck cancer: correlations with parotid dose and quality of life. Oral Oncol 49:42–48
10. Blanco AI, Chao KS, El Naqa I et al (2005) Dose-volume modelling of salivary function in patients with head-and-neck cancer receiving radiotherapy. Int J Radiat Oncol Biol Phys 62:1055–1069
11. Eisbruch A, Harris J, Garden AS et al (2010) Multi-institutional trial of accelerated hypofractionated intensity-modulated radiation therapy for early-stage oropharyngeal cancer (RTOG 00–22). Int J Radiat Oncol Biol Phys 76(5):1333–1338

12. Scrimger R, Kanji A, Parliament M et al (2007) Correlation between saliva production and quality of life measurements in head and neck cancer patients treated with intensity-modulated radiotherapy. Am J Clin Oncol 30:271–277
13. Saarilahti K, Kouri M, Collan J et al (2005) Intensity modulated radiotherapy for head and neck cancer: evidence for preserved salivary gland function. Radiother Oncol 74:251–258
14. Parliament MB, Scrimger RA, Anderson SG et al (2004) Preservation of oral health-related quality of life and salivary flow rates after inverse-planned intensity-modulated radiotherapy (IMRT) for head-and-neck cancer. Int J Radiat Oncol Biol Phys 58:663–673
15. Münter MW, Karger CP, Hoffner SG et al (2004) Evaluation of salivary gland function after treatment of head-and-neck tumors with intensity-modulated radiotherapy by quantitative pertechnetate scintigraphy. Int J Radiat Oncol Biol Phys 58:175–184
16. Eisbruch A, Kim HM, Terrell JE et al (2001) Xerostomia and its predictors following parotid-sparing irradiation of head-and-neck cancer. Int J Radiat Oncol Biol Phys 50:695–704
17. Chao KS, Deasy JO, Markman J et al (2001) A prospective study of salivary function sparing in patients with head-and-neck cancers receiving intensity-modulated or three-dimensional radiation therapy: initial results. Int J Radiat Oncol Biol Phys 49:907–916
18. Lee N, Harris J, Garden AS et al (2009) Intensity-modulated radiation therapy with or without chemotherapy for nasopharyngeal carcinoma: radiation therapy oncology group phase II trial 0225. J Clin Oncol 27(22):3684–3690
19. Pow EH, Kwong DL, McMillan AS et al (2006) Xerostomia and quality of life after intensity-modulated radiotherapy vs. conventional radiotherapy for early-stage nasopharyngeal carcinoma: initial report on a randomized controlled clinical trial. Int J Radiat Oncol Biol Phys 66(4):981–991
20. Kam MK, Leung SF, Zee B et al (2007) Prospective randomized study of intensity-modulated radiotherapy on salivary gland function in early-stage nasopharyngeal carcinoma patients. J Clin Oncol 25:4873–4879
21. Braam PM, Terhaard CH, Roesink JM, Raaijmakers CP (2006) Intensity-modulated radiotherapy significantly reduces xerostomia compared with conventional radiotherapy. Int J Radiat Oncol Biol Phys 66:975–980
22. Feng M, Jabbari S, Lin A et al (2005) Predictive factors of local-regional recurrences following parotid sparing intensity modulated or 3D conformal radiotherapy for head and neck cancer. Radiother Oncol 77:32–38
23. Daly ME, Lieskovsky Y, Pawlicki T et al (2007) Evaluation of patterns of failure and subjective salivary function in patients treated with intensity modulated radiotherapy for head and neck squamous cell carcinoma. Head Neck 29:211–220
24. Eisbruch A, Marsh LH, Dawson LA et al (2004) Recurrences near base of skull after IMRT for head-and-neck cancer: Implications for target delineation in high neck and for parotid gland sparing. Int J Radiat Oncol Biol Phys 59:28–42
25. David MB, Eisbruch A (2007) Delineating neck targets for intensity-modulated radiation therapy of head and neck cancer. What have we learned from marginal recurrences? Front Radiat Ther Oncol 40:193–207
26. Astreinidou E, Dehnad H, Terhaard CH, Raaijmakers CP (2004) Level II lymph nodes and radiation-induced xerostomia. Int J Radiat Oncol Biol Phys 58:124–131
27. Grégoire V, Levendag P, Ang KK et al (2003) CT-based delineation of lymph node levels and related CTVs in the node negative neck: AHANCA, EORTC, GORTEC, RTOG consensus guidelines. Radiother Oncol 69:227–236
28. Grégoire V, Eisbruch A, Hamoir M, Levendag P (2006) Proposal for the delineation of the nodal CTV in the node-positive and the post-operative neck. Radiother Oncol 79:15–20
29. Maes A, Weltens C, Flamen P et al (2002) Preservation of parotid function with uncomplicated conformal radiotherapy. Radiother Oncol 63:203–211
30. Li Y, Taylor J, Ten Haken R et al (2007) The impact of dose on parotid salivary recovery in head and neck cancer patients treated with radiation therapy. Int J Radiat Oncol Biol Phys 67:660–669

31. Robar JL, Day A, Clancey J et al (2007) Spatial and dosimetric variability of organs at risk in head-and-neck intensity-modulated radiotherapy. Int J Radiat Oncol Biol Phys 68:1121–1130
32. Little M et al (2012) Reducing xerostomia after chemo-IMRT for head-and-neck cancer: beyond sparing the parotid glands. Int J Radiat Oncol Biol Phys 83(3):1007–1014
33. Eisbruch A, Ten Haken RK, Kim HM et al (1999) Dose, volume, and function relationships in parotid salivary glands following conformal and intensity-modulated irradiation. Int J Radiat Oncol Biol Phys 45:577–587
34. Barker JL, Garden AS, Ang KK et al (2004) Quantification of volumetric and geometric changes during fractionated radiotherapy. Int J Radiat Oncol Biol Phys 59:960–970
35. Chambers MS, Garden AS, Rosenthal D et al (2005) Intensity-modulated radiotherapy: is xerostomia still prevalent? Curr Oncol Rep 7(2):131–136
36. Deasy JO et al (2010) Radiotherapy dose–volume effects on salivary gland function. Int J Radiat Oncol Biol Phys 76(3):S58–S63
37. Portaluri M, Fucilli F, Castagna R et al (2006) Three-dimensional conformal radiotherapy for locally advanced (stage II and worse) head-and-neck cancer: dosimetric and clinical evaluation. Int J Radiat Oncol Biol Phys 66:1036–1043
38. Tsujii H (1985) Quantitative dose–response analysis of salivary function following radiotherapy using sequential RI-sialography. Int J Radiat Oncol Biol Phys 11:1603–1612
39. Bagesund M, Richter S, Ringden O et al (2007) Longitudinal scintigraphic study of parotid and submandibular gland function after total body irradiation. Int J Paediatr Dent 17:34–40
40. Malpani BL, Samuel AM, Ray S et al (1995) Differential kinetics of parotid and submandibular gland function as demonstrated by scintigraphic means. Nucl Med Commun 16:706–709
41. Raza H, Khan AU, Hameed A et al (2006) Quantitative evaluation of salivary gland dysfunction after radioiodine therapy using salivary gland scintigraphy. Nucl Med Commun 27:495–499
42. Kashima HK, Kirkman WR, Andrews JR (1965) Postradiation sialadenitis: a study following irradiation of human salivary glands. AJR Am J Roentgenol 94:271–291
43. Stephens LC, King GK, Peters LJ et al (1986) Acute and late radiation injury in rhesus parotid glands. Am J Pathol 124:469–478
44. Murdoch-Kinch C et al (2008) Dose effect relationships for the submandibular salivary glands and implications for their sparing by intensity modulated radiotherapy. Int J Radiat Oncol Biol Phys 72(2):373–382
45. Dijkstra PU, Kalk WW, Roodenburg JL (2004) Trismus in head and neck oncology: a systematic review. Oral Oncol 40:879–889
46. Paleri V, Roe JWG, Strojan P, Corry J, Grégoire V, Hamoir M, Eisbruch A, Mendenhall WM, Silver CE, Rinaldo A, Takes RP, Ferlito A (2013) Strategies to reduce long-term postchemoradiation dysphagia in patients with head and neck cancer: an evidence-based review. Head Neck. doi:10.1002/hed.23251
47. Caglar HB, Tishler RB, Othus M et al (2008) Dose to larynx predicts for swallowing complications after intensity-modulated radiotherapy. Int J Radiat Oncol Biol Phys 72:1110–1118
48. Caudell JJ, Schaner PE, Meredith RF, Locher JL, Nabell LM, Carroll WR et al (2009) Factors associated with long-term dysphagia after definitive radiotherapy for locally advanced head-and-neck cancer. Int J Radiat Oncol Biol Phys 73(2):410–415
49. Dirix P, Abbeel S, Vanstraelen B, Hermans R, Nuyts S (2009) Dysphagia after chemoradiotherapy for head-and-neck squamous cell carcinoma: dose–effect relationships for the swallowing structures. Int J Radiat Oncol Biol Phys 75(2):385–392
50. Guadagnolo BA, Liu CC, Cormier JN, Du XL (2010) Evaluation of trends in the use of intensity-modulated radiotherapy for head and neck cancer from 2000 through 2005: socioeconomic disparity and geographic variation in a large population-based cohort. Cancer 116(14):3505–3512

51. Ward EC, van As-Brooks CJ (2007) Head and neck cancer: treatment, rehabilitation, and outcomes. Plural Publishing, , San Diego/Abingdon
52. Van der Molen L, van Rossum MA, Burkhead LM et al (2011) A randomized preventive rehabilitation trial in advanced head and neck cancer patients treated with chemoradiotherapy: feasibility, compliance, and short-term effects. Dysphagia 2:155–170
53. Roe JW, Carding PN, Dwivedi RC et al (2010) Swallowing outcomes following intensity modulated radiation therapy (IMRT) for head & neck cancer – a systematic review. Oral Oncol 46:727–733
54. Levendag PC, Teguh DN, Voet P et al (2007) Dysphagia disorders in patients with cancer of the oropharynx are significantly affected by the radiation therapy dose to the superior and middle constrictor muscle: a dose–effect relationship. Radiother Oncol 85:64–73
55. Van der Molen L, van Rossum MA, Burkhead LM et al (2009) Functional outcomes and rehabilitation strategies in patients treated with chemoradiotherapy for advanced head and neck cancer: a systematic review. Eur Arch Otorhinolaryngol 266:889–900
56. Herb K, Cho S, Stiles MA (2006) Temporomandibular joint pain and dysfunction. Curr Pain Headache Rep 10:408–414
57. Goldstein M, Maxymiw WG, Cummings BJ et al (1999) The effects of antitumor irradiation on mandibular opening and mobility: a prospective study of 58 patients. Oral Surg Oral Med Oral Pathol Oral Radiol Endod 88:365–373
58. Hutcheson KA, Lewin JS, Barringer DA, Lisec A, Gunn GB, Moore MW, Holsinger FC (2012) Late dysphagia after radiotherapy-based treatment of head and neck cancer. Cancer 118(23):5793–5799
59. Eisbruch A, Schwartz M, Rasch C et al (2004) Dysphagia and aspiration after chemoradiotherapy for head-and-neck cancer: which anatomic structures are affected and can they be spared by IMRT? Int J Radiat Oncol Biol Phys 60:1425–1439
60. Snadra N et al (2013) Reduction of the dose to the elective neck in head and neck squamous cell carcinoma, a randomized clinical trial using intensity modulated radiotherapy (IMRT). Dosimetrical analysis and effect on acute toxicity. Radiother Oncol 109(2):323–329
61. Mendenhall WM, Amdur RJ, Morris CG, Kirwan JM, Li JG (2010) Intensity modulated radiotherapy for oropharyngeal squamous cell carcinoma. Laryngoscope 120:2218–2222
62. Cartmill B, Cornwell P, Ward E et al (2011) Emerging understanding of dosimetric factors impacting on dysphagia and nutrition following radiotherapy for oropharyngeal cancer. Head Neck Radiother Oncol 101:394–402
63. Feng FY, Kim HM, Lyden TH et al (2007) IMRT of head and neck cancer aiming to reduce dysphagia: early dose-effect relationships for the swallowing structures. Int J Radiat Oncol Biol Phys 68:1289–1298
64. Jensen K, Lambertsen K, Torkov P, Dahl M, Jensen AB, Grau C (2007) Patient assessed symptoms are poor predictors of objective findings. Results from a cross sectional study in patients treated with radiotherapy for pharyngeal cancer. Acta Oncol 46(8):1159–1168
65. Teguh DN, Levendag PC, Noever I et al (2008) Treatment techniques and site considerations regarding dysphagia-related quality of life in cancer of the oropharynx and nasopharynx. Int J Radiat Oncol Biol Phys 72:1119–1127
66. Feng FY, Kim HM, Lyden TH et al (2010) Intensity-modulated chemoradiotherapy aiming to reduce dysphagia in patients with oropharyngeal cancer: clinical and functional results. J Clin Oncol 28(16):2732–2738
67. Dirix P, Nuyts S (2010) Value of intensity-modulated radiotherapy in stage IV head-and-neck squamous cell carcinoma. Int J Radiat Oncol Biol Phys 78:1373–1380
68. Lee NY, de Arruda FF, Puri DR et al (2006) A comparison of intensity-modulated radiation therapy and concomitant boost radiotherapy in the setting of concurrent chemotherapy for locally advanced oropharyngeal carcinoma. Int J Radiat Oncol Biol Phys 66:966–974
69. Fua TF, Corry J, Milner AD, Cramb J, Walsham SF, Peters LJ (2007) Intensity modulated radiotherapy for nasopharyngeal carcinoma: clinical correlation of dose to the pharyngo-esophageal axis and dysphagia. Int J Radiat Oncol Biol Phys 67:976–981

70. Huguenin P, Beer KT, Allal A et al (2004) Concomitant cisplatin significantly improves locoregional control in advanced head and neck cancers treated with hyperfractionated radiotherapy. J Clin Oncol 22:4665–4673
71. Caudell JJ, Schaner PE, Desmond RA, Meredith RF, Spencer SA, Bonner JA (2010) Dosimetric factors associated with long-term dysphagia after definitive radiotherapy for squamous cell carcinoma of the head and neck. Int J Radiat Oncol Biol Phys 76:403–409
72. Teguh DN, Levendag PC, Sewnaik A et al (2008) Results of fiberoptic endoscopic evaluation of swallowing vs radiation dose in the swallowing muscles after radiotherapy of cancer in the oropharynx. Radiother Oncol 89:57–64
73. Eisbruch A, Lyden T, Bradford CR et al (2002) Objective assessment of swallowing dysfunction and aspiration after radiation concurrent with chemotherapy for head-and-neck cancer. Int J Radiat Oncol Biol Phys 53:23–28
74. Oh YT, Kim CH, Choi JH et al (2004) Sensory neural hearing loss after concurrent cisplatin and radiation therapy for nasopharyngeal carcinoma. Radiother Oncol 72:79–82
75. Merchant TE, Gould CJ, Xiong X et al (2004) Early neuro-otologic effects of three-dimensional irradiation in children with primary brain tumors. Int J Radiat Oncol Biol Phys 58:1194–1207
76. Mendenhall WM, Million RR, Cassisi NJ (1980) Elective neck irradiation in squamous cell carcinoma of the head and neck. Head Neck Surg 3:15–20
77. Van der Laan HP, Christianen ME, Bijl HP, Schilstra C, Langendijk JA (2012) The potential benefit of swallowing sparing intensity modulated radiotherapy to reduce swallowing dysfunction: an in silico planning comparative study. Radiother Oncol 103:76–81
78. van der Molen L et al (2013) Dysphagia and trismus after concomitant chemo-IMRT in advanced head and neck cancer; dose effect relationships for swallowing and mastication structures. Radiother Oncol 106:364–369
79. Wang X, Hu C, Eisbruch A (2011) Organ sparing radiation therapy for head and neck cancer. Nat Rev Clin Oncol 8(11):639–648
80. Bhandare N, Jackson A, Eisbruch A, Pan CC, Flickinger JC, Antonelli P, Mendenhall WM (2010) Radiotherapy and hearing loss. Int J Radiat Oncol Biol Phys 76(3):S50–S57
81. Pacholke HD, Amdur RJ, Schmalfuss IM, Louis D, Mendenhall WM (2005) Contouring the middle and inner ear on radiotherapy planning scans. Am J Clin Oncol 28(2):143–147
82. Hunag E et al (2002) IMRT for pediatric medulloblastoma: early report on the reduction of ototoxicity. Int J Radiat Oncol Biol Phys 52:599–605
83. Sultanem K, Hk S, Xia P et al (2000) Three-dimensional intensity modulated radiotherapy in the treatment of nasopharyngeal carcinoma: the University of California-San Francisco experience. Int J Radiat Oncol Biol Phys 48:711–722
84. Grau C, Møller K, Overgaard M, Overgaard J, Elbrønd O (1991) Sensorineural hearing loss in patients treated with irradiation for nasopharyngeal carcinoma. Int J Radiat Oncol Biol Phys 21:723–728
85. Anteunis LJ, Wanders SL, Hendriks JJ et al (1994) A prospective longitudinal study on radiation-induced hearing loss. Am J Surg 168:408–411
86. Chen WC, Jackson A, Budnick AS et al (2006) Sensorineural hearing loss in combined modality treatment of nasopharyngeal carcinoma. Cancer 106:820–829
87. Honoré HB, Bentzen SM, Møller K, Grau C (2002) Sensori-neural hearing loss after radiotherapy for nasopharyngeal carcinoma: individualized risk estimation. Radiother Oncol 65:9–16
88. Johannesen TB, Rasmussen K, Winther FØ, Halvorsen U, Lote K (2002) Late radiation effects on hearing, vestibular function, and taste in brain tumor patients. Int J Radiat Oncol Biol Phys 53:86–90
89. Pan CC, Eisbruch A, Lee JS et al (2005) Prospective study of inner ear radiation dose and hearing loss in head-and-neck cancer patients. Int J Radiat Oncol Biol Phys 61:1393–1402
90. Herrmann F, Dörr W, Müller R, Herrmann T (2006) A prospective study on radiation-induced changes in hearing function. Int J Radiat Oncol Biol Phys 65:1338–1344

91. Hitchcock YJ, Tward JD, Szabo A, Bentz BG, Shrieve DC (2009) Relative contributions of radiation and cisplatin-based chemotherapy to sensorineural hearing loss in head-and-neck cancer patients. Int J Radiat Oncol Biol Phys 73(3):779–788
92. Van de Putten L, de Bree R, Plukker JT et al (2006) Permanent unilateral hearing loss after radiotherapy for parotid gland tumors. Head Neck 28:902–908
93. Fu KK, Woofhouse RJ, Quivey JM et al (1982) The significance of laryngeal edema following radiotherapy of carcinoma of the vocal cord. Cancer 49:6555–6558
94. Wolden SL, Chen WC, Pfister DG, Kraus DH, Berry SL, Zelefsky MJ (2006) Intensity-modulated radiation therapy (IMRT) for nasopharyngeal cancer: update of the Memorial Sloan-Kettering experience. Int J Radiat Oncol Biol Phys 64:57–62
95. Hirano M (1981) Clinical examination of voice. In: Arnold GE, Winkel F, Wyke BD (eds) Disorders of human communication. Springer, New York, pp 81–84
96. Fung K, Yoo J, Leeper A et al (2001) Vocal function following radiation for non-laryngeal versus laryngeal tumors of the head and neck. Laryngoscope 111:1920–1924
97. Hocevar-Boltezar I, Zargi M, Strojan P (2009) Risk factors for voice quality after radiotherapy for early glottic cancer. Radiother Oncol 93:524–529
98. Cox JD et al (1995) Toxicity criteria of the Radiation Therapy Oncology Group (RTOG) and the European Organization for Research and Treatment of Cancer (EORTC). Int J Radiat Oncol Biol Phys 31(5):1341–1346
99. Sanguineti G, Adapala P, Endres EJ et al (2007) Dosimetric predictors of laryngeal edema. Int J Radiat Oncol Biol Phys 68:741–749
100. Dornfeld K, Simmons JR, Karnell L et al (2007) Radiation doses to structures within and adjacent to the larynx are correlated with long-term diet and speech-related quality of life. Int J Radiat Oncol Biol Phys 68:750–757
101. Rancati T, Sanguineti G, Fiorino C (2007) NTCP modeling of subacute/late laryngeal edema scored by fiberoptic examination: evidence of a large volume effect. Int J Radiat Oncol Biol Phys 69(3):S409–S410
102. Schultheiss TE, Kun LE, Ang KK et al (1995) Radiation response of the central nervous system. Int J Radiat Oncol Biol Phys 31:1093–1112
103. Pallis CA, Louis S, Morgan RL (1961) Radiation myelopathy. Brain 84:460–479
104. Thames HD Jr, Withers HR, Peters LJ, Fletcher GH (1982) Changes in early and late radiation responses with altered dose fractionation: implications for dose-survival relationships. Int J Radiat Oncol Biol Phys 8:219–226
105. Reagan TJ, Thomas JE, Colby MY Jr (1968) Chronic progressive radiation myelopathy. Its clinical aspects and differential diagnosis. JAMA 203:106–110
106. Jones A (1964) Transient radiation myelopathy (with reference to Lhermitte's sign of electrical paraesthesia). Br J Radiol 37:727–744
107. Lim DC, Gagnon PJ, Meranvil S, et al. (2010) Lhermitte's sign developing after IMRT for head and neck cancer. Int J Otolaryngol 2010:907–960
108. St Clair WH, Arnold SM, Sloan AE et al (2003) Spinal cord and peripheral nerve injury: current management and investigations. Semin Radiat Oncol 13:322–332
109. Esik O, Csere T, Stefanits K et al (2003) A review on radiogenic Lhermitte's sign. Pathol Oncol Res 9:115–120
110. Schultheiss TE, Stephens LC, Peters LJ (1986) Survival in radiation myelopathy. Int J Radiat Oncol Biol Phys 12:1765–1769
111. Coderre JA, Morris GM, Micca PL et al (2006) Late effects of radiation on the central nervous system: role of vascular endothelial damage and glial stem cell survival. Radiat Res 166:495–503
112. Philippens ME, Pop LA, Visser AG et al (2007) Dose-volume effects in rat thoracolumbar spinal cord: the effects of nonuniform dose distribution. Int J Radiat Oncol Biol Phys 69:204–213
113. Phillips TL, Buschke F (1969) Radiation tolerance of the thoracic spinal cord. Am J Roentgenol 105:659–664

114. Van der Kogel AJ (1977) Radiation tolerance of the spinal cord. Dependence on fractionation and extended overall times. In: Radiobiological research and radiotherapy, vol 1. International Atomic Energy Agency, Vienna, pp 83–90
115. Abbatucci JS, Delozier T, Quint R, Roussel A, Brune D (1978) Radiation myelopathy of the cervical spinal cord: time, dose and volume factors. Int J Radiat Oncol Biol Phys 4:239–248
116. Kim YH, Fayos JV (1981) Radiation tolerance of the cervical spinal cord. Radiology 139:473–478
117. McCunniff AJ, Liang MJ (1989) Radiation tolerance of the cervical spinal cord. Int J Radiat Oncol Biol Phys 16:675–678
118. Marcus RB, Million RR (1990) The incidence of myelitis after irradiation of the cervical spinal cord. Int J Radiat Oncol Biol Phys 19:3–8
119. Ang KK, Price RE, Stephens LC et al (1993) The tolerance of primate spinal cord to re-irradiation. Int J Radiat Oncol Biol Phys 25:459–464
120. Atkins HL, Tretter P (1966) Time-dose considerations in radiation myelopathy. Acta Radiol Ther Phys Biol 5:79–94
121. Jeremic BJ, Djuric L, Mijatovic L (1991) Incidence of radiation myelitis of the cervical spinal cord at doses of 5500 cGy or greater. Cancer 68:2138–2141
122. Ang KK, Jiang GL, Feng Y et al (2001) Extent and kinetics of recovery of occult spinal cord injury. Int J Radiat Oncol Biol Phys 50:1013–1020
123. Powell S, Cooke J, Parsons C (1990) Radiation-induced brachial plexus injury: follow-up of two different fractionation schedules. Radiother Oncol 18(3):213–220
124. Stoll BA, Andrews JT (1966) Radiation-induced peripheral neuropathy. Br Med J 1:834–837
125. Bowen BC, Verma A, Brandon AH, Fiedler JA (1996) Radiation-induced brachial plexopathy: MR and clinical findings. AJNR Am J Neuroradiol 17(10):1932–1936
126. Johansson S, Svensson H, Denekamp J (2002) Dose response and latency for radiation-induced fibrosis, edema, and neuropathy in breast cancer patients. Int J Radiat Oncol Biol Phys 52(5):1207–1219
127. Schierle C, Winograd JM (2004) Radiation-induced brachial plexopathy: review. Complication without a cure. J Reconstr Microsurg 20:149–152
128. Amini A, Yang J, Williamson R, McBurney ML, Erasmus J Jr, Allen PK, Karhade M, Komaki R, Liao Z, Gomez D, Cox J, Dong L, Welsh J (2012) Dose constraints to prevent radiation-induced brachial plexopathy in patients treated for lung cancer. Int J Radiat Oncol Biol Phys 2012(82):e391–e398
129. Kori SH, Foley KM, Posner JB (1981) Brachial plexus lesions in patients with cancer: 100 cases. Neurology 31:45–50
130. Churn M, Clough V, Slater A (2000) Early onset of bilateral brachial plexopathy during mantle radiotherapy for Hodgkin's disease. Clin Oncol 12:289–291
131. Chen AM, Hall WH, Li J, Beckett L, Farwell DG, Lau DH, Purdy JA (2012) Brachial plexus-associated neuropathy after high-dose radiation therapy for head-and-neck cancer. Int J Radiat Oncol Biol Phys 84:165–169
132. Platteaux N, Dirix P, Hermans R, Nuyts S (2010) Brachial plexopathy after chemoradiotherapy for head and neck squamous cell carcinoma. Strahlenther Onkol. Aug 30. [Epub ahead of print]
133. Truong MT et al (2012) Radiation dose to the brachial plexus in head-and-neck intensity-modulated radiation therapy and its relationship to tumor and nodal stage. Int J Radiat Oncol Biol Phys 84(1):158–164
134. Emami B, Lyman J, Brown A et al (1991) Tolerance of normal tissue to therapeutic irradiation. Int J Radiat Oncol Biol Phys 21:109–122
135. Bajrovic A, Rades D, Fehlauer F et al (2004) Is there a life-long risk of brachial plexopathy after radiotherapy of supraclavicular lymph nodes in breast cancer patients? Radiother Oncol 71(3):297–301

136. Wilson FM (1989) Fundamentals and principles of ophthalmology. Basic and clinical science course. 1987–1988. American Academy of Ophthalmology, San Francisco
137. Lessell S (2004) Friendly fire: neurogenic visual loss from radiation therapy. J Neuroophthalmol 24:243–250
138. Danesh-Meyer HV (2008) Radiation-induced optic neuropathy. J Clin Neurosci 15:95–100
139. Parsons JT (1980) The effect of radiation on normal tissues of the head and neck. In: Million RR, Cassisi NJ (eds) Management of head and neck cancer: a multidisciplinary approach. J. B. Lippincott Company, Philadelphia, pp 173–207
140. Gordon KB, Char DH, Sagerman RH (1995) Late effects of radiation on the eye and ocular adnexa. Int J Radiat Oncol Biol Phys 31:1123–1139
141. Parsons JT, Bova FJ, Fitzgerald CR et al (1994) Radiation optic neuropathy after megavoltage external-beam irradiation: analysis of time–dose factors. Int J Radiat Oncol Biol Phys 30:755–763
142. Kline LB, Kim JY, Ceballos R (1985) Radiation optic neuropathy. Ophthalmology 92:1118–1126
143. Roden D, Bosley TM, Fowble B, Clark J, Savino PJ, Sergott RC, Schatz NJ (1990) Delayed radiation injury to the retrobulbar optic nerves and chiasm. Clinical syndrome and treatment with hyperbaric oxygen and corticosteroids. Ophthalmology 97:346–351
144. Mayo C, Martel MK, Marks LB et al (2010) Radiation dose volume effects of optic nerves and chiasm. Int J Radiat Oncol Biol Phys 76(3 Suppl):S28–S35
145. Goldsmith BJ, Rosenthal SA, Wara WM, Larson DA (1992) Optic neuropathy after irradiation of meningioma. Radiology 185:71–76
146. Weintraub JA, Bennett J, Gaspar LE (2011) Successful treatment of radiation-induced optic neuropathy. Pract Radiat Oncol 1:40–44
147. Mendenhall WM (2004) Mandibular osteoradionecrosis. J Clin Oncol 22:4867–4868
148. Sciubba JJ, Goldenberg D (2006) Oral complications of radiotherapy. Lancet Oncol 7:175–183
149. Vissink A, Jansma J, Spijkervet FKL et al (2003) Oral sequelae of head and neck radiotherapy. Crit Rev Oral Biol Med 14:199–212
150. Reuther T, Schuster T, Mende U, Kubler A (2003) Osteoradionecrosis of the jaws as a side effect of radiotherapy of head and neck tumour patients – a report of a thirty year retrospective review. Int J Oral Maxillofac Surg 32:289–295
151. Curi MM et al (2007) Management of extensive osteoradionecrosis of mandible with radical resection & immediate microvascular reconstruction. J Oral Maxilla Surg 65:434–438
152. Cox JD et al (1995) Toxicity criteria of the Radiotherapy Oncology Group & European Organisation for Research & Treatment of Cancer. Int J Radiat Oncol Biol Phys 31:1314–1346
153. Ben-David MA, Diamante M, Radawski JD et al (2007) Lack of osteoradionecrosis of the mandible after intensity-modulated radiotherapy for head and neck cancer: likely contributions of both dental care and improved dose distributions. Int J Radiat Oncol Biol Phys 68(2):396–402
154. Katsura K, Sasai K, Sato K, Saito M, Hoshina H, Hayashi T (2008) Relationship between oral health status and development of osteoradionecrosis of the mandible: a retrospective longitudinal study. Oral Surg Oral Med Oral Pathol Oral Radiol Endod 105:7381–7382
155. Gomez DR et al (2011) Correlation of osteoradionecrosis & dental events with dosimetric parameters in intensity modulated radiotherapy for head & neck cancer. Int J Radiat Oncol Biol Phys 81(4):e207–e213
156. Studer G, Gratz KW, Glanzmann C (2004) Osteoradionecrosis of the mandibula in patients treated with different fractionations. Strahlenther Onkol 180:233–240
157. Marx RE, Johnson RP (1987) Studies in the radiobiology of osteoradionecrosis and their clinical significance. Oral Surg Oral Med Oral Pathol 64:379–390
158. Glanzmann C, Gratz KW (1995) Radionecrosis of the mandible: a retrospective analysis of the incidence and risk factors. Radiother Oncol 36:94–100

159. Gregoire V, Levendag P, Ang KK, Bernier J et al (2003) CT-based delineation of lymph node levels and related CTVs in the node negative neck: DAHANCA, EORTC, GORTEC, NCIC, RTOG consensus guidelines. Radiother Oncol 69:227–236
160. Claus F, Duthoy W, Boterberg T, De Gersem W, Huys J, Vermeersch H, De Neve W (2002) Intensity modulated radiation therapy for oropharyngeal and oral cavity tumors: clinical use and experience. Oral Oncol 38:597–604
161. van den Broek GB, Balm AJ, van den Brekel MW, Hauptmann M, Schornagel JH, Rasch CR (2006) Relationship between clinical factors and the incidence of toxicity after intra-arterial chemoradiation for head and neck cancer. Radiother Oncol 81:143–150
162. Studer G, Studer SP, Zwahlen RA, Huguenin P, Gratz KW, Lutolf UM, Glanzmann C (2006) Osteoradionecrosis of the mandible: minimized risk profile following intensity-modulated radiation therapy (IMRT). Strahlenther Onkol 182:283–288
163. Mendenhall WM, Morris CG, Amdur RJ, Hinerman RW, Malyapa RS, Werning JW, Lansford CD, Villaret DB (2006) Definitive radiotherapy for tonsillar squamous cell carcinoma. Am J Clin Oncol 29:290–297
164. Jansma JVA et al (1992) Protocol for prevention & treatment of oral sequelae resulting from Head & neck radiotherapy. Cancer 70(8):2171–2180
165. Lyons A, Ghazali N (2008) Osteoradionecrosis of the jaws: current understanding of its pathophysiology and treatment. Br J Oral Maxillofac Surg 46:653–660
166. Marx RE, Johnson RP, Kline SN (1985) (1985) Prevention of osteoradionecrosis: a randomized prospective clinical trial of hyperbaric oxygen versus penicillin. J Am Dent Assoc 111:49–54
167. Goldwaser BR, Chuang SK, Kaban LB, August M (2007) Risk factor assessment for the development of osteoradionecrosis. J Oral Maxillofac Surg 65:2311–2316
168. Assael LA (2004) New foundations in understanding osteoradionecrosis of the jaws. J Oral Maxillofac Surg 62(2):125–126
169. Støre G, Eribe ER, Olsen I (2005) DNA-DNA hybridization demonstrates multiple bacteria in osteoradionecrosis. Int J Oral Maxillofac Surg 34(2):193–196
170. Delanian S, Lefaix JL (2004) The radiation-induced fibroatrophic process: Therapeutic perspective via the antioxidant pathway. Radiother Oncol 73(2):119–131
171. Maurer P, Meyer L (2006) Osteoradionecrosis of the mandible-resection aided by measurement of partial pressure of oxygen (pO2): a technical report. J Oral Maxillofac Surg 64(3):560–562
172. Marx RE, Ehler WJ, Tayapongsak P, Pierce LW (1990) Relationship of oxygen dose to angiogenesis induction in irradiated tissue. Am J Surg 160:519–524
173. Greenwood TW, Gilchrist AG (1973) Hyperbaric oxygen & wound healing in post-irradiation head neck surgery. Br J Surg 60:394–397
174. Elsaleh H, Quoc-Chuong Bui, Michael L, Withers HR et al. (2004) Int J Radiat Oncol Biol Phys 60(3):871–878
175. Winsor T, Winsor D (1985) The noninvasive laboratory-history and future of thermography. Angiology 36(6):341–353
176. Christiansen J, Gerow G (1990) Thermography. Williams and Wilkins, Baltimore, p 200
177. Mainous EG, Boyne J, Hart GB (1973) Elimination of sequestrum & healing of osteoradionecrosis of the mandible after hyperbaric oxygen therapy. J Oral Surg 31:336–339
178. Brown LR, Dreizen S, Daly TE et al (1978) Interrelations of oral microorganisms, immunoglobulins, and dental caries following radiotherapy. J Dent Res 57:882–893
179. Keene HJ, Fleming TJ (1987) Prevalence of caries-associated microflora after radiotherapy in patients with cancer of the head and neck. Oral Surg Oral Med Oral Pathol 64:421–426
180. Llory H, Dammron A, Gioanni M, Frank RM (1972) Some population changes in oral anaerobic microorganisms, Streptococcus mutants and yeasts following irradiation of the salivary glands. Caries Res 6:298–311

181. Regezi JA, Courtney RM, Kerr DA (1976) Dental management of patients irradiated for oral cancer. Cancer 38:994–1000
182. Daly TE, Drane JB (1972) Proceedings: the management of teeth related to the treatment of oral cancer. Proc Natl Cancer Conf 7:147–154
183. Epstein JB, McBride BC, Stevenson-Moore P, Merilees H, Spinelli J (1991) The efficacy of chlorhexidine gel in reduction of streptococcus mutants and lactobacillus species in patients treated with radiation therapy. Oral Surg Oral Med Oral Pathol 71:172–178
184. Joyston-Bechal S, Hayes K, Davenport ES, Hardie JM (1992) Caries incidence, mutants streptococci and lactobacilli in irradiated patients during a 12-month preventive programme using chlorhexidine and fluoride. Caries Res 26:384–390
185. Vissink A, Burlage FR, Spijkervet FK, Jansma J, Coppes RP (2003) Prevention and treatment of the consequences of head and neck radiotherapy. Crit Rev Oral Biol Med 14:213–225
186. Jansma J, Vissink A, Gravenmade EJ, Visch LL, Fidler V, Retief DH (1989) In vivo study on the prevention of postradiation caries. Caries Res 23:172–178
187. Fleming TJ (1983) Use of topical fluoride by patients receiving cancer therapy. Curr Probl Cancer 7:37–41
188. Al-Joburi W, Clark C, Fisher R (1991) A comparison of the effectiveness of two systems for the prevention of radiation caries. Clin Prev Dent 13:15–19
189. Epstein JB, van der Meij EH, Lunn R, Stevenson-Moore P (1996) Effects of compliance with fluoride gel application on caries and caries risk in patients after radiation therapy for head and neck cancer. Oral Surg Oral Med Oral Pathol Oral Radiol Endod 82:268–275
190. Dreizen S, Brown LR, Daly TE, Drane JB (1977) Prevention of xerostomia-related dental caries in irradiated cancer patients. J Dent Res 56:99–104
191. Horiot JC, Bone MC, Ibrahim E, Castro JR (1981) Systematic dental management in head and neck irradiation. Int J Radiat Oncol Biol Phys 7:1025–1029
192. Horiot JC, Schraub S, Bone MC et al (1983) Dental preservation in patients irradiated for head and neck tumours: a 10-year experience with topical fluoride and a randomized trial between two fluoridation methods. Radiother Oncol 1:77–82

Non-small Cell Lung Cancer

12

Quynh-Nhu Nguyen, Ritsuko Komaki, Daniel R. Gomez, and Zhongxing Liao

Keywords
IMRT • Non-small cell lung cancer

12.1 Introduction

In 2013, a statistical fact sheet from the US Surveillance, Epidemiology, and End Results reported that an estimated 228,190 new cases of lung cancer would be diagnosed and that an estimated 159,480 people would die of this disease [1]. Non-small cell lung cancer (NSCLC) remains the predominant variant and the leading cause of mortality worldwide; it represents 13.7 % of all new cancer cases in the United States [1]. Treatment for NSCLC often requires multimodality therapy including surgery, systemic chemotherapy, novel targeted agents, and radiation therapy. High doses of radiation therapy (i.e., those above 60 Gy) have been investigated in attempts to control both local and regional treatment failures in NSCLC. However, delivering higher radiation therapy doses, particularly in combination with chemotherapy, increases the risk of treatment-related toxicity. With the emergence of technologies such as intensity-modulated radiation therapy (IMRT) and particle beam therapy, the goal in radiation therapy is to effectively treat NSCLC while simultaneously minimizing clinically relevant treatment-related toxicity.

The availability of advanced techniques such as IMRT and image-guided radiation therapy has greatly improved the precision of delivering radiation treatments

Q.-N. Nguyen, M.D. (✉) • R. Komaki, M.D. • D.R. Gomez, M.D. • Z. Liao, M.D.
Department of Radiation Oncology, The University of Texas MD Anderson Cancer Center, Unit 1150, 1840 Old Spanish Trail, Houston, TX 77054-1901, USA
e-mail: qnnguyen@mdanderson.org

Y. Nishimura, R. Komaki (eds.), *Intensity-Modulated Radiation Therapy*,
DOI 10.1007/978-4-431-55486-8_12

for patients with lung cancer. IMRT is thought to enhance the therapeutic ratio by using beams of nonuniform intensities to tightly conform the dose to the target. Through inverse treatment planning, a combination of carefully chosen beam arrangements, optimization parameters, and strict adherence to dose limits for normal structures allows the delivery of tightly conformal dose distributions to targets of complex shapes. Because each field contributes a nonuniform intensity pattern, combining the fields can create a uniform target dose distribution. Another advantage of IMRT for treating lung cancer is the potential for dosimetric improvements in terms of delivering high, tightly conformal doses to the tumor while sparing surrounding normal structures such as the healthy lung and spinal cord, in this way improving the therapeutic ratio for lung cancer. This chapter summarizes the current state of the art in the use of IMRT for treating NSCLC.

12.2 Treatment Planning for Thoracic Tumors

12.2.1 Treatment Simulation, Treatment Planning, and Dosimetry

At the authors' institution, treatment planning for all patients involves treatment simulation that is based on the findings from four-dimensional computed tomography (4D CT). Patients are positioned in the intended treatment position on a CT couch and immobilized through the use of a customized, indexed immobilization vacuum device around the upper torso. A respiratory monitoring system is placed on the patient's abdomen, and a series of ten CT datasets representing different points of the respiratory phase are reconstructed. The acquired CT dataset is imported into the treatment planning system, where the average intensity projection is used as the primary dataset for dose calculations, but all datasets are used to determine the internal target volume.

For patients undergoing 4D CT-based treatment simulation, a 5–8 mm margin is added to the internal target volume to create the clinical target volume (CTV), and an additional 5–7 mm margin is added to the CTV to create the planning target volume (PTV) in patients being treated for NSCLC. Treatment plans for IMRT are designed using the inverse planning component of the Pinnacle treatment planning system's software with an optimization algorithm (Philips Healthcare, Inc.). The goals of IMRT planning are to deliver the prescribed dose to the PTV, with a minimum of 95 % of the prescribed dose and a maximum of 110 % of the prescribed dose. The beam configuration for the IMRT plans depends on the location and size of the tumor; however, generally 5–7 beams are sufficient, with gantry angles separated by a minimum of 25–30° (Fig. 12.1).

Normal tissue constraints for radiation therapy used to treat thoracic malignancies are summarized in Table 12.1. At the authors' institution, we attempt to minimize the total lung volume that receives >5 Gy (i.e., the V_5) to the greatest extent possible; we further restrict the mean lung dose (MLD) to 20 Gy or less and the volume of lung that receives >20 Gy (V_{20}) to <40 %. Other dose constraints include

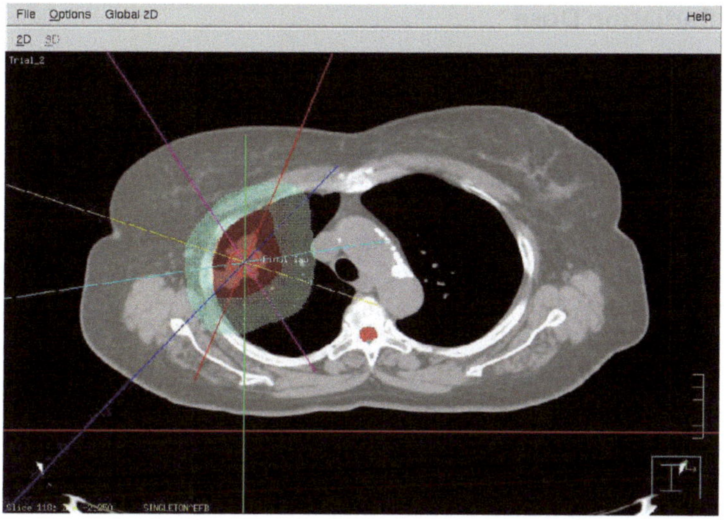

Fig. 12.1 Beam configuration for an intensity-modulated radiation therapy plan to treat a patient with non-small lung cancer. The beams are separated by 25–30° so as to avoid parallel opposed beams

Table 12.1 Dose-volume constraints for normal tissues during standard fractionation radiation therapy

	Radiation only	Radiation with chemotherapy	Radiation with chemotherapy before surgery
Spinal cord	D_{max} <45 Gy	D_{max} <45 Gy	D_{max} <45 Gy
Lung	MLD ≤20 Gy $V_{20} \leq 40$ %	MLD ≤20 Gy $V_{20} \leq 35$ % $V_{10} \leq 45$ % $V_5 \leq 65$ %	MLD ≤20 Gy $V_{20} \leq 30$ % $V_{10} \leq 40$ % $V_5 \leq 55$ %
Heart	$V_{30} \leq 45$ % Mean dose <26 Gy	$V_{30} \leq 45$ % Mean dose <26 Gy	$V_{30} \leq 45$ % Mean dose <26 Gy
Esophagus	D_{max} ≤80 Gy $V_{70} < 20$ % $V_{50} < 50$ % Mean dose <34 Gy	D_{max} ≤80 Gy $V_{70} < 20$ % $V_{50} < 40$ % Mean dose <34 Gy	D_{max} ≤80 Gy $V_{70} < 20$ % $V_{50} < 40$ % Mean dose <34 Gy
Kidney	20 Gy <32 % of bilateral kidney	20 Gy <32 % of bilateral kidney	20 Gy <32 % of bilateral kidney
Liver	$V_{30} \leq 40$ % Mean dose <30 Gy	$V_{30} \leq 40$ % Mean dose <30 Gy	$V_{30} \leq 40$ % Mean dose <30 Gy

minimizing the mean dose and V_{50} to the esophagus, restricting the cardiac V_{30} to <45 %, and limiting the dose to the spinal cord to <45 Gy. We closely follow these and other dose-volume constraints based on summary QUANTEC recommendations for standard fractionation treatment.

12.2.2 Radiation Dose

Principles of basic radiobiology suggest that doses of 80–100 Gy are required to sterilize lung cancer [2]. The Radiation Therapy Oncology Group (RTOG) and other institutions have conducted randomized trials evaluating radiation doses of 60 Gy or more in combination with chemotherapy to treat inoperable NSCLC [3–7]. The optimal dose for definitive radiation therapy for patients with inoperable NSCLC at diagnosis is still controversial. At MD Anderson Cancer Center, such patients are treated with definitive radiation doses of 60–74 Gy with concurrent chemotherapy, if they can tolerate this therapy.

12.2.3 Radiation-Induced Toxicity

12.2.3.1 Radiation Pneumonitis

Patients receiving radiation therapy to the thorax are at risk for developing radiation pneumonitis (RP), which typically manifests within 3–9 months after the completion of radiation therapy. Many studies have demonstrated that MLD [8–15] and the percentage of lung volume receiving more than some threshold dose [8, 11, 13, 16–18] can predict the development of RP; however, other studies have shown that some of these factors are not linked with RP [16, 19, 20], but rather that only a history of smoking [11, 17], chronic obstructive pulmonary disease [9], and receipt of induction chemotherapy with mitomycin [10] predict RP.

Variables typically used in the evaluation of lung dose and risk of RP include the volume of both lungs receiving more than a threshold dose (V_{dose}), the MLD, the lung V_{20}, and normal tissue complication probabilities (NTCPs) given various combinations of dose-volume variables. We and others also evaluate the lung V_5.

In general, in radiation therapy for lung cancer, the total tolerable radiation dose depends on the volume irradiated. In a retrospective analysis, the MLD (Fig. 12.2) and relative V_5–V_{65} in increments of 5 Gy were all found to be associated with the incidence of grade ≥3 RP according to the Common Terminology Criteria for Adverse Events v3.0. Investigators at MD Anderson [21] showed that V_5 was also a significant predictor of RP (Fig. 12.2); in that study, the 1-year incidence of grade ≥3 RP for patients with a relative $V_5 \leq 42\ \%$ was 3 % compared with 38 % for those with $V_5 > 42\ \%$ ($P = 0.001$). This finding suggests that damage to the lung, which has functional subunits in parallel, may depend more on the volume irradiated than on the radiation dose. Gopal et al. similarly demonstrated that exposing normal lung to as little as 13 Gy led to a pronounced decrease in diffusion capacity for carbon monoxide (DLCO), and a loss of DLCO of >30 % was associated with grade ≥2 pulmonary symptoms ($P = 0.003$). Those investigators concluded that such a low threshold for deterioration of DLCO (13 Gy) indicates that it is better to treat a small amount of normal lung to a high dose rather than treating a large volume to a low dose [22]. Similarly, Yorke et al. reported that in patients with NSCLC treated with dose-escalated radiation therapy, the incidence of grade ≥3 pneumonitis correlated with

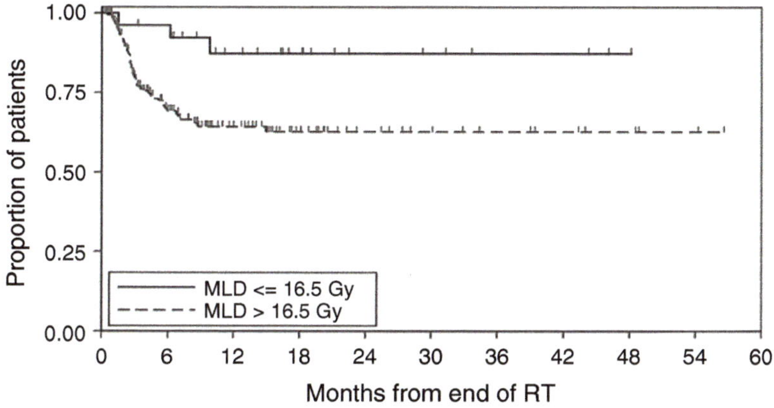

Fig. 12.2 Effect of mean lung dose (MLD) on freedom from grade ≥3 treatment-related pneumonitis. *RT* radiotherapy (Figure republished (with permission) from Wang et al. [21])

MLD ($P \leq 0.05$). The dose response as a function of mean dose to the total lung rises steeply, beginning at approximately 10 Gy [10]. In clinical practice, V_{20} is often used as a surrogate to evaluate total dose to the lung in radiation treatment planning. Graham et al., in their analysis of V_{20} for predicting RP, stratified patients into risk groups and found that the incidence of RP increased steeply when V_{20} levels were 40 % or higher [16].

12.2.3.2 Esophagitis

Another form of radiation-induced toxicity, acute esophagitis, typically occurs within 90 days after the start of radiation therapy, whereas chronic esophagitis occurs after that time. Chronic esophagitis can result in the development of esophageal stricture requiring dilation and, in rare cases, esophageal fistula. Grade 1–2 radiation-related esophagitis is relatively common after treatment for lung cancer, and rates of grade >3 esophagitis range from 10 to 50 % [23–25]. A recent analysis of acute esophagitis in four RTOG trials involving 528 patients reported that 75 % of patients had grade >2 acute esophagitis and 34 % had grade >3 acute esophagitis after radiation therapy. Nineteen percent of these cases had developed within the first month of treatment, 32 % by the second month, and 33 % by the third month [26]. At the authors' institution, we closely monitor acute esophagitis weekly during treatment, and we use aggressive supportive care measures to avoid the need for hospitalization and treatment interruptions.

Reports of potential clinical and dosimetric predictors of esophagitis are many, with substantial variation among studies. Esophagitis has generally been found to be associated with the volume of the esophagus receiving a specific dose, the mean esophageal dose, and the maximum esophageal dose (D_{max}), having a history of esophageal morbidity, having nodal involvement, and receiving twice-daily rather than once-daily irradiation [27–32]. At the authors' institution, we adhere closely to the following dose constraints for patients receiving high-dose radiation for thoracic malignancies: mean esophageal dose <34 Gy, V_{70}<20 %, and D_{max} of 80 Gy. These dose constraints were based in part on the reported experience at Washington University [33]. A group at MD Anderson investigated the potential of IMRT for reducing the volumes of irradiated lung and esophagus during the treatment of NSCLC in a retrospective treatment planning study and found that IMRT produced lower lung V_{20} and MLD than did three-dimensional conformal radiation therapy (3D CRT) in all cases. Notably, IMRT also led to smaller volumes of the esophagus and heart being exposed to radiation doses in excess of 45 Gy [34]. In a similar analysis, Gomez et al. tested the ability of a variety of factors to predict radiation-induced esophagitis in 652 patients with NSCLC treated with 3D CRT, IMRT, or proton beam therapy. In that study, the rate of grade ≥3 esophagitis was highest among patients who had been treated with IMRT (28 % vs. 8 % for 3D CRT and 6 % for proton therapy), leading the authors to conclude that the Lyman-Kutcher-Burman statistical model used in that study seriously underestimated the risk of severe esophagitis among patients treated with IMRT [35].

12.2.3.3 Cardiac Toxicity

Most of the posited effects of radiation-induced cardiotoxicity have been extrapolated from studies in which thoracic irradiation was delivered with older, 2D radiation techniques for breast cancer or lymphoma [36–40]. Findings from these studies, in which patients had been treated many years ago with techniques that could not minimize dose to the heart, are generally not applicable to current technology. Moreover, the reported rates of long-term cardiac morbidity varied considerably across studies, from <1 % to >15 %, although the rates do seem to continue to increase over time.

Several studies have compared the putative dosimetric advantages of IMRT over 3D CRT for sparing normal critical structures such as the heart. In one retrospective treatment planning comparison, Liu et al. investigated whether IMRT could reduce the volumes of lung and other critical structures relative to 3D CRT during radiation therapy for NSCLC. In addition to producing a lower MLD, IMRT led to smaller volumes of the esophagus and heart being exposed to high-dose radiation (>45 Gy). IMRT further allowed an additional safety margin around normal structures including the spinal cord, heart, and esophagus to account for uncertainties related to variations in setup, thereby minimizing the risk of radiation-associated cardiomyopathy [34]. Others at MD Anderson found similar results in their evaluations of IMRT versus 3D CRT for patients with stage III–IV NSCLC. Again, IMRT led to smaller lung V_{10} and V_{20} values as well as smaller MLD and a 10 % absolute reduction in risk of RP; IMRT also reduced the volumes of the heart and esophagus receiving >40–50 Gy. Those investigators concluded that IMRT could significantly improve target coverage and reduce the volume of normal lung irradiated to low doses; they further stated that the extent of low-dose exposure of normal tissues can be controlled in IMRT by choosing appropriate planning parameters [41].

12.3 Clinical Use of IMRT

As noted previously within this chapter, concurrent chemoradiation therapy is usually recommended for patients with locally advanced, inoperable stage IIIA or IIIB NSCLC. Nevertheless, treatment failures are relatively common, and overall survival rates remain relatively low at 5 years. Several studies have demonstrated that improving local disease control can improve overall survival for patients with stage III NSCLC [42–45]. Using IMRT rather than 3D CRT is thought to provide both dosimetric and clinical advantages when sufficiently high doses can be given for locally advanced NSCLC. IMRT enables tighter sculpting of high-dose regions around the target volume; the steep gradients created can reduce the radiation dose to surrounding normal tissues, which presumably could facilitate dose escalation [46].

Govaert et al. assessed survival outcomes and acute pulmonary and esophageal toxicity in 86 patients who received IMRT for stage III NSCLC. The median survival time was 29.7 months after delivery of 66 Gy, with or without chemotherapy. Esophageal toxicity was more pronounced in the group that received concurrent chemoradiation, but no differences were noted in pulmonary toxicity [47]. Long-term clinical outcomes for 165 patients with inoperable stage III–IV NSCLC treated with IMRT to doses >60 Gy were recently reported from MD Anderson; the 3-year overall survival rate was 30 %, the rate of grade ≥3 RP was 14 % at 12 months, and the median time to maximum (grade 3) esophagitis was 6 weeks. Those investigators concluded that IMRT led to low rates of pulmonary and esophageal toxicity and favorable clinical outcomes in terms of survival [48].

The use of 4D CT-based treatment simulation and then IMRT instead of 3D CRT for NSCLC became routine at the authors' institution in 2004. In 2010, Liao et al. published findings on disease control, survival, and toxicity for 496 patients who had been treated with IMRT or 3D CRT, both with concomitant chemotherapy, to a median radiation dose of 63 Gy. Toxicity was considerably lower among patients treated with IMRT, specifically in smaller lung V_{20}, lower rates of grade 3 RP, and improved overall survival, leading the authors to conclude that IMRT was associated with therapeutic gain. Rates of locoregional progression-free and distant metastases-free survival rates were no different between those who received IMRT and those who received 3D CRT (Fig. 12.3). Nevertheless, the advantage of lower toxicity, which presumably would allow effective doses of systemic therapy to be given concurrently, may have been a factor in the improved overall survival in this study [49].

A phase I/II protocol involving the use of image-guided, dose-escalated IMRT for patients with stage II–III NSCLC receiving concurrent chemoradiation is currently underway at MD Anderson. The goal of the study is to determine the maximum tolerated dose to the gross tumor volume, starting at 72 Gy and escalated to the highest dose level of 84 Gy, while keeping the dose to the PTV at 60 Gy (Fig. 12.4). This nonuniform delivery of different radiation dose distributions is

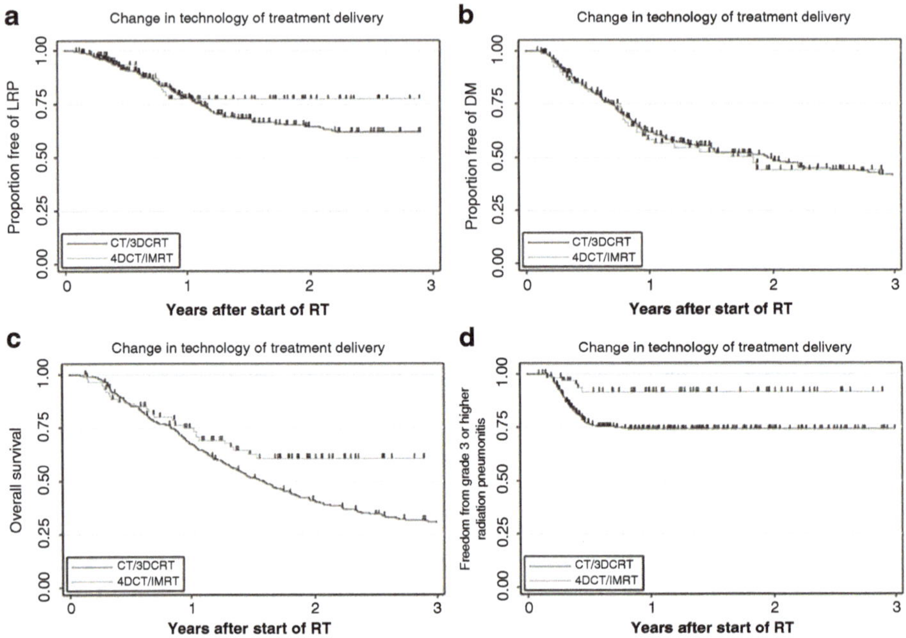

Fig. 12.3 Comparison of (**a**) freedom from locoregional progression (LRP), (**b**) freedom from distant metastases (DM), (**c**) overall survival, and (**d**) freedom from grade ≥3 radiation pneumonitis in patient treated with three-dimensional conformal radiotherapy (3D CRT) or intensity-modulated radiotherapy (IMRT) based on 4D computed tomography simulation (4D CT) (Figure republished (with permission) from Liao et al. [49])

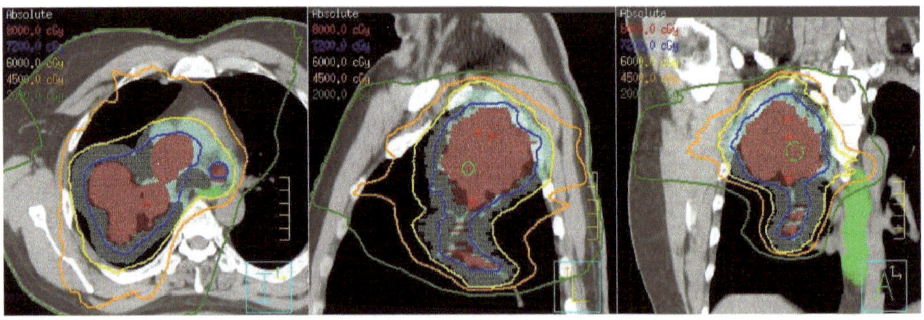

Fig. 12.4 Axial, sagittal, and coronal slices of a treatment plan designed to deliver 72 Gy to the gross tumor volume and 60 Gy to the planning target volume via intensity-modulated radiation therapy with a simultaneous integrated boost

possible with the use of IMRT. The hypothesis of this study is that using a simultaneous integrated boost technique will permit accelerated radiation therapy, with the ultimate goal of improving tumor control without the expected increase in risks of normal tissue toxicity.

12.4 Conclusions

Technologic advances in IMRT and image-guided radiation therapy over the past decade have significantly changed the field of radiation oncology. The availability of daily imaging and more sophisticated treatment delivery systems have allowed the delivery of higher radiation doses to the target volume with tighter conformality, minimizing the dose to normal thoracic structures and thereby improving the therapeutic ratio. Minimizing treatment-related toxicity could expand the number of patients with locally advanced NSCLC who could tolerate concurrent chemotherapy or novel molecular targeted agents, which in turn could lead to improved clinical outcomes.

Acknowledgments Supported in part by Cancer Center Support (Core) Grant CA016672 from the US National Cancer Institute to the University of Texas MD Anderson Cancer Center.

Conflict of Interest The authors declare no conflicts of interest.

References

1. Howlader N, Noone AM, Krapcho M, et al. (eds) (2013) SEER cancer statistics review, 1975–2010. National Cancer Institute. Bethesda. Available at http://seer.cancer.gov/csr/1975_2010/. Based on November 2012 SEER data submission; posted to the SEER web site April 2013
2. Fletcher G (1973) Clinical dose-response curves of human malignant epithelial tumours. Br J Radiol 46:1–12
3. Curran W, Paulus R, Langer CJ et al (2011) Sequential vs concurrent chemoradiation for stage III non-small cell lung cancer: randomized phase III trial RTOG 9410. J Natl Cancer Inst 103(19):1452–1460
4. Dillman RO, Seagren SL, Herndon J et al (1990) A randomized trial of induction chemotherapy plus high-dose radiation versus radiation alone in stage III non-small cell lung cancer. N Engl J Med 323:940–945
5. Sause W, Scott C, Taylor S et al (1995) Radiation Therapy Oncology Group (RTOG) 88-08 and ECOG 4588: preliminary results of a phase III trial in regionally advanced, unresectable non-small cell lung cancer. J Natl Cancer Inst 87:198–205
6. Socinski MA, Blackstock AW, Bogart JA et al (2008) Randomized phase II trial of induction chemotherapy followed by concurrent chemotherapy and dose-escalated thoracic conformal radiotherapy (74 Gy) in stage IIIA and stage IIIB non-small cell lung cancer. J Thorac Oncol 3:250–257
7. Machtay M, Kyounghwa B, Movsas B et al (2012) Higher biologically effective dose of radiotherapy is associated with improved outcomes for locally advanced non-small cell lung carcinoma treated with chemoradiation: an analysis of the Radiation Therapy Oncology Group. Int J Radiat Oncol Biol Phys 82(1):425–434

8. Claude L, Perol D, Ginestet C et al (2004) A prospective study on radiation pneumonitis following conformal radiation therapy in non-small cell lung cancer: clinical and dosimetric factors analysis. Radiother Oncol 71:175–181
9. Rancati T, Ceresoli GL, Gagliardi G et al (2003) Factors predicting radiation pneumonitis in lung cancer patients: a retrospective study. Radiother Oncol 67:275–283
10. Yorke ED, Jackson A, Rosenzweig KE et al (2002) Dose-volume factors contributing to the incidence of radiation pneumonitis in non-small-cell lung cancer patients treated with three-dimensional conformal radiation therapy. Int J Radiat Oncol Biol Phys 54:329–339
11. Hernando ML, Marks LB, Bentel GC et al (2001) Radiation-induced pulmonary toxicity: a dose-volume histogram analysis in 201 patients with lung cancer. Int J Radiat Oncol Biol Phys 51:650–659
12. Kwa SL, Lebesque JV, Theuws JC et al (1998) Radiation pneumonitis as a function of mean lung dose: an analysis of pooled data of 540 patients. Int J Radiat Oncol Biol Phys 42:1–9
13. Oetzel D, Schraube P, Hensley F et al (1995) Estimation of pneumonitis risk in three-dimensional treatment planning using dose-volume histogram analysis. Int J Radiat Oncol Biol Phys 33:455–460
14. Kim TH, Cho KH, Pyo HR et al (2005) Dose-volumetric parameters of acute esophageal toxicity in patients with lung cancer treated with three-dimensional conformal radiotherapy. Int J Radiat Oncol Biol Phys 62:995–1002
15. Willner J, Jost A, Baier K et al (2003) A little to a lot or a lot to a little? An analysis of pneumonitis risk from dose-volume histogram parameters of the lung in patients with lung cancer treated with 3-D conformal radiotherapy. Strahlenther Onkol 179:548–556
16. Graham MV, Purdy JA, Emami B et al (1999) Clinical dose-volume histogram analysis for pneumonitis after 3D treatment for non-small cell lung cancer. Int J Radiat Oncol Biol Phys 45:323–329
17. Jin H, Tucker SL, Liu HH et al (2009) Dose-volume thresholds and smoking status for the risk of treatment-related pneumonitis in inoperable non-small cell lung cancer treated with definitive radiotherapy. Radiother Oncol 91:427–432
18. Roach M 3rd, Gandara DR, Yuo HS et al (1995) Radiation pneumonitis following combined modality therapy for lung cancer: analysis of prognostic factors. J Clin Oncol 13:2606–2612
19. Fu XL, Huang H, Bentel G et al (2001) Predicting the risk of symptomatic radiation-induced lung injury using both the physical and biologic parameters V(30) and transforming growth factor beta. Int J Radiat Oncol Biol Phys 50:899–908
20. Sunyach MP, Falchero L, Pommier P et al (2000) Prospective evaluation of early lung toxicity following three-dimensional conformal radiation therapy in non-small-cell lung cancer: preliminary results. Int J Radiat Oncol Biol Phys 48:459–463
21. Wang S, Liao Z, Wei X et al (2006) Analysis of clinical and dosimetric factors associated with treatment-related pneumonitis in patients with non-small cell lung cancer treated with concurrent chemotherapy and three-dimensional conformal radiotherapy. Int J Radiat Oncol Biol Phys 66(5):1399–1407
22. Gopal R, Tucker SL, Komaki R et al (2003) The relationship between local dose and loss of function for irradiated lung. Int J Radiat Oncol Biol Phys 56(1):102–113
23. Schaake KC, Van den Bogaert W, Dalesio O et al (1992) Effects of concomitant cisplatin and radiotherapy on inoperable non-small cell lung cancer. N Engl J Med 326:524–530
24. Fournal P, Robinet F, Thomas P et al (2005) Randomized phase III trial of sequential chemoradiotherapy compared with concurrent chemoradiotherapy in locally advanced non small cell lung cancer: Groupe Lyon-Saint-Etienne d'Oncologie Thoracique-Groupe Francais de Pneumo-Cancerologie NPC 95-01 Study. J Clin Oncol 23:5910–5917
25. Furuse K, Fukuoka M, Kawahara M et al (1999) Phase III study of concurrent versus sequential thoracic radiotherapy in combination with mitomycin, vindesine, and cisplatin in unresectable stage III non-small cell lung cancer. J Clin Oncol 17:2692–2699
26. Werner-Wasik M, Paulus R, Curran W et al (2011) Acute esophagitis and late lung toxicity in concurrent chemoradiotherapy trials in patients with locally advanced non-small cell lung

cancer: analysis of the Radiation Therapy Oncology Group (RTOG) database. Clin Lung Cancer 12:245–251
27. Kim TH, Cho KH, Pyo HR et al (2005) Dose-volumetric parameters of acute esophageal toxicity in patients with lung cancer treated with three-dimensional conformal radiotherapy. Int J Radiat Oncol Biol Phys 61:995–1002
28. Anh SJ, Kahn D, Zhou S et al (2005) Dosimetric and clinical predictors for radiation-induced esophageal injury. Int J Radiat Oncol Biol Phys 61:335–347
29. Bradley J, Deasy JO, Benzen S et al (2004) Dosimetric correlates for acute esophagitis in patients treated with radiotherapy for lung carcinoma. Int J Radiat Oncol Biol Phys 58:1106–1113
30. Maguire PD, Sibley GS, Zhou SM et al (1999) Clinical and dosimetric predictors of radiation-induced esophageal injury. Int J Radiat Oncol Biol Phys 45:97–103
31. Takeda K, Nemoto K, Saito H et al (2005) Dosimetric correlations of acute esophagitis in lung cancer patients treated with radiotherapy. Int J Radiat Oncol Biol Phys 62:626–629
32. Wei X, Liu HH, Tucker SL et al (2006) Risk factors for acute esophagitis in non-small cell lung cancer patients treated with concurrent chemotherapy and three-dimensional conformal radiotherapy. Int J Radiat Oncol Biol Phys 66:100–107
33. Sing AK, Lockett MA, Bradley JD et al (2003) Predictors of radiation-induced esophageal toxicity in patients with non-small cell lung cancer treated with three-dimensional conformal radiotherapy. Int J Radiat Oncol Biol Phys 55:337–341
34. Liu H, Wang X, Dong L et al (2004) Feasibility of sparing lung and other thoracic structures with intensity-modulated radiotherapy for non-small cell lung cancer. Int J Radiat Oncol Biol Phys 58(4):1208–1279
35. Gomez DR, Tucker SL, Martel MK et al (2012) Predictors of high-grade esophagitis after definitive three-dimensional conformal therapy, intensity-modulated radiation therapy, or proton beam therapy for non-small cell lung cancer. Int J Radiat Oncol Biol Phys 84(4):1010–1016
36. Galper SL, Yu JB, Mauch PM et al (2011) Clinically significant cardiac disease in patients with Hodgkin lymphoma treated with mediastinal irradiation. Blood 117:412–418
37. Hardy D, Liu CC, Cormier JN et al (2010) Cardiac toxicity in association with chemotherapy and radiation therapy in a large cohort of older patients with non-small cell lung cancer. Ann Oncol 21:1825–1833
38. Harris EE, Correa C, Hwang WT et al (2006) Late cardiac mortality and morbidity in early-stage breast cancer patients after breast-conservation treatment. J Clin Oncol 24:4100–4106
39. Hull MC, Morris CG, Pepine CJ et al (2003) Valvular dysfunction and carotid, subclavian, and coronary artery disease in survivors of Hodgkin lymphoma treated with radiation therapy. JAMA 290:2831–2837
40. Vallis KA, Pintilie M, Chong N et al (2002) Assessment of coronary heart disease morbidity and mortality after radiation therapy for early breast cancer. J Clin Oncol 20:1036–1042
41. Murshed J, Liu JJ, Liao Z et al (2004) Dose and volume reduction for normal lung using intensity-modulated radiotherapy for advanced-stage non-small cell lung cancer. Int J Radiat Oncol Biol Phys 58(4):1258–1267
42. Perez CA, Bauer M, Edelstein S et al (1986) Impact of tumor control on survival in carcinoma of the lung treated with irradiation. Int J Radiat Oncol Biol Phys 12:539–547
43. Rengan R, Rosenzweig KE, Venkatraman E et al (2004) Improved local control with higher doses of radiation in large-volume stage III non-small cell lung cancer. Int J Radiat Oncol Biol Phys 60:741–747
44. Wang L, Correa CR, Zhao L et al (2009) The effect of radiation dose and chemotherapy on overall survival in 237 patients with stage III non-small cell lung cancer. Int J Radiat Oncol Biol Phys 73:1383–1390
45. Belderbos JS, Heemsbergen WD, De Jaeger K et al (2006) Final results of a Phase I/II dose escalation trial in non-small cell lung cancer using three-dimensional conformal radiotherapy. Int J Radiat Oncol Biol Phys 66:126–134

46. Christian JA, Jl B, Webb S et al (2007) Comparison of inverse-planned three-dimensional conformal radiotherapy and intensity-modulated radiotherapy for non-small cell lung cancer. Int J Radiat Oncol Biol Phys 67:735–741
47. Govaert SF, Esther GC, Olga CJ et al (2012) Treatment outcome and toxicity of intensity-modulated (chemo) radiotherapy in stage III non-small cell lung cancer patients. Radiat Oncol 150:1–7
48. Jiang S, Yang K, Komaki R et al (2012) Long-term clinical outcome of intensity-modulated radiotherapy for inoperable non-small cell lung cancer: the MD Anderson experience. Int J Radiat Oncol Biol Phys 83(1):332–339
49. Liao ZX, Komaki R, Thames HD Jr et al (2010) Influence of technologic advances on outcomes in patients with unresectable, locally advanced non-small cell lung cancer receiving concomitant chemoradiotherapy. Int J Radiat Oncol Biol Phys 76(3):775–781

Mesothelioma

13

William W. Chance, Neal Rebueno, and Daniel R. Gomez

Keywords
Mesothelioma • Malignant pleural mesothelioma • Multimodality therapy • Hemithoracic radiation • Extrapleural pneumonectomy • Pleurectomy

13.1 Introduction

Mesothelioma is an aggressive and often deadly malignancy for which multimodality therapy remains the standard of care for patients who can tolerate such therapy. For locally advanced disease, treatment that includes surgical resection is the best approach; whether that surgery should be extrapleural pneumonectomy (EPP) or a lung-sparing technique, however, is in debate owing to recent retrospective and prospective evidence that EPP may have greater morbidity but similar disease control [7, 19]. Numerous trials of systemic therapy are also underway in attempts to improve the control of micrometastatic disease [20, 21].

Radiation therapy options for mesothelioma have also evolved in parallel with surgical and systemic approaches over the past 10–15 years. The use of conventional nonconformal techniques did not allow the radiation dose to normal structures to be adequately estimated, and the target volume was frequently underdosed. Computed tomography (CT)-based treatment planning, the ability to calculate and evaluate dose-volume histograms, and ultimately the implementation of inverse planning and fluence modulation with intensity-modulated radiation therapy (IMRT) enabled the evaluation of treatment plans based on adherence to normal

W.W. Chance, M.D. • N. Rebueno, C.M.D. • D.R. Gomez, M.D. (✉)
Department of Radiation Oncology, The University of Texas MD Anderson Cancer Center, Unit 1150, 1840 Old Spanish Trail, Houston, TX 77054-1901, USA
e-mail: DGomez@mdanderson.org

Y. Nishimura, R. Komaki (eds.), *Intensity-Modulated Radiation Therapy*,
DOI 10.1007/978-4-431-55486-8_13

tissue constraints and avoidance of critical structures near the treatment field. However, the use of IMRT to treat mesothelioma, as well as disease at other anatomic sites, involved a "learning curve" in how to use this complex technique effectively yet safely and to establish unique guidelines for its use based on novel features of the treatment planning process. As will become clear from the findings reviewed below, the toxicity associated with the early use of conventional techniques substantially affected how the disease is treated with radiation therapy today. Similarly, IMRT techniques continue to be modified in attempts to improve rates of disease control and expand the number of patients suited for radiation therapy, so that patients can be provided with the best treatment options and the best overall prognosis.

13.2 Conventional Radiation Therapy for Mesothelioma

Because mesothelioma by definition involves the pleura, it is widely accepted that even when imaging shows involvement of discrete regions, the entire pleural cavity is at risk for disease. Therefore, attempts to control localized disease have focused on hemithoracic treatments. In the context of surgical resection, the standard of care involves removal of the entire pleura, with or without the ipsilateral lung and pericardium. Similarly, with respect to radiation therapy, the techniques used as adjuvant therapy encompass a hemithoracic approach that targets the entire pleural bed.

Regardless of initial disease stage, single-modality treatment for mesothelioma, whether that modality is surgery, chemotherapy, or radiation therapy, rarely leads to long-term survival. Typical results after the use of external beam radiation therapy, delivered with conventional techniques as a single modality, are reflected in a retrospective review of 38 patients with mesothelioma treated at the Peter MacCallum Cancer Institute between 1981 and 1985. Twelve of these patients had had definitive radiation therapy to the involved hemithorax to a total dose of 50 Gy. The median survival time for this group was 17 months, and the estimated 2-year survival rate was 17 %. Two patients died of treatment-related causes, one from radiation-induced hepatitis and the other from radiation myelopathy, for a mortality rate of 17 %. The authors of this review concluded that radiation therapy alone is ineffective in prolonging survival in patients with mesothelioma [4].

Given these dismal outcomes, the preferred treatment strategy then shifted toward a multimodality approach in which surgery (EPP) was followed by adjuvant radiation therapy, with or without systemic chemotherapy. An early retrospective review of this approach comprised 49 patients with mesothelioma who underwent EPP at the Brigham and Women's Hospital and Dana-Farber Cancer Institute between 1987 and 1993. Ten of those patients also received adjuvant multi-agent chemotherapy, and 35 patients received adjuvant multi-agent chemotherapy followed by radiation therapy. The radiation therapy involved treating the ipsilateral hemithorax and mediastinum to about 30 Gy, with a boost to 50–55 Gy given to areas of previously bulky disease when possible. The median overall survival time for these patients was 22 months, and the median time to disease progression 19 months. At 3 years, the actuarial overall survival rate was 34 % and the freedom from disease progression rate 33 %. The most common site of first recurrence (35 % of patients) was the ipsilateral hemithorax, and the predominant failure pattern was local, accounting for 67 % of all recurrences. A trend was noted toward improved local control after radiation therapy (local recurrence rates of 31 % with radiation therapy vs. 45 % without radiation therapy) [3].

Collectively, these reviews demonstrated that the combination of surgery and radiation therapy at palliative doses (30–50 Gy) provided modest increases in disease control, but that rates of local recurrence remained high. As a result, subsequent approaches involved delivering higher doses of radiation to the hemithorax in an effort to improve local control. In one such study, Rusch et al. at Memorial Sloan Kettering Cancer Center conducted a prospective phase II trial of surgical resection followed by adjuvant high-dose hemithoracic radiation. From 1995 through 1998, 88 patients underwent surgical resection (70 % with EPP), and 57 received adjuvant radiation to the entire hemithorax, including the thoracotomy and chest tube incision sites. Radiation was given to a dose of 54 Gy via anterior and posterior fields, with the spinal cord protected after 41.4 Gy. Cerrobend blocks were used to limit the dose to the liver, heart, and stomach when necessary. Matched electron fields were used in the blocked regions to prevent underdosing the pleura and diaphragm. The median survival time was 17 months. In contrast to prior studies, only 7 patients (13 %) had locoregional recurrence, which in 3 cases was classified as failures at the margin of the radiation field. Distant metastases were the most common form of relapse [18, 23].

EPP has historically been preferred for patients who are candidates for surgery and have localized epithelioid tumors without nodal involvement. Unfortunately, only a small minority of patients with mesothelioma are suitable candidates for this approach. For patients of advanced age or with inadequate baseline pulmonary function or other comorbid conditions, pleurectomy/decortication is an alternative surgical approach that involves "stripping" the visceral and parietal pleura but not removing the lung. Many comparisons of these techniques have been done, and most studies show that pleurectomy is also associated with high rates of local recurrence (about 60–70 %) but has similar rates of survival and less morbidity and mortality than EPP [19, 7]. Historically, pleurectomy has remained a viable option for patients with mesothelioma who are not candidates for a pneumonectomy.

The use of hemithoracic radiation therapy after pleurectomy presents some unique challenges compared with radiation after EPP. First, patients with functional lung on the ipsilateral side are at risk of radiation pneumonitis from inadvertent irradiation of that lung; indeed, patients can theoretically experience significant functional decline because of the difficulty in sparing the functional ipsilateral lung. Moreover, if much of the ipsilateral lung is deemed nonfunctional after irradiation, a "shunting" effect can be produced whereby perfusion to the ipsilateral lung continues, yet very little air exchange occurs. Thus, radiation treatment for patients with two functional lungs paradoxically can be more challenging than when only one lung is intact.

Studies of hemithoracic radiation given with conventional techniques after pleurectomy are sparse. In 2005, Gupta et al. presented retrospective findings from 123 patients unable to undergo EPP who were treated with pleurectomy at Memorial Sloan Kettering from 1974 through 2003. All patients received adjuvant external beam radiation therapy to the involved hemithorax to a total median dose of 42.5 Gy, and 54 patients also received intraoperative brachytherapy (160 Gy). The radiation technique used was similar to that in the previous Sloan Kettering trial, consisting of standard anteroposterior photon beam geometry with blocks for vital organs and matching electron fields to complete the dose to the target volume. At a median follow-up time of 11 months after treatment, the local control rate for all patients was 44 %. However, for patients with minimal or no residual disease after resection, the median time to failure was 21 months. Two patients (1.6 %) died of treatment-related toxicity, one of cardiac etiology and the other of radiation pneumonitis. These investigators concluded that hemithoracic radiation with conventional techniques was relatively well tolerated but was not effective in reducing local recurrence after pleurectomy [9].

13.2.1 Intensity-Modulated Radiation Therapy After Extrapleural Pneumonectomy

When this chapter was written, IMRT had become a commonly used technique with the potential to tightly conform dose to the target volumes, thereby allowing delivery of higher doses of radiation to the target while reducing the normal tissue toxicity relative to standard conventional treatments. IMRT has been shown to be effective and safe for lung cancer [12], and many of the same principles have been applied in using IMRT for mesothelioma.

Ahamad et al. published an early report on the use of IMRT after EPP for 28 patients treated at MD Anderson Cancer Center. Those patients received IMRT to the surgically violated inner chest wall, insertion of diaphragm, pleural reflections, and deep margin of the incision. The total delivered dose was 45–50 Gy, with focal boost doses to gross disease of up to 60 Gy. At a median follow-up time of 9 months, no in-field failures had taken place. Rates of overall survival and disease-free survival at 1 year were 65 and 88 %. Nausea and vomiting were the most common side effects (65 %, grades 2–3), followed by fatigue (62 %, grades 2–3). No grade 4 or higher toxicity was reported [1]. These findings suggested that IMRT after EPP was well tolerated and that short-term outcomes were encouraging.

However, the limitations of more aggressive dose delivery were illustrated by Allen et al. in 2006. Those authors retrospectively reported outcomes from an initial group of 13 patients treated at the Brigham and Women's Hospital and Dana-Farber Cancer Institute in 2004–2005. All patients had had EPP, and 11 patients had received heated cisplatin during surgery. This treatment was followed by adjuvant IMRT to a dose of 54 Gy to the involved hemithorax. Contralateral lung dose constraints used were similar to those used in treating lung cancer, including a V_{20} of <20 % and a mean lung dose (MLD) of <15 Gy. This approach led to the death of 6 patients (46 %) from pneumonitis, with a median onset of 30 days after the completion of radiation therapy. The V_{20} among those 6 patients ranged from 15.3 to 22.3 %, the V_5 from 81 to 100 %, and the MLD from 13.3 to 17 Gy [2]. Potential contributors to this suboptimal outcome noted in editorials on this study included the use of heated intraoperative cisplatin, but the most significant factor was thought to be the dose to the contralateral (and only functioning) lung. As a result, recommendations based on these outcomes included the lung dose constraint of $V_{20} < 10$ %, with close attention to minimizing V_5 as well [10].

In 2007, an updated retrospective analysis from MD Anderson Cancer Center described 63 patients who had been treated from 2000 through 2005 with EPP followed by adjuvant IMRT to the involved hemithorax to a total dose of 45–50 Gy. At a median follow-up time of 35 months from surgery, the median survival time was 14.2 months, and the 2- and 3-year survival rates were 32 and 21 %. The locoregional failure rate was 13 %, with only 5 % as failures within the treatment field. The rate of fatal pulmonary toxicity was 9.5 %. Additional analysis suggested that patients who received an MLD of <8.5 Gy may have had fewer fatal pulmonary outcomes [15, 16]. Another update from MD Anderson describing 86 patients who had been treated with this technique continued to show relatively low rates of high-grade toxicity and promising improvements in survival. Specifically, rates of grade 3 toxicity were 17 % skin, 12 % lung, 2 % cardiac, and 16 % gastrointestinal. Overall survival rates were almost 90 % at 1 year and 71 % at 2 years, with nodal involvement being associated with both distant metastasis and overall survival [8].

In 2008, Duke University reported their institutional experience in a retrospective study of 13 patients treated with EPP followed by adjuvant IMRT from 2005 through 2007. Patients had been treated to a median dose of 45 Gy to the involved ipsilateral hemithorax, instrument insertion sites, and involved nodal regions, with a focal boost to areas suspected of harboring high-risk disease. Three patients (23 %) experienced grade ≥2 toxicity, including 1 (8 %) with grade 5 pneumonitis after an MLD of 11.4 Gy (compared with 7.6 Gy for the other patients). Six patients (46 %) experienced local recurrence, and 9 patients (69 %) were still alive at the most recent follow-up, at a median 9.5 months after IMRT completion [13].

Even though the use of these more modern techniques has produced encouraging improvements in local control, distant failure rates remain high and limit improvements in long-term survival. An attempt to improve these results with the use of trimodality therapy came in the form of a multicenter phase II trial, the findings of which were published by investigators at Sloan Kettering in 2009. In that trial, 77 patients with medically operable stage T1-3 N0-2 disease were treated with neoadjuvant pemetrexed plus cisplatin followed by EPP and hemithoracic radiation therapy to a dose of 54 Gy. IMRT was allowed but not required. The median overall survival time was about 17 months. Evidence of a radiologic response to induction chemotherapy doubled the survival time to 26 months from 13 months ($P < 0.05$), but a pathologic complete response did not predict long-term survival in two of the three patients with such a response. Among the 40 patients who completed all three treatment modalities, 1 had grade 5 pneumonitis, and 5 had grade ≥3 radiation-related toxicity [11].

13.2.2 Intensity-Modulated Radiation Therapy After Lung-Sparing Techniques

As alluded to above, the superiority of EPP over pleurectomy has been questioned recently, leading some centers away from using EPP, particularly for borderline-operable patients, those with high-risk tumor histology (biphasic or sarcomatoid), or those with nodal involvement. Although studies of conventional radiation used as adjuvant therapy after lung-sparing techniques have generally shown less than optimal outcomes, analyses of IMRT in this circumstance are rare. Most such studies involved patients who had either had pleurectomy and then IMRT or had been considered to have inoperable disease and received only IMRT to the hemithorax. The current literature on the use IMRT in these circumstances is reviewed below.

From 2005 through 2010, 36 patients from Sloan Kettering with mesothelioma who were unable to undergo pneumonectomy underwent induction chemotherapy (89 % of patients) followed by pleurectomy and adjuvant IMRT (56 %) or IMRT alone (44 %). The pleural surface of the involved hemithorax was treated in all cases to a median dose of 46.8 Gy. Seven patients (20 %) developed grade ≥3 pneumonitis (one grade 5). For patients who underwent pleurectomy before IMRT, the 1-year and 2-year survival rates were 75 and 53 %, and the median survival time was 26 months. For patients who did not undergo surgical resection, the 1-year and 2-year survival rates were slightly lower at 69 and 28 %, and the median survival time was slightly shorter at 17 months [17].

Investigators from Australia have also published results on the use of three-dimensional conformal radiation therapy or IMRT to treat mesothelioma in patients with an intact lung. Of the 14 patients in that study treated to 45–60 Gy to the hemithorax, only 1 had undergone EPP. The in-field local control rate was 71 %, with no grade ≥4 toxicity. Median survival times were 25 months after diagnosis and 17 months after starting radiation therapy [6].

A prospective study of pleurectomy and hemithoracic IMRT conducted in Italy was published by Minatel et al. in 2013. In 2009–2010, 20 patients with mesothelioma had radical pleurectomy followed by IMRT to the entire hemithorax, excluding the intact lung, to a prescribed dose of 50 Gy, with a simultaneous integrated boost to 60 Gy to areas suspected of harboring residual disease. Nineteen patients also received cisplatin and pemetrexed chemotherapy. The median follow-up time was 27 months. The median overall survival and progression-free survival times were 33 and 29 months. Overall survival rates at 2 and 3 years were 70 % and 49. Local-regional control rates at 2 and 3 years were 68 % and 59 %. The predominant pattern of failure was distant, with 7 patients developing distant metastases as the first site of relapse. Only 3 patients experienced isolated locoregional recurrences. Five patients had grades 2–3 pneumonitis, but no fatal pulmonary toxicity was noted [14].

A recent report from MD Anderson described 22 patients with mesothelioma treated with pleurectomy followed by adjuvant IMRT to the involved hemithorax (11 prospectively enrolled in an institutional protocol and 11 studied off protocol). Patients were treated to 45 Gy in 25 fractions, and 9 patients received a simultaneous boost to 60 Gy to high-risk areas. Twenty patients received chemotherapy (15 induction, 2 adjuvant, and 3 both). The median follow-up time after surgery was 14.7 months (range 4.1–37.1 months). The therapy was well tolerated, with no episodes of grade 4–5 pulmonary toxicity and one case of grade 4 thrombocytopenia. Progressive decreases in pulmonary function were noted after surgery followed by IMRT. Median percentage of predicted forced vital capacity (FVC), forced expiratory volume at 1 s (FEV1), and DLCO at baseline were 88 %, 83 %, and 87 %. Significant decreases relative to baseline were seen after surgery (PD): FVC 67 % ($P<.01$), FEV1 67 % ($P<.01$), and DLCO 68 % ($P<.01$), and significant additional reductions were noted after IMRT: FVC 57 % ($P=02$), FEV1 58 % ($P<01$), and DLCO 56 % ($P<.01$). Rates of overall survival and progression-free survival were 73 and 60 % at 1 year and 52 and 32 % at 2 years. Outcomes for IMRT after pleurectomy (compared with 22 matched patients who had had IMRT after EPP) included less grade 4–5 toxicity (0/22 vs. 3/22, $P=0.23$) and trends toward improved median overall survival time (28.4 months vs. 14.2 months, $P=0.14$), median disease-free survival time (15.4 months vs. 10.2 months, $P=0.18$), and median time to distant metastasis (not reached vs. 11.8 months, $P=0.15$). Median time to local-regional failure (20.5 months vs. not reached, $P=0.06$) also seemed to be better in the group who had had EPP [5].

Indeed, even when efforts are made to spare the ipsilateral lung in patients undergoing radiation after pleurectomy, many patients ultimately experience significant declines in function on the treated side, which can develop into a "functional" pneumonectomy. However, the clinical effects of such changes, manifested by high rates of grade 3 radiation pneumonitis or high-grade dyspnea, are often disproportionally low. Possible reasons for this discrepancy are as follows. First, the ipsilateral lung is typically substantially smaller than the uninvolved lung, with much less perfusion on the involved side. As a result, treating the ipsilateral hemithorax is rarely synonymous with delivering radiation to 50 % of the lung volume. Second, IMRT allows the dose to the target to be made more homogeneous, and hot spots within normal lung parenchyma can be avoided to a greater extent than is possible with conventional techniques. If the amount of lung receiving high-dose radiation is important in the development of radiation pneumonitis, as some have suggested [22], then even though almost all of the ipsilateral lung receives a low dose of radiation in most cases, sparing this region from high doses that would result from the use of 2D techniques may also reduce the rates of pulmonary toxicity.

13.3 Intensity-Modulated Radiation Therapy Technique

13.3.1 Before Treatment Simulation

All patients undergoing hemithoracic IMRT should undergo a nuclear perfusion renal scan and measurements of blood urea nitrogen and creatinine levels before treatment to assess the relative perfusion in each kidney, because covering the entire pleura and pleural bed will typically extend inferiorly to this region. The treating physician should ensure that kidney function is normal and that perfusion in the contralateral kidney, which should receive almost no radiation dose, is roughly equal to that of the kidney on the involved side. If these two criteria are not met, then a nephrologist should be consulted.

Because hemithoracic IMRT for patients with an intact lung is not yet considered an established technique, it should be performed only in the context of a clinical trial or at centers with experience in this type of treatment. The method used at the authors' institution is as follows. Before treatment simulation, patients undergo pulmonary function testing and a quantitative perfusion study to determine the anticipated functional loss after treatment to the hemithorax. For the purposes of pretreatment evaluation, it is appropriate to estimate a "worst-case scenario" in which the patient would experience a long-term functional pneumonectomy and therefore depend on pulmonary function from the contralateral side. We use the following parameters to identify patients who are candidates for this approach: (1) expected postradiation FEV1 >30 % predicted {defined as [pretreatment FEV1 × (% perfusion to contralateral lung)]} and (2) pretreatment DLCO >30 % predicted.

13.3.2 Treatment Simulation

The simulation technique for adjuvant IMRT for mesothelioma is the same regardless of whether the lung is intact or not. In either situation, before the patient is immobilized, scars can be marked with wire and drain sites with BBs and a bolus placed on these regions so that the treatment volume can be extended laterally toward the skin. At the authors' institution, if a bolus is placed, the material is 5 mm thick and encompasses the scar and drain sites with a 2.5- to 3-cm margin circumferentially. However, because rates of treatment failures in drain sites and scars have been low in most series and because skin reactions are expected to increase when this approach is used, it is reasonable to reserve bolus placement for patients at high risk for recurrence in these regions, as indicated by operative or imaging findings.

Treatment is simulated while the patient is supine with arms over the head to maximize the number of beam angles for treatment planning. Patients are immobilized with a customized upper-body cradle, and 4D CT scanning is done to account for tumor motion. Respiratory management techniques such as breath hold or gating are not typically used.

13.3.3 Radiation Treatment Volumes

13.3.3.1 IMRT After Extrapleural Pneumonectomy

For patients who have undergone an EPP, we recommend covering the entire hemithorax, from approximately the thoracic inlet to L1, to ensure that all ribs are encompassed in the internal target volume (ITV) (Fig. 13.1). We do not recommend coverage of the mediastinal lymph nodes, although the ipsilateral hilar lymph nodes are often included in the treatment volume when the pleural bed is contoured. The

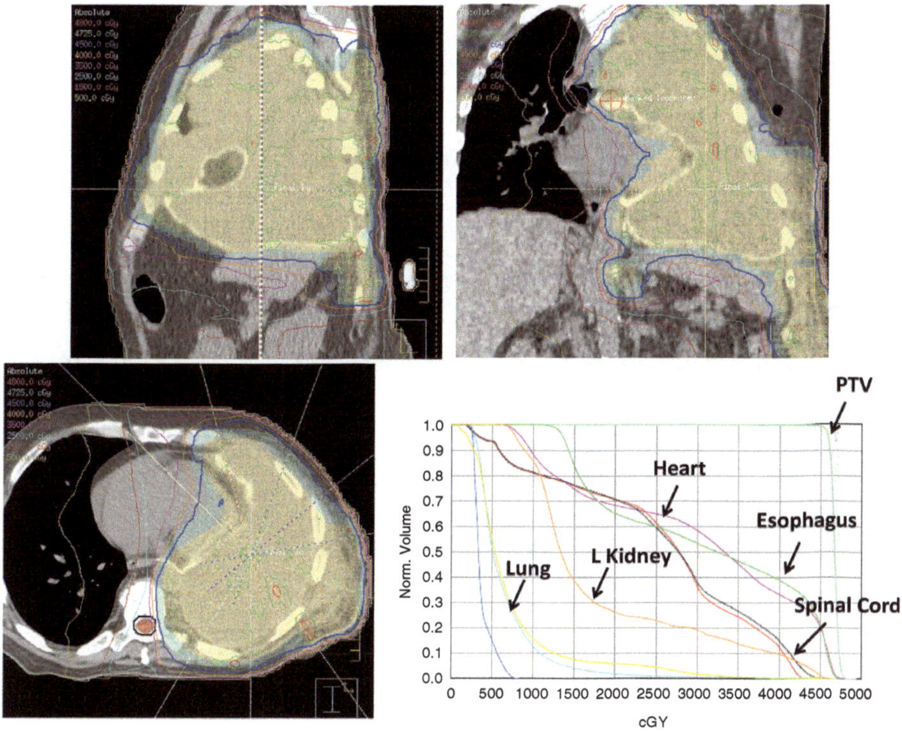

Fig. 13.1 Scans for intensity-modulated radiation therapy to be given after an extrapleural pneumonectomy for mesothelioma, with representative sagittal, coronal, and axial slices and the dose-volume histogram (DVH)

pleura should be covered medially such that the crus and the reconstructed diaphragm are within the treatment volume. Lateral expansion to drain sites and scars should also be included in the ITV if a bolus was placed during simulation. If daily kilovoltage imaging is used, the ITV can be expanded by 5 mm to encompass the planning target volume.

13.3.3.2 IMRT with an Intact Lung

As noted above, we recommend that this technique be used only in the context of clinical protocols or at institutions with experience in using it. Our treatment method is as follows. Consistent with reported methods from other centers, our treatment volumes include an ITV "rind," that is, a 5-mm expansion off the pleural bed (Fig. 13.2). These volumes are then expanded similarly to those for IMRT after EPP as described above, with sparing of the ipsilateral lung when feasible. The target is adjusted to account for respiratory motion and to include all at-risk pleural regions; the target is also expanded to include the scar and drain sites laterally if applicable. For patients with gross residual disease after surgery, limited studies have suggested

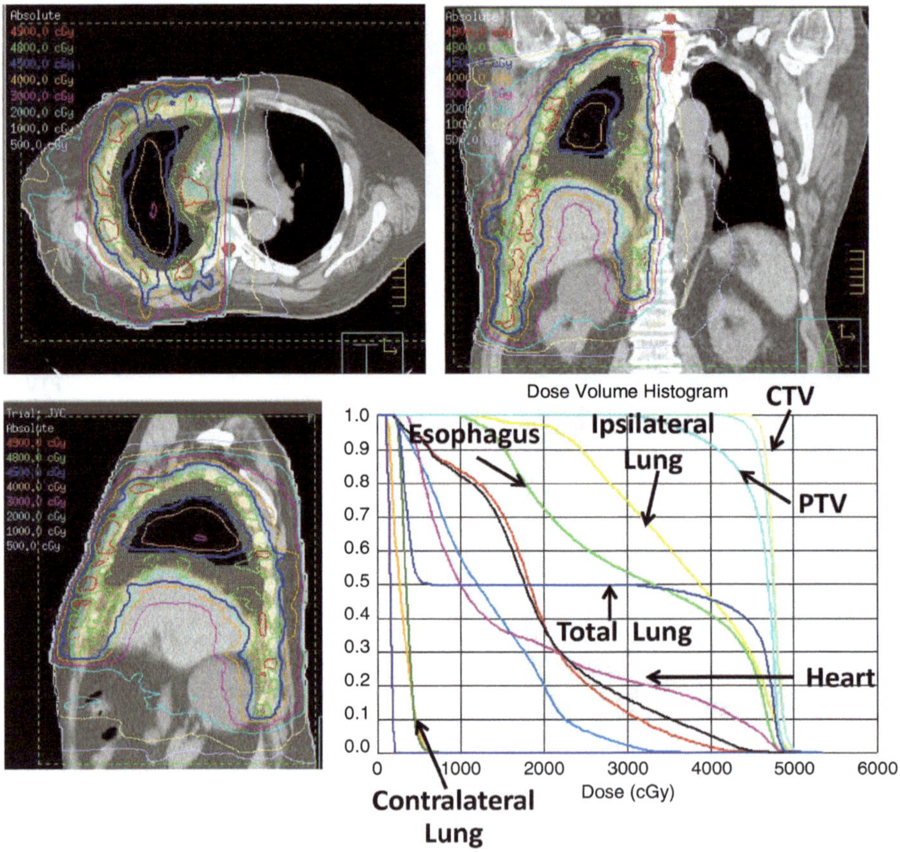

Fig. 13.2 Scans for intensity-modulated radiation therapy to be given after a pleurectomy or decortication, with lung intact for mesothelioma, with representative sagittal, coronal, and axial slices and the dose-volume histogram (DVH). The normal tissue complication probability in this case was 24 %

using a concomitant boost, although again this technique must be considered experimental at this time and should not be attempted except in the context of a clinical trial or at a center with experience in its use.

13.3.4 Radiation Dose

Doses to microscopic disease when IMRT is to be used as adjuvant treatment are typically 45–50.4 Gy, given in 1.8-Gy fractions. A concomitant boost has been used to treat gross disease in limited studies, but this approach is not considered standard and its safety and efficacy have not been established.

13.3.5 Normal Tissue Constraints

We recommend standard thoracic normal tissue constraints to the esophagus (mean dose <34 Gy), heart (V_{40} < 40 % or mean dose <20 Gy), and spinal cord (maximum dose <45 Gy). For left-sided disease, meeting heart constraints can be difficult while still covering the pericardium and anterior pericardial pleural bed and lymph nodes, and so treatment beyond these constraints is at the treating physician's discretion. The current literature on this technique does not describe high rates of heart toxicity. Because the fields extend inferiorly to the kidneys, and for right-sided tumors to the liver, we recommend adhering to normal tissue constraints for these organs as well (liver V_{30} < 40 %, mean dose <30 Gy; kidney V_{20} < 33 %). As for lung constraints, for patients who have had an EPP, we recommend limiting the dose to the remaining lung to an MLD of <8.5 Gy and a V_{20} of <10 %. When treating patients with an intact lung, we use a normal tissue complication probability limit of <25 % to the total lung, and we also attempt to limit the contralateral lung to an MLD of <8.5 Gy and a V_{20} of <10 %.

Acknowledgments Supported in part by Cancer Center Support (Core) Grant CA016672 from the US National Cancer Institute to the University of Texas MD Anderson Cancer Center. We would like to sincerely thank Christine Wogan, MS, for her valuable assistance in editing this manuscript.

Conflict of Interest The authors declare no conflicts of interest.

References

1. Ahamad A, Stevens CW, Smythe WR (2003) Promising early local control of malignant pleural mesothelioma following postoperative intensity modulated radiotherapy (IMRT) to the chest. Cancer J 9:476–484
2. Allen AM, Czerminska M, Janne PA, Sugarbaker DJ, Bueno R, Harris JR, Court L, Baldini EH (2006) Fatal pneumonitis associated with intensity-modulated radiation therapy for mesothelioma. Int J Radiat Oncol Biol Phys 65(3):640–645
3. Baldini EH, Recht A, Strauss GM (1997) Patterns of failure after trimodality therapy for malignant pleural mesothelioma. Ann Thorac Surg 63:334–338
4. Ball D, Cruickshank DG (1990) The treatment of malignant mesothelioma of the pleura: review of a 5-year experience, with special reference to radiotherapy. Am J Clin Oncol 13:4–9
5. Chance WW, Rice DC, Allen PK, Tsao AS, Fontanilla HP, Liao Z, Chang JY, Tang C, Pan HY, Welsh JW, Mehran RJ, Gomez, DR (2014) Hemithoracic intensity modulated radiation therapy after pleurectomy/decortication for malignant pleural mesothelioma: toxicity, patterns of failure, and a matched survival analysis. Int J Radiat Oncol Biol Phys (Epub ahead of print)
6. Feigen M, Lee ST, Lawford C, Churcher K, Zupan E, Scott AM, Hamilton C (2011) Establishing locoregional control of malignant pleural mesothelioma using high-dose radiotherapy and (18) F-FDG PET/CT scan correlation. J Med Imaging Radiat Oncol 55(3):320–332

7. Flores RM, Pass HI, Seshan VE, Dycoco J, Zakowski M, Carbone M, Bains MS, Rusch VW (2008) Extrapleural pneumonectomy versus pleurectomy/decortication in the surgical management of malignant pleural mesothelioma: results in 663 patients. J Thorac Cardiovasc Surg 135(3):620–626, 626 e621-623
8. Gomez DR, Hong DS, Allen PK, Welsh JS, Mehran RJ, Tsao AS, Liao Z, Bilton SD, Komaki R, Rice DC (2013) Patterns of failure, toxicity, and survival after extrapleural pneumonectomy and hemithoracic intensity-modulated radiation therapy for malignant pleural mesothelioma. J Thorac Oncol 8(2):238–245
9. Gupta V, Mychalczak B, Krug L, Flores R, Bains M, Rusch VW, Rosenzweig KE (2005) Hemithoracic radiation therapy after pleurectomy/decortication for malignant pleural mesothelioma. Int J Radiat Oncol Biol Phys 63(4):1045–1052
10. Komaki R, Liao Z, Liu H, Tucker S, Rice D (2006) Fatal pneumonitis associated with intensity-modulated radiation therapy for mesothelioma: in regard to Allen et al. (Int J Radiat Oncol Biol Phys 2006;65:640–645). Int J Radiat Oncol Biol Phys 66(5):1595–1596, author reply 1596
11. Krug LM, Pass HI, Rusch VW, Kindler HL, Sugarbaker DJ, Rosenzweig KE, Flores R, Friedberg JS, Pisters K, Monberg M, Obasaju CK, Vogelzang NJ (2009) Multicenter phase II trial of neoadjuvant pemetrexed plus cisplatin followed by extrapleural pneumonectomy and radiation for malignant pleural mesothelioma. J Clin Oncol 27(18):3007–3013
12. Liao ZX, Komaki RR, Thames HD Jr, Liu HH, Tucker SL, Mohan R, Martel MK, Wei X, Yang K, Kim ES, Blumenschein G, Hong WK, Cox JD (2010) Influence of technologic advances on outcomes in patients with unresectable, locally advanced non-small-cell lung cancer receiving concomitant chemoradiotherapy. Int J Radiat Oncol Biol Phys 76(3):775–781
13. Miles EF, Larrier NA, Kelsey CR, Hubbs JL, Ma J, Yoo S, Marks LB (2008) Intensity-modulated radiotherapy for resected mesothelioma: the Duke experience. Int J Radiat Oncol Biol Phys 71(4):1143–1150
14. Minatel E, Trovo M, Polesel J, Baresic T, Bearz A, Franchin G, Gobitti C, Rumeileh IA, Drigo A, Fontana P, Pagan V, Trovo MG (2014) Radical pleurectomy/decortication followed by high dose of radiation therapy for malignant pleural mesothelioma. Final results with long-term follow-up. Lung Cancer 83(1):78–82
15. Rice DC, Smythe WR, Liao Z, Guerrero T, Chang JY, McAleer MF, Jeter MD, Correa A, Vaporciyan AA, Liu HH, Komaki R, Forster KM, Stevens CW (2007) Dose-dependent pulmonary toxicity after postoperative intensity-modulated radiotherapy for malignant pleural mesothelioma. Int J Radiat Oncol Biol Phys 69(2):350–357
16. Rice DC, Stevens CW, Correa AM, Vaporciyan AA, Tsao A, Forster KM, Walsh GL, Swisher SG, Hofstetter WL, Mehran RJ, Roth JA, Liao Z, Smythe WR (2007) Outcomes after extrapleural pneumonectomy and intensity-modulated radiation therapy for malignant pleural mesothelioma. Ann Thorac Surg 84(5):1685–1692, discussion 1692–1683
17. Rosenzweig KE, Zauderer MG, Laser B, Krug LM, Yorke E, Sima CS, Rimner A, Flores R, Rusch V (2012) Pleural intensity-modulated radiotherapy for malignant pleural mesothelioma. Int J Radiat Oncol Biol Phys 83(4):1278–1283
18. Rusch VW, Rosenzweig K, Venkatraman E, Leon L, Raben A, Harrison L, Bains MS, Downey RJ, Ginsberg RJ (2001) A phase II trial of surgical resection and adjuvant high-dose hemithoracic radiation for malignant pleural mesothelioma. J Thorac Cardiovasc Surg 122(4):788–795
19. Treasure T, Lang-Lazdunski L, Waller D, Bliss JM, Tan C, Entwisle J, Snee M, O'Brien M, Thomas G, Senan S, O'Byrne K, Kilburn LS, Spicer J, Landau D, Edwards J, Coombes G, Darlison L, Peto J (2011) Extra-pleural pneumonectomy versus no extra-pleural pneumonectomy for patients with malignant pleural mesothelioma: clinical outcomes of the Mesothelioma and Radical Surgery (MARS) randomised feasibility study. Lancet Oncol 12(8):763–772
20. Tsao AS, Wistuba I, Roth JA, Kindler HL (2009) Malignant pleural mesothelioma. J Clin Oncol 27(12):2081–2090

21. Tsao A, He D, Saigal B, Liu S, Lee JJ, Bakkannagari S, Ordonez NG, Hong WK, Wistuba II, Johnson FM (2007) Activated Src kinase is expressed in malignant pleural mesothelioma tumors; dasatinib inhibition leads to cytotoxicity, cell cycle inhibition, and prevention of invasion and migration (abstract). J Clin Oncol 25(18s):7713
22. Willner J, Jost A, Baier K, Flentje M (2003) A little to a lot or a lot to a little? An analysis of pneumonitis risk from dose-volume histogram parameters of the lung in patients with lung cancer treated with 3-D conformal radiotherapy. Strahlenther Onkol 179(8):548–556
23. Yajnik S, Rosenzweig KE, Mychalczak B, Krug L, Flores R, Hong L, Rusch VW (2003) Hemithoracic radiation after extrapleural pneumonectomy for malignant pleural mesothelioma. Int J Radiat Oncol Biol Phys 56(5):1319–1326

14 Breast Cancer

Gregory M. Chronowski

Keywords
Breast • IMRT • Forward-planned IMRT • Inverse-planned IMRT • Contouring

14.1 Introduction

Radiation therapy has a central role in the management of breast cancer after either breast-conserving surgery or mastectomy, with attendant improvements in local control [8] and survival [5, 21, 22]. Numerous trials showing a benefit from adjuvant irradiation of the breast [2, 4, 8, 30] used conventional two-dimensional techniques involving a beam of uniform intensity, often simply modulated with a physical wedge. Such techniques routinely resulted in areas of the breast receiving up to 115 % of the prescribed dose.

With the development of commercially available treatment planning systems that model the radiation dose in three dimensions, it became possible to homogenize the dose to the breast by progressively blocking "hot spots" within the breast by using Cerrobend blocks or multileaf collimators. This form of intensity modulation of the radiation beam is often referred to as "field-in-field" or "step-and-shoot" technique. These techniques are often categorized as "forward-planned," in that a radiation dose is not specified to a target volume (typically the breast with or without regional lymphatics); instead, in an iterative process, the radiation beam is modulated by progressively blocking hot spots by viewing them on digitally reconstructed radiographs (DRRs) along the axis of the radiation beam. In addition, this technique typically involves the use of only two tangentially oriented gantry positions to treat

G.M. Chronowski, M.D. (✉)
Department of Radiation Oncology, The University of Texas MD Anderson Cancer Center, Unit 1639, 1515 Holcombe Blvd, Houston, TX 77030, USA
e-mail: gchronowski@mdanderson.org

Y. Nishimura, R. Komaki (eds.), *Intensity-Modulated Radiation Therapy*,
DOI 10.1007/978-4-431-55486-8_14

the breast. The end result is a homogeneous plan in which "hot spots" typically exceed the prescribed dose by no more than 7 %.

More recently, the term intensity-modulated radiation therapy (IMRT) has come to refer to a treatment planning process in which targets and structures to be avoided are contoured on axial images, typically computed tomography (CT); a dose to the target and the structures to be avoided are specified within the planning computer; and the planning system then generates a plan involving multiple shaped beams and multiple, often non-tangential, gantry positions. This technique can result in more complex dose distributions than those that can be achieved with a field-in-field technique and can be used to treat irregular target volumes or avoid critical structures, such as the heart, that are close to the target volume.

The role of IMRT in the treatment of breast cancer is rapidly evolving, but its adoption and development have lagged somewhat compared with the use of IMRT to treat disease at other sites such as prostate cancer. Some of this lag may reflect the typically modest radiation doses used to treat breast cancer, which are decreasing still further in clinical practice [27]. However, with the recognition that excess radiation dose to the heart from radiation therapy for breast cancer can result in long-term morbidity and mortality [9, 11], interest is increasing in using IMRT to limit even modest radiation doses to the heart. In addition, the evolving evidence showing increasing benefits from regional nodal irradiation [28], in particular irradiation of the internal mammary chain, has further increased interest in using IMRT to treat regional lymphatics in breast cancer [3].

14.2 Field-in-Field or "Forward-Planned" IMRT

Forward-planned IMRT [17] usually results in a radiation treatment plan with a more homogeneous dose distribution than can typically be achieved with a conventional plan using physical or dynamic wedges (Fig. 14.1). The use of forward planning requires that the planning system be capable of generating a dose cloud that is visible along a beam's eye view on a DRR (Fig. 14.2). The process begins by setting up two tangential photon beams in a typical oblique arrangement on the chest wall. Various dose clouds are then visualized along a beam's eye view (Fig. 14.3), and new fields using the identical gantry positions are created. Then monitor units for each new "field within a field" are subtracted from the open field and added to the new field, which blocks high-dose clouds until the isodose line of interest disappears. The process then begins anew until the 5–7 % isodose cloud is reduced as much as possible. Once mastered, such plans can be generated and delivered rapidly. Centers without access to linear accelerators equipped with multileaf collimators can use fabricated Cerrobend blocks for each individual subfield, with the attendant increase in treatment time owing to the need to switch out blocks for each blocked subfield.

Limitations of forward-planned IMRT are the constraints associated with using just two gantry positions with medial and lateral tangentially oriented fields.

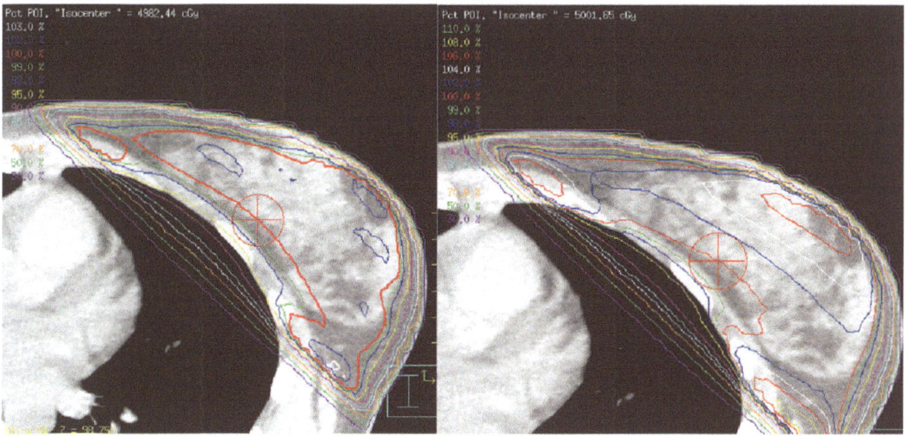

Fig. 14.1 A forward-planned intensity-modulated radiation therapy (IMRT) plan (*left*) vs. a conventional wedge plan (*right*)

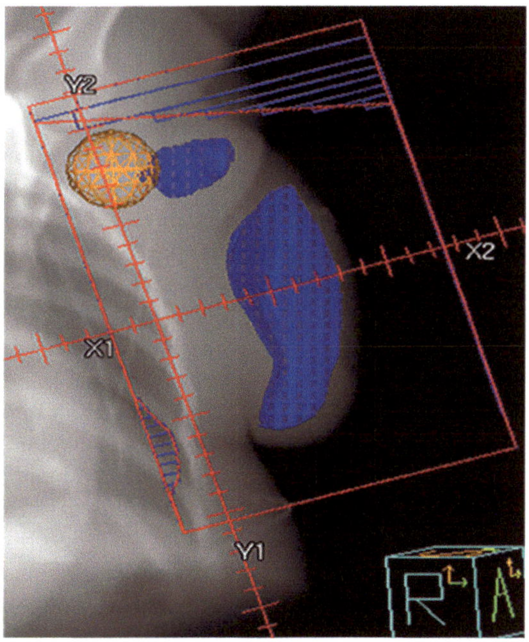

Fig. 14.2 A 107 % isodose cloud (*blue*) visible on a digitally reconstructed radiograph (DRR)

Comprehensive irradiation of the regional lymphatics of the internal mammary, axillary, and supraclavicular regions is often not possible with this technique, and it is ideally suited for treatment that includes only the breast.

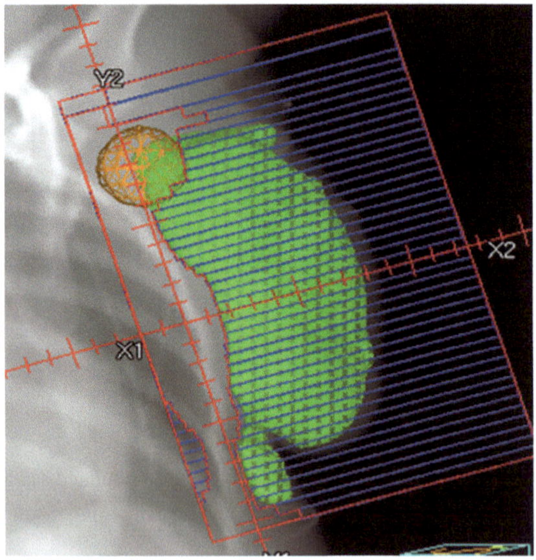

Fig. 14.3 A 105 % isodose cloud (*green*) visible on a digitally reconstructed radiograph (DRR)

14.3 Inverse-Planned IMRT

Over the past decade, the general term IMRT has been used to refer to inverse planning of radiation therapy involving the use of either multiple, non-coplanar fixed gantry positions or, more recently, a continuous photon arc. Such techniques can result in far more complex and homogenous dose distributions than forward-planned IMRT, which offers some advantages when treating lymphatics in the internal mammary, axillary, and supraclavicular regions owing to the non-coplanar orientation of the beams or arcs used. A disadvantage of inverse-planned IMRT is that the multiple-beam geometry results in higher doses to both superficial and deep tissues outside the target volume, such as the lung and contralateral breast. The clinical implications of these high-dose exposures are unclear, but scattered dose to the uninvolved breast during radiation therapy could result in the development of breast cancer [1, 10] or lung cancer [19, 31] in smokers. As a result, the benefits of the improved conformality and homogeneity of inverse-planned IMRT should be balanced against the risk of additional scattered radiation to normal structures when inverse-planned IMRT is used. Inverse-planned IMRT is best reserved for specific cases where more conventional techniques such as forward-planned IMRT are unable to cover regions at risk and should probably not be used routinely, especially in cases where only the breast is being treated.

14.3.1 Avoiding the Heart

Inverse-planned IMRT can result in treatment plans that can avoid the heart in patients with left-sided breast cancer (Fig. 14.4). A deep inspiratory breath-hold technique [23] should be used to minimize target motion during therapy, in addition

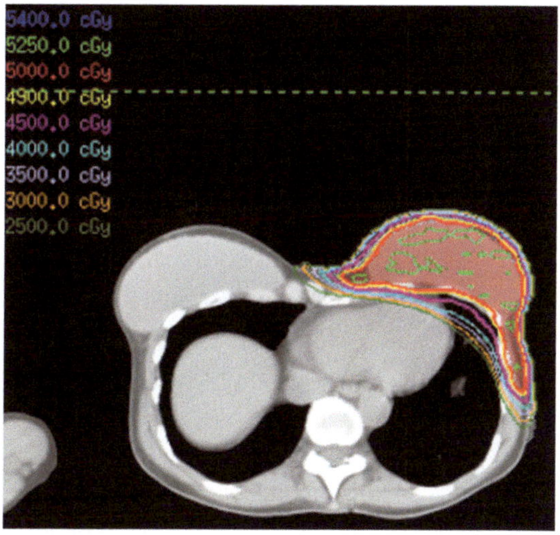

Fig. 14.4 An inverse-planned IMRT plan for a patient with a reconstructed breast that avoids the heart while treating the internal mammary chain of lymphatics

to pulling the chest wall away from the heart as much as possible. The dose to the heart should be thoroughly minimized. Although the optimal dose-volume histogram parameters have not been clearly defined, a reasonable guideline would be a mean dose to the entire muscular heart of <5 Gy [6].

14.3.2 Treatment of Regional Lymphatics

The lymphatic regions at risk in patients with breast cancer include the axillary, infraclavicular, supraclavicular, and internal mammary chains. Evidence is evolving that comprehensive treatment of the lymphatics may be of benefit for patients with high-risk disease [28]. Traditionally the lymphatics of the breast were treated with multiple abutting photon or electron fields (Fig. 14.5), usually matched on the skin. These techniques often resulted in underdosing at the field junctions. IMRT can eliminate underdosing at these regions (Figs. 14.6 and 14.7), at the expense of higher doses to the lung and contralateral breast in some cases. Several detailed contouring atlases have been published [20, 29], and some general principles of nodal contouring are described below.

14.3.2.1 Internal Mammary Lymphatics

The internal mammary artery and vein should be contoured within the first three ipsilateral rib interspaces, starting at the superior aspect of the ipsilateral, medial first rib and extending to the cranial aspect of the ipsilateral, medial, fourth rib (Fig. 14.8).

14.3.2.2 Level I of the Axilla

Level I of the axilla should be contoured starting at the level where the axillary vessels cross the lateral edge of the pectoralis minor muscle superiorly and extending inferiorly to the insertion of the pectoralis minor muscle on the ribs. The

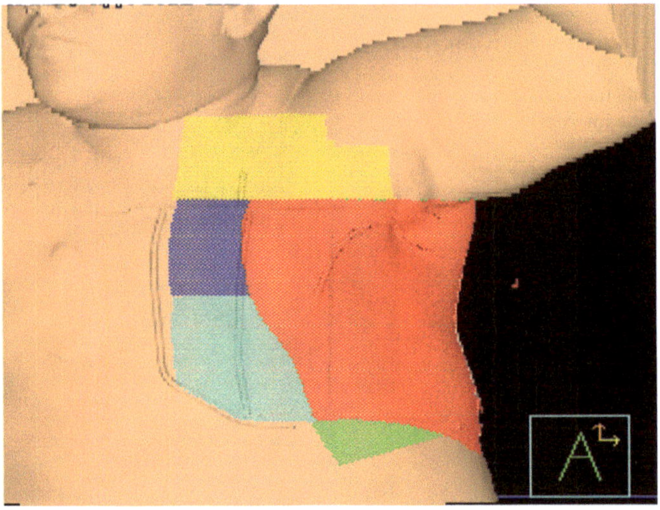

Fig. 14.5 Abutting photon and electron fields used to treat a patient after mastectomy

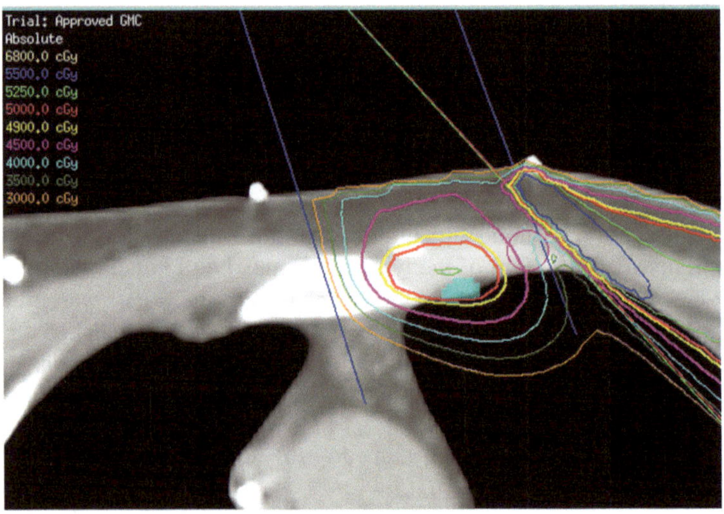

Fig. 14.6 An underdosed region in the internal mammary chain

medial and lateral borders extend between the medial border of the latissimus dorsi muscle and the lateral border of the pectoralis minor muscle, respectively. The anterior border is the anterior surface of the pectoralis major muscle, and the posterior border should be the anterior surface of the subscapularis muscle (Fig. 14.9).

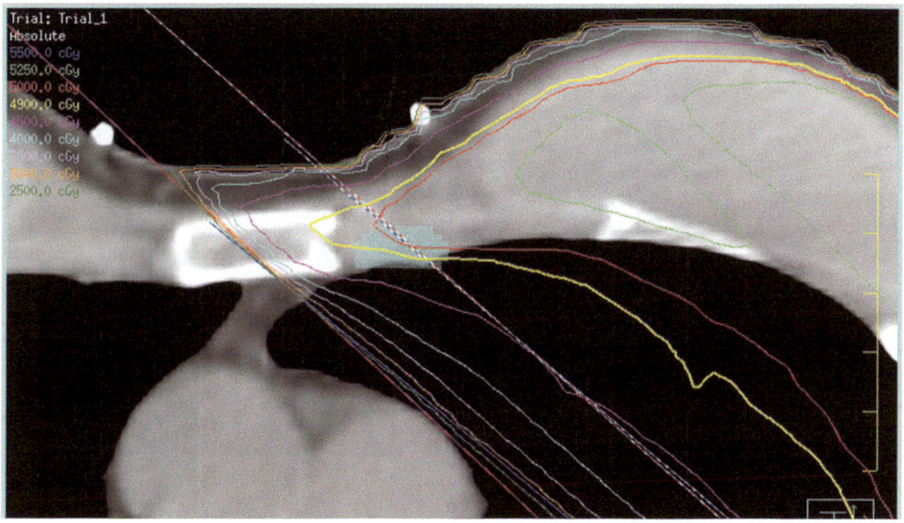

Fig. 14.7 Elimination of the underdosed region in the internal mammary chain with IMRT

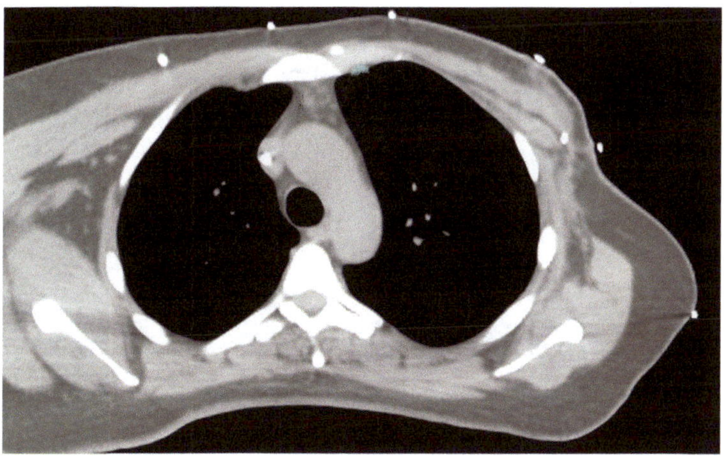

Fig. 14.8 The location of the internal mammary lymphatics

14.3.2.3 Level II of the Axilla

Level II of the axilla should be contoured at the level where the axillary vessels cross the medial edge of the pectoralis minor muscle superiorly, extending inferiorly to the cranial edge of level I of the axilla (where the axillary vessels cross the lateral edge of the pectoralis minor muscle). The medial and lateral borders extend between the medial border of the pectoralis minor muscle and the lateral border of the latissimus dorsi muscle, respectively. The anterior border is the anterior surface of the pectoralis minor muscle and the posterior borders are the ribs and intercostal muscles (Fig. 14.10).

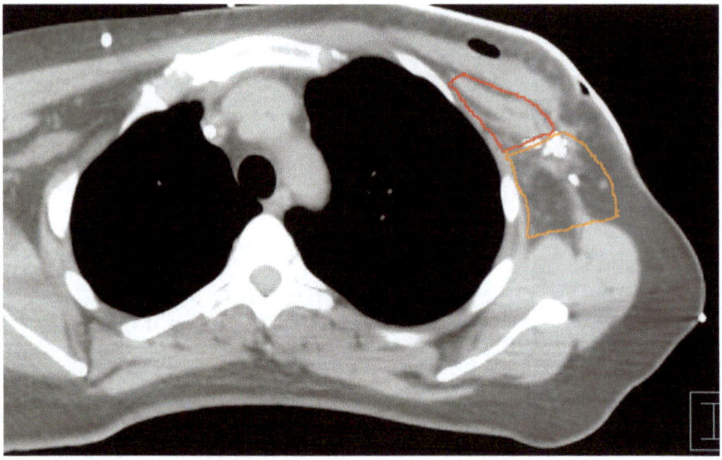

Fig. 14.9 Axillary contours. *Red*, level II; *orange*, level I

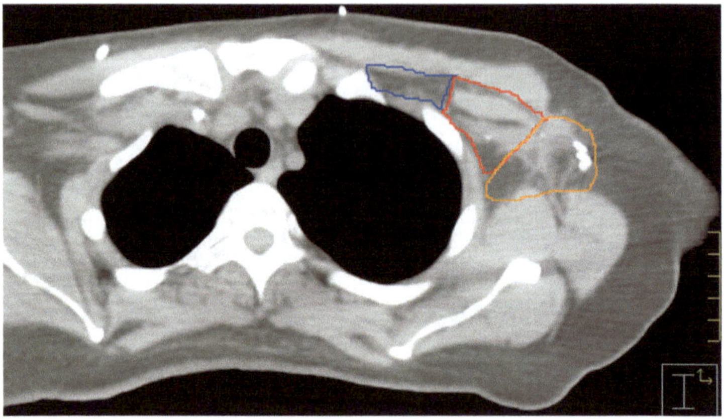

Fig. 14.10 Axillary contours. *Blue*, level III; *red*, level II; *orange*, level I

14.3.2.4 Level III of the Axilla

Level III of the axilla should be contoured at the level of the insertion of the pectoralis minor muscle on the coronoid process superiorly, extending inferiorly to the cranial edge of level II of the axilla (where the axillary vessels cross the medial edge of the pectoralis minor muscle). The medial and lateral borders extend from the thoracic inlet to the medial border of the pectoralis minor muscle, respectively. The anterior border is the posterior surface of the pectoralis major muscle, and the posterior borders are the ribs and intercostal muscles (Fig. 14.11).

14.3.2.5 Supraclavicular Lymphatics

The lymphatics of the supraclavicular region should be contoured at the inferior border of the cricoid cartilage superiorly, extending inferiorly to the caudal edge of

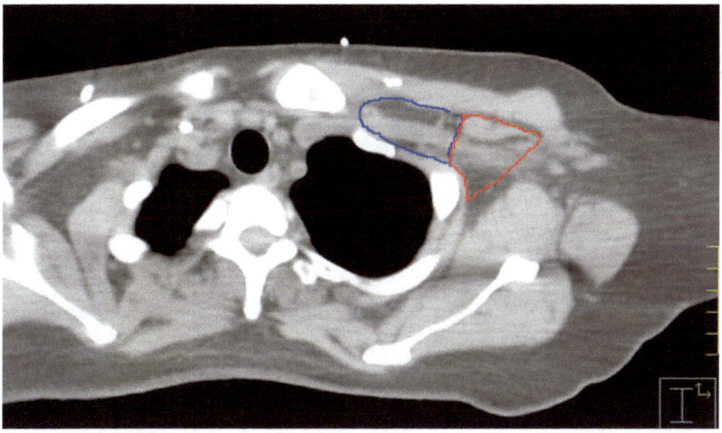

Fig. 14.11 Axillary contours. *Blue*, level III; *red*, level II

the clavicular head. The medial and lateral borders should be the edge of the thyroid gland and trachea and the lateral edge of the sternocleidomastoid muscle, respectively. The anterior border is the posterior surface of the sternocleidomastoid muscle, and the posterior border is the anterior surface of the scalene muscles (Fig. 14.12).

14.3.3 Treatment of the Breast and Tumor Bed

Treating the intact breast with inverse-planned IMRT involves contouring the entire breast mound on CT images. Breast mounds vary considerably between patients depending on body habitus and positioning; common sense should be used when contouring this structure, but some general guidelines are as follows: The breast mound often extends from the region between the clavicular head and second rib insertion superiorly and extends inferiorly to a variable degree. The medial border is typically at the junction of the sternum and the ribs, and the lateral junction should be determined clinically. In patients with early-stage disease who have not received neoadjuvant therapy and who have tumors that do not approximate the pectoralis muscles, the posterior border should be at the anterior border of the pectoralis major muscle. Otherwise, the posterior border should be at the pleural surface and include the ribs and pectoralis muscles. The skin surface should constitute the anterior border. The tumor cavity should comprise the entire visible postoperative seroma seen on CT and should be included in its entirety within the breast contours. The placement of surgical clips at the time of lumpectomy can assist in delineating the tumor cavity if the seroma is not visible on CT and constitutes best practice when using IMRT to treat the breast. Use of an integrated boost [24, 26] can shorten treatment time through the administration of a higher dose per fraction to the tumor bed concurrently with treatment of the breast, with acceptable cosmetic results [18].

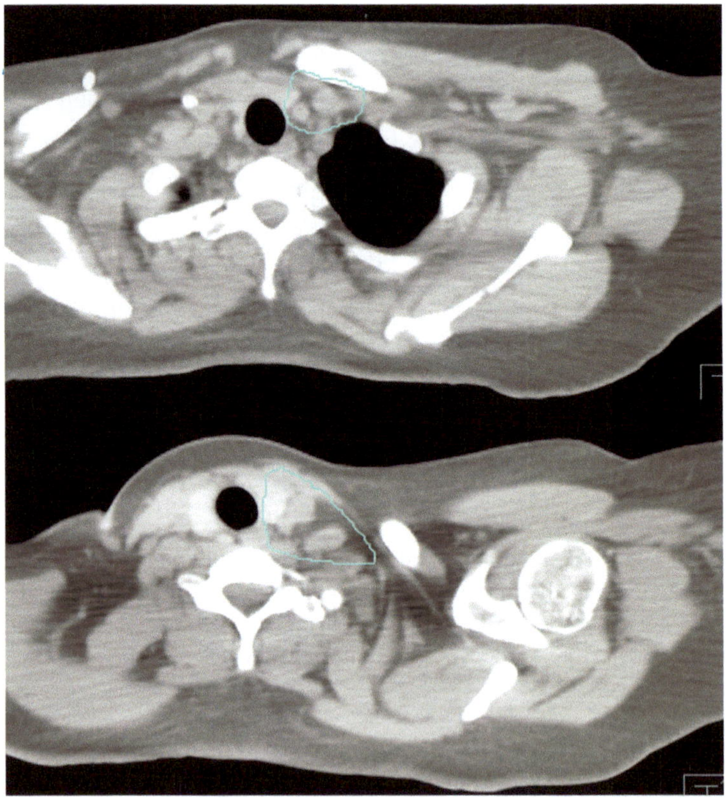

Fig. 14.12 Contours of the supraclavicular region

14.3.4 Treatment of the Chest Wall

For patients with locally advanced disease, treatment of the chest wall after mastectomy is accompanied by comprehensive irradiation of the draining lymphatics in most cases. In some patients found to be node-negative after mastectomy, irradiating the chest wall alone, without regional nodal irradiation, may be appropriate [13]. Care should be taken to include the entire surgical scar when irradiating the chest wall. The superior border is usually placed at the inferior border of the clavicular head; the contralateral breast is used as a guide in determining the inferior border. If the contralateral breast is surgically absent, preoperative imaging should be used to determine the inferior border. The posterior border should be the pleural surface and include the ribs and pectoralis muscles. The skin surface should constitute the anterior border.

14.3.5 Normal Tissue Constraints

Normal tissue constraints for a variety of organs have been published [7]. An ongoing trial by the National Surgical Adjuvant Breast and Bowel Project and Radiation Therapy Oncology Group (NSABP B-51/RTOG 1304) [16] is investigating the utility of regional nodal irradiation for patients with lymph node-positive disease that responds completely to neoadjuvant chemotherapy. This trial allows the use of inverse-planned breast IMRT. The normal tissue constraints used in this trial are listed below, with the caveat that finding from the trial has not yet been published; indeed, when this chapter was written, the trial was only in the early stages of accrual.

14.3.5.1 Normal Tissue Constraints from NSABP B-51/RTOG 1304

Contralateral Breast

- <5 % receives 3 Gy.

Ipsilateral Lung

- ≤ 30 % of the ipsilateral lung should receive ≥ 20 Gy.
- ≤ 50 % of the ipsilateral lung should receive ≥ 10 Gy.
- ≤ 65 % of the ipsilateral lung should receive ≥ 5 Gy.

Contralateral Lung

- ≤ 10 % of the contralateral lung should receive ≥ 5 Gy.

Heart

- ≤ 5 % of the whole heart should receive ≥ 25 Gy for left-sided breast cancers, and 0 % of the heart should receive ≥ 25 Gy for right-sided breast cancers.
- ≤ 30 % of the whole heart should receive ≥ 15 Gy for left-sided breast cancers, and ≤ 10 % of the heart should receive ≥ 15 Gy for right-sided breast cancers.
- The mean heart dose should be ≤ 4 Gy.

14.4 Accelerated Partial-Breast Irradiation

Accelerated partial-breast irradiation (APBI) is a relatively new approach being investigated in ongoing randomized trials [25]. APBI typically involves the irradiation of the tumor bed alone to doses of 34–38 Gy administered in 10 fractions twice daily over 5 treatment days. The ongoing NSABP B-39/RTOG 0413 trial randomly assigned patients with early-stage breast cancer to receive either whole-breast radiation therapy or APBI using either brachytherapy or 3D conformal radiation therapy. The findings from this trial continue to mature; however, early reports [12, 15] have suggested unacceptable cosmetic outcomes, particularly subcutaneous fibrosis and fat necrosis, in patients treated with APBI with 3D conformal radiation therapy. In addition, findings from single-institution studies have also suggested that cosmetic results are unacceptable when IMRT is used for APBI [14]. At this time, external-beam modalities, including IMRT, seem to result in inferior cosmetic outcomes when used for APBI and should be avoided outside the context of clinical trials.

14.5 Conclusions

When only the breast is to be treated, forward-planned IMRT results in excellent dose homogeneity compared with conventional techniques involving the use of physical wedges. With modern multileaf collimator-equipped linear accelerators, forward-planned IMRT can be administered quickly and safely, without the increased costs associated with inverse-planned IMRT. On the other hand, inverse-planned IMRT is an attractive option for patients who require extensive irradiation of the regional lymphatics. Inverse-planned IMRT typically involves a trade-off between increased dose conformality and homogeneity at the expense of higher radiation doses to normal structures such as the contralateral breast, heart, and lungs. Although experience with the technique can allow minimization of such doses to normal structures, the question remains as to whether the increased doses to normal structures from inverse-planned IMRT are clinically significant. Reports of improved long-term clinical outcomes with inverse-planned IMRT are lacking but should become available as the technique becomes more widely accepted.

References

1. Baral E, Larsson LE, Mattsson B (1977) Breast cancer following irradiation of the breast. Cancer 40(6):2905–2910
2. Bartelink H, Horiot JC, Poortmans P, Struikmans H, Van den Bogaert W, Barillot I, Fourquet A, Borger J, Jager J, Hoogenraad W, Collette L, Pierart M (2001) Recurrence rates after treatment of breast cancer with standard radiotherapy with or without additional radiation. N Engl J Med 345(19):1378–1387
3. Cho BC, Hurkmans CW, Damen EM, Zijp LJ, Mijnheer BJ (2002) Intensity modulated versus non-intensity modulated radiotherapy in the treatment of the left breast and upper internal mammary lymph node chain: a comparative planning study. Radiother Oncol 62(2):127–136
4. Clark RM, Whelan T, Levine M, Roberts R, Willan A, McCulloch P, Lipa M, Wilkinson RH, Mahoney LJ (1996) Randomized clinical trial of breast irradiation following lumpectomy and axillary dissection for node-negative breast cancer: an update. Ontario Clinical Oncology Group. J Natl Cancer Inst 88(22):1659–1664
5. Clarke M, Collins R, Darby S, Davies C, Elphinstone P, Evans E, Godwin J, Gray R, Hicks C, James S, MacKinnon E, McGale P, McHugh T, Peto R, Taylor C, Wang Y (2005) Effects of radiotherapy and of differences in the extent of surgery for early breast cancer on local recurrence and 15-year survival: an overview of the randomised trials. Lancet 366(9503):2087–2106

6. Darby SC, Ewertz M, McGale P, Bennet AM, Blom-Goldman U, Bronnum D, Correa C, Cutter D, Gagliardi G, Gigante B, Jensen MB, Nisbet A, Peto R, Rahimi K, Taylor C, Hall P (2013) Risk of ischemic heart disease in women after radiotherapy for breast cancer. N Engl J Med 368(11):987–998
7. Emami B, Lyman J, Brown A, Cola L, Goitein M, Munzenrider JE, Shank B, Solin LJ, Wesson M (1991) Tolerance of normal tissue to therapeutic irradiation. Int J Radiat Oncol Biol Phys 21(1):109–122
8. Fisher B, Anderson S, Bryant J, Margolese RG, Deutsch M, Fisher ER, Jeong JH, Wolmark N (2002) Twenty-year follow-up of a randomized trial comparing total mastectomy, lumpectomy, and lumpectomy plus irradiation for the treatment of invasive breast cancer. N Engl J Med 347(16):1233–1241
9. Giordano SH, Kuo YF, Freeman JL, Buchholz TA, Hortobagyi GN, Goodwin JS (2005) Risk of cardiac death after adjuvant radiotherapy for breast cancer. J Natl Cancer Inst 97(6):419–424
10. Hankey BF, Curtis RE, Naughton MD, Boice JD Jr, Flannery JT (1983) A retrospective cohort analysis of second breast cancer risk for primary breast cancer patients with an assessment of the effect of radiation therapy. J Natl Cancer Inst 70(5):797–804
11. Harris EE, Correa C, Hwang WT, Liao J, Litt HI, Ferrari VA, Solin LJ (2006) Late cardiac mortality and morbidity in early-stage breast cancer patients after breast-conservation treatment. J Clin Oncol 24(25):4100–4106
12. Hepel JT, Tokita M, MacAusland SG, Evans SB, Hiatt JR, Price LL, DiPetrillo T, Wazer DE (2009) Toxicity of three-dimensional conformal radiotherapy for accelerated partial breast irradiation. Int J Radiat Oncol Biol Phys 75(5):1290–1296
13. Jagsi R, Raad RA, Goldberg S, Sullivan T, Michaelson J, Powell SN, Taghian AG (2005) Locoregional recurrence rates and prognostic factors for failure in node-negative patients treated with mastectomy: Implications for postmastectomy radiation. Int J Radiat Oncol Biol Phys 62(4):1035–1039
14. Jagsi R, Ben-David MA, Moran JM, Marsh RB, Griffith KA, Hayman JA, Pierce LJ (2010) Unacceptable cosmesis in a protocol investigating intensity-modulated radiotherapy with active breathing control for accelerated partial-breast irradiation. Int J Radiat Oncol Biol Phys 76(1):71–78
15. Leonard KL, Hepel JT, Hiatt JR, Dipetrillo TA, Price LL, Wazer DE (2013) The effect of dose-volume parameters and interfraction interval on cosmetic outcome and toxicity after 3-dimensional conformal accelerated partial breast irradiation. Int J Radiat Oncol Biol Phys 85(3):623–629
16. Mamounas E, White J (2013) NSABP PROTOCOL B-51 RTOG PROTOCOL 1304. A randomized phase III clinical trial evaluating post-mastectomy chestwall and regional nodal xrt and post-lumpectomy regional nodal XRT in patients with positive axillary nodes before neoadjuvant chemotherapy who convert to pathologically negative axillary nodes after neoadjuvant chemotherapy. http://www.cancer.gov/clinicaltrials/search/view?cdrid=750327&version=HealthProfessional
17. McDonald MW, Godette KD, Butker EK, Davis LW, Johnstone PAS (2008) Long-term outcomes of IMRT for breast cancer: a single-institution cohort analysis. Int J Radiat Oncol Biol Phys 72(4):1031–1040
18. McDonald MW, Godette KD, Whitaker DJ, Davis LW, Johnstone PAS (2010) Three-year outcomes of breast intensity-modulated radiation therapy with simultaneous integrated boost. Int J Radiat Oncol Biol Phys 77(2):523–530
19. Neugut AI, Murray T, Santos J, Amols H, Hayes MK, Flannery JT, Robinson E (1994) Increased risk of lung cancer after breast cancer radiation therapy in cigarette smokers. Cancer 73(6):1615–1620
20. Nielsen MH, Berg M, Pedersen AN, Andersen K, Glavicic V, Jakobsen EH, Jensen I, Josipovic M, Lorenzen EL, Nielsen HM, Stenbygaard L, Thomsen MS, Vallentin S, Zimmermann S, Offersen BV, Danish Breast Cancer Cooperative Group Radiotherapy C (2013) Delineation of target volumes and organs at risk in adjuvant radiotherapy of early breast cancer: national guidelines and contouring atlas by the Danish Breast Cancer Cooperative Group. Acta Oncol 52(4):703–710

21. Overgaard M, Jensen MB, Overgaard J, Hansen PS, Rose C, Andersson M, Kamby C, Kjaer M, Gadeberg CC, Rasmussen BB, Blichert-Toft M, Mouridsen HT (1999) Postoperative radiotherapy in high-risk postmenopausal breast-cancer patients given adjuvant tamoxifen: Danish Breast Cancer Cooperative Group DBCG 82c randomised trial. Lancet 353(9165):1641–1648
22. Ragaz J, Jackson SM, Le N, Plenderleith IH, Spinelli JJ, Basco VE, Wilson KS, Knowling MA, Coppin CM, Paradis M, Coldman AJ, Olivotto IA (1997) Adjuvant radiotherapy and chemotherapy in node-positive premenopausal women with breast cancer. N Engl J Med 337(14):956–962
23. Remouchamps VM, Letts N, Vicini FA, Sharpe MB, Kestin LL, Chen PY, Martinez AA, Wong JW (2003) Initial clinical experience with moderate deep-inspiration breath hold using an active breathing control device in the treatment of patients with left-sided breast cancer using external beam radiation therapy. Int J Radiat Oncol Biol Phys 56(3):704–715
24. Singla R, King S, Albuquerque K, Creech S, Dogan N (2006) Simultaneous-integrated boost intensity-modulated radiation therapy (SIB-IMRT) in the treatment of early-stage left-sided breast carcinoma. Med Dosim 31(3):190–196
25. Smith BD, Arthur DW, Buchholz TA, Haffty BG, Hahn CA, Hardenbergh PH, Julian TB, Marks LB, Todor DA, Vicini FA, Whelan TJ, White J, Wo JY, Harris JR (2009) Accelerated partial breast irradiation consensus statement from the American Society for Radiation Oncology (ASTRO). Int J Radiat Oncol Biol Phys 74(4):987–1001
26. van der Laan HP, Dolsma WV, Maduro JH, Korevaar EW, Hollander M, Langendijk JA (2007) Three-dimensional conformal simultaneously integrated boost technique for breast-conserving radiotherapy. Int J Radiat Oncol Biol Phys 68(4):1018–1023
27. Whelan T, MacKenzie R, Julian J, Levine M, Shelley W, Grimard L, Lada B, Lukka H, Perera F, Fyles A, Laukkanen E, Gulavita S, Benk V, Szechtman B (2002) Randomized trial of breast irradiation schedules after lumpectomy for women with lymph node-negative breast cancer. J Natl Cancer Inst 94(15):1143–1150
28. Whelan TJ, Olivotto I, Ackerman I, Chapman JW, Chua B, Nabid A, Vallis KA, White JR, Rousseau P, Fortin A, Pierce LJ, Manchul L, Craighead P, Nolan MC, Bowen J, McCready DR, Pritchard KI, Levine MN, Parulekar W (2011) NCIC-CTG MA.20: an intergroup trial of regional nodal irradiation in early breast cancer. J Clin Oncol (Meeting Abstracts) 29 (18 Suppl):LBA 1003
29. White J, Tai A, Arthur D, Buchholz TA, MacDonald S, Marks LB, Pierce LJ, Recht A, Rabinovitch RA, Taghian AG, Vicini F, Woodward W, Li XA RTOG breast cancer contouring atlas. http://www.rtog.org/LinkClick.aspx?fileticket=vzJFhPaBipE%3d&tabid=236
30. Winzer KJ, Sauer R, Sauerbrei W, Schneller E, Jaeger W, Braun M, Dunst J, Liersch T, Zedelius M, Brunnert K, Guski H, Schmoor C, Schumacher M (2004) Radiation therapy after breast-conserving surgery; first results of a randomised clinical trial in patients with low risk of recurrence. Eur J Cancer 40(7):998–1005
31. Zablotska LB, Neugut AI (2003) Lung carcinoma after radiation therapy in women treated with lumpectomy or mastectomy for primary breast carcinoma. Cancer 97(6):1404–1411

Clinical Application of IMRT for Cervical Esophageal Cancer

15

Satoshi Itasaka

Keywords
Cervical esophageal cancer • IMRT • QOL

15.1 Anatomy

The esophagus is divided into cervical and thoracic components. The American Joint Committee on Cancer report divides the esophagus into four regions: cervical, upper thoracic, midthoracic, and lower thoracic. The cervical esophagus begins at the cricopharyngeus muscle and extends to the thoracic inlet. According to the Japanese classification of carcinoma of the esophagus, the cervical esophagus is defined as beginning at the esophageal orifice and extending to the level of the sternal notch.

The esophagus is characterized by a rich, longitudinal, lymphatic drainage network. Regarding thoracic esophageal cancer, extended lymph node metastases are often observed. As compared with thoracic esophageal cancer, a higher incidence of cervical lymph node metastasis is observed in cervical esophageal cancer, and mediastinal lymph node metastases are often limited to the upper mediastinal lymph nodes. Fujita et al. [1] reported on the incidence of lymph node metastases from cervical esophageal cancer in 36 patients who had undergone curative resection and divided them into two groups with or without tumor

S. Itasaka, M.D. (✉)
Department of Radiation Oncology and Image-Applied Therapy, Graduate School of Medicine, Kyoto University, 54 Shogoin Kawahara-cho, Sakyo-ku, Kyoto 606-8507, Japan
e-mail: sitasaka@kuhp.kyoto-u.ac.jp

Y. Nishimura, R. Komaki (eds.), *Intensity-Modulated Radiation Therapy*,
DOI 10.1007/978-4-431-55486-8_15

invasion into the pharynx. Cervical esophageal cancer with invasion into the pharynx had a high incidence of cervical paraesophageal and deep cervical node metastases, while cervical esophageal cancer without invasion into the pharynx had a higher incidence of cervical paraesophageal and recurrent nerve node metastases; resection of these nodes was related to better prognosis [1].

According to the 7th UICC staging system, supraclavicular node metastasis is classified as distant metastasis regardless of the primary tumor location. However, in the case of cervical esophageal cancer, even in the presence of supraclavicular node metastasis, surgery or radiation therapy is performed with curative intent. According to the Japanese classification of carcinoma of the esophagus, the supraclavicular nodes are defined as a regional lymph node station (group 2, N2) and definitive treatment is selected.

15.2 Risk Factors

Alcohol and smoking are well-known major risk factors for hypopharyngeal and esophageal cancer [2, 3]. This is a major problem in East Asia, because approximately 36 % of East Asians show low tolerability to alcohol because of an inherited deficiency in the enzyme aldehyde dehydrogenase 2 (ALDH2) [3]. However, cervical esophageal cancer is a rare disease, because the incidence of cervical esophageal cancer among all esophageal cancers is low. According to the Japanese *Comprehensive Registry of Esophageal Cancer* published in 2006 [4], the incidence of cervical esophageal cancer was only 4.2 % among all esophageal cancers.

15.3 Surgery

Cervical esophageal cancer is often discovered as locally advanced disease at diagnosis and is frequently associated with local invasion to other organs and lymph node metastases. However, lymph node metastases are often limited to the cervical region and are candidates for surgery. According to the Japanese guidelines for the diagnosis and therapy of esophageal cancer (2012), larynx-preserving therapy is recommended for cases in which the tumor does not invade into the larynx or trachea and does not reach the esophageal orifice. If the tumor extends to the pharynx, trachea, or hypopharynx, or there is not enough space for anastomosis of the cervical esophagus, laryngoesophagopharyngectomy is recommended and a permanent tracheostomy is performed. If the tumor directly invades the internal carotid artery, the left subclavian artery, or the cervical vertebral body, it is unresectable. The role of neoadjuvant chemotherapy or chemoradiation therapy has not yet been established.

Results of surgery for cervical esophageal cancer have often been reported together with hypopharyngeal cancer (Table 15.1). In recent reports regarding

Table 15.1 Outcomes of patients treated using surgery

Study	Year	Ce/Hp	5ySR (%)	Morbidity *n* (%)	Hospital mortality *n* (%)
Wei et al. [5]	1998	32/37	24	34(49)	6(9)
Triboulet et al. [6]	2001	78/131	24	42(33.1)	10(4.8)
Wang et al. [7]	2006	15/26	31.5	19(46.3)	4(9.8)
Daiko et al. [8]	2007	74/0	33	25(34 %)	3(4 %)
Tong et al. [9]	2011	43/25	37.6(2y)		5(7.1)

surgical treatments, 5-year overall survival rates were in the range of 24–37.6 %, and high morbidity (33.1–49 %) and mortality (4–9.8 %) have also been reported. In a Japanese study, Daiko et al. [8] reported that 74 cases of cervical esophageal cancer (UICC 1997 pStage I, 6; pStage II, 30; pStage III, 38 (T4, 19)) were treated using surgery, and the 3- and 5-year overall survival rates were 42 % and 33 %, respectively. Fujita et al. reported a 5-year overall survival rate of 31 % for 29 cases of cervical esophageal cancer (UICC 1997 pStage I, 1; pStage II, 8; pStage III, 13; pStage IV, 7).

Daiko et al. [8] reported that the first recurrence after surgery was mostly locoregional (82 %) and that the rate of distant metastasis was 14 %. They reported that pathological staging, namely, pT, pN, pM1 distant lymph node metastases, lymphatic invasion, and extracapsular invasion, was prognostic factors identified using univariate analysis. Triboulet et al. [6] reported that the pT3 and pT4 histopathological stages were poor prognostic factors in surgical cases of cervical esophageal cancer. Thus, surgery can be applied for locally advanced disease, but outcome is not satisfactory and a worse quality of life (QOL) is a problem.

15.4 Radiation Therapy

15.4.1 Results of Radiation Therapy

Radiation therapy for cervical esophageal cancer has a major positive impact on the QOL of patients as a result of the preservation of the larynx and esophagus [10]. Therefore, radiation therapy is often preferred over surgery in the clinic. The optimal radiation dose and fractionation have not yet been established; a wide range (50–70 Gy) of prescribed doses regarding the doses, which are used for either thoracic esophageal cancer or head and neck cancer, have been reported [9, 11–19]. Although no large prospective trials have been carried out because of the small number of cases involved, the outcome of radiation therapy has been reported in several studies (Table 15.2). Burmeister et al. [13] reported on 34 patients (Stage I, 4; Stage II, 27; Stage III, 3) treated using chemoradiation therapy at a total

Table 15.2 Outcomes of patients treated using radiation therapy

Study	Year	*n*	Dose (Gy)	CCRT (%)	IMRT (%)	LRC (%)	2y-OS (%)	5y-OS (%)
Stuschke et al. [12]	1999	17	60–66	100	0	33(2y)	24	NA
Burmeister et al. [13]	2000	34	50.4–65 (mean 61.2)	100	0	NA	NA	55
Yamada et al. [14]	2005	27	44–73.7 (mean 66)	85.2	0	13(5y)[a]	38	38
Wang et al. [15]	2006	35[b]	24.5–64.8 (median 50.4)	100	0	47.7(5y)	NA	18.6
Wang (MDA) [20]	2006	7	59.4–66 (median 64.8)	100	100	NA	NA	NA
Uno et al. [18]	2007	21	60–74 (median 64)	90	0	NA	41	27
Huang et al. [16]	2008	71		52		37(2y)[c]	35	18.6
		(29)	54Gy/20fr.	45	0	NA	41	NA
		(42)	70Gy/35fr.	57	Some	NA	32	NA
Tong et al. [9]	2011	21	60–68	100	0	NA	46.9	NA
Cao et al. [17]	2013	115	59.4–80	30.4	73.6	68.3(2y)[d]	47.6	NA

[a]Disease-free survival
[b]Including 13 upper thoracic esophageal cases
[c]Locoregional relapse-free survival
[d]Local failure-free survival

radiation dose of 61.2 Gy (mean dose); 3- and 5-year overall survival rates were 60 % and 55 %, respectively. Yamada et al. [14] reported on 27 patients (UICC 1997 Stage I, 5; Stage II, 6; Stage III, 12; Stage IV, 4) treated using definitive radiation therapy at a mean total dose of 66 Gy; both 3- and 5-year overall survival rates were 37.9 %. Uno et al. [18] reported on 21 patients (UICC 1997 Stage I, 1; Stage II, 5; Stage III, 8; Stage IV, 7) treated using definitive radiation therapy (median dose 64 Gy), and the 2- and 5-year overall survival rates were 41 % and 27 %, respectively. For 11 patients who did not have grade T4 tumors, the 5-year overall survival rate was 41 %. Gkika et al. [21] reported on 55 patients (UICC 1992 Stage II, 20; Stage III, 35) treated by means of definitive chemoradiation therapy (median dose 60 [range, 50–70] Gy). Overall survival rates at 2, 3, 5, and 10 years were 35 %, 29 %, 25 %, and 10 %, respectively. Although the above reports represent only a small number of retrospective studies involving differences in doses and chemotherapy regimens, the results of radiation therapy have been promising. A few studies regarding treatment results using intensity-modulated radiation therapy (IMRT) have been used as a basis for the treatment of cervical esophageal cancer [22]. Huang et al. [16] reported that patients treated from

2001 at a total radiation dose of 70 Gy delivered in 35 fractions, using either three-dimensional conformal radiation therapy (3D CRT) or IMRT together with concurrent cisplatin, achieved a 2-year overall survival rate of 32 %. Wang et al. [20] reported on the initial experience of seven patients treated with IMRT at the MD Anderson Cancer Center, at total doses of 59.4–66 Gy (median 64.8 Gy) given in combination with concurrent chemotherapy. Several preliminary studies regarding IMRT for both cervical and upper thoracic esophageal cancer have also been published [22, 23]. Overall, whether or not better coverage of the planning target volume (PTV) by IMRT leads to an improvement in local tumor control and patient survival is unknown and a prospective study is needed.

15.4.2 IMRT

15.4.2.1 Merits of IMRT

Classically, the AP-PA field, the so-called short-T field, is the initial field; cord cut opposing oblique fields has been used for a radiation boost. The distance between the esophagus and spinal cord is short at the level of the neck, so it is often difficult to deliver an adequate dose to the tumor and reduce the dose to the spinal cord with 3D CRT. In addition, differences in the thickness of the body at the neck and the chest make it difficult to achieve an even dose distribution; this is especially true in patients with bulky tumors, a bilateral neck, or supraclavicular nodes. Several treatment planning studies have clearly demonstrated the usefulness of IMRT for cervical esophageal cancer, with superior target volume coverage and conformity, and a reduced dose to the spinal cord and lungs [24–26]. Comparisons between fixed-beam IMRT and volumetric modulated arc therapy (VMAT) [27] have also been reported. Both methods basically achieve abundant dose coverage of the PTV; VMAT has exhibited slight improvements in the radiation dose distribution and a large reduction in the number of monitor units (MU) required for treatment, but a slight increase in low-dose lung radiation exposures.

15.4.2.2 Target Delineation

Usually, the type of primary tumor is determined by means of multiple studies including barium studies, endoscopy, computed tomography, magnetic resonance imaging, and FDG-PET. Because Lugol chromoendoscopy cannot be used in the head and neck region because of the risk of aspiration, narrow-band imaging esophagogastroduodenoscopy is useful for determining superficial squamous cell carcinoma extension in cervical esophageal cancer [28]. The radiation field for cervical esophageal cancer (Fig. 15.1a–d) usually includes prophylactic lymph node stations including the bilateral supraclavicular, mid-deep cervical, paraesophageal, and recurrent nerve extending as far as the subcarina. When tumors have invaded the pharynx, the upper deep cervical station is also incorporated into the radiation fields.

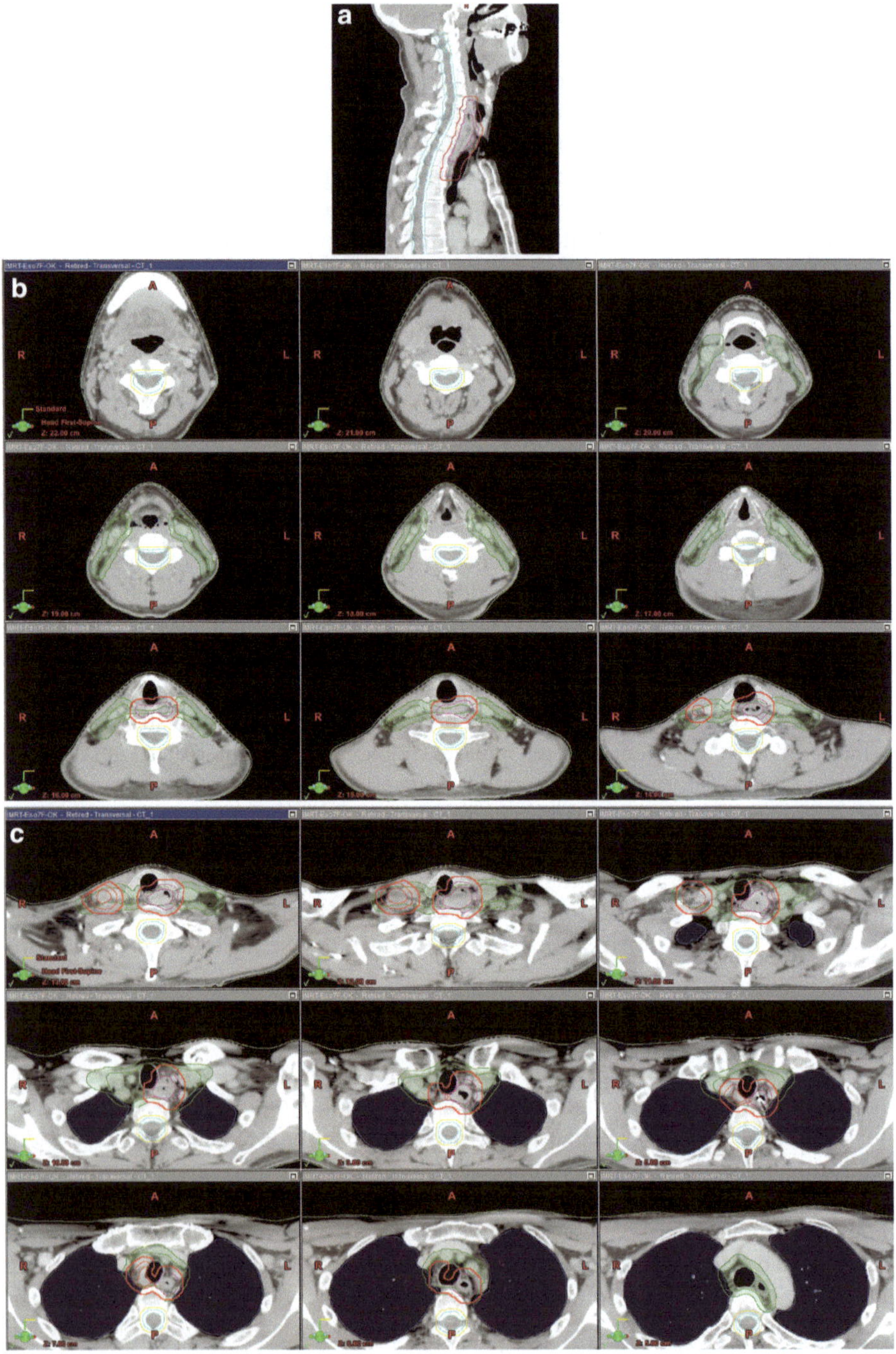

Fig. 15.1 (**a**–**d**): An example of contouring for a cervical esophageal tumor. (**a**) The clinical target volume (CTV) for the primary tumor is defined as the gross tumor volume plus a 2-cm craniocaudal margin with a 0.5-cm radial margin. The CTV over the trachea and bone is cut unless there is direct invasion of the tumor. (**b**–**d**) Axial images of contouring. Planning target volume (PTV) for prophylactic lymph node stations (*green*); PTV for primary tumor and metastatic lymph nodes (*red*)

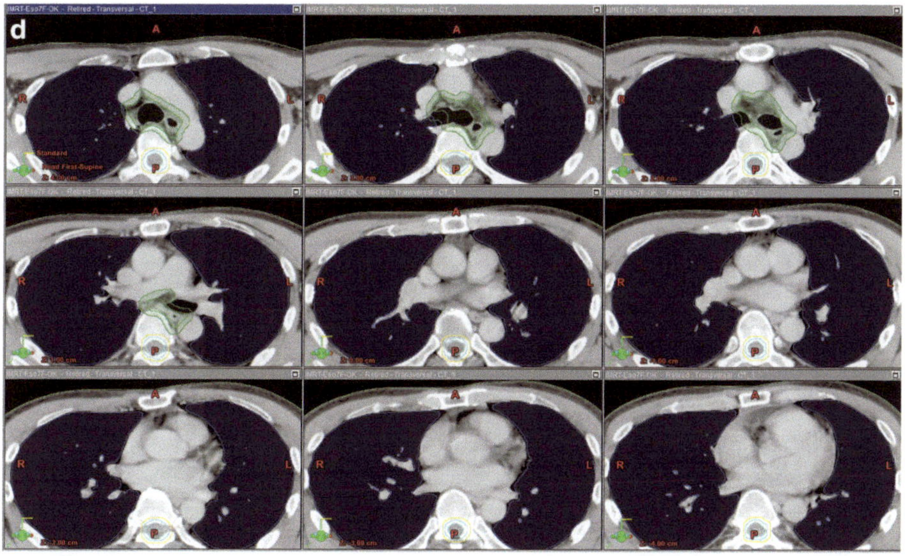

Fig. 15.1 (continued)

15.4.2.3 Case Reports

Examination of a 66-year-old male patient who suffered from dysphagia using laryngoscopy revealed no abnormalities. Two months later, the patient received upper gastrointestinal endoscopy, and a 5-cm, localized, ulcerative tumor that started at a distance of 17 cm from the incisor teeth was found. Squamous cell carcinoma was determined by means of biopsy, and CT and FDG-PET revealed extensive lymph node swellings at the bilateral supraclavicular and mediastinal nodes. Gastrostomy was performed prior to chemoradiation therapy. The total dose delivered was 60 Gy in 30 fractions for the PTV and 51 Gy for the prophylactic lymph node region delivered by means of seven fields involving a simultaneous integrated boost using IMRT (Fig. 15.2a–d); cisplatin and 5FU were administered concurrently in two courses. After chemoradiation, metastatic lymph nodes showed a complete response (CR), but ulceration remained at the primary tumor site. Adjuvant chemotherapy with cisplatin and 5FU was repeated in four courses. Eight months after treatment, the CR of the primary tumor was confirmed using endoscopy. At the same time as adjuvant chemotherapy, balloon dilation of the esophagus was repeated for stenosis of the esophagus, and gastrostomy was terminated at 8 months after the completion of radiation therapy. The patient has continued to exhibit a CR over a 4-year period.

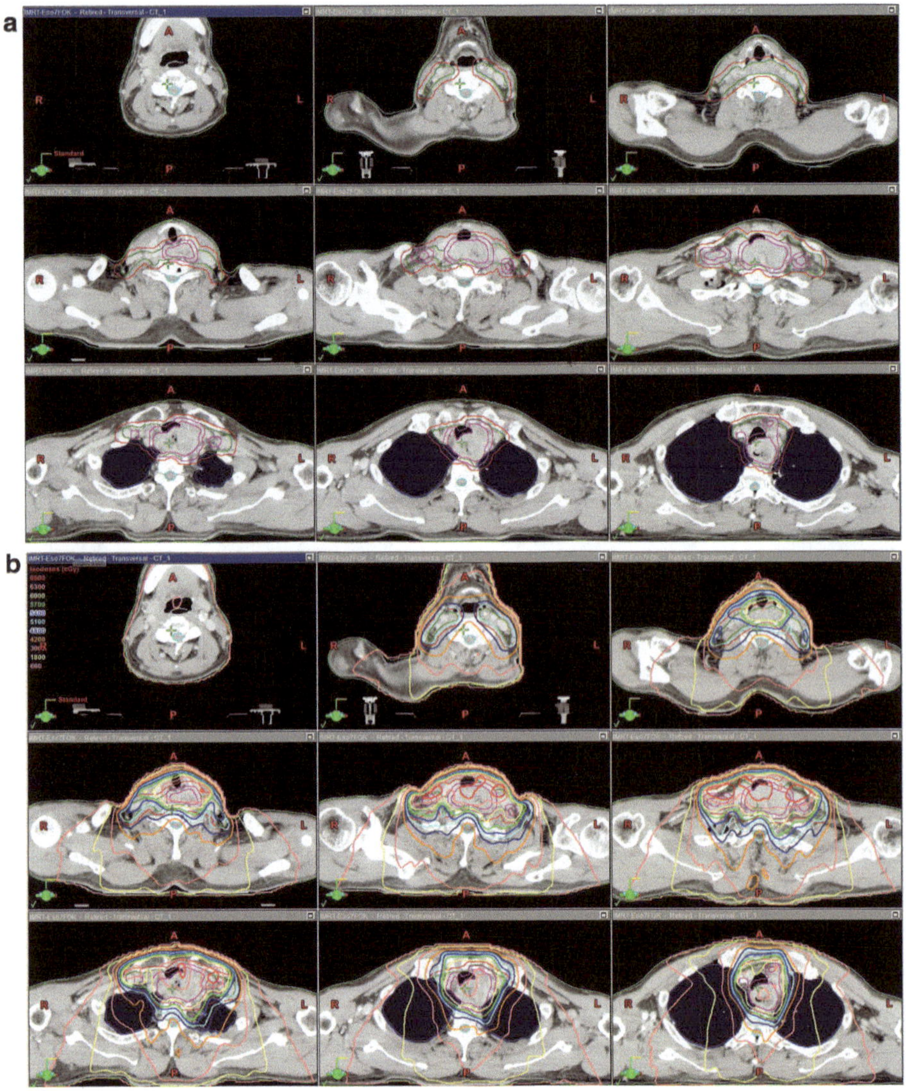

Fig. 15.2 (**a–d**): Case report. A 66-year-old male patient with cT4b (trachea and thyroid) N2M1(LYM) cervical esophageal cancer. A total dose of 60 Gy was delivered in 30 fractions to the planning target volume and 51 Gy to the prophylactic lymph node region using seven fields of simultaneous integrated boost intensity-modulated radiation therapy (IMRT); cisplatin and 5FU were concurrently administered in two courses. Four courses of adjuvant chemotherapy with cisplatin and 5FU were added. (**a**) Contouring. Note the huge tumor located in both the cervical and upper thoracic esophagus. Bilateral supraclavicular and mediastinal lymph node swellings can also be observed. (**b**) Dose distribution of IMRT in the axial plane. (**c**) Dose distribution of IMRT in three-dimensional view. (**d**) Computed tomography scans showing dramatic improvement in esophageal wall thickness and the disappearance of bilateral supraclavicular lymph node metastases at 10 months after chemoradiation therapy

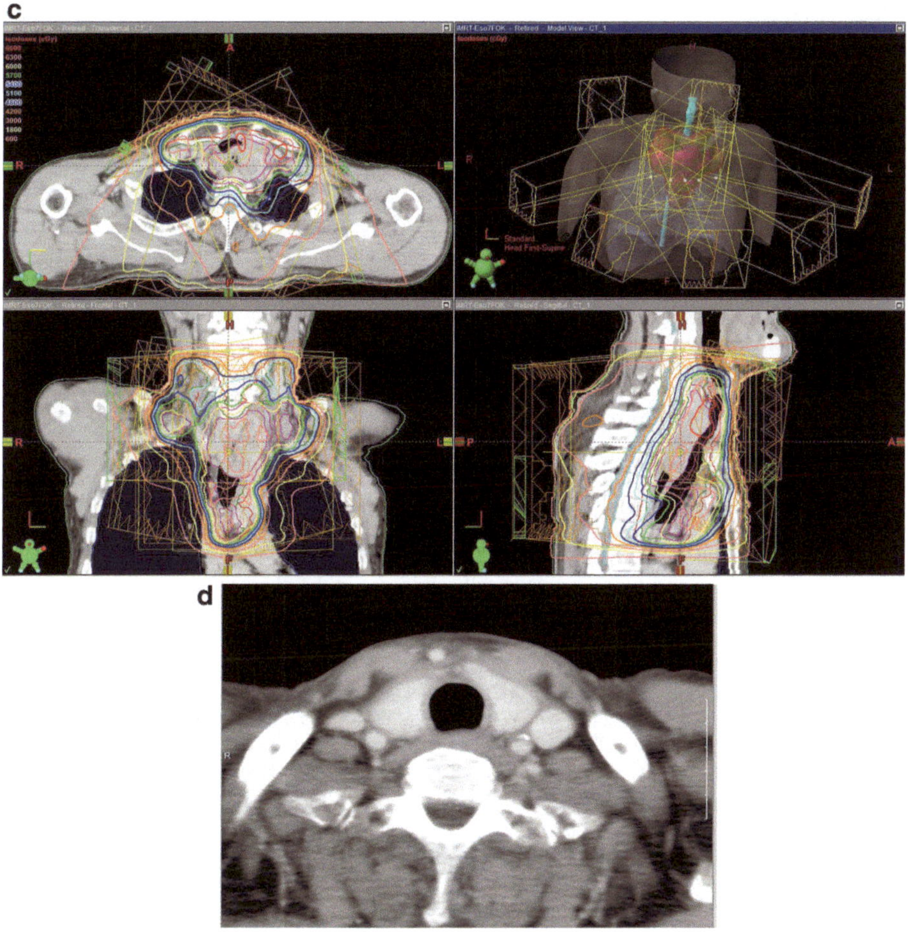

Fig. 15.2 (continued)

15.5 Future Considerations

Although cervical esophageal cancer is a rare disease, radiation therapy plays an important role in its treatment, and IMRT has clear advantages over 3D CRT. To establish the role of IMRT in the treatment of cervical esophageal cancer, the Japanese Radiation Oncology Study Group is now running a multi-institutional phase II study of chemoradiation therapy using IMRT for cervical esophageal cancer to confirm the efficacy and safety of IMRT.

Acknowledgments This study was supported in part by a Grant-in-Aid for Cancer Research (H23-009) from the Ministry of Health, Labour and Welfare of Japan.

References

1. Fujita H, Sueyoshi S, Tanaka T, Tanaka Y, Matono S, Mori N, Tsubuku T, Nishimura K, Shirouzu K (2008) A new N category for cancer of the cervical esophagus based on lymph node compartments. Esophagus 5:19–26. doi:10.1007/s10388-007-0134-8
2. Lee CH, Wu DC, Lee JM, Wu IC, Goan YG, Kao EL, Huang HL, Chan TF, Chou SH, Chou YP, Lee CY, Chen PS, Ho CK, He J, Wu MT (2007) Carcinogenetic impact of alcohol intake on squamous cell carcinoma risk of the oesophagus in relation to tobacco smoking. Eur J Cancer 43(7):1188–1199. doi:10.1016/j.ejca.2007.01.039
3. Brooks PJ, Enoch MA, Goldman D, Li TK, Yokoyama A (2009) The alcohol flushing response: an unrecognized risk factor for esophageal cancer from alcohol consumption. PLoS Med 6(3):e50. doi:10.1371/journal.pmed.1000050
4. Tachimori Y, Ozawa S, Fujishiro M, Matsubara H, Numasaki H, Oyama T, Shinoda M, Toh Y, Udagawa H, Uno T (2014) Comprehensive registry of esophageal cancer in Japan, 2006. Esophagus 11(1):21–47. doi:10.1007/s10388-013-0393-5
5. Wei WI, Lam LK, Yuen PW, Wong J (1998) Current status of pharyngolaryngo-esophagectomy and pharyngogastric anastomosis. Head Neck 20(3):240–244
6. Triboulet JP, Mariette C, Chevalier D, Amrouni H (2001) Surgical management of carcinoma of the hypopharynx and cervical esophagus: analysis of 209 cases. Arch Surg 136(10):1164–1170
7. Wang HW, Chu PY, Kuo KT, Yang CH, Chang SY, Hsu WH, Wang LS (2006) A reappraisal of surgical management for squamous cell carcinoma in the pharyngoesophageal junction. J Surg Oncol 93(6):468–476. doi:10.1002/jso.20472
8. Daiko H, Hayashi R, Saikawa M, Sakuraba M, Yamazaki M, Miyazaki M, Ugumori T, Asai M, Oyama W, Ebihara S (2007) Surgical management of carcinoma of the cervical esophagus. J Surg Oncol 96(2):166–172. doi:10.1002/jso.20795
9. Tong DK, Law S, Kwong DL, Wei WI, Ng RW, Wong KH (2011) Current management of cervical esophageal cancer. World J Surg 35(3):600–607. doi:10.1007/s00268-010-0876-7
10. Chou SH, Li HP, Lee JY, Huang MF, Lee CH, Lee KW (2010) Radical resection or chemoradiotherapy for cervical esophageal cancer? World J Surg 34(8):1832–1839. doi:10.1007/s00268-010-0595-0
11. Ma JB, Song YP, Yu JM, Zhou W, Cheng EC, Zhang XQ, Kong L (2011) Feasibility of involved-field conformal radiotherapy for cervical and upper-thoracic esophageal cancer. Onkologie 34(11):599–604. doi:10.1159/000334194
12. Stuschke M, Stahl M, Wilke H, Walz MK, Oldenburg AR, Stuben G, Jahnke K, Seeber S, Sack H (1999) Induction chemotherapy followed by concurrent chemotherapy and high-dose radiotherapy for locally advanced squamous cell carcinoma of the cervical oesophagus. Oncology 57(2):99–105. doi:10.1159/000012015
13. Burmeister BH, Dickie G, Smithers BM, Hodge R, Morton K (2000) Thirty-four patients with carcinoma of the cervical esophagus treated with chemoradiation therapy. Arch Otolaryngol Head Neck Surg 126(2):205–208
14. Yamada K, Murakami M, Okamoto Y, Okuno Y, Nakajima T, Kusumi F, Takakuwa H, Matsusue S (2006) Treatment results of radiotherapy for carcinoma of the cervical esophagus. Acta Oncol 45(8):1120–1125. doi:10.1080/02841860600609768
15. Wang S, Liao Z, Chen Y, Chang JY, Jeter M, Guerrero T, Ajani J, Phan A, Swisher S, Allen P, Cox JD, Komaki R (2006) Esophageal cancer located at the neck and upper thorax treated with concurrent chemoradiation: a single-institution experience. J Thorac Oncol 1(3):252–259
16. Huang SH, Lockwood G, Brierley J, Cummings B, Kim J, Wong R, Bayley A, Ringash J (2008) Effect of concurrent high-dose cisplatin chemotherapy and conformal radiotherapy on cervical esophageal cancer survival. Int J Radiat Oncol Biol Phys 71(3):735–740. doi:10.1016/j.ijrobp.2007.10.022. S0360–3016(07)04421-5 [pii]
17. Cao C, Luo J, Gao L, Xu G, Yi J, Huang X, Wang K, Zhang S, Qu Y, Li S, Xiao J, Zhang Z (2013) Definitive radiotherapy for cervical esophageal cancer. Head Neck. doi:10.1002/hed.23572

18. Uno T, Isobe K, Kawakami H, Ueno N, Shimada H, Matsubara H, Okazumi S, Nabeya Y, Shiratori T, Kawata T, Ochiai T, Ito H (2007) Concurrent chemoradiation for patients with squamous cell carcinoma of the cervical esophagus. Dis Esophagus 20(1):12–18. doi:10.1111/j.1442-2050.2007.00632.x. DES632 [pii]
19. Mendenhall WM, Parsons JT, Vogel SB, Cassisi NJ, Million RR (1988) Carcinoma of the cervical esophagus treated with radiation therapy. Laryngoscope 98(7):769–771. doi:10.1288/00005537-198807000-00017
20. Wang SL, Liao Z, Liu H, Ajani J, Swisher S, Cox JD, Komaki R (2006) Intensity-modulated radiation therapy with concurrent chemotherapy for locally advanced cervical and upper thoracic esophageal cancer. World J Gastroenterol 12(34):5501–5508
21. Gkika E, Gauler T, Eberhardt W, Stahl M, Stuschke M, Pottgen C (2013) Long-term results of definitive radiochemotherapy in locally advanced cancers of the cervical esophagus. Dis Esophagus. doi:10.1111/dote.12146
22. Tu L, Sun L, Xu Y, Wang Y, Zhou L, Liu Y, Zhu J, Peng F, Wei Y, Gong Y (2013) Paclitaxel and cisplatin combined with intensity-modulated radiotherapy for upper esophageal carcinoma. Radiat Oncol 8:75. doi:10.1186/1748-717X-8-75
23. Zhu WG, Zhou K, Yu CH, Han JH, Li T, Chen XF (2012) Efficacy analysis of simplified intensity-modulated radiotherapy with high or conventional dose and concurrent chemotherapy for patients with neck and upper thoracic esophageal carcinoma. Asian Pac J Cancer Prevent 13(3):803–807
24. Fu WH, Wang LH, Zhou ZM, Dai JR, Hu YM, Zhao LJ (2004) Comparison of conformal and intensity-modulated techniques for simultaneous integrated boost radiotherapy of upper esophageal carcinoma. World J Gastroenterol 10(8):1098–1102
25. Chen J, Shrieve DC, Hitchcock YJ (2008) Comparison of cervical esophagus dose-volumes for three radiotherapy techniques for head and neck cancer. Radiother Oncol 87(2):274–280. doi:10.1016/j.radonc.2007.12.014
26. Fenkell L, Kaminsky I, Breen S, Huang S, Van Prooijen M, Ringash J (2008) Dosimetric comparison of IMRT vs. 3D conformal radiotherapy in the treatment of cancer of the cervical esophagus. Radiother Oncol 89(3):287–291. doi:10.1016/j.radonc.2008.08.008. S0167-8140(08)00432-5 [pii]
27. Yin L, Wu H, Gong J, Geng JH, Jiang F, Shi AH, Yu R, Li YH, Han SK, Xu B, Zhu GY (2012) Volumetric-modulated arc therapy vs. c-IMRT in esophageal cancer: a treatment planning comparison. World J Gastroenterol 18(37):5266–5275. doi:10.3748/wjg.v18.i37.5266
28. Muto M, Minashi K, Yano T, Saito Y, Oda I, Nonaka S, Omori T, Sugiura H, Goda K, Kaise M, Inoue H, Ishikawa H, Ochiai A, Shimoda T, Watanabe H, Tajiri H, Saito D (2010) Early detection of superficial squamous cell carcinoma in the head and neck region and esophagus by narrow band imaging: a multicenter randomized controlled trial. J Clin Oncol 28(9):1566–1572. doi:10.1200/JCO.2009.25.4680

Thoracic Esophageal Cancer 16

Steven H. Lin

Keywords
Esophageal cancer • IMRT • 3D conformal radiation therapy • Chemoradiation • Trimodality therapy

16.1 Introduction

Esophageal cancer is the eighth most common cancer worldwide, with an estimated 456,000 cases in 2012, and is the sixth most common cause of cancer death, accounting for over 400,000 deaths [15]. The prognosis for patients with esophageal cancer is poor, with mortality nearly matching incidence in most countries and an overall 5-year survival rate of less than 20 %. Esophageal cancer is most prevalent in Eastern Asia and Southern and Eastern Africa. The two histologic types of esophageal cancer, esophageal squamous cell carcinoma (CC) and esophageal adenocarcinoma, occur in distinct geographic areas, with esophageal SCC being most common in the Eastern Hemisphere and esophageal adenocarcinoma predominating in the Western Hemisphere. This difference in distribution reflects the difference in etiology of the two subtypes, with esophageal SCC associated with chronic damage from smoking and alcohol consumption and esophageal adenocarcinoma linked with squamous metaplasia from chronic reflux disease and obesity [13].

Historically, thoracic esophageal cancer has been managed with surgical resection or, for patients who are not candidates for surgery, definitive radiation therapy. Two reviews published in 1980 of more than 80,000 patients concluded that either approach produced poor treatment outcomes, with 5-year survival rates ranging

S.H. Lin, M.D., Ph.D. (✉)
Department of Radiation Oncology, The University of Texas MD Anderson Cancer Center, Unit 97, 1515 Holcombe Blvd, Houston, TX 77030, USA
e-mail: shlin@mdanderson.org

Y. Nishimura, R. Komaki (eds.), *Intensity-Modulated Radiation Therapy*,
DOI 10.1007/978-4-431-55486-8_16

from 0 to 10 % [11, 12]. Improvements in surgical techniques and postoperative care have improved the morbidity and mortality associated with surgery, and the 5-year survival rates after surgery alone are generally about 20 % regardless of whether the surgical technique is transhiatal or open thoracotomy [22, 23]. Radiation therapy used as monotherapy produces limited cure rates; however, the addition of chemotherapy (given alone or concurrently with radiation before surgery) has substantially improved cure rates. In one randomized trial comparing radiation alone with chemoradiation, the corresponding 5-year survival rates were 0 and 25 % [20]. Chemoradiation also produced outcomes comparable to those of surgical resection in another randomized trial [7]. The benefit of adding surgery to chemoradiation is still somewhat controversial because the two randomized trials conducted to date showed a benefit in disease-free survival but not in overall survival, and the high perioperative mortality rate (8–12 %) may have offset any survival benefit from preoperative therapy [3, 42]. Most retrospective studies have shown a benefit from adding surgery to chemoradiation, but selection bias precludes the findings from those studies from being considered definitive.

The belief that outcomes could be improved by adding neoadjuvant therapy (either chemotherapy or chemoradiation) to surgery has led to several randomized trials, all of which have shown survival benefits from neoadjuvant therapy relative to surgery alone, a finding confirmed by a recent meta-analysis involving 4,188 patients [41]. Specifically, preoperative chemotherapy was associated with a hazard ratio (HR) of 0.87 ($p=0.005$), but the benefit was confined to adenocarcinomas. The HR for preoperative chemoradiation, by contrast, was 0.78 ($p<0.0001$), and benefits were similar for both SCC and adenocarcinoma. An indirect comparison of neoadjuvant chemotherapy versus neoadjuvant chemoradiation had an HR of 0.88, but this was not statistically significant ($p=0.07$).

That meta-analysis did not include a recently published phase III randomized trial from the Dutch CROSS group [47]. This trial, the largest conducted to date, involved 366 patients randomized to receive preoperative chemoradiation or surgery alone. Unlike previous neoadjuvant chemoradiation trials, the chemotherapy was carboplatin and paclitaxel (unlike the more common regimen of cisplatin with fluorouracil), and the radiation dose was lower than in other trials at 41.4 Gy in 23 fractions. Chemoradiation was found to substantially enhance median survival time, being 49 months versus 24 months in the surgery-only group, and improve 3-year overall survival rates (59 % vs. 48 %). The benefit seems to have been greater for the 25 % of patients who had SCC in that study. Moreover, chemoradiation did not increase perioperative mortality, and the pathologic complete response rate was 29 %, which was no different than findings from previous studies involving higher radiation doses (45–50.4 Gy). This high-quality trial, conducted using modern-day treatment approaches, has shifted the paradigm for managing esophageal cancer worldwide. Even countries in which preoperative chemotherapy has been the standard of care are now comparing preoperative chemotherapy versus preoperative chemoradiation. The Japan Clinical Oncology Group is conducting a phase III clinical trial, JCOG 1109, comparing three treatment approaches—preoperative cisplatin with fluorouracil versus docetaxel with cisplatin and fluorouracil versus cisplatin, fluorouracil, and radiation therapy—for locally advanced esophageal cancer. Future meta-analyses that include the CROSS trial will likely demonstrate a benefit from preoperative chemoradiation over preoperative chemotherapy.

16.2 Adverse Events After Radiation Therapy

Although radiation therapy is a key component of the current management of esophageal cancer, the safe delivery of radiation is challenging owing to the proximity of the heart and lung, particularly for adenocarcinomas that develop in the mid- to distal regions of the esophagus. Inadvertent exposure of the heart and lung can lead to significant morbidity, and several studies have linked dosimetric variables with pleural and pericardial effusion. In one such study, Wei et al. evaluated data from 101 patients with inoperable disease who received chemoradiation using 3-dimensional conformal radiation therapy (3D CRT) in 2000–2003 [54]. The crude rate of pericardial effusion in that study was 27.7 %, with median time to onset of 5.3 months. On multivariable analysis, a pericardial V_{30} of>46 % was the most significant predictor of pericardial effusion. In a separate study of 167 patients, symptomatic pericardial effusion (grade ≥3 on the Common Terminology Criteria for Adverse Events scale) was observed in 14 patients (8.4 %), and the authors found that the cutoff values for predicting symptomatic pericardial effusion were a mean pericardial dose of 36.5 Gy and a pericardial V_{45} of 58 % [16]. In another study evaluating dose-volume variables associated with pleural effusions, the authors found 35 % of the 43 patients in that study developed pleural effusions. Multivariable analysis revealed the heart V_{50} to be the strongest predictor of pleural effusions, with the risk being <6 % if the heart V_{50} was <20 % but 64 % if the heart V_{50} was >40 % [40]. In a study evaluating long-term toxicity after definitive chemoradiation for thoracic esophageal SCC, more than 25 % of patients experienced late grade ≥3 cardiopulmonary toxicity [24]. This late toxicity was considerably more common among patients older than 75 years (29 %) than among younger patients (3 %) ($p=0.005$) [34].

Although trimodality therapy is currently the standard approach for managing nonmetastatic esophageal cancer for patients who can tolerate surgery [10, 44, 47, 50], esophagectomy carries a high risk of complications [8–10, 48] that can lead to poor prognosis and decrements in the quality of life [9, 21, 39]. Bosset and colleagues [4] reported higher rates of postoperative death among patients who received multimodality therapy than among those given surgery alone (17 % vs. 5 %), although the risk of death was related to both the surgical volume of the institution and the skill of the surgeons [38, 43]. Adelstein and colleagues [2] reported a postoperative mortality rate of 18 % in their phase II study of preoperative concurrent chemotherapy and accelerated fractionated radiation therapy for esophageal carcinoma. The morbidity associated with esophagectomy procedures is even more alarming. In several large prospective trials examining neoadjuvant chemoradiation followed by surgical resection [44, 46, 47, 50], the rate of postoperative pulmonary complications ranged from 33 to 46 %, with anastomotic leaks occurring in 4 to 22 % of cases. Large retrospective analyses [8, 9] also showed high rates of postoperative pulmonary complications (up to 45 %) and gastrointestinal complications (up to 24 %). The lung dose-volume relationship has also been linked with postoperative pulmonary complications [28, 51]. In one series of patients who had preoperative chemoradiation in 1998–2002, 18 % had pulmonary complications, most of which were postoperative pneumonia and some of which were fatal. The incidence of pulmonary complications correlated significantly with a total lung V_{10} of ≥40 % (35 %) and a lung V_{15} of ≥ 30 % (33 %) and may have been associated with a V_{20} of ≥20 % as well (≥20 % [32 %] vs. <20 % [10 %]) ($p = 0.079$) [28]. Subsequent analyses seem to indicate that other dosimetric variables such as mean lung dose (MLD) can also predict postoperative pulmonary complications [45].

16.3 Dosimetric Benefits from IMRT Versus Conformal Techniques

Even though the complex beam arrangements used for IMRT can produce highly conformal treatment plans [55], particularly for tumors in the lung and esophagus, any dosimetric benefit from IMRT relative to other conformal techniques depends on the way the beams are arranged. Simply modulating the beam arrangements used for conformal techniques or creating a 9-field plan that spreads out the dose will not substantially improve dosimetric sparing of nearby normal tissues [37] and could even enhance pulmonary toxicity [26]. However, more often than not, IMRT improves dosimetric sparing of adjacent tissues, particularly the heart, compared with 3D CRT (Fig. 16.1). In an early treatment planning comparison involving 10 patients with distal esophageal tumors that had been treated with 3D CRT, IMRT plans created with 4, 7, or 9 beams could reduce the total lung volume exposed to >10 Gy and >20 Gy as well as the MLD, but there was no benefit in terms of doses to the spinal cord, heart, or liver or integral doses to the total body [5]. This improvement in lung dose was noted not only for mid- to distal esophageal tumors but also for proximally located tumors in the cervical esophagus, with IMRT providing better target volume coverage and conformality and lower doses to the spinal cord and parotids [14]. This improved coverage and conformality should allow further dose escalation and, presumably, improved tumor control [56]. For distal esophageal tumors, the heart is immediately anterior to the tumor. Given traditional beam arrangements, the anterior-posterior entering beam is often complemented by 2 or 3 other beams, either as two posterior obliques or as a cross-like distribution. Although this beam arrangement improves lung dosimetry, the heart dose often is not substantially reduced with IMRT. Grosshans et al. [18] recently analyzed IMRT beam arrangements that could optimize heart sparing without drastically increasing lung dose. Removing the anterior entrance beams and placing all 5 beams in a posterior-left lateral beam arrangement significantly reduced mean cardiac doses (from 33 Gy to 23 Gy, $p < 0.05$) as well as the heart V_{20} and V_{30}. This arrangement did increase the MLD and lung V_5 and V_{20} slightly, but those variables remained within acceptable

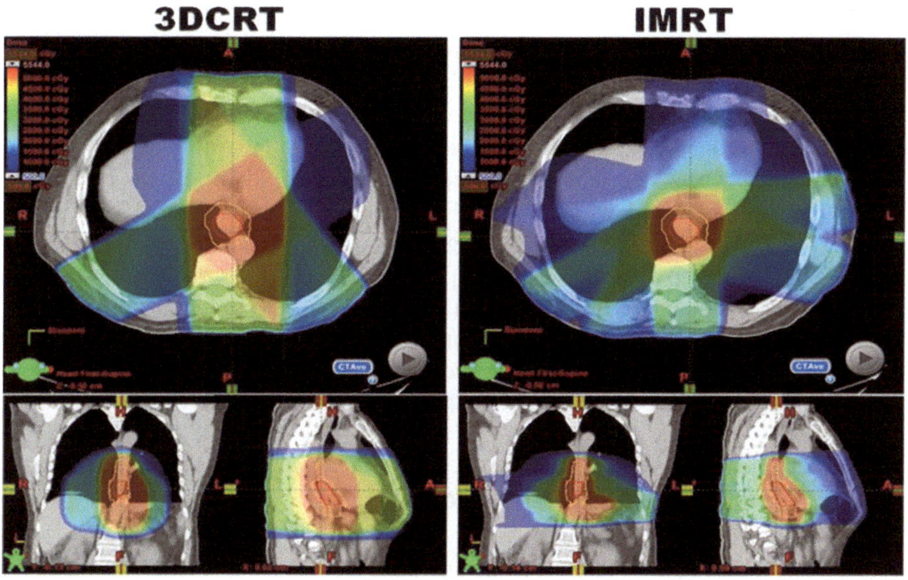

3DCRT: 4-field static photons; **IMRT**: 5-field modulated photons

Fig. 16.1 The relative distribution of radiation dose in the treatment of esophageal cancer with 3-dimensional conformal radiation therapy (3D CRT) versus intensity-modulated radiation therapy (IMRT)

limits. Target coverage, homogeneity, and conformality were similar to or improved with the alternative beam arrangements. Another dosimetric planning study comparing heart and coronary vessel exposure between IMRT and 4-field 3D CRT showed better conformality (conformality index 1.3 for IMRT vs. 1.56 for 3D CRT) and a significant reduction in the mean heart dose, heart V_{30}, and sparing of the right coronary artery with the use of IMRT [25].

Thus, when optimal beam arrangements are used, IMRT can substantially improve sparing of organs at risk (OARs) relative to conformal techniques. Moreover, the technology used to deliver IMRT can produce further improvements in conformality and dose delivery. In one treatment planning comparison study of 10 patients with upper thoracic esophageal tumors, the use of "micro" 4-mm-wide multileaf collimators (MLCs) was compared to the use of standard 10-mm-wide MLCs. The 4-mm MLCs were more efficient, using fewer monitor units (MUs) than did the standard 10-mm MLCs (mean 703 vs. 833 MUs). Dose coverage was also improved, as was sparing of the spinal cord and lung [17]. Other studies have explored the use of volumetric modulated arc therapy (VMAT, also known as RapidArc), a specialized IMRT delivery system in which the radiation is delivered while the gantry of the linear accelerator is being rotated through one or more arcs. This application can be used to further improve OAR sparing over that which is possible with 3D CRT or fixed-field IMRT and significantly reduces the "beam-on" time [1, 19, 31, 36, 49, 57]. One treatment planning study that compared single-arc and double-arc VMAT plans to those for 4-field IMRT or 3D CRT showed that the IMRT and VMAT plans improved both OAR sparing and dose conformality over that possible with the 3D CRT plans; also, the double-arm VMAT plan further enhanced OAR sparing while also reducing beam-on time compared with the single-arc VMAT plan [49]. These improvements may be further enhanced with the use of flattening filter-free RapidArc plans. In a study comparing this technique with standard RapidArc therapy, IMRT, and 3D CRT, all of the IMRT plans showed better sparing of the lung compared with 3D CRT, but the flattening filter-free RapidArc therapy showed a 20 % increase in MU/Gy, a 90 % increase in average dose rate, and a 20 % reduction in beam-on time [36]. The use of IMRT delivered by helical tomotherapy has also been reported in several treatment planning studies, and the dose distributions are quite similar to those of the double-arc RapidArc plans [6, 33, 35].

16.4 Clinical Evidence in Support of IMRT

As alluded to in the previous section, many treatment planning studies have demonstrated potential dosimetric advantages from IMRT in terms of improving avoidance of normal tissues and the efficiency of radiation treatment; however, clinical evidence of such benefits is only recently emerging from a few single-institution reports. No large randomized trials have been undertaken to directly compare IMRT with 3D CRT, and indeed, such studies are unlikely. Although 3D CRT remains the standard of care in many parts of the world, IMRT is increasingly being used at large academic centers and in many community practices. Several small studies have shown that IMRT for esophageal tumors is feasible and has promising long-term outcomes and good tolerability [27, 52]. A small randomized trial was conducted in China comparing IMRT versus 3D CRT in 60 patients [29]; the radiation dose was 64 Gy in 30 fractions and was given with concurrent cisplatin and docetaxel. Although the 1- to 3-year survival rates seemed to favor IMRT, the numbers were too small to reach statistical significance. MD Anderson Cancer Center reported a large retrospective analysis that spanned 1998–2010; 3D CRT was more common in 1998–2002, and IMRT was more common in 2002–2010 [30]. Because of the substantial potential for bias in this nonrandomized comparison, propensity-matched analysis was used with an inverse probability weight approach to balance the two treatment era groups for 13 potential confounding patient- and treatment-related characteristics. Imbalances noted were that patients given IMRT were less likely to have received induction chemotherapy (35.7 % in the IMRT group vs. 46.7 % in the 3D CRT group), had worse performance status scores (≤80 in 66 % of IMRT vs. 50 % of 3D CRT), were more likely to have undergone positron emission tomography (PET) with fluorodeoxyglucose (FDG) as part of the initial disease-staging workup (95 % vs. 55 %), and had higher levels of predicted forced expiratory volume in 1 second (FEV1) (96 % vs. 90 %). Clinical outcomes, evaluated after inverse probability weight adjustment and Cox logistic regression, showed that IMRT produced significantly better overall survival and locoregional control

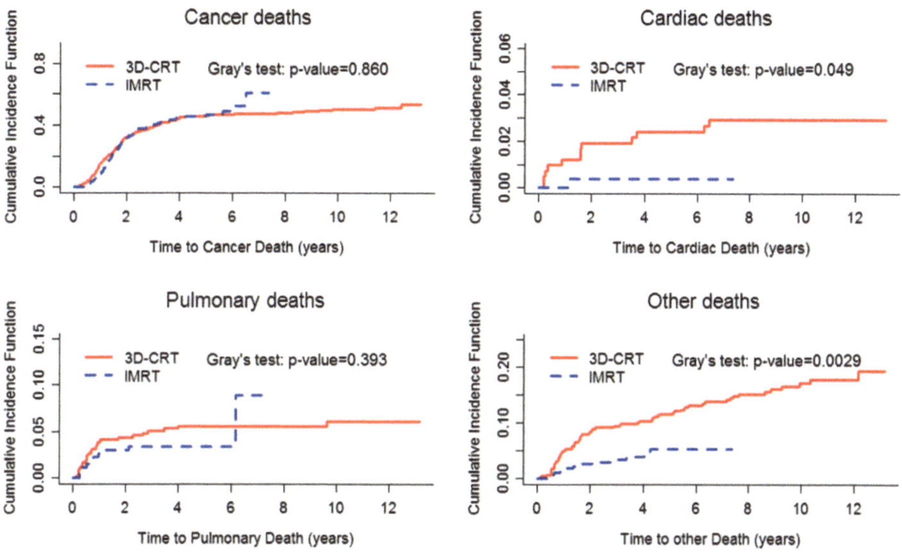

Fig. 16.2 The cumulative risk of cause-specific death after 3D CRT versus IMRT for esophageal cancer. Cancer and pulmonary deaths were not significantly different between IMRT and 3D CRT; however, for cardiac and other deaths (undefined deaths), rates of deaths were significantly higher in patients treated with 3D CRT (Reprinted with permission from [30])

but not cancer-specific survival or distant metastasis-free survival. To determine why overall survival but not cancer-specific survival was better in the IMRT group, cause-specific survival was compared between groups, specifically evaluating deaths from cancer, pulmonary or cardiac disease, or other causes (defined as deaths that were not related to cancer but were unknown). Cancer-related deaths accounted for most of the deaths in these patients, accounting for >40 % of deaths during the long-term follow-up period (median 40.3 months for the IMRT group and 82.4 months for the 3D CRT group). No differences were found in cancer-specific death or pulmonary death between groups, but the cumulative risks of cardiac-related mortality and other death were strikingly higher in the 3D CRT group than in the IMRT group, with nearly three times the risk of dying from causes not related to cancer (Fig. 16.2). To determine if this observation could be generalized to other populations, these investigators evaluated two large populational databases, the United States Surveillance, Epidemiology, and End Results (SEER) and the Texas Cancer Registry-Medicare databases. Inclusion criteria included having nonmetastatic esophageal cancer, being older than 65 years, and having been treated in 2002–2009 with either conformal radiation or IMRT. A total of 2,578 patients were identified who met these criteria, 2,265 who received conformal radiation and 313 IMRT. The two groups were well balanced in age, sex, race, disease stage (localized or regional), tumor grade, practice location, Charlson comorbidity index (CCI), use of chemotherapy, use of surgery, and various underlying cardiac and pulmonary diseases; the only differences were in marital status and SEER geographic region.

On average, 85 % of patients had had chemotherapy with radiation, and 18 % had surgical resection. The rate of IMRT use increased over time from 2.6 % in 2002 to 30 % in 2009, and the use of 2D or 3D radiation therapy decreased from 97.4 % in 2002 to 69 % in 2009. All-cause mortality was significantly lower in the IMRT group (52.4 % vs. 74.5 %, $p<0.0001$). Cause-specific mortality also revealed that IMRT led to lower rates of esophageal cancer-specific death (40.3 % vs. 55.6 %, $p<0.0001$) and cardiac-related mortality (1.6 % vs. 5.3 %, $p=0.0043$), but not pulmonary-specific death (0.96 % vs. 1.55 %, $p=0.419$) or all other non-cancer deaths (9.6 % vs. 12.1 %, $p=0.204$). On multivariable analysis, not being married (HR 1.17, $p=0.02$), having regional vs. localized disease (HR 1.29, $p<0.0001$), having a Charlson comorbidity index of 2+ (HR 1.21, $p=0.01$), receiving surgery (HR 0.52, $p<0.0001$), receiving chemotherapy (HR 0.57, $p<0.0001$), and living in the New Jersey region (HR 0.73, $p=0.002$) predicted cancer-specific mortality, but receipt of IMRT (rather than 3D CRT) did not. However, other multivariable analyses showed that IMRT (HR 0.39, $p=0.047$) and comorbidity index 2+ (HR 3.9, $p<0.0001$) were the only significant predictors of for cardiac-related mortality [32].

Because postoperative complications are well known to be related to radiation exposure of OARs, MD Anderson investigators further studied whether postoperative complications could be improved with the use of IMRT relative to 3D CRT. In the subgroup of 444 patients who received preoperative chemoradiation followed by surgical resection between 1998 and 2011, 208 had received 3D CRT, 164 patients IMRT, and 72 patients proton beam therapy (PBT; use of PBT began at MD Anderson in 2006). Postoperative complications evaluated were pulmonary, gastrointestinal, cardiac, and wound-healing complications [53]. Although radiation was not associated with cardiac or wound-healing complications, significant differences were seen in pulmonary and gastrointestinal complications: IMRT was associated with a substantial reduction in risk of both gastrointestinal (odds ratio [OR] 0.57, 95 % CI [confidence interval] 0.34–0.93) and pulmonary (OR 0.50, 95 % CI 0.27–0.91) complications compared with 3D CRT, but no differences were found between IMRT and PBT (with the possible exception of IMRT being more likely to produce pulmonary complications [OR 1.56, 95 % CI 0.68–3.60]). Actuarial rates of pulmonary complications were 34 % for the 3D CRT group, 23 % for the IMRT group, and 14 % for the PBT group. The Lyman-Kutcher-Burman modeling showed that the MLD was the strongest predictor of pulmonary toxicity (Fig. 16.3a); indeed, after MLD was added to the multivariate model, the link between radiation modality and pulmonary toxicity disappeared. Superimposing the MLD of each patients treated with each modalities revealed that pulmonary complications were driven not by the type of radiation used but by the ability of each of the modalities to achieve lower MLD (Fig. 16.3b). PBT produced the lowest MLD and IMRT a low-to-intermediate range MLD, and 3D CRT was associated with intermediate-to-high MLD. These findings strongly link MLD with the percent probability of pulmonary complications [53]. The rates of pulmonary and gastrointestinal complications associated with 3D CRT in this study were similar to those reported in the literature, but were much lower among patients treated with either IMRT or PBT (Table 16.1).

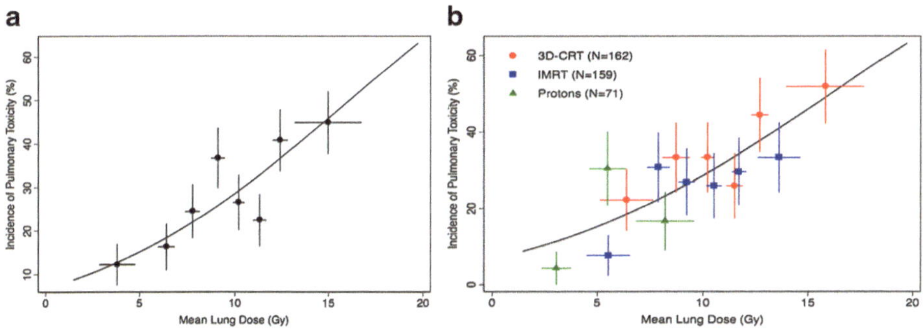

Fig. 16.3 LKB modeling of pulmonary toxicity in relation to mean lung dose and radiation modality. (**a**) The probability of developing postoperative pulmonary toxicity is related to the mean lung dose (MLD). (**b**) The relationship between incidence of postoperative pulmonary toxicities and MLD in the context of radiation modality (Reprinted with permission from [53])

Table 16.1 Studies of pulmonary and gastrointestinal toxicity after surgery for esophageal cancer

Study	Pulmonary	Gastrointestinal
[50]	46.40 %	3.5 % †
[8]	13.6 %*	24.40 %
[46]	Not reported	14.9 % †
[44]	33.30 %	20.80 %
[9]	45.00 %	14.0 %†
[47]	46.00 %	22.3 %†
[53]	25.2 % (overall)	23.0 % (overall)
	30.3 % (3D CRT)	28.4 % (3D CRT)
	23.8 % (IMRT)	18.3 % (IMRT)
	13.9 % (PBT)	18.1 % (PBT)

3D CRT 3-dimensional conformal radiation therapy, *IMRT* intensity-modulated radiation therapy, *PBT* proton beam therapy

*Did not include respiratory insufficiency

†The only gastrointestinal toxicity was an anastomotic leak

16.5 Treatment Planning for Chemoradiation

Investigators at MD Anderson take a multidisciplinary approach to treating esophageal cancer. Each case is discussed at multidisciplinary tumor board meetings, and eligibility for various ongoing clinical trials is considered. After a complete staging workup that includes FDG-PET/CT and esophagoduodenoscopy and endoscopic ultrasonography with biopsy, all patients with nonmetastatic disease are treated with chemoradiation, to be followed by an evaluation to determine surgical resectability. That workup also includes FDG-PET to rule out metastatic disease and endoscopy and biopsy to identify any local residual disease.

For treatment planning, 4D CT-based treatment simulation is used for all patients. This technique assesses diaphragmatic motion of the gastroesophageal junction, where most tumors are located, and that motion is accounted for in the treatment planning, along with findings from pretreatment PET scans and ultrasonography. Outlines of the gross tumor volume (GTV) and the path of its motion (iGTV) are contoured on the axial slices of maximum intensity projection (MIP) images, with the margin expanded by 4 cm superiorly and by 3 cm inferiorly, into the lesser curvature of the stomach. If nodal disease is present in the left gastric nodes along the lesser curvature of the stomach, generally, those nodes are included in the iGTV, but the internal clinical target volume (iCTV) will extend down to encompass the celiac axis nodes because of the risk of regional recurrence in those nodes. Laterally, the iCTV encompasses all of the soft tissues surrounding the esophagus up to the parietal pleural surface and avoids all anatomic boundaries (e.g., aorta, vertebral body, and the inferior border of the heart). The planning target volume (PTV) is a uniform expansion of 0.5 cm around the iCTV to account for setup errors and is used only if daily kV or MV projection imaging is used for daily image-guided setup. If daily imaging is not performed, the PTV should be a 1-cm expansion from the iCTV. The standard radiation dose used is 50.4 Gy in 28 fractions, although other regimens (e.g., 40 Gy in 20 fractions, 41.4 Gy in 23 fractions, or 45 Gy in 25 fractions) are also acceptable. The most common field arrangement is a 5-field IMRT approach, with 5 posteriorly placed beams without any beams directly entering anteriorly through the heart. In light of the potential advantages of VMAT or RapidArc IMRT in terms of workflow efficiency, these techniques are also being implemented. In terms of dosimetric (dose-volume histogram or DVH) limits, the total lung V_{10} is kept at <40 %, lung $V_{15}<30$ %, and lung $V_{20}<20$ %, particularly for patients who will go on to undergo surgery, to avoid postoperative pulmonary complications. The heart V_{40} is generally kept at <40 %, but considerably lower doses can generally be achieved. The most common chemotherapy regimen used concurrently with IMRT is cisplatin with fluorouracil, although other regimens can also be used, as none have been shown to have particular advantages. Per the CROSS trial, carboplatin (AUC 2.0) and paclitaxel (50 mg/m^2) can also be considered a standard chemotherapy regimen.

16.6 Conclusions

Radiation is a crucial component of the management of esophageal cancer. Based on findings from recent randomized trials and meta-analyses, trimodality therapy (chemotherapy, radiation, and surgery) can be considered the standard of care. For patients who cannot tolerate surgery, chemoradiation is the definitive treatment, with long-term cure rates of around 20 %. Treatment of esophageal tumors, particularly those that arise in the mid- to distal locations, can cause significant morbidity and postoperative complications such as pleural and pericardial effusions owing to the proximity of the heart and lung. The use of IMRT can improve dose conformality to the target and also reduce the dose to the OARs. The clinical experience at MD Anderson, in which large numbers of patients have been treated with IMRT or 3D CRT, has demonstrated significant improvements for IMRT in overall survival and reduced cardiac mortality as well as in postoperative pulmonary and gastrointestinal complications. A recent population-based analysis of SEER-Medicare data further supports a benefit from IMRT in terms of improved survival, not from a reduction in cancer-specific death but rather from a reduction in risk of cardiac mortality compared with 3D CRT. Given the unlikeliness of a large randomized trial that directly compares IMRT with 3D CRT, the findings reviewed in this chapter should provide sufficient evidence to support the use of IMRT for esophageal cancer.

Acknowledgments Supported in part by Cancer Center Support (Core) Grant CA016672 from the US National Cancer Institute to the University of Texas MD Anderson Cancer Center.

Conflict of Interest The author declares no conflicts of interest.

References

1. Abbas AS, Moseley D, Kassam Z, Kim SM, Cho C (2013) Volumetric-modulated arc therapy for the treatment of a large planning target volume in thoracic esophageal cancer. J Appl Clin Med Phys 14(3):4269
2. Adelstein DJ, Rice TW, Becker M, Larto MA, Kirby TJ, Koka A, Tefft M, Zuccaro G (1997) Use of concurrent chemotherapy, accelerated fractionation radiation, and surgery for patients with esophageal carcinoma. Cancer 80(6):1011–1020
3. Bedenne L, Michel P, Bouchart O, Milan C, Mariette C, Conroy T, Pezet D, Roullet B, Seitz JF, Herr JP, Paillot B, Arveux P, Bonnetain P, Binquet C (2007) Chemoradiation followed by surgery compared with chemoradiation alone in squamous cancer of the esophagus: FFCD 9102. J Clin Oncol 25(10):1160–1168
4. Bosset JF, Gignoux M, Triboulet JP, Tiret E, Mantion G, Elias D, Lozach P, Ollier JC, Pavy JJ, Mercier M, Sahmoud T (1997) Chemoradiotherapy followed by surgery compared with surgery alone in squamous-cell cancer of the esophagus. N Engl J Med 337(3):161–167
5. Chandra A, Guerrero TM, Liu HH, Tucker SL, Liao Z, Wang X, Murshed H, Bonnen MD, Garg AK, Stevens CW, Chang JY, Jeter MD, Mohan R, Cox JD, Komaki R (2005) Feasibility of using intensity-modulated radiotherapy to improve lung sparing in treatment planning for distal esophageal cancer. Radiother Oncol 77(3):247–253
6. Chen YJ, Liu A, Han C, Tsai PT, Schultheiss TE, Pezner RD, Vora N, Lim D, Shibata S, Kernstine KH, Wong JYC (2007) Helical tomotherapy for radiotherapy in esophageal cancer: a preferred plan with better conformal target coverage and more homogeneous dose distribution. Med Dosim 32(3):166–171
7. Chiu PW, Chan AC, Leung SF, Leong HT, Kwong KH, Li MK, Au-Yeung AC, Chung SC, Ng EK (2005) Multicenter prospective randomized trial comparing standard esophagectomy with chemoradiotherapy for treatment of squamous esophageal cancer: early results from the Chinese University Research Group for Esophageal Cancer (CURE). J Gastrointest Surg 9(6):794–802
8. Daly JM, Fry WA, Little AG, Winchester DP, McKee RF, Stewart AK, Fremgen AM (2000) Esophageal cancer: results of an American College of Surgeons Patient Care Evaluation Study. J Am Coll Surg 190(5):562–572
9. Derogar M, Orsini N, Sadr-Azodi O, Lagergren P (2012) Influence of major postoperative complications on health-related quality of life among long-term survivors of esophageal cancer surgery. J Clin Oncol 30(14):1615–1619

10. Doty JR, Salazar JD, Forastiere AA, Heath EI, Kleinberg L, Heitmiller RF (2002) Postesophagectomy morbidity, mortality, and length of hospital stay after preoperative chemoradiation therapy. Ann Thorac Surg 74(1):227–231
11. Earlam R, Cunha-Melo JR (1980) Oesophageal squamous cell carcinoma: I. A critical review of surgery. Br J Surg 67(6):381–390
12. Earlam R, Cunha-Melo JR (1980) Oesophogeal squamous cell carcinoms: II. A critical view of radiotherapy. Br J Surg 67(7):457–461
13. Enzinger PC, Mayer RJ (2003) Esophageal cancer. N Engl J Med 349(23):2241–2252
14. Fenkell L, Kaminsky I, Breen S, Huang S, Van Prooijen M, Ringash J (2008) Dosimetric comparison of IMRT vs. 3D conformal radiotherapy in the treatment of cancer of the cervical esophagus. Radiother Oncol 89(3):287–291
15. Ferlay J, Soerjomataram I, Ervik M, Dikshit R, Eser S, Mathers C, Rebelo M, Parkin DM, Forman D, Bray F (2013) GLOBOCAN 2012 v1.0, Cancer incidence and mortality worldwide: IARC CancerBase No. 11. International Agency for Research on Cancer, Lyon
16. Fukada J, Shigematsu N, Takeuchi H, Ohashi T, Saikawa Y, Takaishi H, Hanada T, Shiraishi Y, Kitagawa Y, Fukuda K (2013) Symptomatic pericardial effusion after chemoradiation therapy in esophageal cancer patients. Int J Radiat Oncol Biol Phys 87(3):487–493
17. Gong Y, Wang S, Zhou L, Liu Y, Xu Y, Lu Y, Bai S, Fu Y, Xu Q, Jiang Q (2010) Dosimetric comparison using different multileaf collimeters in intensity-modulated radiotherapy for upper thoracic esophageal cancer. Radiat Oncol 5(1):65
18. Grosshans DR, Boehing NS, Palmer M, Spicer C, Erice R, Cox JD, Komaki R, Chang JY (2012) Improving cardiac dosimetry: alternative beam arrangements for intensity modulated radiation therapy planning in patients with carcinoma of the distal esophagus. Pract Radiat Oncol 2(1):41–45
19. Hawkins MA, Bedford JL, Warrington AP, Tait DM (2012) Volumetric modulated arc therapy planning for distal oesophageal malignancies. Br J Radiol 85(1009):44–52
20. Herskovic A, Martz K, Al-Sarraf M, Leichman L, Brindle J, Vaitkevicius V, Cooper J, Byhardt R, Davis L, Emami B (1992) Combined chemotherapy and radiotherapy compared with radiotherapy alone in patients with cancer of the esophagus. N Engl J Med 326(24):1593–1598
21. Hirai T, Yamashita Y, Mukaida H, Kuwahara M, Inoue H, Toge T (1998) Poor prognosis in esophageal cancer patients with postoperative complications. Surg Today 28(6):576–579
22. Hulscher JBF, Tijssen JGP, Obertop H, Van Lanschot JJB (2001) Transthoracic versus transhiatal resection for carcinoma of the esophagus: a meta-analysis. Ann Thorac Surg 72(1):306–313
23. Hulscher JBF, Van Sandick JW, De Boer AGEM, Wijnhoven BPL, Tijssen JGP, Fockens P, Stalmeier PFM, Ten Kate FJW, Van Dekken H, Obertop H, Tilanus HW, Lanschot J (2002) Extended transthoracic resection compared with limited transhiatal resection for adenocarcinoma of the esophagus. N Engl J Med 347(21):1662–1669
24. Ishikura S, Nihei K, Ohtsu A, Boku N, Hironaka S, Mera K, Muto M, Ogino T, Yoshida S (2003) Long-term toxicity after definitive chemoradiotherapy for squamous cell carcinoma of the thoracic esophagus. J Clin Oncol 21(14):2697–2702
25. Kole TP, Aghayere O, Kwah J, Yorke ED, Goodman KA (2012) Comparison of heart and coronary artery doses associated with intensity-modulated radiotherapy versus three-dimensional conformal radiotherapy for distal esophageal cancer. Int J Radiat Oncol Biol Phys 83(5):1580–1586
26. Kumar G, Rawat S, Puri A, Sharma MK, Chadha P, Babu AG, Yadav G (2012) Analysis of dose-volume parameters predicting radiation pneumonitis in patients with esophageal cancer treated with 3D-conformal radiation therapy or IMRT. Jpn J Radiol 30(1):18–24
27. La TH, Minn AY, Su Z, Fisher GA, Ford JM, Kunz P, Goodman KA, Koong AC, Chang DT (2010) Multimodality treatment with intensity modulated radiation therapy for esophageal cancer. Dis Esophagus 23(4):300–308
28. Lee HK, Vaporciyan AA, Cox JD, Tucker SL, Putnam JB Jr, Ajani JA, Liao Z, Swisher SG, Roth JA, Smythe WR, Walsh GL, Mohan R, Liu HH, Mooring D, Komaki R (2003) Postoperative pulmonary complications after preoperative chemoradiation for esophageal

carcinoma: correlation with pulmonary dose-volume histogram parameters. Int J Radiat Oncol Biol Phys 57(5):1317–1322

29. Lin XD, Shi XY, Zhou TC, Zhang WJ (2011) Intensity-modulated or 3-D conformal radiotherapy combined with chemotherapy with docetaxel and cisplatin for locally advanced esophageal carcinoma. Nan Fang Yi Ke Da Xue Xue Bao 31(7):1264–1267
30. Lin SH, Wang L, Myles B, Thall PF, Hofstetter WL, Swisher SG, Ajani JA, Cox JD, Komaki R, Liao Z (2012) Propensity score-based comparison of long-term outcomes with 3-dimensional conformal radiotherapy vs intensity-modulated radiotherapy for esophageal cancer. Int J Radiat Oncol Biol Phys 84(5):1078–1085
31. Lin CY, Huang WY, Jen YM, Chen CM, Su YF, Chao HL, Lin CS (2013) Dosimetric and efficiency comparison of high-dose radiotherapy for esophageal cancer: volumetric modulated arc therapy versus fixed-field intensity-modulated radiotherapy. Dis Esophagus October 18 [ePub ahead of print]
32. Lin SH, Zhang N, Godby J, Wang J, Marsh GD, Liao Z, Komaki R, Ho L, Hofstetter WL, Swisher SG, Mehran RJ, Buchholz TA, Elting LS, Giordano SH (2015) Radiation modality and cardiopulmonary mortality risk in the elderly with esophageal cancer (Submitted)
33. Martin S, Chen JZ, Rashid Dar A, Yartsev S (2011) Dosimetric comparison of helical tomotherapy, RapidArc, and a novel IMRT & Arc technique for esophageal carcinoma. Radiother Oncol 101(3):431–437
34. Morota M, Gomi K, Kozuka T, Chin K, Matsuura M, Oguchi M, Ito H, Yamashita T (2009) Late toxicity after definitive concurrent chemoradiotherapy for thoracic esophageal carcinoma. Int J Radiat Oncol Biol Phys 75(1):122–128
35. Nguyen NP, Krafft SP, Vinh-Hung V, Vos P, Almeida F, Jang S, Ceizyk M, Desai A, Davis R, Hamilton R, Modarresifar H, Abraham D, Smith-Raymond L (2011) Feasibility of tomotherapy to reduce normal lung and cardiac toxicity for distal esophageal cancer compared to three-dimensional radiotherapy. Radiother Oncol 101(3):438–442
36. Nicolini G, Ghosh-Laskar S, Shrivastava SK, Banerjee S, Chaudhary S, Agarwal JP, Munshi A, Clivio A, Fogliata A, Mancosu P, Vanetti E, Cozzi L (2012) Volumetric modulation arc radiotherapy with flattening filter-free beams compared with static gantry IMRT and 3D conformal radiotherapy for advanced esophageal cancer: a feasibility study. Int J Radiat Oncol Biol Phys 84(2):553–560
37. Nutting CM, Bedford JL, Cosgrove VP, Tait DM, Dearnaley DP, Webb S (2001) A comparison of conformal and intensity-modulated techniques for oesophageal radiotherapy. Radiother Oncol 61(2):157–163
38. Patti MG, Corvera CU, Glasgow RE, Way LW (1998) A hospital's annual rate of esophagectomy influences the operative mortality rate. J Gastrointest Surg 2(2):186–192
39. Rizk NP, Bach PB, Schrag D, Bains MS, Turnbull AD, Karpeh M, Brennan MF, Rusch VW (2004) The impact of complications on outcomes after resection for esophageal and gastroesophageal junction carcinoma. J Am Coll Surg 198(1):42–50
40. Shirai K, Tamaki Y, Kitamoto Y, Murata K, Satoh Y, Higuchi K, Nonaka T, Ishikawa H, Katoh H, Takahashi T, Nakano T (2011) Dose-volume histogram parameters and clinical factors associated with pleural effusion after chemoradiotherapy in esophageal cancer patients. Int J Radiat Oncol Biol Phys 80(4):1002–1007
41. Sjoquist KM, Burmeister BH, Smithers BM, Zalcberg JR, Simes RJ, Barbour A, Gebski V (2011) Survival after neoadjuvant chemotherapy or chemoradiotherapy for resectable oesophageal carcinoma: an updated meta-analysis. Lancet Oncol 12(7):681–692
42. Stahl M, Stuschke M, Lehmann N, Meyer HJ, Walz MK, Seeber S, Klump B, Budach W, Teichmann R, Schmitt M, Schmitt G, Franke C, Wilke H (2005) Chemoradiation with and without surgery in patients with locally advanced squamous cell carcinoma of the esophagus. J Clin Oncol 23(10):2310–2317
43. Swisher SG, DeFord L, Merriman KW, Walsh GL, Smythe R, Vaporicyan A, Ajani JA, Brown T, Komaki R, Roth JA, Putnam JB (2000) Effect of operative volume on morbidity, mortality, and hospital use after esophagectomy for cancer. J Thorac Cardiovasc Surg 119(6):1126–1134

44. Tepper J, Krasna MJ, Niedzwiecki D, Hollis D, Reed CE, Goldberg R, Kiel K, Willett C, Sugarbaker D, Mayer R (2008) Phase III trial of trimodality therapy with cisplatin, fluorouracil, radiotherapy, and surgery compared with surgery alone for esophageal cancer: CALGB 9781. J Clin Oncol 26(7):1086–1092
45. Tucker SL, Liu HH, Wang S, Wei X, Liao Z, Komaki R, Cox JD, Mohan R (2006) Dose-volume modeling of the risk of postoperative pulmonary complications among esophageal cancer patients treated with concurrent chemoradiotherapy followed by surgery. Int J Radiat Oncol Biol Phys 66(3):754–761
46. Urba SG, Orringer MB, Turrisi A, Iannettoni M, Forastiere A, Strawderman M (2001) Randomized trial of preoperative chemoradiation versus surgery alone in patients with locoregional esophageal carcinoma. J Clin Oncol 19(2):305–313
47. van Hagen P, Hulshof MC, van Lanschot JJ, Steyerberg EW, van Berge Henegouwen MI, Wijnhoven BP, Richel DJ, Nieuwenhuijzen GA, Hospers GA, Bonenkamp JJ, Cuesta MA, Blaisse RJ, Busch OR, ten Kate FJ, Creemers GJ, Punt CJ, Plukker JT, Verheul HM, Spillenaar Bilgen EJ, van Dekken H, van der Sangen MJ, Rozema T, Biermann K, Beukema JC, Piet AH, van Rij CM, Reinders JG, Tilanus HW, van der Gaast A (2012) Preoperative chemoradiotherapy for esophageal or junctional cancer. N Engl J Med 366(22):2074–2084
48. Viklund P, Lindblad M, Lu M, Ye W, Johansson J, Lagergren J (2006) Risk factors for complications after esophageal cancer resection: a prospective population-based study in Sweden. Ann Surg 243(2):204–211
49. Vivekanandan N, Sriram P, Kumar SA, Bhuvaneswari N, Saranya K (2012) Volumetric modulated arc radiotherapy for esophageal cancer. Med Dosim 37(1):108–113
50. Walsh TN, Noonan N, Hollywood D, Kelly A, Keeling N, Hennessy TPJ (1996) A comparison of multimodal therapy and surgery for esophageal adenocarcinoma. N Engl J Med 335(7):462–467
51. Wang SL, Liao Z, Vaporciyan AA, Tucker SL, Liu H, Wei X, Swisher S, Ajani JA, Cox JD, Komaki R (2006) Investigation of clinical and dosimetric factors associated with postoperative pulmonary complications in esophageal cancer patients treated with concurrent chemoradiotherapy followed by surgery. Int J Radiat Oncol Biol Phys 64(3):692–699
52. Wang J, Han C, Li XN, Gao C, Jia JH, Cai BN, Zhang X, Xiao AQ (2009) Short-term efficacy of intensity-modulated radiotherapy on esophageal carcinoma. Ai Zheng 28(11):1138–1142
53. Wang J, Wei C, Tucker SL, Myles B, Palmer M, Hofstetter WL, Swisher SG, Ajani JA, Cox JD, Komaki R, Liao Z, Lin SH (2013) Predictors of postoperative complications after trimodality therapy for esophageal cancer. Int J Radiat Oncol Biol Phys 86(5):885–891
54. Wei X, Liu HH, Tucker SL, Wang S, Mohan R, Cox JD, Komaki R, Liao Z (2008) Risk factors for pericardial effusion in inoperable esophageal cancer patients treated with definitive chemoradiation therapy. Int J Radiat Oncol Biol Phys 70(3):707–714
55. Wu VWC, Kwong DLW, Sham JST (2004) Target dose conformity in 3-dimensional conformal radiotherapy and intensity modulated radiotherapy. Radiother Oncol 71(2):201–206
56. Wu VWC, Sham JST, Kwong DLW (2004) Inverse planning in three-dimensional conformal and intensity-modulated radiotherapy of mid-thoracic oesophageal cancer. Br J Radiol 77(919):568–572
57. Yin L, Wu H, Gong J, Geng JH, Jiang F, Shi AH, Yu R, Li YH, Han SK, Xu B, Zhu GY (2012) Volumetric-modulated arc therapy vs. c-IMRT in esophageal cancer: a treatment planning comparison. World J Gastroenterol 18(37):5266–5275

Pancreatic Cancer

17

Keiko Shibuya, Takehiro Shiinoki, and Akira Nakamura

Keywords
Pancreatic cancer • Chemoradiation • Locally advanced pancreatic cancer • Borderline resectable tumor

17.1 Introduction

Pancreatic cancer is often fatal. The only curative treatment is complete resection of the tumor, but even patients who are able to undergo surgery have a 5-year survival rate of less than 20 %. Most patients have metastatic or locally advanced, unresectable tumors at diagnosis. For all disease stages combined, the 5-year survival rate is less than 5 % [1–4].

17.1.1 Definitive Treatments for Unresectable Locally Advanced Pancreatic Cancer

For locally advanced, unresectable pancreatic cancer, chemoradiation therapy is the standard treatment based on the results of two randomized trials conducted by the Gastrointestinal Tumor Study Group (GITSG). These trials demonstrated a survival

K. Shibuya, M.D., Ph.D. (✉) • T. Shiinoki, Ph.D.
Department of Therapeutic Radiology, Graduate School of Medicine, Yamaguchi University, 1-1-1, Minamikogushi, Ube, Yamaguchi 755-8505, Japan
e-mail: kshibuya@yamaguchi-u.ac.jp

A. Nakamura, M.D., Ph.D.
Department of Radiation Oncology and Image-applied Therapy, Graduate School of Medicine, Kyoto University, Kyoto, Japan

Y. Nishimura, R. Komaki (eds.), *Intensity-Modulated Radiation Therapy*,
DOI 10.1007/978-4-431-55486-8_17

advantage for the combination of radiation therapy plus fluorouracil over either radiation therapy alone [5] or chemotherapy (streptozotocin, mitomycin-C, and fluorouracil) alone [6]. However, the survival benefits were very modest. Recently developed chemotherapeutic agents showing clinical benefits in the treatment of pancreatic cancer include gemcitabine, capecitabine, and the fluoropyrimidine derivative S-1 [7–11]. Some clinicians have raised concerns about the need for radiation therapy for locally advanced pancreatic cancer from the perspective of toxicity versus the small survival benefit. Results of two randomized trials directly comparing chemoradiation therapy with chemotherapy alone were recently reported. In one of these studies, the Fédération Francophone de Cancérologie Digestive/ Société Francophone de Radiothérapie Oncologique (FFCD/SFRO) randomly assigned 119 patients into one of two treatment groups, one receiving intensive chemoradiation therapy (60 Gy with infused fluorouracil and intermittent cisplatin) followed by maintenance gemcitabine and the other receiving gemcitabine alone [12]. In that study, the overall survival time was shorter in the chemoradiation group than the gemcitabine group (median 8.6 months, 99 % confidence interval [CI] 7.1–11.4 vs. 13 months, 99 % CI 8.7–18.1, $P=0.03$). However, in this study, only 42 % received at least 75 % of both the planned radiation dose and concomitant chemotherapy because of hematologic toxicity. In addition, the rate of grade 3/4 nonhematologic toxicity was higher in the chemoradiation group (43.6 %) than in the chemotherapy group (10 %). The high incidence of toxicity in the chemoradiation group was thought to be one reason why the survival times were much shorter than in previous reports of chemoradiation therapy. By way of comparison, the other study, by the Eastern Cooperative Oncology Group, randomly assigned 74 patients to receive gemcitabine or gemcitabine plus radiation [13]. The primary end point was the survival time, which was 9.2 months (95 % CI 7.9–11.4) with gemcitabine alone and 11.1 months (95 % CI 7.6–15.5) with chemoradiation ($P=0.017$). In this study, patients in the chemoradiation group had a greater incidence of grade 4 hematologic toxicities than those in the gemcitabine-alone group, but rates of grade 3 and 4 toxicities were similar in the two groups. No statistical differences were seen in quality of life measurements. The radiation dose was 50.4 Gy in 28 fractions, with a required field reduction after 39.6 Gy. Three-dimensional (3D) treatment planning was encouraged in the protocol. The Quality Assurance Review Center conducted two separate reviews of the 3D benchmark, including dose-volume histograms for the critical normal tissues. Although this trial did not reach the planned accrual goals, the results were highly suggestive that well-planned radiation therapy could control toxicity, even in combination with novel cytotoxic agents, and could contribute to prolonging survival time.

Results of several recent phase I/II studies of chemoradiation combined with novel chemotherapeutic agents have been promising. Wilkowski et al. reported the results of a prospective study of 32 patients with locally advanced pancreatic cancer treated with low-dose (300 mg/m^2) gemcitabine and fluorouracil as a continuous infusion at 350 mg/m^2/day with concurrent radiation (45–50 Gy) to the tumor and regional lymph nodes [14]. The median survival time was 13.6 months (95 % CI 12.7–14.6) for all patients versus 16.4 months (95 % CI 13.4–19.4) for those undergoing secondary resection. The actuarial 1-, 2-, and 3-year survival rates were 67.2, 20.0, and 8.0 %. On the basis of these findings and those from other phase I/II studies of gemcitabine with radiation, gemcitabine-based chemoradiation is currently considered a standard therapy for locally advanced pancreatic cancer [4].

S-1 is an oral fluoropyrimidine derivative that has shown anticancer activity in various types of solid tumors. Several clinical trials of radiation therapy combined with S-1 for locally advanced pancreatic cancer have been conducted in Japan. In one of those studies, Sudo et al. completed a phase II trial of a regimen comprising radiation therapy delivered in 1.8-Gy daily fractions to a total dose of 50.4 Gy, with S-1 administered orally twice a day at a dose of 80 mg/m^2/day on days 1–14 and 22–35, followed by maintenance chemotherapy with S-1 [15]. In the 34 patients treated with this regimen, the median survival time was 16.8 months (95 % CI 12.9–20.7) and the 1-year survival rate was 70.6 %. Another phase II study reported by Shinchi et al. enrolled 50 patients, of whom 43 completed the scheduled course of chemoradiation therapy: oral S-1 at a dose of 80 mg/m^2/day twice daily on days 1–21 and radiation therapy delivered in 1.25-Gy twice-daily fractions for 4 weeks (total dose 50 Gy in 40 fractions) [16]. The median survival time in that trial was 14.3 months (95 % CI 10.8–20.8), and the survival rates at 1, 2, 3, and 4 years were 62, 27, 15, and 12 %. In 2012, Ikeda et al. reported the results of a multicenter phase II trial of radiation therapy at a dose of 50.4 Gy in 28 fractions combined with S-1 at a dose of 80 mg/m^2 twice daily on the day of irradiation, followed by a 2- to 8-week break and a maintenance dose of S-1 (80 mg/m^2/days for 28 consecutive days and a 14-day rest period) [17]. In all, 61 patients were enrolled and 60 patients were treated with this regimen. The median survival time was 16.2 months (95 % CI 13.5–21.3) and the survival rates were 72 % at 1 year and 26 % at 2 years.

Capecitabine is an oral fluorouracil prodrug used to treat pancreatic cancer in the United States and Europe. Some clinical trials testing treatment with radiation and capecitabine have suggested that this may be an alternative to the classic fluorouracil-based chemoradiation regimens. In 2013, the results of a multicenter, randomized, phase II trial of gemcitabine or capecitabine-based chemoradiation therapy for locally advanced pancreatic cancer (SCALOP) were reported [18]. In this regimen, induction therapy was given with gemcitabine and capecitabine (four cycles of gemcitabine [1,000 mg/m^2 on days 1, 8, 15 of a 28-day cycle] and capecitabine [830 mg/m^2 twice daily on days 1–21 of a 28-day cycle]), followed by either gemcitabine (300 mg/m^2 once per week) or capecitabine (830 mg/m^2 twice daily, Monday to Friday only) combined with radiation (50.4 Gy in 28 fractions). That study registered 114 patients, 74 of whom were randomly assigned to one of the two treatment groups. The median overall survival time was 15.2 months (95 % CI 13.9–19.2) in the capecitabine group versus 13.4 months (95 % CI 11.0–15.7) in the gemcitabine group (adjusted hazard ratio [HR] 0.39, 95 % CI 0.18–0.81; $P=0.012$). The 1-year overall survival rate was 79.2 % (95 % CI 61.1–89.5) in the capecitabine group and 64.2 % (95 % CI 46.4–77.5) in the gemcitabine group.

To summarize, the survival data varied markedly in these studies. Overall, higher proportions of patients who survived for longer than 2 or 3 years were found among those treated with chemoradiation than among those treated with chemotherapy alone. These results indicate that at least some subset of patients get a survival benefit from intensive local treatments, i.e., radiation therapy.

Based on a limited number of randomized controlled trials, the American National Comprehensive Cancer Network (NCCN) guidelines recommend both chemoradiation and chemotherapy alone as standard treatments for locally advanced pancreatic cancer. Unlike other solid tumors, pancreatic cancer is a disease for which chemotherapy and chemoradiation therapy share positions as standard therapy.

17.1.2 Neoadjuvant Radiation Therapy for Resectable or Borderline Resectable Pancreatic Cancer

Numerous studies of factors that could influence outcomes after surgery for pancreatic cancer have shown that resection margin status is significantly associated with survival [19–21]: the survival outcomes of patients with positive margins after resection were similar to those of patients with unresectable disease. However, no randomized controlled study has shown any improvement from more radical procedures (such as an extended retroperitoneal lymphadenectomy) as compared with a standard pancreaticoduodenectomy [22–27]. Thus, interest is increasing in optimizing the use of multimodality therapy in neoadjuvant or adjuvant settings.

Neoadjuvant chemoradiation has been investigated in several studies [28–31], but no conclusions were reached regarding effects on survival. A recent retrospective analysis of the Surveillance, Epidemiology, and End Results (SEER) registry database (1994–2003) compared survival of patients who received preoperative radiation therapy with that of patients who underwent surgical resection without radiation therapy or with that of patients who received surgery followed by adjuvant radiation therapy [32]. The median survival time for patients receiving neoadjuvant radiation therapy was 23 months vs. 12 months for those with no radiation therapy and 17 months for those with adjuvant radiation therapy. Adjuvant and neoadjuvant radiation therapy were both associated with significantly higher survival rates than those for no radiation therapy (for neoadjuvant radiation therapy HR for $=0.49$, $P=0.00$; for adjuvant radiation, HR $=0.71$, $P=0.00$). A recent meta-analysis found that 40 % of patients with initially unresectable pancreatic cancer who underwent neoadjuvant chemoradiotherapy could ultimately undergo resection and achieved similar survival outcomes to those of patients with resectable disease [33]. In that study, neoadjuvant chemoradiation therapy was not associated with a statistically significant increase in the rate of total complications.

Neoadjuvant therapy has also been investigated for the purpose of downstaging or downsizing borderline resectable tumors. The definitions of borderline resectable tumors in the NCCN guidelines (version 1.2014), adopted from a consensus report from Callery et al. [34], are as follows: tumors with no metastasis, no venous involvement of the superior mesenteric vein or portal vein, no gastroduodenal artery encasement up to the hepatic artery with either short segment encasement or direct abutment of the hepatic artery without extension to the celiac axis, or no tumor abutment of the superior mesenteric artery (SMA) that exceeds > 180° of the circumference of the vessel wall. Criteria for resectable or borderline resectable tumors differ among institutions and countries; for example, whether portal vein infiltration should be included in the resectable category has been found to correlate with the experience of the operating surgeon [35]. However, tumor abutment of the arteries that is less than half the circumference of those arteries is generally accepted to meet the criteria of borderline resectable disease. Neoadjuvant radiation therapy, using highly conformal techniques, is thought to be a reasonable approach to downsizing borderline resectable tumors and debulking tumors away from involved vessels, possibly resulting in a margin-free resection.

So far, no standard preoperative treatment regimen exists for resectable or borderline resectable pancreatic cancer. The NCCN guidelines (version 1.2014) recommend that neoadjuvant therapy for patients with resectable tumors should be conducted in the context of a clinical trial.

17.1.3 Adjuvant Radiation Therapy After Resection

Although adjuvant chemoradiation after resection has been shown to be effective in several types of gastrointestinal cancer [36, 37], the role of adjuvant chemoradiation in pancreatic cancer is still controversial. In 1985, a prospective randomized study by the GITSG showed improved overall survival with adjuvant fluorouracil-based chemoradiation, in which the median survival of the treatment group (20 months) was significantly longer than that of the control group (11 months) [38]. However, the results of a clinical trial conducted by the European Study Group for Pancreatic Cancer (ESPAC-1) suggested that the effect of adjuvant chemoradiation on survival was equal to the effects of adjuvant chemotherapy or surgery alone [39]. Despite much discussion about possible shortcomings of this study including its design, the radiation delivery methods, and radiation quality control, the conclusions of the ESPAC-1 trial raised doubts about the validity of adjuvant chemoradiation therapy for pancreatic cancer. On the other hand, several US study groups have reported findings supporting the use of adjuvant chemoradiation therapy. A review of Johns Hopkins' data prospectively collected on 908 patients (seen from 1993 through 2005) included examination of the efficacy of adjuvant chemoradiation therapy after pancreaticoduodenectomy or total pancreatectomy. Excluding patients with metastatic disease, those who died within 60 days of surgery, those who received preoperative therapy or an experimental vaccine, and those treated with adjuvant chemotherapy or radiation alone, the final cohort included 616 patients, of whom 271 patients received adjuvant chemoradiation therapy, whereas 345 patients elected to receive no therapy, after being fully informed of the potential risks and benefits of such therapy. In that study, patients who received adjuvant chemoradiation therapy had better median survival time (21.2 vs. 14.4 months, $P<0.001$), 2-year survival rates (43.9 % vs.

31.9 %), and 5-year survival rates (20.1 % vs. 15.4 %) compared with patients given no chemoradiation therapy [40]. Also, a retrospective review of 472 consecutive patients treated from 1975 through 2005, who underwent complete resection with negative margins (R0) at Mayo Clinic, revealed that they experienced a median overall survival time after adjuvant chemoradiation therapy of 25.2 months vs. 19.2 months for those who had had no adjuvant therapy ($P=0.001$) [41]. The respective 2-year overall survival rates were 50 % vs. 39 %, and the 5-year survival rates were 28 % vs. 17 %. Multivariate analysis identified several factors that predicted adverse prognosis: no adjuvant therapy (risk ratio [RR] = 1.3, $P<0.001$), positive lymph nodes (RR = 1.3, $P<0.001$), and high histologic grade (RR = 1.2, $P<0.001$). Subsequently, investigators at Johns Hopkins and the Mayo Clinic compared the efficacy of adjuvant fluorouracil-based chemoradiation therapy for resected pancreatic adenocarcinoma versus surgery alone. In that study, all patients who underwent pancreaticoduodenectomy at Johns Hopkins from 1993 through 2005 ($n=794$) and at the Mayo Clinic from 1985 through 2005 ($n=592$) were prospectively evaluated. The results of matched-pair analysis of 496 patients by treatment group (1:1, with 248 patients per group) in terms of institutional, marginal, and nodal positivity showed that overall survival improved when chemoradiation therapy was added to surgery versus surgery (median survival times 21.9 months vs. 14.3 months; 2-year overall survival rate 45.5 % vs. 31.4 %; 5-year overall survival rate 25.4 % vs.12.2 %, all P values <0.001) [42].

In 2012, an interesting report explored whether failure to adhere to specified radiation therapy guidelines could have influenced survival or toxicity in the Radiation Therapy Oncology Group (RTOG) trial 9704 [43], a randomized phase III trial that compared the use of either continuous fluorouracil or gemcitabine before and after concurrent chemoradiation therapy with fluorouracil in patients with resected pancreatic cancer. In this analysis, patients were divided into two groups using a cutoff score confirmed by radiation therapy quality assurance (QA) review. These were "per protocol" (PP) patients. They also defined "less than PP (<PP) patients as those exhibiting both acceptable and unacceptable deviations [44]. Median survival time of PP patients was significantly better than that of < PP patients (1.74 vs. 1.46 years, $P=0.0077$). Upon multivariate analysis of factors affecting survival, nodal status ($P=0.043$), tumor size ($P=0.0036$), and QA score ($P=0.016$) were significant. Toxicity in the fluorouracil arm was not affected by QA score. However, in the gemcitabine arm, a trend toward increased toxicity was evident in < PP patients; the toxicities were of hematologic grade 4 ($P=0.08$) and non-hematologic grades 4/5 ($P=0.065$).

In summary, studies that compare adjuvant chemotherapy with adjuvant chemotherapy combined with quality-controlled radiation therapy are needed to determine the most effective combination with surgery to control both systemic and locoregional lesions.

17.2 IMRT for Pancreatic Cancer

17.2.1 Why IMRT?

As mentioned above, despite evidence that chemoradiation can result in long-term survival in some subsets of patients with pancreatic cancer, evidence is still limited regarding the superiority of combined radiation and chemotherapy over chemotherapy alone.

One reason why a survival benefit from chemoradiation therapy cannot be demonstrated clearly is the very high rate of metastasis in pancreatic cancer. For patients with latent but rapidly progressive systemic disease, treatment with highly toxic local therapy is obviously inappropriate. The key to controlling such tumors is to reduce the toxicity and achieve a balance between local and systemic intervention. Another critical issue is that the optimal radiation dose for pancreatic cancer has not been determined. In many studies, the occurrence of normal tissue toxicity consistently limited the radiation dose that could be given, because the pancreas is surrounded by several critical organs with a very low tolerance for radiation, including the stomach, duodenum, small and large bowels, kidneys, and liver. The average radiation doses in these studies were about 50–54 Gy, but generally no more than 60-Gy doses that seem weak considering the radiosensitivity of common adenocarcinomas derived from other organs. A phase I trial conducted at the University of Michigan examined the maximum tolerated dose of radiation combined with gemcitabine [45]. This study was revolutionary in several ways. First, it combined radiation dose escalation (given as 3D conformal radiation therapy) with a fixed, full dose of gemcitabine, with goals of retaining systemic antitumor activity and exploiting the radiosensitizing effects of gemcitabine during radiation therapy. Second, it omitted elective nodal irradiation (ENI), narrowing the radiation field to the gross tumor volume (GTV). Nevertheless, gastrointestinal dose-limiting toxicity was noted at 42 Gy; hence, the recommended dose for the phase of this study was 36 Gy in 15 fractions for 3 weeks, which was still insufficient for tumor control.

Increasing the effectiveness of pancreatic cancer treatments will require the use of advanced, highly conformal irradiation techniques that can facilitate intensification of the radiation dose without enhancing normal tissue toxicity. Intensity-modulated radiation therapy (IMRT) techniques have the potential to deliver more highly conformal dose distributions to pancreatic tumors than a 3D-CRT. In one treatment plan comparison, Landry et al. compared plans for IMRT and for 3D-CRT for pancreatic cancer [46]. Planning data for 10 randomly selected patients led them to predict probability of small bowel complications of 9.3 % ± 6 % with IMRT compared with 24.4 % ± 8.9 % with 3D-CRT ($P = 0.021$) when 61.2 Gy was delivered to the GTV and 45 Gy to the clinical target volume (CTV). Fuss et al. reported that ultrasound-guided IMRT is feasible clinically and that this form of image-guided IMRT reduced the required safety margin and allowed a moderate dose escalation [47]. Eppinga et al. showed that volumetric modulated arc therapy achieved a superior conformity index compared with IMRT and led to modest reductions in the dose to organs at risk (OAR) in the dosimetric analysis [48].

17.2.2 Clinical Assessments

Clinical assessments of IMRT for pancreatic cancer are just beginning. Ben-Josef et al. retrospectively analyzed 15 patients (7 patients after curative resection and 8 with unresectable disease) treated with IMRT given concurrently with capecitabine at 1,600 mg/m^2/day, 5 days per week during the radiation period [49]. Two target volumes were set: target 1 consisted of the GTV or the tumor bed after resection and target 2 consisted of the draining lymph nodes. The total dose to target 1 was 45–54 Gy (median 54 Gy) for postoperative therapy and 54–55 Gy (median 54 Gy) for unresectable disease; the dose to target 2 was 45 Gy for all patients, and all radiation was given in 25 fractions, 5 days a week. This treatment regimen was well tolerated, with only 1 patient experiencing grade >3 toxicity. The median follow-up time in this study was 8.5 months, and no deaths occurred in the group with resectable disease. Two patients with initially unresectable disease were reclassified as having resectable disease after treatment. These results led to a phase I/II trial of IMRT dose escalation with concurrent fixed-dose-rate gemcitabine at the University of Michigan [50]. In that trial, the planning target volume (PTV) consisted of the GTV plus 1 cm and did not include the elective nodal region. The IMRT dose was delivered in 25 fractions over 5 weeks and was escalated from 50 to 60 Gy. The gemcitabine dose was 1,000 mg/m^2 over 100 min on days 1 and 8 of a 21-day cycle. One cycle of run-in chemotherapy was given before IMRT, followed by two cycles concurrently with IMRT, and four cycles of gemcitabine were recommended after IMRT. The study accrued 50 patients, and dose-limiting toxicity was observed in 11 patients. The recommended dose was 55 Gy in 25 fractions. The 2-year freedom from local progression rate in that study was 59 % (95 % CI 32–79), the median survival time was 14.8 months (95 % CI 12.6–22.2), and the 2-year survival rate was 30 % (95 % CI 17–45). Twelve patients underwent resection, ten with negative resection margins (R0), and two with microscopically positive margins (R1); the median survival time was 32 months.

Some clinical investigations examining IMRT combined with novel systemic therapies in adjuvant or neoadjuvant settings are summarized in Table 17.1 [51–54]. In another such study, Yovino et al. [55] retrospectively analyzed patterns of the first failure among 71 patients treated with IMRT for resected pancreatic cancer. At a median follow-up time of 24 months, treatment failure had occurred in 49 patients (69 %); 14 patients (19 %) had developed locoregional failure (in the tumor bed alone in 5 patients, regional nodes in 4 patients, and concurrently with metastases in 5 patients). In that study, median overall survival time was 25 months, and late small bowel obstruction occurred in four patients (6 %). Yovino et al. also evaluated acute gastrointestinal toxicity after IMRT in patients with pancreatic and ampullary cancers [56] treated similarly to those in the US intergroup trial of adjuvant chemoradiation (RTOG 97-04 [43]), in which all patients had been treated with conventional 3-D planning techniques. In that comparison, Yovino et al. found that IMRT produced significantly lower rates of grade 3–4 upper- and lower-gastrointestinal toxicity than did 3D-CRT.

Table 17.1 Clinical investigations of IMRT with novel systemic therapies in neoadjuvant and adjuvant settings

Reference and year of publication	Study design	Number of patients	IMRT dose/ fraction size, Gy	Chemotherapy	% with resected disease	Median progression-free survival time, months	Median overall survival time, months
Neoadjuvant chemoradiation therapy							
Pipas et al. [51]	Phase II	37 (12 % resectable, 70 % borderline resectable, 18 % unresectable)	54/2.0 with ENI	Before RT: cetuximab Concurrently: cetuximab gemcitabine	76 % (R0 from total resected: 92 %)		24.3
Patel et al. 2011 [52]	Retrospective	17 (100 % borderline resectable)	45–50/1.8–2.0 without ENI	Before RT: gemcitabine docetaxel capecitabine Concurrently: fluorouracil	52 % (R0 from total resected: 89 %)	10.5	15.6
Adjuvant chemoradiation therapy							
Herman et al. [53]	Phase II	48 (40 negative margins, 8 positive margins)	50.4/1.8	Concurrently: erlotinib capecitabine After RT: gemcitabine erlotinib		15.6	24.4
Abelson et al. [54]	Retrospective	29 (22 negative margins, 7 positive margins)	50.4/1.8	Concurrently: fluorouracil		N/A Rates at 1 year = 79 % 2 years = 40 %	20.4

IMRT (intensity-modulated radiation therapy), *ENI* (elective nodal irradiation), *N/A* (not applicable)

17.2.3 IMRT Techniques

17.2.3.1 Treatment Volume

The GTV generally comprises the primary tumor and any grossly involved lymph nodes (minimum diameter >1 cm) based on computed tomography (CT), magnetic resonance imaging (MRI), and (optionally) ^{18}F-fluorodeoxyglucose positron emission tomography (PET) [57]. The clinical target volume (CTV) is the area thought to harbor micrometastasis and is usually covered by a 0.5–1.5-cm margin around the GTV. The need for ENI for unresectable or neoadjuvant/borderline resectable cases is controversial, but ENI is typically used even though its benefits have not been proven in a planned prospective randomized study. Findings from some recent trials suggest that omitting ENI can both be effective and allow dose escalation of radiation or chemotherapeutic agents [45, 58, 59]. However, if the dose constraints for normal tissues permit, the retropancreatic space between the root of the celiac and superior mesenteric arteries could be reasonably included in the CTV, because microinvasion of the neural plexus in this region is common and causes severe back pain that reduces the patients' quality of life. When IMRT is used for adjuvant therapy, the CTV is the preoperative tumor bed and the region at highest risk for residual subclinical disease. Clinical studies have also involved targeting the following volumes as specific regions of interest (ROIs), according to the guidelines used to construct the 3D fields used in RTOG 9704: the celiac artery (the most proximal 1.0–1.5 cm of that artery from where it splits from the aorta and including up to its first branching), the superior mesenteric artery (the most proximal 2.5–3.0 cm of that artery from where it splits from the aorta), the portal vein, the site of pancreaticojejunostomy (may be omitted if clinically appropriate), and the aorta (from the cephalad contour of the celiac artery, and portal vein or pancreaticojejunostomy site [whichever is most cephalad] to the bottom of the L2 vertebral body). The PTV is the CTV plus a setup margin and the internal target volume (ITV). ITV is generated via management of respiratory motion, as described below.

17.2.3.2 Respiratory Motion Management

The delivery of highly conformal radiotherapy is usually problematic because the pancreas moves with respiration. According to fluoroscopic, ultrasonographic, MRI, and four-dimensional (4D) CT studies [60–67], the peak-to-peak magnitude of movement associated with respiratory motion can be as large as 20–30 mm [68]. Using cine MRI, Feng et al. [65] found that the motion of the borders of pancreatic tumors was highly variable among patients and reported that a margin of 20 mm inferiorly was required to provide 99 % geometric coverage. Mori et al. [66] used a 256-multislice 4D CT scanner and verified a displacement of the GTV by >10 mm in the superior-inferior direction.

Traditionally, radiation treatment planning for pancreatic cancer involves adding large margins around the CTV to generate the PTV [67] to account for respiratory motion during free breathing; however, the use of large margins can result in high-grade gastrointestinal toxicity [68], as the PTV often contains large volumes of organs at risk.

Respiratory motion of the pancreas is also a critical factor limiting IMRT. The interplay between motion of the multileaf collimator and motion of the target could lead to degradation of the planned dose distribution [69, 70]. Several investigators have proposed various dose delivery methods to resolve problems related to respiratory motion. The use of breath-hold (BH) techniques can reduce the impact of respiratory motion and has been used successfully in the treatment of cancer at various sites [71–76]. Dawson et al. showed the reproducibility of organ position using active breathing control during liver radiotherapy [74]. Hanley et al. applied a deep inspiration BH technique for the treatment of thoracic cancer [75]. Respiratory gating has been used for the treatment of lung and liver cancer [76–79]. Taniguchi et al. showed in a dosimetric analysis that expiratory gating might be preferable to inspiratory BH and free breathing for avoiding normal tissue toxicity [80]. Dynamic tumor tracking [81–84] has reduced the impact of respiratory motion and interfraction motion during radiation treatment. Gibbs et al. found the CyberKnife system, which includes dynamic tumor tracking, to be one of the most effective techniques for treating pancreatic tumors that move with respiration [85].

At Kyoto University, the use of a visual feedback technique during end-exhalation (EE)-BH conditions led to highly reproducible positioning of pancreatic tumors [67, 86]. IMRT in combination with EE-BH could be a reliable method to facilitate dose escalation in locally advanced pancreatic tumors with small margins. This technique and its application in the treatment of locally advanced pancreatic cancer are described in the following sections:

17.2.3.2.1 Simulation and Planning for IMRT with the End-Expiration Breath-Hold Technique: The Kyoto University Regimen

CT-Based Simulation

For CT-based treatment simulation, patients fast for at least 3 h and are positioned supine on individualized vacuum-molded pillows to ensure immobilization (BodyFIX, Medical Intelligence, Schwabmünchen, Germany) with both arms raised (Fig. 17.1). A marker block with two infrared-reflecting dots is placed on the anterior abdominal surface. A real-time position management (RPM) system (Varian Medical Systems, Palo Alto, CA) is used to monitor the abdominal skin-surface displacement in the anterior-posterior direction. Patients are given video goggles that display the extent of abdominal displacement acquired by the RPM system; the abdominal motion signal gives visual feedback to the patient. CT scans are obtained during EE-BH at a slice thickness of 2.5 mm.

IMRT Planning

The GTV includes the primary tumor, and the CTV is defined as the GTV plus a 5-mm isotropic margin. The retropancreatic space between the root of the celiac trunk and superior mesenteric artery is included in the CTV. The PTV is the CTV surrounded by another 5-mm isotropic margin. The GTV, stomach, duodenum, small intestine, liver, kidneys, and spinal cord are delineated. The planning OAR volume is determined from the BH-CT for the dose-limiting organs (the stomach, duodenum, and small intestine). The PTV is divided into two regions to satisfy the

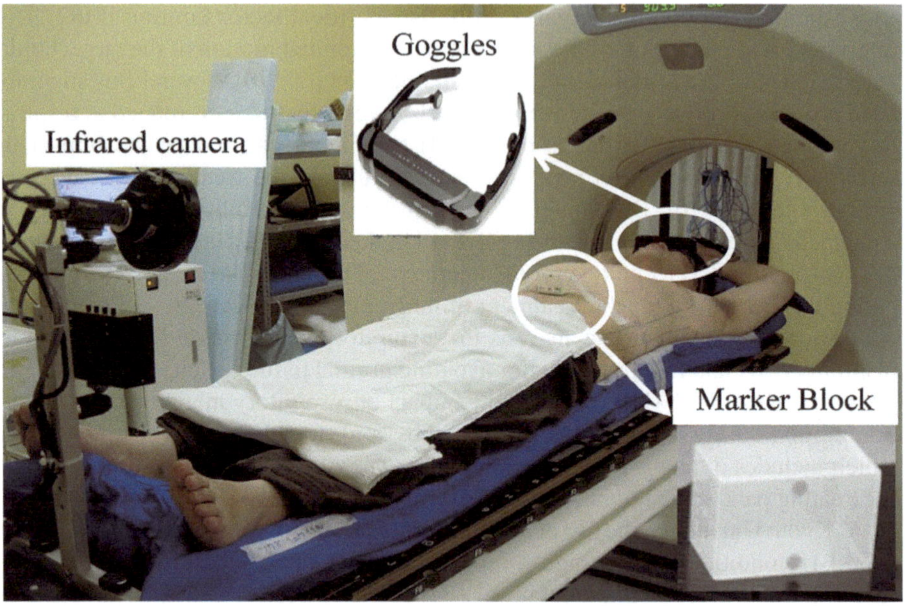

Fig. 17.1 Patient setup. All patients were placed supine on an individualized vacuum pillow with both arms raised, and an infrared-reflective marker block was placed on the anterior abdominal surface. Patients wore goggles on which the extent of abdominal displacement was displayed for visual feedback

dose constraints for the PTV and planning OAR volume, i.e., the volume after deducting the planning OAR volume from the PTV and the volume where the PTV and planning OAR volume overlap.

IMRT plans are based on the BH-CT set produced by inverse planning by a commercially available planning system (Eclipse version 8.6; Varian Medical Systems). The prescribed dose is 39 Gy in 15 fractions, based on the clinical results from McGinn et al. [45] and Allen et al. [87]. Five coplanar ports with gantry angles of 40°, 100°, 180°, 260°, and 320° are selected. Radiation is delivered with a 15-MV photon beam at a dose rate of 600 monitor units (MU)/min. The radiation dose is calculated by using the analytical anisotropic algorithm (version 8.6.15) with heterogeneity correction. The calculation grid size is 2.5 mm×2.5 mm. Dose-volume constraints are summarized in Table 17.2.

After the optimization process, the intensity fluences are converted into continuous sets of multileaf collimator segments for dynamic delivery. MUs for each port are kept at <150 MU so that the duration of the EE-BH is kept at 15 s or less, with a dose rate of 600 MU/min. Typically the IMRT technique involves using 5–7 ports to deliver the entire prescribed set of MUs, and one set of MUs in one port can be delivered during a 15-s EE-BH by using the highest dose rate. Therefore, one fraction requires a number of BHs that equal the number of ports. An IMRT dose distribution on a CT image obtained during EE-BH while the patient uses the visual feedback technique is shown in Fig. 17.2.

Table 17.2 Dose-volume constraints in intensity-modulated radiotherapy planning in the Kyoto University regimen

Structure	Dose-volume constraints
PTV	Maximum dose ≤110 %[a]
	D98% >36 Gy
PTV minus Planning OAR Volume	D95% >95 %[a]
Stomach and duodenum	V42<0.5 mL
	V39<1.0 mL
	V36<10.0 mL
Liver	Mean dose <30 Gy
Kidney	V20<30 % (in each kidney)
Spinal cord	Maximum dose <36 Gy

PTV (planning target volume), *OARs* (organs at risk), *Dxx%* (dose covering ≥ xx% of the structure's volume), *Vxx* (volume of the structure receiving > xx Gy)

[a]Dose relative to the prescription dose

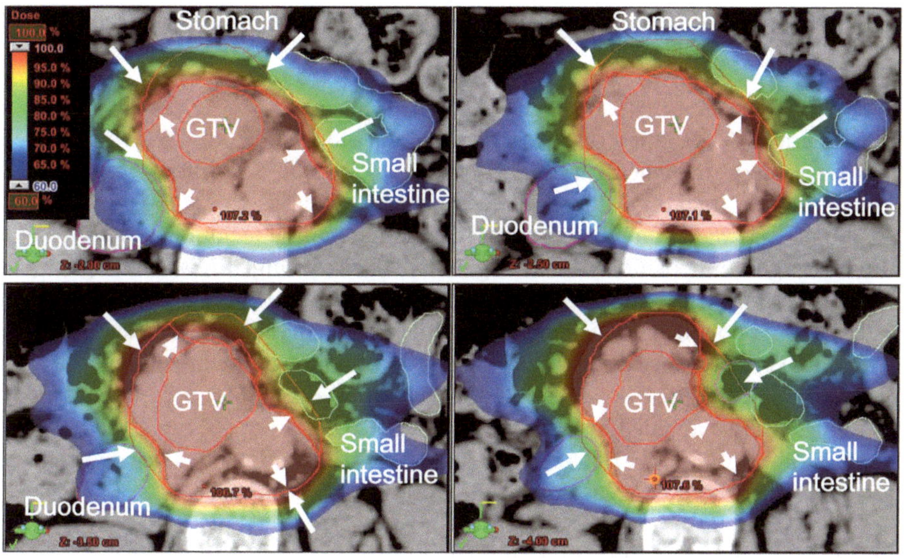

Fig. 17.2 Dose distribution on a CT scan obtained during an end-exhalation breath-hold with a visual feedback technique during IMRT, delivered using the Kyoto University regimen. The CTV is defined as the GTV plus a 5-mm isotropic margin and the retropancreatic space between the root of the celiac trunk and the superior mesenteric artery. PTV1 is the CTV plus another 5 mm of isotropic margin (long arrow). A prescribed dose of 39 Gy in 2.6-Gy daily fractions was specified for PTV2 (small arrow), which is the volume remaining after subtraction of the planning OAR volume from PTV1. The OAR volume includes the volume of the stomach plus a 10-mm isotropic margin, and the volume of the duodenum, the small intestine, and the spinal cord, plus 5-mm isotropic margins. The dose to the PTV1 was to be at least 36 Gy in 15 fractions. A dose-escalation study based on this regimen is ongoing at Kyoto University

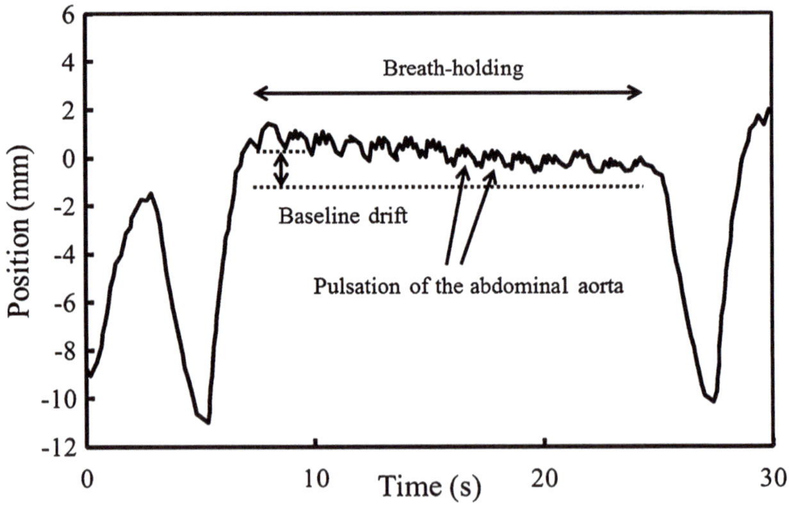

Fig. 17.3 An abdominal motion pattern as documented by an external respiratory monitoring system during end-exhalation breath-hold. Abdominal motion varied with pulsation of the abdominal aorta and with baseline drift

Quality Assurance

The American Association of Physicists in Medicine Task Group 76 recommended that respiratory management techniques such as BH, respiratory gating, and dynamic tumor tracking techniques be considered in cases where organ motion of >5 mm is observed in any direction [88]. Although the BH technique is applied to a moving target, the position of that target can still vary (Fig. 17.3). Investigators at Kyoto University evaluated the source and effects of variations in respiratory pattern under EE-BH conditions, such as pulsation from the abdominal aorta and baseline drift, on BH/IMRT dose distribution. That group found that the effect of aortic arch pulsation on BH/IMRT dose distribution was often negligible; a baseline drift of >5 mm should be avoided in using the BH/IMRT [89].

An example of the EE-BH technique used for IMRT for locally advanced pancreatic cancer is given in the case report that follows.

17.2.3.2.2 Case Presentation: Using the Kyoto University Regimen to Treat Locally Advanced Pancreatic Cancer with IMRT

A 43-year-old man presented with a 1-month history of epigastric and back pain. CT (Fig. 17.4a) and sonographic imaging showed a mass in the pancreatic head. At laparotomy, the tumor was deemed unresectable because of invasion of major vessels including the superior mesenteric artery. The diagnostic workup revealed a 4-cm pancreatic adenocarcinoma that was diagnosed as stage III (cT4N0M0). The recommended treatment was gemcitabine-based induction chemotherapy followed by consolidative chemoradiation therapy; the patient is elected to participate in a dose-escalation clinical trial involving BH/IMRT that was underway at Kyoto University Hospital.

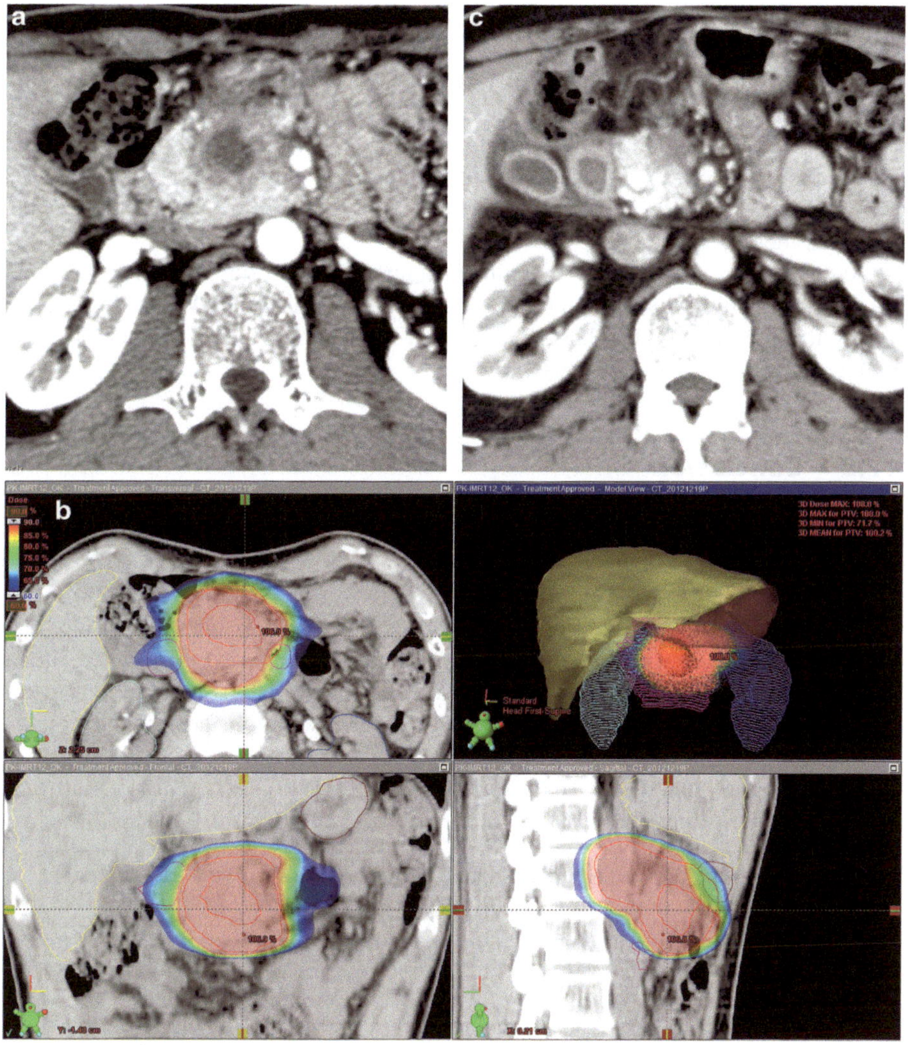

Fig. 17.4 A 43-year-old man presented with epigastric and back pain. (**a**) CT scans showed a 4-cm mass in the pancreatic head, which was diagnosed as stage III (cT4N0M0) adenocarcinoma. Breath-hold IMRT was administered concurrently with once-weekly full-dose (1,000 mg/m^2) gemcitabine. (**b**) IMRT dose distribution: the prescribed dose was 48 Gy in 3.2-Gy daily fractions for the volume after subtraction of the planning OAR volume from the PTV and 36 Gy in 2.4-Gy daily fractions for the entire PTV. (**c**) CT scans of the tumor 2 months after treatment

CT simulation and IMRT planning were done according to the Kyoto University regimen, as described in the previous section. The prescribed dose of 48 Gy in 3.2-Gy daily fractions was specified to the PTVboost region (the volume after subtracting the planning OAR volume from the PTV), and the minimum dose to the entire

PTV was to be at least 36 Gy in 15 fractions (Fig. 17.4b). Dose constraints for each OAR are shown in Table 17.2.

BH/IMRT took place concurrently with once-weekly full-dose (1,000 mg/m^2) gemcitabine. The patients fasted for at least 3 h before receiving each daily fraction. During the treatment, slight nausea and vomiting (<3 times a day) were observed during the acute phase, as was mild abdominal pain. These symptoms were controlled completely with medication, and the patient was able to complete the chemoradiation therapy as planned.

At 3 months after IMRT, the tumor diameter had shrunk by >20 % tumor (Fig. 17.4c), and tumor markers had returned to normal levels. When this chapter was written, the patient had had 12 months of maintenance systemic chemotherapy with no evidence of tumor relapse.

17.3 Conclusion

Both local and systemic progressions of pancreatic cancer are very rapid. The key to control of such tumors via radiation therapy is to reduce toxicity. This allows delivery of intensive systemic therapy and optimization of radiation doses. IMRT can potentially deliver highly conformal doses to pancreatic tumors and facilitate intensification of local therapy without enhancing normal tissue toxicities. Clinical assessment of IMRT as a treatment for pancreatic cancer is in its infancy, but several promising results have already emerged. Further investigations are expected.

Acknowledgments This writing is supported by Research Infrastructure Fund of Yamaguchi University.

Conflicts of Interest: Ultrasonographic MRI The authors declare no conflicts of interest.

References

1. Hidalgo M (2010) Pancreatic cancer. N Engl J Med 362(17):1605–1617
2. Vincent A, Herman J, Schulick R, Hruban RH, Goggins M (2011) Pancreatic cancer. Lancet 378(9791):607–620
3. Hartwig W, Werner J, Jager D, Debus J, Buchler MW (2013) Improvement of surgical results for pancreatic cancer. Lancet Oncol 14(11):e476–e485
4. Huguet F, Girard N, Guerche CS, Hennequin C, Mornex F, Azria D (2009) Chemoradiotherapy in the management of locally advanced pancreatic carcinoma: a qualitative systematic review. J Clin Oncol 27(13):2269–2277
5. Moertel CG, Frytak S, Hahn RG, O'Connell MJ, Reitemeier RJ, Rubin J, Schutt AJ, Weiland LH, Childs DS, Holbrook MA, Lavin PT, Livstone E, Spiro H, Knowlton A, Kalser M, Barkin J, Lessner H, Mann-Kaplan R, Ramming K, Douglas HO Jr, Thomas P, Nave H, Bateman J, Lokich J, Brooks J, Chaffey J, Corson JM, Zamcheck N, Novak JW (1981) Therapy of locally unresectable pancreatic carcinoma: a randomized comparison of high dose (6000 rads) radiation alone, moderate dose radiation (4000 rads+5-fluorouracil), and high dose radiation+5-fluorouracil: The Gastrointestinal Tumor Study Group. Cancer 48(8):1705–1710

6. Gastrointestinal Tumor Study Group (1988) Treatment of locally unresectable carcinoma of the pancreas: comparison of combined-modality therapy (chemotherapy plus radiotherapy) to chemotherapy alone. J Natl Cancer Inst 80(10):751–755
7. Burris HA 3rd, Moore MJ, Andersen J, Green MR, Rothenberg ML, Modiano MR, Cripps MC, Portenoy RK, Storniolo AM, Tarassoff P, Nelson R, Dorr FA, Stephens CD, Von Hoff DD (1997) Improvements in survival and clinical benefit with gemcitabine as first-line therapy for patients with advanced pancreas cancer: a randomized trial. J Clin Oncol 15(6):2403–2413
8. Okusaka T, Funakoshi A, Furuse J, Boku N, Yamao K, Ohkawa S, Saito H (2008) A late phase II study of S-1 for metastatic pancreatic cancer. Cancer Chemother Pharmacol 61(4):615–621
9. Ueno H, Ioka T, Ikeda M, Ohkawa S, Yanagimoto H, Boku N, Fukutomi A, Sugimori K, Baba H, Yamao K, Shimamura T, Sho M, Kitano M, Cheng AL, Mizumoto K, Chen JS, Furuse J, Funakoshi A, Hatori T, Yamaguchi T, Egawa S, Sato A, Ohashi Y, Okusaka T, Tanaka M (2013) Randomized phase III study of gemcitabine plus S-1, S-1 alone, or gemcitabine alone in patients with locally advanced and metastatic pancreatic cancer in Japan and Taiwan: GEST study. J Clin Oncol 31(13):1640–1648
10. Von Hoff DD, Bearss D (2002) New drugs for patients with pancreatic cancer. Curr Opin Oncol 14(6):621–627
11. Smith DB, Neoptolemos JP (2007) Capecitabine: an evidence-based review of its effectiveness in the treatment of carcinoma of the pancreas. Core Evid 2(2):111–119
12. Chauffert B, Mornex F, Bonnetain F, Rougier P, Mariette C, Bouche O, Bosset JF, Aparicio T, Mineur L, Azzedine A, Hammel P, Butel J, Stremsdoerfer N, Maingon P, Bedenne L (2008) Phase III trial comparing intensive induction chemoradiotherapy (60 Gy, infusional 5-FU and intermittent cisplatin) followed by maintenance gemcitabine with gemcitabine alone for locally advanced unresectable pancreatic cancer. Definitive results of the 2000-01 FFCD/SFRO study. Ann Oncol 19(9):1592–1599
13. Loehrer PJ Sr, Feng Y, Cardenes H, Wagner L, Brell JM, Cella D, Flynn P, Ramanathan RK, Crane CH, Alberts SR, Benson AB 3rd (2011) Gemcitabine alone versus gemcitabine plus radiotherapy in patients with locally advanced pancreatic cancer: an Eastern Cooperative Oncology Group trial. J Clin Oncol 29(31):4105–4112
14. Wilkowski R, Thoma M, Bruns C, Wagner A, Heinemann V (2006) Chemoradiotherapy with gemcitabine and continuous 5-FU in patients with primary inoperable pancreatic cancer. J Pancr 7(4):349–360
15. Sudo K, Yamaguchi T, Ishihara T, Nakamura K, Hara T, Denda T, Tawada K, Imagumbai T, Araki H, Sakai M, Hatano K, Kawakami H, Uno T, Ito H, Yokosuka O (2011) Phase II study of oral S-1 and concurrent radiotherapy in patients with unresectable locally advanced pancreatic cancer. Int J Radiat Oncol Biol Phys 80(1):119–125
16. Shinchi H, Maemura K, Mataki Y, Kurahara H, Sakoda M, Ueno S, Hiraki Y, Nakajo M, Natsugoe S, Takao S (2012) A phase II study of oral S-1 with concurrent radiotherapy followed by chemotherapy with S-1 alone for locally advanced pancreatic cancer. J Hepatobiliary Pancreat Sci 19(2):152–158
17. Ikeda M, Ioka T, Ito Y, Yonemoto N, Nagase M, Yamao K, Miyakawa H, Ishii H, Furuse J, Sato K, Sato T, Okusaka T (2013) A multicenter phase II trial of S-1 with concurrent radiation therapy for locally advanced pancreatic cancer. Int J Radiat Oncol Biol Phys 85(1):163–169
18. Mukherjee S, Hurt CN, Bridgewater J, Falk S, Cummins S, Wasan H, Crosby T, Jephcott C, Roy R, Radhakrishna G, McDonald A, Ray R, Joseph G, Staffurth J, Abrams RA, Griffiths G, Maughan T (2013) Gemcitabine-based or capecitabine-based chemoradiotherapy for locally advanced pancreatic cancer (SCALOP): a multicentre, randomised, phase 2 trial. Lancet Oncol 14(4):317–326
19. Neoptolemos JP, Stocken DD, Dunn JA, Almond J, Beger HG, Pederzoli P, Bassi C, Dervenis C, Fernandez-Cruz L, Lacaine F, Buckels J, Deakin M, Adab FA, Sutton R, Imrie C, Ihse I, Tihanyi T, Olah A, Pedrazzoli S, Spooner D, Kerr DJ, Friess H, Buchler MW (2001) Influence of resection margins on survival for patients with pancreatic cancer treated by adjuvant chemoradiation and/or chemotherapy in the ESPAC-1 randomized controlled trial. Ann Surg 234(6):758–768

20. Kuhlmann KF, de Castro SM, Wesseling JG, ten Kate FJ, Offerhaus GJ, Busch OR, van Gulik TM, Obertop H, Gouma DJ (2004) Surgical treatment of pancreatic adenocarcinoma; actual survival and prognostic factors in 343 patients. Eur J Cancer 40(4):549–558
21. Millikan KW, Deziel DJ, Silverstein JC, Kanjo TM, Christein JD, Doolas A, Prinz RA (1999) Prognostic factors associated with resectable adenocarcinoma of the head of the pancreas. Am Surg 65(7):618–623; discussion 623–614
22. Pedrazzoli S, DiCarlo V, Dionigi R, Mosca F, Pederzoli P, Pasquali C, Kloppel G, Dhaene K, Michelassi F (1998) Standard versus extended lymphadenectomy associated with pancreatoduodenectomy in the surgical treatment of adenocarcinoma of the head of the pancreas: a multicenter, prospective, randomized study. Lymphadenectomy Study Group. Ann Surg 228(4):508–517
23. Yeo CJ, Cameron JL, Lillemoe KD, Sohn TA, Campbell KA, Sauter PK, Coleman J, Abrams RA, Hruban RH (2002) Pancreaticoduodenectomy with or without distal gastrectomy and extended retroperitoneal lymphadenectomy for periampullary adenocarcinoma, part 2: randomized controlled trial evaluating survival, morbidity, and mortality. Ann Surg 236(3):355–366, discussion 366–358
24. Nguyen TC, Sohn TA, Cameron JL, Lillemoe KD, Campbell KA, Coleman J, Sauter PK, Abrams RA, Hruban RH, Yeo CJ (2003) Standard vs. radical pancreaticoduodenectomy for periampullary adenocarcinoma: a prospective, randomized trial evaluating quality of life in pancreaticoduodenectomy survivors. J Gastrointest Surg 7(1):1–9, discussion 9–11
25. Stojadinovic A, Hoos A, Brennan MF, Conlon KC (2002) Randomized clinical trials in pancreatic cancer. Surg Oncol Clin N Am 11(1):207–229
26. Farnell MB, Pearson RK, Sarr MG, DiMagno EP, Burgart LJ, Dahl TR, Foster N, Sargent DJ (2005) A prospective randomized trial comparing standard pancreatoduodenectomy with pancreatoduodenectomy with extended lymphadenectomy in resectable pancreatic head adenocarcinoma. Surgery 138(4):618–628, discussion 628–630
27. Nimura Y, Nagino M, Takao S, Takada T, Miyazaki K, Kawarada Y, Miyagawa S, Yamaguchi A, Ishiyama S, Takeda Y, Sakoda K, Kinoshita T, Yasui K, Shimada H, Katoh H (2012) Standard versus extended lymphadenectomy in radical pancreatoduodenectomy for ductal adenocarcinoma of the head of the pancreas: long-term results of a Japanese multicenter randomized controlled trial. J Hepatobiliary Pancreat Sci 19(3):230–241
28. Snady H, Bruckner H, Cooperman A, Paradiso J, Kiefer L (2000) Survival advantage of combined chemoradiotherapy compared with resection as the initial treatment of patients with regional pancreatic carcinoma. An outcomes trial. Cancer 89(2):314–327
29. Breslin TM, Hess KR, Harbison DB, Jean ME, Cleary KR, Dackiw AP, Wolff RA, Abbruzzese JL, Janjan NA, Crane CH, Vauthey JN, Lee JE, Pisters PW, Evans DB (2001) Neoadjuvant chemoradiotherapy for adenocarcinoma of the pancreas: treatment variables and survival duration. Ann Surg Oncol 8(2):123–132
30. White RR, Xie HB, Gottfried MR, Czito BG, Hurwitz HI, Morse MA, Blobe GC, Paulson EK, Baillie J, Branch MS, Jowell PS, Clary BM, Pappas TN, Tyler DS (2005) Significance of histological response to preoperative chemoradiotherapy for pancreatic cancer. Ann Surg Oncol 12(3):214–221
31. Lind PA, Isaksson B, Almstrom M, Johnsson A, Albiin N, Bystrom P, Permert J (2008) Efficacy of preoperative radiochemotherapy in patients with locally advanced pancreatic carcinoma. Acta Oncol 47(3):413–420
32. Stessin AM, Meyer JE, Sherr DL (2008) Neoadjuvant radiation is associated with improved survival in patients with resectable pancreatic cancer: an analysis of data from the surveillance, epidemiology, and end results (SEER) registry. Int J Radiat Oncol Biol Phys 72(4):1128–1133
33. Laurence JM, Tran PD, Morarji K, Eslick GD, Lam VW, Sandroussi C (2011) A systematic review and meta-analysis of survival and surgical outcomes following neoadjuvant chemoradiotherapy for pancreatic cancer. J Gastrointest Surg 15(11):2059–2069
34. Callery MP, Chang KJ, Fishman EK, Talamonti MS, William Traverso L, Linehan DC (2009) Pretreatment assessment of resectable and borderline resectable pancreatic cancer: expert consensus statement. Ann Surg Oncol 16(7):1727–1733

35. Sargent M, Boeck S, Heinemann V, Jauch KW, Seufferlein T, Bruns CJ (2011) Surgical treatment concepts for patients with pancreatic cancer in Germany–results from a national survey conducted among members of the "Chirurgische Arbeitsgemeinschaft Onkologie" (CAO) and the "Arbeitsgemeinschaft Internistische Onkologie" (AIO) of the Germany Cancer Society (DKG). Langenbeck's Arch Surg 396(2):223–229
36. Macdonald JS, Smalley SR, Benedetti J, Hundahl SA, Estes NC, Stemmermann GN, Haller DG, Ajani JA, Gunderson LL, Jessup JM, Martenson JA (2001) Chemoradiotherapy after surgery compared with surgery alone for adenocarcinoma of the stomach or gastroesophageal junction. N Engl J Med 345(10):725–730
37. Kapiteijn E, Marijnen CA, Nagtegaal ID, Putter H, Steup WH, Wiggers T, Rutten HJ, Pahlman L, Glimelius B, van Krieken JH, Leer JW, van de Velde CJ (2001) Preoperative radiotherapy combined with total mesorectal excision for resectable rectal cancer. N Engl J Med 345(9):638–646
38. Kalser MH, Ellenberg SS (1985) Pancreatic cancer. Adjuvant combined radiation and chemotherapy following curative resection. Arch Surg 120(8):899–903
39. Neoptolemos JP, Stocken DD, Friess H, Bassi C, Dunn JA, Hickey H, Beger H, Fernandez-Cruz L, Dervenis C, Lacaine F, Falconi M, Pederzoli P, Pap A, Spooner D, Kerr DJ, Buchler MW (2004) A randomized trial of chemoradiotherapy and chemotherapy after resection of pancreatic cancer. N Engl J Med 350(12):1200–1210
40. Herman JM, Swartz MJ, Hsu CC, Winter J, Pawlik TM, Sugar E, Robinson R, Laheru DA, Jaffee E, Hruban RH, Campbell KA, Wolfgang CL, Asrari F, Donehower R, Hidalgo M, Diaz LA Jr, Yeo C, Cameron JL, Schulick RD, Abrams R (2008) Analysis of fluorouracil-based adjuvant chemotherapy and radiation after pancreaticoduodenectomy for ductal adenocarcinoma of the pancreas: results of a large, prospectively collected database at the Johns Hopkins Hospital. J Clin Oncol 26(21):3503–3510
41. Corsini MM, Miller RC, Haddock MG, Donohue JH, Farnell MB, Nagorney DM, Jatoi A, McWilliams RR, Kim GP, Bhatia S, Iott MJ, Gunderson LL (2008) Adjuvant radiotherapy and chemotherapy for pancreatic carcinoma: the Mayo Clinic experience (1975–2005). J Clin Oncol 26(21):3511–3516
42. Hsu CC, Herman JM, Corsini MM, Winter JM, Callister MD, Haddock MG, Cameron JL, Pawlik TM, Schulick RD, Wolfgang CL, Laheru DA, Farnell MB, Swartz MJ, Gunderson LL, Miller RC (2010) Adjuvant chemoradiation for pancreatic adenocarcinoma: the Johns Hopkins Hospital-Mayo Clinic collaborative study. Ann Surg 17(4):981–990
43. Regine WF, Winter KA, Abrams RA, Safran H, Hoffman JP, Konski A, Benson AB, Macdonald JS, Kudrimoti MR, Fromm ML, Haddock MG, Schaefer P, Willett CG, Rich TA (2008) Fluorouracil vs gemcitabine chemotherapy before and after fluorouracil-based chemoradiation following resection of pancreatic adenocarcinoma: a randomized controlled trial. JAMA 299(9):1019–1026
44. Abrams RA, Winter KA, Regine WF, Safran H, Hoffman JP, Lustig R, Konski AA, Benson AB, Macdonald JS, Rich TA, Willett CG (2012) Failure to adhere to protocol specified radiation therapy guidelines was associated with decreased survival in RTOG 9704–a phase III trial of adjuvant chemotherapy and chemoradiotherapy for patients with resected adenocarcinoma of the pancreas. Int J Radiat Oncol Biol Phys 82(2):809–816
45. McGinn CJ, Zalupski MM, Shureiqi I, Robertson JM, Eckhauser FE, Smith DC, Brown D, Hejna G, Strawderman M, Normolle D, Lawrence TS (2001) Phase I trial of radiation dose escalation with concurrent weekly full-dose gemcitabine in patients with advanced pancreatic cancer. J Clin Oncol 19(22):4202–4208
46. Landry JC, Yang GY, Ting JY, Staley CA, Torres W, Esiashvili N, Davis LW (2002) Treatment of pancreatic cancer tumors with intensity-modulated radiation therapy (IMRT) using the volume at risk approach (VARA): employing dose-volume histogram (DVH) and normal tissue complication probability (NTCP) to evaluate small bowel toxicity. Med Dosim 27(2):121–129
47. Fuss M, Wong A, Fuller CD, Salter BJ, Fuss C, Thomas CR (2007) Image-guided intensity-modulated radiotherapy for pancreatic carcinoma. Gastrointest Cancer Res 1(1):2–11

48. Eppinga W, Lagerwaard F, Verbakel W, Slotman B, Senan S (2010) Volumetric modulated arc therapy for advanced pancreatic cancer. Strahlenther Onkol 186(7):382–387
49. Ben-Josef E, Shields AF, Vaishampayan U, Vaitkevicius V, El-Rayes BF, McDermott P, Burmeister J, Bossenberger T, Philip PA (2004) Intensity-modulated radiotherapy (IMRT) and concurrent capecitabine for pancreatic cancer. Int J Radiat Oncol Biol Phys 59(2):454–459
50. Ben-Josef E, Schipper M, Francis IR, Hadley S, Ten-Haken R, Lawrence T, Normolle D, Simeone DM, Sonnenday C, Abrams R, Leslie W, Khan G, Zalupski MM (2012) A phase I/II trial of intensity modulated radiation (IMRT) dose escalation with concurrent fixed-dose rate gemcitabine (FDR-G) in patients with unresectable pancreatic cancer. Int J Radiat Oncol Biol Phys 84(5):1166–1171
51. Pipas JM, Zaki BI, McGowan MM, Tsapakos MJ, Ripple GH, Suriawinata AA, Tsongalis GJ, Colacchio TA, Gordon SR, Sutton JE, Srivastava A, Smith KD, Gardner TB, Korc M, Davis TH, Preis M, Tarczewski SM, Mackenzie TA, Barth RJ Jr (2012) Neoadjuvant cetuximab, twice-weekly gemcitabine, and intensity-modulated radiotherapy (IMRT) in patients with pancreatic adenocarcinoma. Ann Oncol 23(11):2820–2827
52. Patel M, Hoffe S, Malafa M, Hodul P, Klapman J, Centeno B, Kim J, Helm J, Valone T, Springett G (2011) Neoadjuvant GTX chemotherapy and IMRT-based chemoradiation for borderline resectable pancreatic cancer. J Surg Oncol 104(2):155–161
53. Herman JM, Fan KY, Wild AT, Hacker-Prietz A, Wood LD, Blackford AL, Ellsworth S, Zheng L, Le DT, De Jesus-Acosta A, Hidalgo M, Donehower RC, Schulick RD, Edil BH, Choti MA, Hruban RH, Pawlik TM, Cameron JL, Laheru DA, Wolfgang CL (2013) Phase 2 study of erlotinib combined with adjuvant chemoradiation and chemotherapy in patients with resectable pancreatic cancer. Int J Radiat Oncol Biol Phys 86(4):678–685
54. Abelson JA, Murphy JD, Minn AY, Chung M, Fisher GA, Ford JM, Kunz P, Norton JA, Visser BC, Poultsides GA, Koong AC, Chang DT (2012) Intensity-modulated radiotherapy for pancreatic adenocarcinoma. Int J Radiat Oncol Biol Phys 82(4):e595–e601
55. Yovino S, Maidment BW 3rd, Herman JM, Pandya N, Goloubeva O, Wolfgang C, Schulick R, Laheru D, Hanna N, Alexander R, Regine WF (2012) Analysis of local control in patients receiving IMRT for resected pancreatic cancers. Int J Radiat Oncol Biol Phys 83(3):916–920
56. Yovino S, Poppe M, Jabbour S, David V, Garofalo M, Pandya N, Alexander R, Hanna N, Regine WF (2011) Intensity-modulated radiation therapy significantly improves acute gastrointestinal toxicity in pancreatic and ampullary cancers. Int J Radiat Oncol Biol Phys 79(1):158–162
57. Schellenberg D, Quon A, Minn AY, Graves EE, Kunz P, Ford JM, Fisher GA, Goodman KA, Koong AC, Chang DT (2010) 18Fluorodeoxyglucose PET is prognostic of progression-free and overall survival in locally advanced pancreas cancer treated with stereotactic radiotherapy. Int J Radiat Oncol Biol Phys 77(5):1420–1425
58. Small W Jr, Berlin J, Freedman GM, Lawrence T, Talamonti MS, Mulcahy MF, Chakravarthy AB, Konski AA, Zalupski MM, Philip PA, Kinsella TJ, Merchant NB, Hoffman JP, Benson AB, Nicol S, Xu RM, Gill JF, McGinn CJ (2008) Full-dose gemcitabine with concurrent radiation therapy in patients with nonmetastatic pancreatic cancer: a multicenter phase II trial. J Clin Oncol 26(6):942–947
59. Yamazaki H, Nishiyama K, Koizumi M, Tanaka E, Ioka T, Uehara H, Iishi H, Nakaizumi A, Ohigashi H, Ishikawa O (2007) Concurrent chemoradiotherapy for advanced pancreatic cancer: 1,000 mg/m2 gemcitabine can be administered using limited-field radiotherapy. Strahlenther Onkol 183(6):301–306
60. Bryan PJ, Custar S, Haaga JR, Balsara V (1984) Respiratory movement of the pancreas: an ultrasonic study. J Ultrasound Med 3(7):317–320
61. Suramo I, Paivansalo M, Myllyla V (1984) Cranio-caudal movements of the liver, pancreas and kidneys in respiration. Acta Radiol Diagn 25(2):129–131
62. Murphy MJ, Martin D, Whyte R, Hai J, Ozhasoglu C, Le QT (2002) The effectiveness of breath-holding to stabilize lung and pancreas tumors during radiosurgery. Int J Radiat Oncol Biol Phys 53(2):475–482

63. Gierga DP, Chen GT, Kung JH, Betke M, Lombardi J, Willett CG (2004) Quantification of respiration-induced abdominal tumor motion and its impact on IMRT dose distributions. Int J Radiat Oncol Biol Phys 58(5):1584–1595
64. Bussels B, Goethals L, Feron M, Bielen D, Dymarkowski S, Suetens P, Haustermans K (2003) Respiration-induced movement of the upper abdominal organs: a pitfall for the three-dimensional conformal radiation treatment of pancreatic cancer. Radiother Oncol 68(1):69–74
65. Feng M, Balter JM, Normolle D, Adusumilli S, Cao Y, Chenevert TL, Ben-Josef E (2009) Characterization of pancreatic tumor motion using cine MRI: surrogates for tumor position should be used with caution. Int J Radiat Oncol Biol Phys 74(3):884–891
66. Mori S, Hara R, Yanagi T, Sharp GC, Kumagai M, Asakura H, Kishimoto R, Yamada S, Kandatsu S, Kamada T (2009) Four-dimensional measurement of intrafractional respiratory motion of pancreatic tumors using a 256 multi-slice CT scanner. Radiother Oncol 92(2):231–237
67. Shiinoki T, Shibuya K, Nakamura M, Nakamura A, Matsuo Y, Nakata M, Sawada A, Mizowaki T, Itoh A, Hiraoka M (2011) Interfractional reproducibility in pancreatic position based on four-dimensional computed tomography. Int J Radiat Oncol Biol Phys 80(5):1567–1572
68. Langen KM, Jones DT (2001) Organ motion and its management. Int J Radiat Oncol Biol Phys 50(1):265–278
69. ICRU Report 50. Bethesda MI (1993) International Commission on Radiation Units and Measurements (ICRU). Prescribing, recording and reporting photon beam therapy
70. Murphy JD, Adusumilli S, Griffith KA, Ray ME, Zalupski MM, Lawrence TS, Ben-Josef E (2007) Full-dose gemcitabine and concurrent radiotherapy for unresectable pancreatic cancer. Int J Radiat Oncol Biol Phys 68(3):801–808
71. Bortfeld T, Jokivarsi K, Goitein M, Kung J, Jiang SB (2002) Effects of intra-fraction motion on IMRT dose delivery: statistical analysis and simulation. Phys Med Biol 47(13):2203–2220
72. Jiang SB, Pope C, Al Jarrah KM, Kung JH, Bortfeld T, Chen GT (2003) An experimental investigation on intra-fractional organ motion effects in lung IMRT treatments. Phys Med Biol 48(12):1773–1784
73. Wong JW, Sharpe MB, Jaffray DA, Kini VR, Robertson JM, Stromberg JS, Martinez AA (1999) The use of active breathing control (ABC) to reduce margin for breathing motion. Int J Radiat Oncol Biol Phys 44(4):911–919
74. Dawson LA, Brock KK, Kazanjian S, Fitch D, McGinn CJ, Lawrence TS, Ten Haken RK, Balter J (2001) The reproducibility of organ position using active breathing control (ABC) during liver radiotherapy. Int J Radiat Oncol Biol Phys 51(5):1410–1421
75. Hanley J, Debois MM, Mah D, Mageras GS, Raben A, Rosenzweig K, Mychalczak B, Schwartz LH, Gloeggler PJ, Lutz W, Ling CC, Leibel SA, Fuks Z, Kutcher GJ (1999) Deep inspiration breath-hold technique for lung tumors: the potential value of target immobilization and reduced lung density in dose escalation. Int J Radiat Oncol Biol Phys 45(3):603–611
76. Kimura T, Hirokawa Y, Murakami Y, Tsujimura M, Nakashima T, Ohno Y, Kenjo M, Kaneyasu Y, Wadasaki K, Ito K (2004) Reproducibility of organ position using voluntary breath-hold method with spirometer for extracranial stereotactic radiotherapy. Int J Radiat Oncol Biol Phys 60(4):1307–1313
77. Kubo HD, Len PM, Minohara S, Mostafavi H (2000) Breathing-synchronized radiotherapy program at the University of California Davis Cancer Center. Med Phys 27(2):346–353
78. Ohara K, Okumura T, Akisada M, Inada T, Mori T, Yokota H, Calaguas MJ (1989) Irradiation synchronized with respiration gate. Int J Radiat Oncol Biol Phys 17(4):853–857
79. Vedam SS, Keall PJ, Kini VR, Mohan R (2001) Determining parameters for respiration-gated radiotherapy. Med Phys 28(10):2139–2146
80. Taniguchi CM, Murphy JD, Eclov N, Atwood TF, Kielar KN, Christman-Skieller C, Mok E, Xing L, Koong AC, Chang DT (2013) Dosimetric analysis of organs at risk during expiratory gating in stereotactic body radiation therapy for pancreatic cancer. Int J Radiat Oncol Biol Phys 85(4):1090–1095
81. Keall PJ, Colvill E, O'Brien R, Ng JA, Poulsen PR, Eade T, Kneebone A, Booth JT (2014) The first clinical implementation of electromagnetic transponder-guided MLC tracking. Med Phys 41(2):020702

82. Lang S, Zeimetz J, Ochsner G, Schmid Daners M, Riesterer O, Klock S (2014) Development and evaluation of a prototype tracking system using the treatment couch. Med Phys 41(2):021720
83. Bahig H, Campeau MP, Vu T, Doucet R, Beliveau Nadeau D, Fortin B, Roberge D, Lambert L, Carrier JF, Filion E (2013) Predictive parameters of CyberKnife fiducial-less (XSight Lung) applicability for treatment of early non-small cell lung cancer: a single-center experience. Int J Radiat Oncol Biol Phys 87(3):583–589
84. Depuydt T, Poels K, Verellen D, Engels B, Collen C, Haverbeke C, Gevaert T, Buls N, Van Gompel G, Reynders T, Duchateau M, Tournel K, Boussaer M, Steenbeke F, Vandenbroucke F, De Ridder M (2013) Initial assessment of tumor tracking with a gimbaled linac system in clinical circumstances: a patient simulation study. Radiother Oncol 106(2):236–240
85. Gibbs IC (2006) Frameless image-guided intracranial and extracranial radiosurgery using the Cyberknife robotic system. Cancer Radiother 10(5):283–287
86. Nakamura M, Shibuya K, Shiinoki T, Matsuo Y, Nakamura A, Nakata M, Sawada A, Mizowaki T, Hiraoka M (2011) Positional reproducibility of pancreatic tumors under end-exhalation breath-hold conditions using a visual feedback technique. Int J Radiat Oncol Biol Phys 79(5):1565–1571
87. Allen AM, Zalupski MM, Robertson JM, Eckhauser FE, Simone D, Brown D, Hejna G, Normolle D, Lawrence TS, McGinn CJ (2004) Adjuvant therapy in pancreatic cancer: phase I trial of radiation dose escalation with concurrent full-dose gemcitabine. Int J Radiat Oncol Biol Phys 59(5):1461–1467
88. Keall PJ, Mageras GS, Balter JM, Emery RS, Forster KM, Jiang SB, Kapatoes JM, Low DA, Murphy MJ, Murray BR, Ramsey CR, Van Herk MB, Vedam SS, Wong JW, Yorke E (2006) The management of respiratory motion in radiation oncology report of AAPM Task Group 76. Med Phys 33(10):3874–3900
89. Nakamura M, Kishimoto S, Iwamura K, Shiinoki T, Nakamura A, Matsuo Y, Shibuya K, Hiraoka M (2012) Dosimetric investigation of breath-hold intensity-modulated radiotherapy for pancreatic cancer. Med Phys 39(1):48–54

Anal Canal Cancer

18

Bradford A. Perez, Christopher G. Willett, Brian G. Czito, and Manisha Palta

Keywords
Anal cancer • Chemoradiation • IMRT

18.1 Introduction

Historically, abdominoperineal resection (APR) was the standard of care in the management of anal cancer [1, 2]. Although this procedure resulted in cure for many patients, there were significant drawbacks: a permanent colostomy and high rates of morbidity and mortality. In 1974, Nigro et al. introduced chemotherapy and radiation as a novel treatment approach in the management of anal cancer [3]. Since Nigro's publications, phase III randomized trials have examined different strategies of chemotherapy and radiation administration [4–9] (Table 18.1). In 2014, radiation therapy with concurrent 5-fluorouracil (5FU) and mitomycin C (MMC) is the standard treatment of patients with localized squamous cell carcinoma of the anal canal. While this approach offers high rates of disease control with sphincter preservation, treatment is associated with significant acute and late toxicity. Radiation therapy is delivered to large target volumes including the anal canal, mesorectum, pelvic lymph nodes, and inguinal lymph nodes. Conventional radiation therapy using 2D or even 3D planning techniques treats large volumes of nontarget tissue which can injure the bowel, bladder, genitalia, femoral heads, and bone marrow. Acute toxicity may also lead to treatment interruption resulting in poorer outcomes. In RTOG 92-08 a scheduled 2-week rest interval permitted recovery from acute toxicity;

B.A. Perez • C.G. Willett • B.G. Czito • M. Palta (✉)
Department of Radiation Oncology, Duke University Medical Center,
Morris Clinic Building, DUMC 3085, Research Drive, NC 27710 Durham, USA
e-mail: manisha.palta@duke.edu

Y. Nishimura, R. Komaki (eds.), *Intensity-Modulated Radiation Therapy*,
DOI 10.1007/978-4-431-55486-8_18

Table 18.1 Phase III randomized conventional RT studies and phase II/institutional IMRT trials in anal cancer patients

Trial	Initial publication year	# Pts enrolled	Radiation technique	Median treatment duration (days)	Treatment breaks (%)	Median treatment break duration (days)	Acute G3+ heme toxicity (%)	Acute G3+ derm toxicity (%)	Acute G3+ GI toxicity (%)	Acute G3+ GU toxicity (%)	Colostomy free survival or colostomy rate (%)	Overall survival (%)
ACT I (5FU/MMC arm)	1996	292	2D	NA	19	NA	11	17	48	1	CFS-50 (3 YR)	65 (3 YR)
EORTC (5FU/MMC arm)	1997	51	2D	NA	NA	NA	NA	29	10 (diarrhea)	NA	CFS-72 (5 YR)	56 (5 YR) (both arms)
RTOG 87–04 (5FU/MMC arm)	1996	146	2D	NA	NA	NA	G4+ 18	NA	NA	NA	CFS-71 (4 YR)	76 (4 YR)
RTOG 98–11 (5FU/MMC arm)	2008	341	2D	49	62	9	58	45	34	3	CFS-72 (5 YR)	84 (3 YR)
ACCORD 3 (CIS/5FU, no induction arms)	2012	157	3D	NA	NA	NA	13	NA	12 (diarrhea)	NA	CFS-76 (3 YR)	71 (5 YR)
ACT II (5FU/MMC arms)	2013	472	3D	38	14	NA	26	48	16	1	CFS-74 (3 YR)	79 (5 YR)
RTOG 05-29	2013	52	IMRT (DP)	43	49	0	58	23	21	2	CFS-86 (2 YR)	88 (2 YR)
Salama et al.	2007	53	IMRT (DP or SEQ)	42	42	4	59	38	15	0	CFS-84 (1.5 YR)	93 (1.5 YR)

Pepek et al.	2010	19/45 (SCC)	IMRT (SEQ)	NA	18	5	24	0	13	2	CFS-91 (2 YR)	100 (2 YR)
Bazan et al. (IMRT group)	2011	29	IMRT (SEQ)	40	35	1.5	21	21	7	NA	CFS-86 (3 YR)	88 (3 YR)
DeFoe et al	2011	78	IMRT (DP)	NA	66	2	43	29	28	0	CFS-81 (2 YR)	87 (2 YR)
Kachnic et al.	2012	43	IMRT (DP)	NA	40	2	61	10	7	7	CFS-90 (2 YR)	94 (2 YR)
Vieillot et al.	2012	39	IMRT (DP)	NA	15	6	25	42	10	5	CFS-85 (2 YR)	89 (2 YR)
Dasgupta et al.	2013	45	IMRT (DP)	40	NA	NA	NA	NA	NA	NA	CFS-97 (2 YR)	93 (2 YR)
Mitchell et al.	2013	65	IMRT (DP)	NA	9	1.5	3	17	9	2	CR-3 (2 YR)	96 (2 YR)

ACT Anal Cancer Trial, *EORTC* European Organisation for Research and Treatment of Cancer, *RTOG* Radiation Therapy Oncology Group, *ACCORD*, *2D* 2 dimensional, *3D* 3 dimensional, *IMRT* intensity-modulated radiation therapy, *DP* dose painting, *SEQ* sequential boosts, *Heme* hematologic, *Derm* dermatologic, *GI* gastrointestinal, *GU* genitourinary, *YR* year, *SCC* squamous cell carcinoma however, results from this trial showed a higher colostomy rate (30 % 2-year colostomy rate) compared to historical controls [10]. Other studies have also shown detrimental effects on tumor control with prolonged treatment intervals [11–15]. By reducing radiation dose to normal structures, IMRT minimizes toxicities and treatment interruptions. Institutional experiences and a single phase II, multi-institution prospective study have shown the feasibility of IMRT with improvements in acute toxicity compared to historical studies. Due to the high precision of IMRT, planning and delivery of radiation therapy requires a thorough understanding of the local and regional progression patterns to define planning target volumes and surrounding normal organs. A knowledgeable treatment planning team is required to optimally utilize IMRT planning algorithms ensuring homogeneous dose to target areas while reducing dose to normal tissues. In this chapter we will highlight studies examining the use of IMRT in anal cancer and describe our therapeutic approach for patients with this malignancy.

18.2 Published Data Using IMRT

In 2005, Milano et al. published the first institutional experience of IMRT in the treatment of 17 patients with anal cancer at the University of Chicago. Patients treated with IMRT were compared to those treated with simple AP/PA plans for dosimetric and clinical outcomes comparison. IMRT reduced mean and threshold doses to the small bowel, bladder, and genitals. In the 17 patients treated with IMRT, no grade 3+ acute non-hematologic toxicity was seen, and there were no treatment breaks secondary to dermatologic or gastrointestinal toxicity [16].

Subsequently, Salama et al. published a pooled, multi-institution, retrospective analysis of 53 patients treated with IMRT and chemotherapy for anal cancer from the University of Chicago, University of Illinois, and the Mayo Clinic [17]. Median dose was 45 Gy to the pelvis and inguinal nodes and 51.5 Gy to the primary tumor and involved nodes. Fifteen percent of patients experienced grade 3 GI toxicity, and 37 % of patients experienced grade 3 dermatologic toxicity. All grade 4 toxicity was hematologic. Tumor control outcomes after IMRT were comparable to historical controls utilizing 2D or 3D techniques with 92.5 % of patients achieving complete clinical response after treatment. Since these initial reports, others have published IMRT results showing similar tumor control and colostomy-free rates comparable to historical results of studies using conventional radiation and chemotherapy. This treatment is well tolerated with favorable rates of acute dermatologic, gastrointestinal (GI), and genitourinary (GU) toxicity (Table 18.1) [18–24]. Dosimetric analyses also indicate that the use of IMRT allows for significantly lower dose to organs at risk (OARs) [25–28].

Inherently, retrospective series have limitations, especially in toxicity scoring. The most compelling data indicating that IMRT is an alternative to 2D or 3D conformal radiation therapy is from the recently published RTOG 05-29 trial. This prospective, multi-institution phase II study of 63 patients evaluated dose-painted IMRT with 5FU and MMC [29]. The primary study hypothesis was that IMRT would reduce grade 2+ GI/GU toxicity by at least 15 % compared to the 5FU/MMC arm of the RTOG 98-11 study utilizing conventional radiotherapy techniques with 5FU/MMC. The study did not show any difference with respect to the primary endpoint with grade 2+ GI/GU toxicity rates of 77 % in both RTOG 05-29 and RTOG 98-11. Importantly, the study did show lower grade 3+ GI toxicity rates (21 % vs. 36 %, $p = 0.0082$), lower grade 3+dermatologic toxicity rates (23 % vs. 49 %, $p < 0.0001$), and lower grade 2+ hematologic toxicity rates (73 % vs. 85 %, $p = 0.032$). At 2 years the rates of locoregional and colostomy failure, DFS, and OS of RTOG 05-19 were similar to disease outcomes from RTOG 98-11 [30].

An additional finding of RTOG 05-29 is the importance of real-time quality assurance processes. Eighty-one percent of the cases required at least 1 planning revision on central, rapid pretreatment plan review with 46 % of cases requiring multiple resubmissions and re-reviews [29]. In 21 % of the patients, the gross tumor volume was inaccurately delineated. Errors in contouring of the elective nodal volumes were also common especially in definition of the mesorectum, presacrum, inguinal fossa, and iliac nodes in 55 %, 43 %, 33 %, and 31 % of cases, respectively. Normal structures, including small bowel and large bowel, were inappropriately delineated in 60 % and 45 % of cases, respectively. We have outlined previously published phase III randomized studies using conventional radiation as well as retrospective series and RTOG 05-29 using IMRT to allow comparison of tumor control rates, colostomy-free survival rates, and acute toxicity rates in Table 18.1.

Toxicity with chemoradiotherapy for anal canal cancer is significant. In RTOG 98-11, rates of chronic grade 3+ toxicity were 50 % with conventional radiation and 5FU/MMC [31]. Patients followed for a median of 66 months after conventional chemoradiotherapy were reported to have significant long-term impairment of health-care-related quality of life in a Norwegian study [32]. Lower rates of acute toxicity and dosimetric analyses indicate that IMRT may also decrease rates of late toxicity; however, these data are not mature. It will be important to continue to follow patients treated with IMRT closely to assess these outcomes.

18.3 Evaluation and Staging

We recommend that patients with newly diagnosed anal cancer be evaluated in a multidisciplinary setting by a surgeon, medical oncologist, and radiation oncologist with expertise in the treatment of anal cancer [33]. Patients who are being considered for radiation therapy and chemotherapy should undergo complete physical examination including thorough clinical evaluation of inguinal lymph nodes, digital rectal examination, and proctoscopy to determine tumor extent. In females, a pelvic examination to rule out vaginal extension is indicated as well as identification of synchronous gynecological cancers (HPV related). All patients should have a biopsy to confirm invasive malignancy at the primary site. In patients with suspicious inguinal adenopathy, biopsy can be performed to clarify the diagnosis as it may alter radiation treatment volumes and doses. Axial CT imaging of the chest, abdomen, and pelvis is critical for appropriate staging. Anal cancer most commonly spreads by local and lymphatic invasion: however, in about 13 % of cases, distant metastases are present at initial presentation [34]. Distant metastases are most commonly seen in the liver and lungs [35].

An alternative to CT or MRI is PET/CT, which can be useful for assessing disease extent and often assists in treatment planning. Multiple, small studies have shown that PET/CT alters staging in about 20 % of patients compared to conventional imaging due to improved sensitivity in detection of primary tumor, involved regional lymph nodes, and distant metastases [36–40]. Winton et al. reported that initial diagnostic PET/CT altered design of radiation treatment fields in 8 of 61 (13 %) of patients with anal cancer [36]. Similarly a series of 50 patients from Australia reported 19 % of cases underwent treatment planning revision based on pelvic or nodal inguinal involvement on PET/CT [39].

18.4 Patterns of Spread

An understanding of local and lymphatic patterns of invasion is critical for accurate radiation treatment planning. Approximately 50 % of patients with squamous cell carcinoma of the anal canal will invade into the rectum and/or perianal skin [41]. Careful physical exam and attention to axial imaging as described above should rule out T4 disease with local invasion into nearby structures such as the pelvic bones, prostate, vagina, bladder, or urethra.

Lymphatic involvement is common in anal cancer with a reported 31 % incidence of regional lymphatic spread at presentation [34]. Tumors arising from the anal canal, anal verge, or perianal skin can have similar lymphatic drainage as rectal malignancies due to a rich lymphatic plexus between the anal canal and rectum. Tumors below the dentate line predominately drain to the superficial inguinal and femoral nodes [42]. For distal tumors which are deeply infiltrating or poorly differentiated, the reported rate of inguinal node metastases is 63 % [2]. Tumors within the anal canal at or above the dentate line are more likely to drain to the perirectal and internal iliac lymph nodes with rare extension to the external iliac or common iliac nodes except in advanced cases [42].

18.5 Simulation

After clinical and radiological staging, CT-based simulation is performed for radiation treatment planning. If available, PET/CT at the time of simulation may be helpful to define local and regional target structures. Patients can be simulated in the supine or prone position, and there are benefits to each approach in the appropriate clinical setting. Prone setup with a false tabletop allows for improved small bowel avoidance and may be useful in individuals with a large pannus and pelvic node involvement. Supine setup is usually more reproducible with less setup variability, potentially allowing for reduced PTV margins and smaller treatment fields. We typically simulate patients for anal cancer IMRT planning in the supine position with legs slightly abducted (frog legged) with semirigid immobilization in vacuum-locked bag or alpha-cradle. Patients are instructed to maintain a full bladder for simulation and treatment. In males, the external genitalia are typically positioned inferiorly such that setup is reproducible. In females, a vaginal dilator can be placed to help delineate the genitalia and displace the vulva, anterior vagina, and urethra away from the primary tumor. A radio-opaque marker should be placed at the anal verge, and perianal skin involvement can be outlined with radio-opaque catheters. It may be helpful to place a catheter with rectal contrast in the anal canal at the time of simulation for tumor delineation. In patients with adequate renal function, IV contrast facilitates identification of the pelvic and groin vasculature (which approximates at-risk nodal regions). Oral contrast identifies small bowel as an avoidance structure during treatment planning. For tumors involving the perianal skin or superficial inguinal nodes, bolus should be placed as necessary for adequate dosing of gross disease in these areas. The routine use of bolus may not be necessary as the tangential effect of IMRT may minimize skin sparing. In situations where adequate dosing of superficial targets is uncertain, in vivo diode dosimetry with the first treatment fraction can ensure appropriate dose at the skin surface.

18.6 Target Volume Definition

Target volume definition should be performed per ICRU 50 recommendations [43]. Gross tumor volume (GTV) should include all primary tumor and involved lymph nodes, utilizing information from physical examination, endoscopic findings, diagnostic imaging, and simulation planning study for delineation. Clinical target volumes (CTV) should include the gross tumor volume plus areas at risk for microscopic spread from the primary tumor and at-risk nodal areas. If the primary tumor cannot be determined with available information (such as after local excision), the anal canal may be used as a surrogate target. Ortholan et al. [44] published a study of 181 patients with anal cancer without inguinal nodal involvement at presentation. Seventy-five patients received elective inguinal irradiation compared to 106 patients who did not receive elective inguinal irradiation. With a median follow-up of 61 months, rates of inguinal recurrence were 2 % in patients receiving inguinal irradiation compared to 16 % in patients without inguinal irradiation. Given these recurrence rates, our general policy is that pelvic and inguinal nodes should be routinely treated in all patients. When using IMRT, a separate CTV volume for each planned treatment dose tier is contoured. Our approach has been to define three tiers: a gross disease only volume, a high-risk elective nodal volume (including gross disease), and low-risk elective nodal volume (including gross disease) [18, 19]. These volumes are determined by the presence or absence of tumor based on physical exam, biopsy, diagnostic and planning studies, and risk of nodal spread depending on tumor stage at presentation. The rationale for this approach is based on the shrinking fields technique [6, 7]. At other institutions using IMRT and in RTOG 05-29, a gross disease volume with a single elective nodal volume is used to deliver the prescribed course (dose painting) [16, 20–24, 29, 45].

In defining the gross disease CTV around the primary tumor, an approximately 2.5 cm margin around GTV should be used with manual editing to avoid muscle or bone at low risk for tumor infiltration. To define the gross disease CTV around involved nodes, a 1 cm expansion should be made beyond the contoured involved lymph node with manual editing to exclude areas at low risk for tumor infiltration. Additionally, in our practice, the entire mesorectum below the level of small bowel is included within the volume defined as gross disease CTV.

At-risk nodal regions include mesorectal, presacral, internal and external iliac, and inguinal nodes [42]. The mesorectal volume encompasses the rectum and surrounding lymphatic tissue. The presacral nodal volume is typically defined as an approximately 1 cm strip over the anterior sacral prominence. To contour the internal and external iliac nodes, our practice is to generally contour the iliac arteries and veins with approximately 0.7 cm margin (1–1.5 cm anteriorly on external iliac vessels) to include adjacent lymph nodes. In order to include the obturator lymph nodes, external and internal iliac volume contours should be joined parallel to the pelvic sidewall. The inguinal node volume extends beyond the external iliac contour along the femoral artery from approximately the upper edge of the superior pubic rami to approximately 2 cm caudad to saphenofemoral artery junction. The medial and lateral borders may be defined by the adductor longus and sartorius muscles, respectively. Several recently published atlases are helpful to review when defining elective nodal CTVs [46–48]. The above descriptions are generalizations, and each plan should be individual based on the anatomy of each patient and tumor distribution.

The high-risk elective nodal volume typically includes the gross disease CTV plus the entire mesorectum, presacral nodes, and bilateral internal and external iliac lymph nodes inferior to the sacroiliac joint. In patients with gross inguinal nodal involvement, the bilateral or unilateral inguinal nodes may be included in the high-risk elective nodal volume. The low-risk elective nodal volume should include the gross disease CTV, high-risk elective nodal CTV, as well as presacral, bilateral internal and external iliac nodes above the inferior border of the sacroiliac joint to the bifurcation of the internal and external iliac vessels at approximately L5/S1 vertebral body junction. If there is no obvious involvement of the bilateral inguinal nodes, these are included in the low-risk elective nodal volume.

The planning target volume (PTV) should account for effects of organ and patient movement and inaccuracies in beam and patient setup. PTV expansions should typically be about 0.5–1.0 cm depending on the use of image guidance and physician practice with treatment setup for each defined CTV. To account for differences in bladder and rectal filling, a more generous CTV to PTV margin is applied in these regions. These volumes may be manually edited to limit the borders to the skin surface for treatment planning purposes.

18.7 Dose Prescription

With IMRT treatment planning, doses are typically prescribed to PTVs. The dose of radiation required to control disease is extrapolated from historical studies which show excellent rates of control with concurrent radiation and chemotherapy [7, 49].

Table 18.2 RTOG 05-29 dose prescriptions according to stage and target volume

Stage	# Daily fractions	Elective nodal volume	Metastatic nodal volume (<3 cm)	Metastatic nodal volume (≥3 cm)	Primary tumor volume
T2N0	28	42 Gy (1.5 Gy/day)	NA	NA	50.4 Gy (1.8 Gy/day)
T3–4 N0–3	30	45 Gy (1.5 Gy/day)	50.4 Gy (1.8 Gy/day)	54 Gy (1.8 Gy/day)	54 Gy (1.8 Gy/day)

Typically, prescribed dose varies by the size of tumor and risk of microscopic spread in elective nodal areas. At our institution, the low-risk elective nodal PTV volume is typically prescribed 30.6–36 Gy in 1.8 Gy daily fractions. The high-risk elective nodal PTV is sequentially prescribed an additional 9–14.4 Gy in 1.8 Gy daily fractions for a total prescribed dose of 45 Gy. Finally, an additional 5.4–9 Gy in 1.8 Gy daily fractions is again sequentially prescribed to the gross disease PTV volume (total dose 50.4–54 Gy).

In RTOG 05-29, the prescription parameters are different due to the use of only a single elective nodal volume and slightly different dose prescriptions depending on tumor stage. Furthermore, the delivery of escalating dose to different target volumes was performed using a simultaneous integrated boost (SIB) dose painting technique with a maximum dose of 1.8 Gy per fraction to the primary tumor and large volume gross nodal involvement and 1.5 Gy per daily fraction to elective nodal areas. Table 18.2 outlines dose prescriptions by TNM stage according to the RTOG 05-29 protocol. The SIB approach offers the convenience of developing a single treatment plan with reduced planning complexity. The utilization of SIB dose painting is a relatively new approach in the treatment of anal cancer, and the implications of 1.5 Gy per fraction to the elective nodal region are not well studied in this disease.

18.8 Organs at Risk and IMRT Constraints

It is important to accurately define OARs so that the dose to these structures can be minimized during treatment. In anal cancer, 2D and 3D treatment planning techniques are limited in their ability to spare most pelvic normal tissues due to the location of the target. With IMRT, dose to the small bowel, bladder, pelvic/femoral bones, and external genitalia can be sculpted and minimized despite close proximity of these organs to target volumes. When contouring these structures, it is typically best to demarcate normal tissues on axial CT at least 2 cm above and below the PTV. Oral contrast is helpful to delineate small bowel. While there is significant variability in how to contour the small bowel, one approach entails contouring the entire volume of peritoneal space in which the small bowel can move. As with elective nodal volume delineation, contouring atlases offer guidance on defining organs at risk [46, 47]. Once the OARs have been identified, the chief aim of IMRT planning is to limit the dose to these structures without compromising PTV coverage.

The extent to which OARs can be avoided largely depends on the location and extent of tumor involvement at presentation as well as the extent to which the bowel extends into the lower pelvis and a given individual's anatomy.

Devisetty et al. published a multi-institution, dosimetric analysis which found that limiting the volume of bowel receiving 30 Gy (V30 Gy) to less than 450 cm^3 significantly reduced acute GI toxicity (8 vs. 33 %) [50]. Defoe et al. have published retrospective data correlating lower rates of acute GI toxicity with V30 Gy less than 310 cm^3 and V40 Gy less than 70 cm^3 [51]. To minimize late small bowel toxicity, the Quantitative Analysis of Normal Tissue Effects in the Clinic (QUANTEC) analysis reviewed available clinical data for small bowel dose and recommends minimizing volume receiving greater than 45 Gy to less than 195 cm^3 when contouring the entire potential peritoneal space [52]. Emami et al. estimate a dose of 50 Gy to 1/3 of the small bowel is associated with 5 % likelihood of obstruction or perforation at 5 years [53].

In general, the risk of grade 3+ GU toxicity is low in anal cancer (Table 18.1). There are limited clinical data correlating bladder dose–volume relationships with increased GU toxicity in anal cancer. Extrapolating from other cancer treatment sites, normal tissue complication probability models suggest that the risk of serious GU complications with bladder doses below 65 Gy is low [54]. The risk of clinical complications appears to increase with larger volumes of bladder receiving high dose. Marks et al. estimate that limiting 50 % of the bladder to less than 40–50 Gy will limit complications to less than 5–10 % (based on cervical cancer clinical literature) [55]. Despite limited available data for anal cancer, many patients experience acute grade 1–2 GU toxicity during radiation therapy, and limiting dose to the bladder, without compromising PTV coverage, may help minimize these symptoms further.

Reduction in radiation dose to the proximal femoral and pelvic bones is also important. A retrospective cohort study using Surveillance, Epidemiology, and End Results (SEER) cancer registry data linked to Medicare claims reported that in women treated for anal cancer, the cumulative 5-year pelvic fracture rate was 14 % among women receiving radiation treatment versus 7.5 % among women who did not receive radiation therapy [56]. There is limited empirical data on dose–response relationships for femoral neck complications. Emami et al. reported a tolerance dose of 52 Gy to the entire femoral neck to limit the risk of complication to less than 5 % [53]. Bedford et al. have recommended limiting the volume of femoral neck receiving 52 Gy to less than 10 % [57].

Studies also indicate that sexual dysfunction is a late effect of radiation therapy for anal cancer with significant impact on quality of life [32, 58]. Pelvic radiotherapy has been associated with high rates of impotence, sterility in young men, as well as dyspareunia, vaginal bleeding, and vaginal dryness in women. Limited data on dose–response relationships between late sexual toxicity and genitalia dose are available. As with other pelvic OARs where data are limited, it is advisable to minimize genitalia dose without compromising PTV coverage.

Given patient variation with respect to OAR position and areas of tumor involvement, practical dose constraint guidelines are challenging. In tumors without gross nodal involvement, it is often possible to limit OAR doses even further. Alternatively,

Table 18.3 RTOG 05-29 dose-painted intensity-modulated radiation therapy dose constraints for normal tissues

Organ	Dose (Gy) at <5 % volume	Dose (Gy) at <35 % volume	Dose (Gy) at <50% volume
Small bowel	45 (<20 cm^3)	35 (<150 cm^3)	30 (<200 cm^3)
Femoral heads	44	40	30
Illiac crest	50	40	30
External genitalia	40	30	20
Bladder	50	40	35
Large bowel	45 (<20 cm^3)	35 (<150 cm^3)	30 (<200 cm^3)

in tumors with gross nodal involvement within the pelvis, compromise of PTV coverage may be necessary to limit doses to the small bowel. Example normal tissue dose constraints from RTOG 05-29 are shown in Table 18.3.

18.9 Quality Assurance and Image-Guided Treatment Delivery

Due to the sophistication and complexity of IMRT planning for anal cancer, comprehensive quality assurance measures must be implemented to ensure minimal variability between the designed and delivered treatment plans. Each institution should have a quality assurance program in place for the treatment of anal cancer patients. Quality assurance measures should include the ability to generate verification plans for each field to replicate the number of monitor units and multi-leaf collimator sequence on an acrylic phantom containing a measurement device (ion chamber, diodes, film, etc.) to calculate absolute dose and dose distribution. At our institution, after the approval of each treatment plan by the prescribing physician, 2D measurements are collected using portal dosimetry to ensure less than 3 % variability between the designed and delivered plans prior to the delivery of the first treatment fraction.

The use of image guidance for radiation treatment delivery has significantly improved confidence in daily treatment setup. This has allowed for shrinking CTV to PTV expansions during the treatment planning process, which in turn further minimizes dose to OARs. In prostate cancer, the use of image guidance has been shown to improve toxicity associated with radiation therapy [59]. We often perform cone beam computed tomography (CBCT) for the first few days of radiation treatment, and then weekly as needed thereafter to ensure treatment setup positioning of bony anatomy and soft tissue structures anatomy is similar to images obtained at the time of simulation. The CBCT can then be imported into the radiation planning software where overlay of treatment isodose lines can be performed to ensure appropriate dose delivery. In addition to CBCT, onboard kilovoltage imaging is used to ensure alignment of bony anatomy. Our practice is to typically use daily image guidance to assist with treatment delivery.

18.10 Case Studies

In order to assist with IMRT treatment planning, we have included two case studies.

18.10.1 Case Study 1

A 50-year-old female patient with T3N2 squamous cell carcinoma of the anal canal was evaluated in our multidisciplinary clinic. A diagnostic PET-CT was performed and indicated FDG avidity concerning for tumor involvement in an external iliac lymph node and the anal canal (Fig. 18.1). A recommendation for combined radiation and chemotherapy with infusional 5FU and MMC was made. The patient was treated with IMRT to a primary volume including the low- and high-risk elective nodes as well as the primary tumor and involved metastatic left external iliac node using an 11-field IMRT plan to a dose of 30.6 Gy. An additional 14.4 Gy was delivered sequentially to the first boost volume including high-risk elective and involved nodes at the level and below the sacroiliac joints and the primary tumor also using an 11-field IMRT plan. The second and final boost volume was sequentially delivered with an additional 9 Gy to the primary tumor and involved left external iliac node using an 8-field IMRT plan. Representative axial CT images from the composite plan summary including the composite isodose lines are outlined in Fig. 18.2.

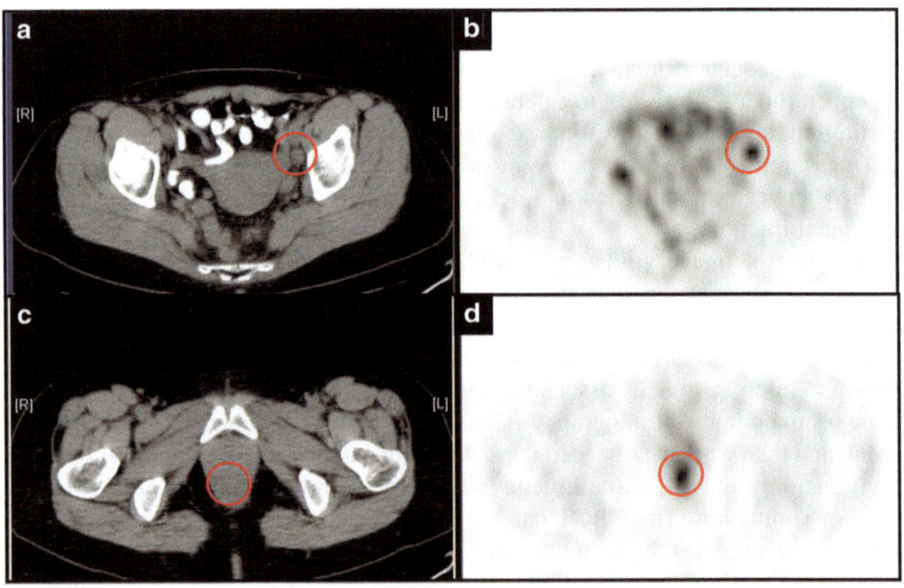

Fig. 18.1 PET-CT indicates FDG avidity involving left external iliac lymph node (**a**, **b**) and anal canal (**c**, **d**). *Red circles* indicate areas of disease involvement

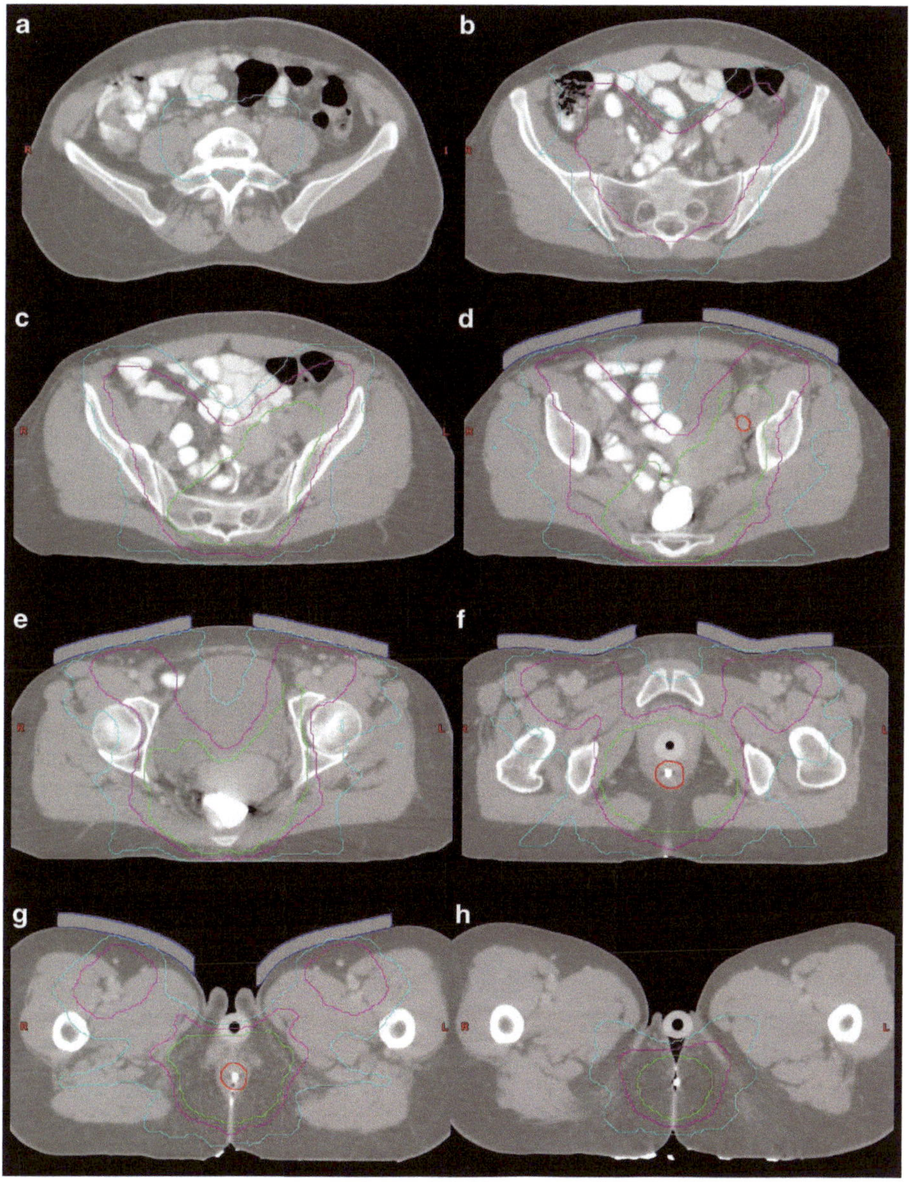

Fig. 18.2 IMRT plan delivered to a patient with squamous cell carcinoma of the anal canal involving a left pelvic lymph node. (**a–h**) Serial axial CT slices including gross tumor volume (*Red*) and summary plan isodose lines (30.6 Gy, *cyan*; 45.0 Gy, *magenta*; 54.0 Gy, *green*)

18.10.2 Case Study 2

A 60-year-old male patient with T2N2 squamous cell carcinoma of the anal canal was evaluated in our multidisciplinary clinic. Diagnostic PET-CT indicated FDG avidity concerning for malignancy in the anal canal and in a single left inguinal lymph node which was also palpable on clinical exam (Fig. 18.3). A recommendation for combined radiation and chemotherapy with infusional 5FU and MMC was made. The patient was treated with IMRT to a primary volume including the low- and high-risk elective nodes as well as the primary tumor and the involved left inguinal metastatic lymph node using a 13-field IMRT plan to a dose of 36 Gy in 1.8 Gy daily fractions. An additional 9 Gy was delivered sequentially to the first boost volume including the high-risk elective nodes inferior to the sacroiliac joint, primary tumor, and involved left inguinal node using a 13-field IMRT plan. The second and final boost volume was sequentially delivered with an additional 9 Gy to the primary tumor and involved left inguinal node using a 12-field IMRT plan. Representative axial CT images from the composite plan summary including the primary treatment course as well as the two boost courses with composite isodose lines are outlined in Fig. 18.4.

18.11 Conclusions

The primary treatment of anal cancer with combined modality radiation and chemotherapy allows for sphincter preservation in most patients. Compared to conventional 2D or 3D conformal radiotherapy, IMRT reduces toxicity rates without compromising disease-related outcomes. Published series indicate acute toxicity is

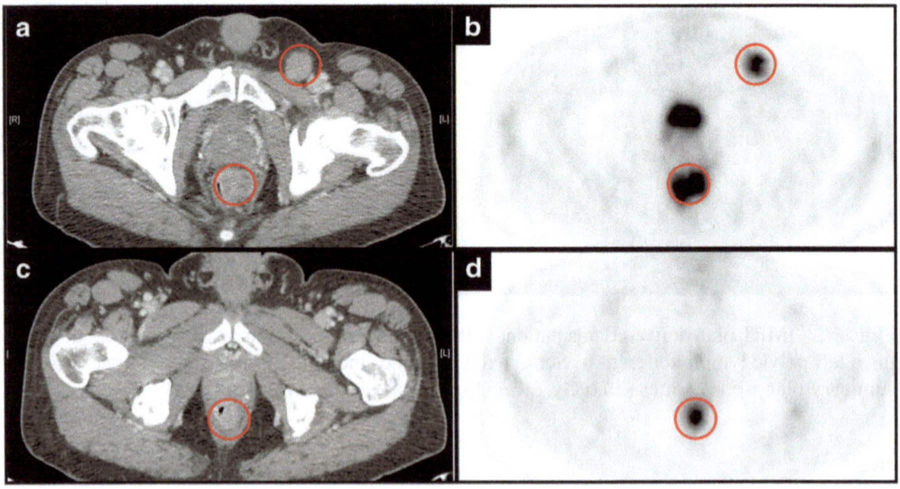

Fig. 18.3 PET-CT indicates indicates FDG avidity involving left inguinal lymph node (**a**, **b**) and anal canal (**a**–**d**). *Red circles* indicate areas of disease involvement

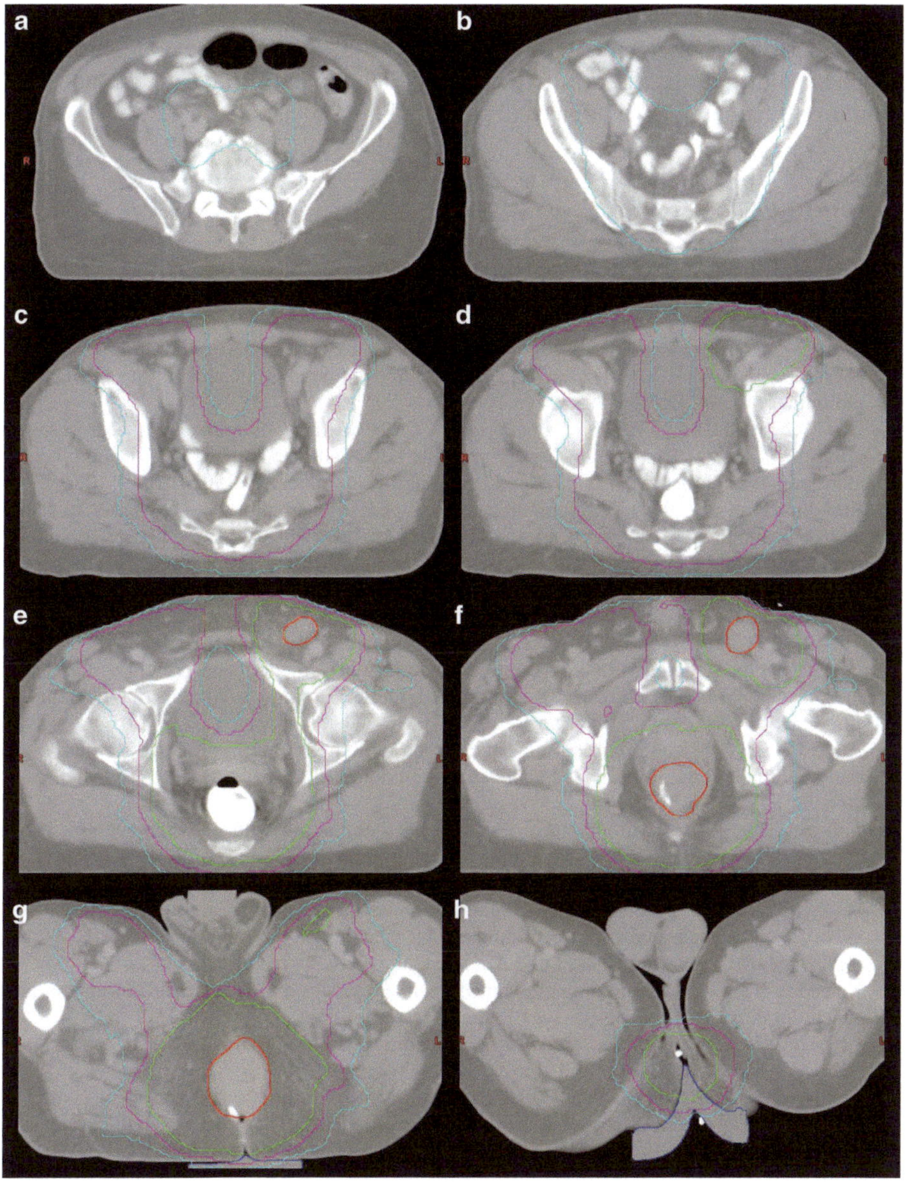

Fig. 18.4 IMRT plan delivered to a patient with squamous cell carcinoma of the anal canal involving a left inguinal lymph node. (**a–h**) Serial axial CT slices including gross tumor volume (*Red*) and summary plan isodose lines (36.0 Gy, *cyan*; 45.0 Gy, *magenta*; 54.0 Gy, *green*)

reduced with IMRT compared to historical data. Long-term analysis of late effects is necessary and forthcoming. Dosimetrically, IMRT allows for lower dose to OARs which should improve rates of both acute and late toxicity and, as a consequence, minimize treatment interruptions and potentially improve outcomes.

References

1. Klotz R, Pamukcoglu T, Souilliard D (1967) Transitional cloacogenic carcinoma of the anal canal. Clinicopathologic study of three hundred seventy-three cases. Cancer 20(10):1727–1745
2. Frost D et al (1984) Epidermoid cancer of the anorectum. Cancer 53(6):1285–1293
3. Nigro N, Vaitkevicius V, Considine B (1974) Combined therapy for cancer of the anal canal: a preliminary report. Dis Colon Rectum 17(3):354–356
4. UKCCCR Anal Cancer Trial Working Party. UK Co-ordinating Committee on Cancer Research (1996) Epidermoid anal cancer: results from the UKCCCR randomised trial of radiotherapy alone versus radiotherapy, 5-fluorouracil, and mitomycin. Lancet 348(9034):1049–1054
5. Bartelink H et al (1997) Concomitant radiotherapy and chemotherapy is superior to radiotherapy alone in the treatment of locally advanced anal cancer: results of a phase III randomized trial of the European Organization for Research and Treatment of Cancer Radiotherapy and Gastrointestinal Cooperative Groups. J Clin Oncol Off J Am Soc Clin Oncol 15(5):2040–2049
6. Flam M et al (1996) Role of mitomycin in combination with fluorouracil and radiotherapy, and of salvage chemoradiation in the definitive nonsurgical treatment of epidermoid carcinoma of the anal canal: results of a phase III randomized intergroup study. J Clin Oncol Off J Am Soc Clin Oncol 14(9):2527–2539
7. Ajani J et al (2008) Fluorouracil, mitomycin, and radiotherapy vs fluorouracil, cisplatin, and radiotherapy for carcinoma of the anal canal: a randomized controlled trial. J Am Med Assoc 299(16):1914–1921
8. James R et al (2013) Mitomycin or cisplatin chemoradiation with or without maintenance chemotherapy for treatment of squamous-cell carcinoma of the anus (ACT II): a randomised, phase 3, open-label, 2×2 factorial trial. Lancet Oncol 14(6):516–524
9. Peiffert D et al (2012) Induction chemotherapy and dose intensification of the radiation boost in locally advanced anal canal carcinoma: final analysis of the randomized UNICANCER ACCORD 03 trial. J Clin Oncol Off J Am Soc Clin Oncol 30(16):1941–1948
10. Konski A et al (2008) Evaluation of planned treatment breaks during radiation therapy for anal cancer: update of RTOG 92–08. Int J Radiat Oncol Biol Phys 72(1):114–118
11. Constantinou E et al (1997) Time-dose considerations in the treatment of anal cancer. Int J Radiat Oncol Biol Phys 39(3):651–657
12. Ben-Josef E et al (2010) Impact of overall treatment time on survival and local control in patients with anal cancer: a pooled data analysis of Radiation Therapy Oncology Group trials 87–04 and 98–11. J Clin Oncol Off J Am Soc Clin Oncol 28(34):5061–5066
13. Graf R et al (2003) Impact of overall treatment time on local control of anal cancer treated with radiochemotherapy. Oncology 65(1):14–22
14. Huang K et al (2007) Higher radiation dose with a shorter treatment duration improves outcome for locally advanced carcinoma of anal canal. World J Gastroenterol 13(6):895–900
15. Roohipour R et al (2008) Squamous-cell carcinoma of the anal canal: predictors of treatment outcome. Dis Colon Rectum 51(2):147–153
16. Milano M et al (2005) Intensity-modulated radiation therapy (IMRT) in the treatment of anal cancer: toxicity and clinical outcome. Int J Radiat Oncol Biol Phys 63(2):354–361
17. Salama J et al (2007) Concurrent chemotherapy and intensity-modulated radiation therapy for anal canal cancer patients: a multicenter experience. J Clin Oncol Off J Am Soc Clin Oncol 25(29):4581–4586

18. Pepek J et al (2010) Intensity-modulated radiation therapy for anal malignancies: a preliminary toxicity and disease outcomes analysis. Int J Radiat Oncol Biol Phys 78(5):1413–1419
19. Bazan J et al (2011) Intensity-modulated radiation therapy versus conventional radiation therapy for squamous cell carcinoma of the anal canal. Cancer 117(15):3342–3351
20. Vieillot S et al (2012) IMRT for locally advanced anal cancer: clinical experience of the Montpellier Cancer Center. Radiat Oncol (Lond) 7:45
21. DeFoe S et al (2012) Concurrent chemotherapy and intensity-modulated radiation therapy for anal carcinom--linical outcomes in a large National Cancer Institute-designated integrated cancer centre network. Clin Oncol (R Coll Radiol (G B)) 24(6):424–431
22. Kachnic L et al (2012) Dose-painted intensity-modulated radiation therapy for anal cancer: a multi-institutional report of acute toxicity and response to therapy. Int J Radiat Oncol Biol Phys 82(1):153–158
23. Dasgupta T et al (2013) Intensity-modulated radiotherapy vs. conventional radiotherapy in the treatment of anal squamous cell carcinoma: a propensity score analysis. Radiother Oncol J Eur Soc Ther Radiol Oncol 107(2):189–194
24. Mitchell M et al (2014) Intensity-modulated radiation therapy with concurrent chemotherapy for anal cancer: outcomes and toxicity. Am J Clin Oncol 37(5):461–466
25. Chen Y-J et al (2005) Organ sparing by conformal avoidance intensity-modulated radiation therapy for anal cancer: dosimetric evaluation of coverage of pelvis and inguinal/femoral nodes. Int J Radiat Oncol Biol Phys 63(1):274–281
26. Lin A, Ben-Josef E (2007) Intensity-modulated radiation therapy for the treatment of anal cancer. Clin Colorectal Cancer 6(10):716–719
27. Menkarios C et al (2007) Optimal organ-sparing intensity-modulated radiation therapy (IMRT) regimen for the treatment of locally advanced anal canal carcinoma: a comparison of conventional and IMRT plans. Radiat Oncol (Lond) 2:41
28. Wright J et al (2010) Squamous cell carcinoma of the anal canal: patterns and predictors of failure and implications for intensity-modulated radiation treatment planning. Int J Radiat Oncol Biol Phys 78(4):1064–1072
29. Kachnic L et al (2013) RTOG 0529: a phase 2 evaluation of dose-painted intensity modulated radiation therapy in combination with 5-fluorouracil and mitomycin-C for the reduction of acute morbidity in carcinoma of the anal canal. Int J Radiat Oncol Biol Phys 86(1):27–33
30. Kachnic LA, Winter KA, Myerson RJ, Goodyear MD, Willins J, Esthappan J, Haddock MG, Rotman M, Parikh PJ, Willett CG (2011) Two-year outcomes of RTOG 0529: a phase II evaluation of dose-painted IMRT in combination with 5-fluorouracil and mitomycin-C for the reduction of acute morbidity in carcinoma of the anal canal. J Clin Oncol 29(4):368
31. Gunderson L et al (2012) Long-term update of US GI intergroup RTOG 98–11 phase III trial for anal carcinoma: survival, relapse, and colostomy failure with concurrent chemoradiation involving fluorouracil/mitomycin versus fluorouracil/cisplatin. J Clin Oncol Off J Am Soc Clin Oncol 30(35):4344–4351
32. Bentzen A et al (2013) Impaired health-related quality of life after chemoradiotherapy for anal cancer: late effects in a national cohort of 128 survivors. Acta oncologica (Stockh Swed) 52(4):736–744
33. Benson A et al (2012) Anal Carcinoma, Version 2.2012: featured updates to the NCCN guidelines. J Natl Compr Cancer Netw 10(4):449–454
34. Howlader N, Noone AM, Krapcho M, Garshell J, Neyman N, Altekruse SF, Kosary CL, Yu M, Ruhl J, Tatalovich Z, Cho H, Mariotto A, Lewis DR, Chen HS, Feuer EJ, Cronin KA (eds) (2013) SEER cancer statistics review, 1975–2010. National Cancer Institute, Bethesda. http://seer.cancer.gov/csr/1975_2010/. Based on November 2012 SEER data submission, posted to the SEER web site, April 2013
35. Hoppe R, Phillips TL, Roach M (2010) Leibel and Phillips textbook of radiation oncology: expert consult. Elsevier Health Sciences, London
36. Winton E et al (2009) The impact of 18-fluorodeoxyglucose positron emission tomography on the staging, management and outcome of anal cancer. Br J Cancer 100(5):693–700
37. Cotter S et al (2006) FDG-PET/CT in the evaluation of anal carcinoma. Int J Radiat Oncol Biol Phys 65(3):720–725

38. Trautmann T, Zuger J (2005) Positron emission tomography for pretreatment staging and post-treatment evaluation in cancer of the anal canal. Mol Imag Biol Off Publ Acad Mol Imag 7(4):309–313
39. Nguyen B et al (2008) Assessing the impact of FDG-PET in the management of anal cancer. Radiother Oncol J Eur Soc Ther Radiol Oncol 87(3):376–382
40. Krengli M et al (2010) FDG-PET/CT imaging for staging and target volume delineation in conformal radiotherapy of anal carcinoma. Radiat Oncol (Lond) 5:10
41. Boman B et al (1984) Carcinoma of the anal canal. A clinical and pathologic study of 188 cases. Cancer 54(1):114–125
42. Godlewski G, Prudhomme M (2000) Embryology and anatomy of the anorectum. Basis of surgery. Surg Clin N Am 80(1):319–343
43. Jones D (1994) ICRU report 50 – prescribing, recording and reporting photon beam therapy. Med Phys 21(6):833–834
44. Ortholan C et al (2012) Anal canal cancer: management of inguinal nodes and benefit of prophylactic inguinal irradiation (CORS-03 Study). Int J Radiat Oncol Biol Phys 82(5):1988–1995
45. Brooks C et al (2013) Organ-sparing Intensity-modulated radiotherapy for anal cancer using the ACTII schedule: a comparison of conventional and intensity-modulated radiotherapy plans. Clin Oncol (R Coll Radiol (G B)) 25(3):155–161
46. Ng M et al (2012) Australasian Gastrointestinal Trials Group (AGITG) contouring atlas and planning guidelines for intensity-modulated radiotherapy in anal cancer. Int J Radiat Oncol Biol Phys 83(5):1455–1462
47. Gay H et al (2012) Pelvic normal tissue contouring guidelines for radiation therapy: a Radiation Therapy Oncology Group consensus panel atlas. Int J Radiat Oncol Biol Phys 83(3):62
48. Myerson R et al (2009) Elective clinical target volumes for conformal therapy in anorectal cancer: a radiation therapy oncology group consensus panel contouring atlas. Int J Radiat Oncol Biol Phys 74(3):824–830
49. Aggarwal A et al (2012) Clinical target volumes in anal cancer: calculating what dose was likely to have been delivered in the UK ACT II trial protocol. Radiother Oncol 103(3):341–346
50. Devisetty K et al (2009) A multi-institutional acute gastrointestinal toxicity analysis of anal cancer patients treated with concurrent intensity-modulated radiation therapy (IMRT) and chemotherapy. Radiother Oncol J Eur Soc Ther Radiol Oncol 93(2):298–301
51. DeFoe S et al (2013) Dosimetric parameters predictive of acute gastrointestinal toxicity in patients with anal carcinoma treated with concurrent chemotherapy and intensity-modulated radiation therapy. Oncology 85(1):1–7
52. Kavanagh BD et al (2010) Radiation dose-volume effects in the stomach and small bowel. Int J Radiat Oncol Biol Phys 76(3 Suppl):S101–S107
53. Emami B et al (1991) Tolerance of normal tissue to therapeutic irradiation. Int J Radiat Oncol Biol Phys 21(1):109–122
54. Viswanathan AN et al (2010) Radiation dose-volume effects of the urinary bladder. Int J Radiat Oncol Biol Phys 76(3 Suppl):S116–S122
55. Marks LB et al (1995) The response of the urinary bladder, urethra, and ureter to radiation and chemotherapy. Int J Radiat Oncol Biol Phys 31(5):1257–1280
56. Baxter NN et al (2005) Risk of pelvic fractures in older women following pelvic irradiation. J Am Med Assoc 294(20):2587–2593
57. Bedford JL et al (2000) Optimization of coplanar six-field techniques for conformal radiotherapy of the prostate. Int J Radiat Oncol Biol Phys 46(1):231–238
58. Das P et al (2010) Long-term quality of life after radiotherapy for the treatment of anal cancer. Cancer 116(4):822–829
59. Zelefsky MJ et al (2012) Improved clinical outcomes with high-dose image guided radiotherapy compared with non-IGRT for the treatment of clinically localized prostate cancer. Int J Radiat Oncol Biol Phys 84(1):125–129

Early Prostate Cancer (T1–2N0M0) 19

Michael Scott, Amber Orman, and Alan Pollack

Keywords
Risk classifications • Androgen deprivation therapy • Interfraction motion • Intrafraction motion • Image-guided treatment delivery

19.1 Introduction

The definitive treatment of early stage prostate cancer with radiation therapy has progressed dramatically over the past two decades primarily due to the development and implementation of intensity-modulated radiation therapy (IMRT) techniques and better definition of the role of androgen deprivation therapy (ADT). IMRT has permitted the escalation of radiotherapy dose to the target tissues, namely, the prostate and proximal seminal vesicles, in an effort to improve tumor cell killing and local tumor control, while also reducing dose to nearby organs at risk (OARs) including the bladder, rectum, bowel, femoral heads, and penile bulb. Advances in image-guided radiotherapy (IGRT) have improved the accuracy of the delivery of IMRT, reduced PTV margins, and consequently decreased acute and long-term side effects. This chapter will review the clinical evidence for the use of IMRT for early stage, clinically localized (T1–2N0M0) prostate cancer and will outline the processes involved in designing and implementing a safe and effective IMRT treatment plan.

M. Scott • A. Orman • A. Pollack (✉)
Department of Radiation Oncology, Sylvester Comprehensive Cancer Center, University of Miami Miller School of Medicine,
1475 NW 12th Ave, Suite 1500, Miami, FL 33136, USA
e-mail: APollack@med.miami.edu

Y. Nishimura, R. Komaki (eds.), *Intensity-Modulated Radiation Therapy*,
DOI 10.1007/978-4-431-55486-8_19

19.2 Staging/Risk Classifications

Prostate cancer is the leading cause of non-skin cancer in the US adult male population [1] and is the second most common cause of cancer mortality in American men according to the Centers for Disease Control and Prevention. The American Joint Committee on Cancer (AJCC) anatomic staging system for prostate cancer is outlined in Table 19.1 and the prognostic grouping based on TNM stage, PSA, and Gleason score in Table 19.2. Men with early stage, localized prostate cancer who are candidates for primary therapy most commonly fall into the low-risk (group I: cT1–2a, Gleason 6, PSA <10) and intermediate-risk (group IIa: cT2b and/or Gleason 7 and/or PSA 10–20) categories, which are similar to the NCCN risk groups, the main difference being that cT2c is considered high risk in the AJCC grouping (IIb) and intermediate risk in the NCCN grouping.

Table 19.1 American Joint Committee on Cancer 2010

TNM staging system for prostate cancer	
Primary tumor(T)	
TX	Primary tumor cannot be assessed
T0	No evidence of primary tumor
T1	Clinically inapparent tumor neither palpable nor visible by imaging
T1a	Tumor incidental histologic finding in ≤5 % of tissue resected
T1b	Tumor incidental histologic finding in >5 % of tissue resected
T1c	Tumor identified by needle biopsy (e.g., because of elevated PSA)
T2	Tumor confined within prostate
T2a	Tumor involves one lobe
T2b	Tumor involves more than one half of a lobe but not both lobes
T2c	Tumor involves both lobes
T3	Tumor extends through the prostate capsule
T3a	Extracapsular extension
T3b	Tumor invades seminal vesicle(s)
T4	Tumor is fixed or invades adjacent structures other than seminal vesicles
T4a	Tumor invades bladder neck, external sphincter, or rectum
T4b	Tumor invades levator muscles or is fixed to pelvic wall, or both
Regional lymph nodes (N)	
NX	Regional lymph nodes cannot be assessed
N0	No regional node metastasis
N1	Metastasis in single lymph node, ≤2 cm
N2	Metastasis in a single node, >2 cm but not >5 cm
N3	Metastasis in a node >5 cm
Distant metastasis (M)	
MX	Presence of metastasis cannot be assessed
M0	No distant metastasis
M1	Distant metastasis
M1a	Nonregional lymph node(s)
M1b	Metastasis in bone(s)
M1c	Metastasis in other sites

Table 19.2 American Joint Committee on Cancer 2010

Anatomic stage prognostic groups					
Group	T stage	N stage	M stage	PSA	Gleason
I	T1c	N0	M0	<10	≤6
	T2a	N0	M0	<10	≤6
	T1–2a	N0	M0	X	X
IIA	T1a–c	N0	M0	<20	7
	T1a–c	N0	M0	≥10, <20	≤6
	T2a	N0	M0	≥10, <20	≤6
	T2b	N0	M0	<20	≤7
	T2b	N0	M0	X	X
IIB	T2c	N0	M0	Any	Any
	T1–2	N0	M0	≥20	Any
	T1–2	N0	M0	Any	≥8
III	T3a–b	N0	M0	Any	Any
IV	T4	N0	M0	Any	Any
	Any T	N1	M0	Any	Any
	Any T	Any N	M1	Any	Any

Used with permission of the American Joint Committee on Cancer (AJCC), Chicago, Illinois. The original and primary source for this information is the AJCC Cancer Staging Manual, Seventh Edition (2010) published by Springer Science + Business Media

19.3 Treatment Options

The most established treatment options for favorable to intermediate-risk prostate cancer include radical prostatectomy, external beam radiotherapy, or brachytherapy. There are no published contemporary phase III randomized trials that have directly compared these treatment options to determine which therapy, if any, is superior with regard to outcome and toxicity, and such comparisons may not be feasible [2].

19.4 Historical Review of RT Treatment Options and Outcomes

Conventional 2D RT was initially used to treat early stage prostate cancer, but efficacy at ≤70 Gy was found to be less than previously thought when PSA became available for assessing duration of response [3]. When using 2D techniques to achieve doses >70 Gy, the genitourinary and gastrointestinal side effects increase considerably [4]. However, several retrospective analyses indicated that the delivery of doses higher than 70 Gy improved local control of the tumor [5–7]. Further technological innovations introduced CT-based 3D conformal radiotherapy (3D CRT) delivery systems. These 3D systems enabled radiation oncologists to more safely escalate radiotherapy to doses >70 Gy before reaching the upper limit of acceptable bladder and rectal toxicities [8]. An early study utilizing 3D CRT for the prostate boost portion of treatment demonstrated on DVH analysis that the volume of bladder and rectum receiving the prescribed dose could be reduced to nearly one half of the volume treated with 2D RT delivery [9]. Eade et al. described the dose response for over 1,500 men treated with 3D conformal or IMRT at Fox Chase Cancer Center from 1998 to 2002, demonstrating that there are continued gains of approximately 2.2 % freedom from biochemical failure (FFBF) for every 1 Gy increase, even beyond 80 Gy using either the nadir+2 or ASTRO definitions of biochemical failure [10].

Several randomized, controlled trials also compared higher dose radiotherapy (>70 Gy) to conventionally dosed radiotherapy (70 Gy) with respect to outcomes in men with prostate cancer [11–15]. In a randomized dose escalation trial of 301 patients, Kuban et al. found an improvement in freedom from biochemical or clinical failure at a median follow-up of 8.7 years of 78 % in the group receiving 78 Gy compared to 59 % in patients receiving 70 Gy [11]. However, gastrointestinal toxicity of grade 2 or greater occurred in 26 % of patients in the high-dose 78 Gy arm compared to 13 % in the conventional 70 Gy arm in treatment delivered by 3D CRT. A meta-analysis that included 7 trials found a significant FFBF benefit for dose escalation when all of the trials ($p<0.0001$) were considered [16]. This benefit was significant for NCCN low-risk ($p=0.007$), intermediate-risk ($p<0.0001$), and high-risk ($p<0.0001$) groups, although no difference was found for overall survival ($p=0.69$) or disease-specific survival ($p=0.41$). All of the trials had less than 10 years of follow-up, which is not sufficient for survival endpoints.

19.5 Clinical Evidence for IMRT

While the 3D CRT prostate cancer dose escalation trials were accruing, intensity-modulated radiotherapy (IMRT) was introduced, revolutionizing the radiation treatment of numerous tumor types, including prostate cancer. In 1996, Ling et al. published one of the first descriptions of prostate cancer inversely planned intensity-modulated photon beams using dynamic multi-leaf collimation [17]. The ability to collimate the treatment fields in a dynamic way during the delivery of radiation resulted in more conformal dose distributions with the potential to reduce toxicity.

When compared directly to 3D CRT plans in dosimetric studies, IMRT proved superior in target volume coverage at the prescribed dose and at reducing the volume of normal tissues treated to specified constraint doses [18–22]. Ailleres reported that 95 % of the PTV1 received 5 Gy more with IMRT when compared to 3D CRT planning without compromising dose limits on the bladder and rectal walls [23]. Using an endorectal balloon for prostate immobilization, Ashman compared sequential IMRT plans to 3D CRT in delivering whole pelvic radiotherapy and found that IMRT reduced the volume of bowel receiving 45 Gy by 60 % when compared to 3D CRT delivery [21]. Without placing intentional constraints on the penile bulb dose, Kao designed IMRT and 3D CRT plans to deliver 74 Gy to the target for ten patients with clinically organ-confined prostate cancer; IMRT reduced the mean penile bulb dose, the percentage of bulb receiving >40 Gy, and the dose received by >95 % of the penile bulb; however, the maximum penile bulb dose was higher to a very small volume [22]. Subsequent studies have confirmed the reduction in dose to critical structures including the penile bulb with IMRT while achieving improved coverage to target structures including the pelvic lymph nodes [24, 25].

Investigators at Memorial Sloan Kettering Cancer Center were leaders in the application of IMRT to prostate cancer. Zelefsky et al. [26] described the early experience where 61 men with localized disease were treated with 3D CRT and 171 with IMRT to a prescribed dose of 81 Gy. When comparing the plans of 20 randomly selected patients, IMRT showed a significant improvement in the coverage of the clinical target volume (CTV) to the prescribed dose while lowering the volume of the bladder and rectal walls that received this dose. The analysis also demonstrated a reduction in the 2-year risk of rectal bleeding from 10 % with 3D CRT to 2 % for IMRT [26]. In a larger analysis, Zelefsky analyzed 1,100 patients with cT1c–T3N0 prostate cancer treated with 3D CRT or IMRT and reported statistically significant improvements in the 5-year PSA relapse-free survival in the favorable, intermediate, and unfavorable risk groups. For the patients who received a prescription dose of 81 Gy, IMRT resulted in significantly less late grade 2 rectal toxicity (14 % vs. 2 %) with no impact on GU toxicity [27]. In the Dutch randomized dose escalation trial CKVO 96-10, a subset analysis comparing the toxicities for patients who received 78 Gy via sequential 3D CRT or via simultaneous integrated boost, with IMRT, IMRT resulted in significantly lower acute grade 2+ GI toxicity (61 % vs. 20 %), while the rates of 5-year FFBF were comparable in both groups (61 % vs. 70 %). Acute GU and late GI and GU toxicities were similar in both groups [28]. Several other reports confirmed that IMRT is a safer delivery method for high-dose RT to the prostate [18, 23, 29], which eventually resulted in the adoption of IMRT as the standard of care.

IMRT is associated with an increased volume of low-dose radiation to normal tissues as a result of the multiple gantry angles by which IMRT is delivered to create the highly conformal dose distribution to target structures. Kry reported that IMRT required 3.5–4.9 times more monitor units (MU) when compared to more conventional treatments. According to the analysis, the calculated conservative maximal lifetime risk of fatal radiation-induced malignancy was 1.7 % for conventional RT, 2.1 % for IMRT with 10-MV X-rays, and 5.1 % for IMRT using 18-MV X-rays [30]. There are conflicting data on the observed risk of second malignancies after prostate cancer radiotherapy [31–35], and the overwhelming majority of clinical and dosimetric evidence support IMRT as the standard of care over 3D CRT.

19.6 IMRT Technique

Close collaboration between physicians, nurses, radiation therapists, dosimetrists, and medical physicists facilitates coordination that can reduce uncertainties in IMRT treatment delivery. In our institution, prior to diagnostic/planning MRI and CT simulation, the patient receives instructions on bowel and bladder preparation. Table 19.3 outlines the simulation technique and sequence that is utilized at the University of Miami.

All patients who are candidates for an MRI with contrast undergo diagnostic 3.0 T MRI of the prostate and pelvis using a body coil. Multiparametric MRI that includes T2w, T1 non-contrast, T1 dynamic contrast-enhanced (DCE)-MRI,

Table 19.3 Simulation technique and sequence

1. Diagnostic MRI. A 3.0 T MRI using a pelvic body coil is acquired when possible with the following sequences: T2, T1 plain, and T1 with dynamic contrast enhancement (DCE) and diffusion-weighted imaging (DWI) in 2–2.5 mm slices throughout the pelvis. This is a diagnostic staging exam that is fused to the planning CT. The same bowel preparation is recommended for the MRI and CT simulation.
   (a) Bowel prep: the patient is encouraged to abstain from drinking carbonated beverages and gas-producing foods starting the day before imaging. The evening before, magnesium citrate is consumed to empty the bowel. An enema or two are administered within 2 h of imaging.
2. Placement of four gold fiducial markers (tracking beacons or gold seed fiducials) transrectally.
3. Wait a minimum of 7 days for fiducial marker position to stabilize.
4. On our clinical trials, we have also performed a second limited 2–2.5 mm slice thickness MRI simulation using T2w, T2*, and DWI sequences. An MRI after fiducial marker placement aids in aligning the MRI and CT simulation images. This is not usually reimbursed when attempted outside of a clinical trial, but is very helpful in image fusion.
5. CT simulation at 2.0–2.5 mm slices throughout the pelvis in supine position with legs in an immobilization device.
6. CT-MR fusion.

and diffusion-weighted imaging (DWI) sequences (our standard sequences), as well as MR proton spectroscopy (MRS), has been shown to improve the sensitivity and specificity of tumor localization [36–39]. T2w-MRI provides an excellent depiction of prostate and pelvic anatomy with regions of healthy peripheral zone prostate tissue demonstrating higher signal intensity than prostate cancer. The observed reduction in MRI image signal intensity is due to a loss of the normal glandular (ductal) morphology in regions of prostate cancer. However, other benign pathologies such as inflammation, benign prostatic hyperplasia (BPH), blood, and prior radiation treatment also cause a loss of ductal morphology and low signal intensity on T2w-MRI. Additionally, infiltrating prostate cancer does not always cause a reduction in normal glandular morphology and therefore may not be hypointense on T2w-MRI. Due to these confounding factors, T2w-MRI alone can localize cancer larger than 0.5 cm^3 in volume with only a 65–74 % sensitivity and low specificity [40]. Other studies also report quite variable sensitivity (50–83 %) and specificity (21–88 %). Utilizing an endorectal coil improves MRI's sensitivity to 78 %, but the specificity still remains poor (55 %).

MRS, DWI, and DCE-MRI have been used to improve the sensitivity and specificity of MRI and to determine tumor location and extent. Localization accuracies of above 80 % may be achieved by combining these methods [41]. Each has shown promise when combined with T2w-MRI. DWI is sensitive to random thermal movement of water molecules and provides a determination of stiffness. DWI is used to calculate apparent diffusion coefficient (ADC) values, which are significantly lower in tumor than in normal prostate due to the restriction of water displacement. Lower ADC values are associated with Gleason 7 or above disease [42].

DCE-MRI also discriminates between normal and cancerous tissues [43, 44]. Greater and earlier enhancement with washout over time is seen in tumor tissue vs. delayed enhancement in normal tissue. DCE-MRI is a measure of tissue vascularity and angiogenesis. Specificity has been notably high at close to or over 90 % in a number of studies [45, 46], while sensitivity has been 65 to >80 %. At the University of Miami, we routinely use DCE-MRI combined with DWI (and ADC maps) and T2w-MRI in our assessments and planning of radiotherapy. These sequences aid in the determination of tumor location and extraprostatic extension, thereby informing the CTV. In addition, there is an emerging literature on delivering higher daily doses using dose painting to the dominant prostatic lesion(s) based on MRI criteria [47–54], and we are testing this approach in clinical trials (Fig. 19.1). In some high-risk patients with well-defined MRI lesions, we have boosted the dominant intraprostatic lesion(s) slightly (2.15 Gy per fraction to 86 Gy) if not a candidate for a clinical trial.

Patients are given bowel preparation and bladder filling instructions for the diagnostic MRI to mimic treatment bladder and rectal filling. If possible, the MRI is performed on a flat tabletop to most accurately reproduce the patient's treatment position. After the MRI, 3–4 gold fiducial markers are implanted transrectally in the prostate at a distance of 1 cm apart [55], to reduce interfraction setup error [56–59]. Seed migration is not a major issue [57, 58, 60, 61]. We typically wait a week after fiducial marker placement before simulation to allow for the stabilization of fiducial position.

Gold seed fiducials are currently the most commonly employed method to correct for interfraction motion. Although patients do require an extra outpatient procedure, they tolerate the insertion well [62]. The most frequent reported side effects are hematuria (15 %) and rectal bleeding (4 %) [63]. Patients report their pain is less than that of diagnostic biopsy, even without local anesthesia. Daily fiducial marker alignment prior to each treatment via orthogonal portal imaging, cone beam CT, or

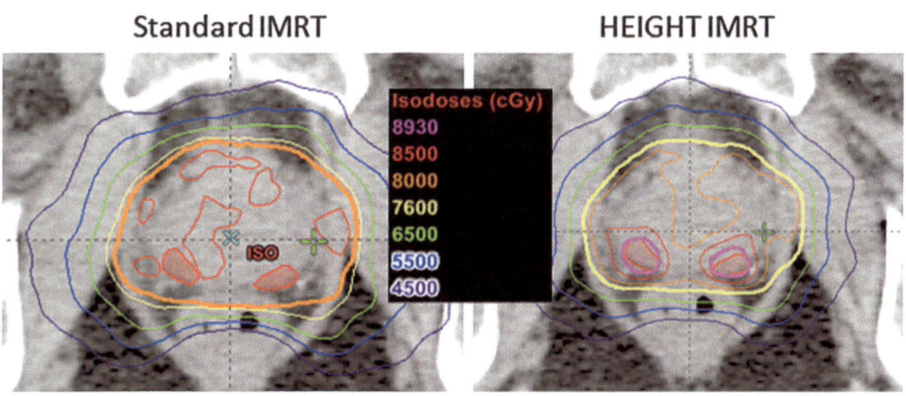

Fig. 19.1 Dose distribution on a single axial slice for a dose-painting IMRT plan prescribing a boost to dominant intraprostatic lesion(s) on multiparametric MRI for patient on an inhouse (UM) clinical trial

in-room CT reduces uncertainties from interfraction motion [56], allowing for reduced margins [58, 59, 64, 65].

Daily alignment to gold seed fiducials reduces the subjectivity of the alignment process over cone beam CT (CBCT) alone. With fiducial correction for interfraction motion and arc-based therapy that typically takes <5 min (2–3 arcs), we use a PTV expansion of 5 mm everywhere, except posteriorly where the margin is 3–4 mm. Langen et al. have shown that the chance for prostate displacement increases over time during patient setup [66]; arc-based therapy reduces the chance of a portion of the CTV falling outside the PTV.

Real-time tracking of fiducials using fluoroscopy, electromagnetic transponders, or transperineal ultrasound [67] limits effects from intrafraction motion, which takes on considerable importance. When using transponders, three are placed in the prostate to define prostate position [68]. The system continuously monitors and compares the triangulated center of mass of the transponders to the planned isocenter. This information is used to stop the treatment if a predetermined threshold is exceeded [69]. The threshold that has been used in our institution is 3 mm. Of note, transponders result in significant artifacts on MRI that preclude assessments of tumor location or recurrence in the prostate.

After waiting 7 days to allow for fiducial marker position to stabilize, CT simulation with 2 mm slices is performed throughout the pelvis in supine position with arms on chest and legs supported or in an immobilization device such as a Vac-Lok™. On the same day, a limited 2.5 mm slice thickness MRI simulation using T1, DWI, T2*, and T2w sequences has been used for patients on protocols in which a GTV boost is administered in order to facilitate more accurate MRI-CT fusion based on the position of the implanted markers. While the simulation MRI is of value in fusing the diagnostic MRI to the planning CT, it is not usually reimbursed.

Bladder and rectal volumes should be thoroughly analyzed prior to approval of the CT simulation for IMRT planning. At our institution, we have a low threshold for repeating the CT simulation after the patient has adjusted the bladder filling or rectal emptying if these normal organs do not conform to our preferred volume and anatomic orientation. The smaller the rectal volume at simulation, the more the plan will reflect a worst case dosimetric assessment during planning.

19.7 Contouring of Target and Normal Anatomy

The IMRT planning process begins with the fusion of the CT and MRI images followed by delineation of the target structures and normal tissue volumes by the treating radiation oncologist. As background, the prostate is an exocrine gland that surrounds the prostatic urethra. The normal adult gland measures approximately 4×3×3 cm (transverse, AP, craniocaudal) and weighs 15–20 g [70]. The size of the prostate generally increases with age, and according to Zackrisson et al., up to 85 % of healthy males older than 40 years have prostate volumes higher than 20 cm^3 [71]. Five anatomical prostatic zones are recognized: (1)

anterior fibromuscular stroma, (2) periurethral glandular tissue, (3) transition zone, (4) central zone, and (5) peripheral zone [70].

Inconsistencies in CT contouring are the result of poor definition of the prostate relative to adjacent structures and wide variation in anatomic position relative to the pelvic bones [72]. The prostate tends to blend in with muscle, making it unclear where the prostate borders are located relative to the levator ani muscle, the rectum (particularly inferiorly near the prostatic apex), and the bladder wall superiorly. The inferior aspect of the prostatic apex is not recognizable on CT. The rectum blends into the apex posteriorly, and the urogenital diaphragm blends into the apex inferiorly. These structures are more clearly visualized on T2w-MRI [73]. Using CT alone, the apex location may be estimated to be about 1.0–1.5 cm superior to the bulb of the penis. It is best to overestimate the inferior location of the apex in the absence of MRI. MRI considerably resolves these boundaries and makes accurate contouring of the prostate more consistent [74, 75]. MRI prostate volumes are more aligned with ultrasound volumes [76]; CT prostate volumes are about 30–40 % larger than MRI volumes [77, 78]. However, current planning algorithms are based on CT, making CT-MR fusion the best approach to define the prostate, seminal vesicles, and pelvic lymph node regions.

CT-MR fusion is fraught with potential problems that could lead to significant errors if not performed appropriately. The CT should be used primarily and the MRI only used as a reference because of the inherent problems with the accuracy of CT-MR fusion. The position of the prostate on MRI may be substantially different from the CT because of bladder and rectal filling. These differences are accentuated when the MRI is being performed at a different time (not in sequence with the CT) and on a concave tabletop (instead of a flat table on the CT simulator). Patients should be given instructions on diet to minimize gas and should perform an enema before going for both the MRI and CT. Ideally, a radiation simulation therapist will be present at the MRI to confirm optimal bladder and bowel filling at the time of MRI. An MRI equipped with lasers to ensure accuracy of patient positioning will also help to optimize the subsequent CT-MR fusion. If there are considerable differences in bladder and rectal filling, the fusion will be inaccurate. The random error for CT-MR registration along the three spatial directions was estimated to be on the order of 0.5 mm and around 0.4° in rotation (standard deviation) for each axis [79]. Fusion error is minimized when a second limited MRI simulation (T1, DWI, T2, and T2* MRI sequences) with fiducials in place on the same day of CT simulation is performed and the fiducial markers are considered in the fusion process [54].

The seminal vesicles (SV) are paired organs located in the connective tissue lodged between the urinary bladder and the rectum lateral to the ampulla of the vas deferens. Seminal vesicles can vary in size, and differences in dimension between the right and left seminal vesicle have been reported [80, 81]. According to surgical specimen reports, the length is about 31 ± 10.3 mm [80], which is concordant with reported results on ultrasound [71]. The angle between the seminal vesicles and the horizontal plane (normally 50 to 60°) changes with bladder and rectal filling [82]. Seminal vesicle contouring, similar to that of prostate contouring, is better delineated on MRI because of enhanced anatomic detail [83]. We include the proximal SVs in the CTV with the prostate.

The bulb of the penis is formed by the elongation of the corpus spongiosum after the separation of the corpora cavernosa to form the crura of the penis. The bulb of the penis is attached superiorly to the inferior surface of the urogenital diaphragm [84]. The penile bulb is best visualized on T2w-MRI as an oval-shaped, hyperintense midline structure [25, 84]. Although the penile bulb can also be identified on CT imaging and transverse transrectal ultrasound, MRI is best for the superior and inferior aspects. Contouring should stop inferiorly when the bulb loses the lateral bulging aspects of the corpus spongiosum. As summarized by Van der Wielen et al. [85], the sparing of the penile bulb, corporal bodies, and neurovascular structures has sometimes been associated with increased preservation of erectile function, but results have been mixed [25, 86]. IMRT reduces the dose received by the penile bulb, as shown by some authors and reported above [22, 87]. Regarding the relationship between the penile bulb dose and the development of erectile dysfunction, some studies have not shown any significant association [88], whereas Merrick et al. have [89]. Roach et al. advocate keeping the penile bulb dose to <52.5 Gy [90].

19.8 IMRT Planning and Dosimetry

Table 19.4 outlines common dose constraints utilized in IMRT planning for prostate cancer. Figure 19.2 demonstrates a typical IMRT treatment plan with corresponding dose-volume histograms when prescribing 80 Gy in 40 fractions to the PTV, the standard dose used at the University of Miami. At the University of Miami, particular attention is paid to the 30 Gy isodose line, and optimization structures are created to ensure that this line is anterior to the posterior rectal wall, as well as excluding the lateral rectal wall as much as possible (Fig. 19.3). The bladder volume at 30 Gy is also reduced by designing a plan that pulls in dose tightly in the anterior. The resultant optimization of dose across the rectal and bladder is accomplished using optimization structures that result in an increase in doses to the lateral soft tissues and femoral heads while still maintaining constraints to the femoral heads (Fig. 19.4).

19.9 IMRT Delivery Methods

Two common treatment delivery techniques include standard fixed gantry IMRT, otherwise known as step-and-shoot, and volumetric modulated arc therapy (VMAT). The step-and-shoot technique results in increased treatment time compared to VMAT as the gantry must be repositioned between the deliveries of each small, irregularly shaped field. The VMAT technique typically requires 2–3 arcs for the delivery of one standard daily fraction. VMAT reduces the overall treatment time on average by 1.5–3 min for each 2 Gy fraction by delivering the dose during a continuous gantry rotation and also reduces the calculated monitor units by

Table 19.4 Common dose constraints for prostate IMRT

Dose (Gy) or dose-volume parameters			
Organ	UM: 80 Gy/40 fx	RTOG 0126 arm 2: 79.2 Gy/44 fx	MSKCC: 86.4 Gy/48 fx
PTV	V80 Gy ≥95 %	V79.2≥98 %	V86.4≥95 %
	V92 Gy(115 %)=0 %	V84.7≤2 %	V95.04=0 %
Bladder	V65 Gy<25 %	V80≤15 %	Bladder wall
	V40 Gy <50 %	V75≤25 %	V47<53 %
		V70≤35 %	V75.6<30 %
		V65≤50 %	
Bowel	V40 Gy<150 cm³		V60=0 %
Femoral heads	V50 Gy=0 %	V50=0 %	
Penile bulb		Mean≤52.5	
Rectum/anus	V65 Gy <17 %	V75≤15 %	Rectal wall
	V40 Gy <35 %	V70≤25 %	V47<53 %
		V65≤35 %	V75.6<30 %
		V60≤50 %	V85.5=0 %

PTV(UM) prostate+proximal seminal vesicles (extending 10 mm from the origin) with a 3 mm margin posteriorly and a 5 mm margin in all other directions; bowel is contoured as a bowel bag and includes both small and large bowel proximal to rectum; rectum/anus includes the anus at the level of the ischial tuberosities to the rectosigmoid flexure; femoral heads include femoral neck to midway between the greater and lesser trochanter; *MSKCC* Memorial Sloan Kettering Cancer Center

approximately 50 % [91]. Treatment time is reduced more significantly when the pelvic lymph nodes are treated for high-risk disease. Additionally, VMAT may offer dosimetric advantages over IMRT in some cases, especially in high-dose regions [92]. Another variant of continuous rotational treatment is TomoTherapy®, which offers at least equivalent, if not slightly improved, dose distributions [93].

19.10 Image-Guided Treatment Delivery (IGRT): Interfraction Motion

Interfraction motion occurs during the daily setup of patients and can result in decreased dose to the PTV and increased toxicity due to increased dose to normal tissues. Uncertainties in prostate position are related to both setup error and prostate motion between daily fractions. Setup error could be systematic or random. Systematic errors do not change, are reproducible, and always occur in the same direction and magnitude. Random errors vary daily and are not reproducible [94].

The prostate is not a fixed organ, and its location can vary significantly from day to day [95–99]. The mean prostate shift was found to be 1.0+1.5 mm in the lateral direction, 0.9+2.1 mm in the AP, and 1.9+2.1 mm in the craniocaudal direction in an Italian study. After DVH recalculation, CTV coverage was maintained despite organ motion, whereas rectal DVHs were often dramatically different [95]. The amount of prostate motion varies from patient to patient, and the reproducibility of daily prostate positioning has become increasingly important with decreased PTV margins and dose escalation permitted by IMRT.

The impact of rectal distension on prostate target volume variability, treatment dose, and patient outcomes has been well studied [100–104]. The effects of rectal distension at simulation described by de Crevoisier et al. [103] can be mitigated by the use of daily image guidance [105]. There are many methods to correct the observed interfraction motion. These methods are daily CT bony pelvis alignment, daily cone beam CT or ultrasound soft tissue alignment, definitive isocenter calculation, and daily online fiducial correction. If gold seed markers are not placed within the prostate, daily CT localization using bony anatomy could be used with reasonable results [98]. The increase in PTV margin would consist of the uncertainty of the location of the prostate with respect to the bone in addition to intrafractional motion.

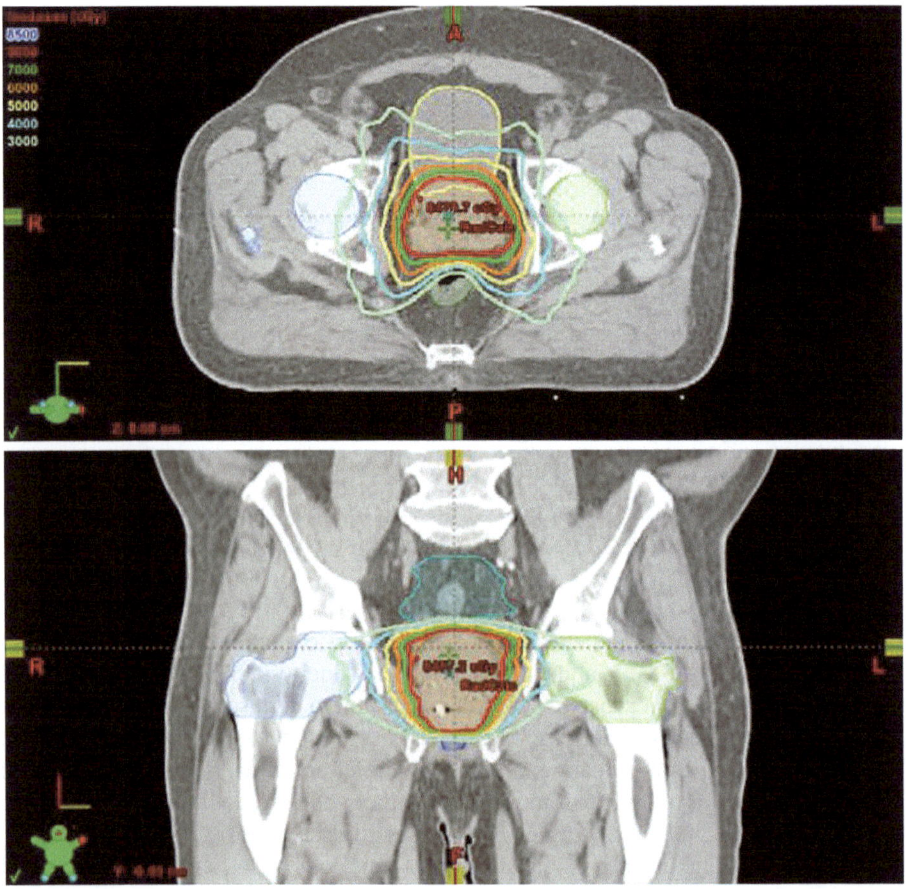

Fig. 19.2 Dose distribution and DVH for a typical IMRT plan prescribing 80 Gy in 40 fractions to the PTV

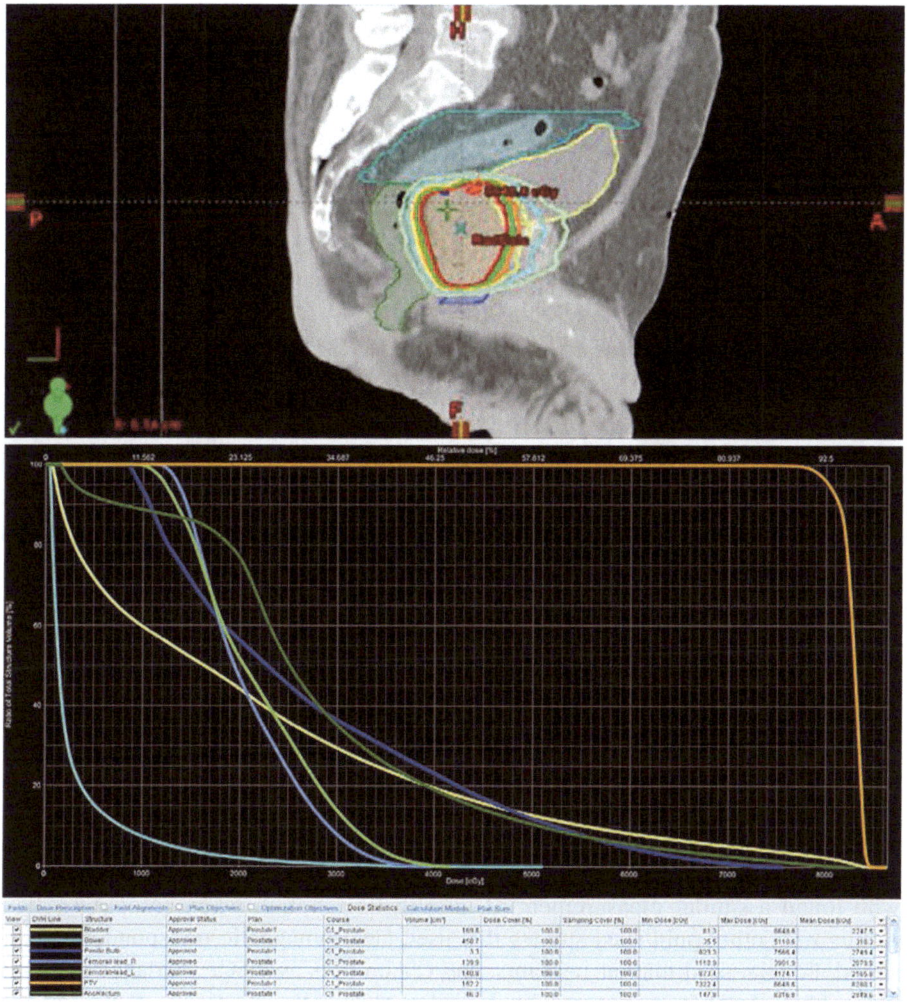

Fig. 19.2 (continued)

Hanna et al. also described how centrally located intraprostatic calcifications (IPC) can be used as natural fiducials with similar pattern of displacement compared to implanted fiducials [106]. The use of IPC in IGRT for prostate cancer can eliminate an invasive procedure to implant fiducials which comes with both financial cost and unnecessary morbidity when IPCs are identified.

Cone beam CT (CBCT) 3D volume reconstruction images have improved over the years and are often used for correction of prostate interfraction motion. The measures obtained by the use of CBCT are highly accurate, with the greatest displacement usually observable in the AP axis [107, 108].

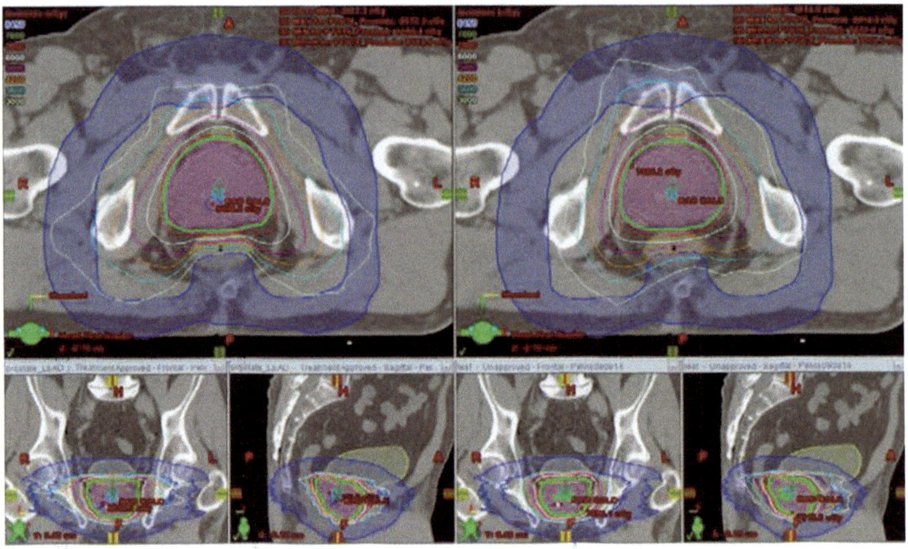

Fig. 19.3 Example of optimization structure (*shaded blue*) utilized to pull dose in anteriorly to spare the rectum and posteriorly to better spare the bladder while pushing out lower dose regions into lateral soft tissues. The optimized plan is on the *left*

CBCT appears to be associated with less variability than ultrasound [109]. However, therapists must still make subjective decisions concerning prostate anatomy. We combine daily fiducial marker assessment with CBCT to reduce the subjectiveness of the alignment and assess bladder and rectal filling. A broad comparison of MV CBCT with other imaging modalities was performed by Bylund et al. [108], who found a similar measurement of interfraction motion when comparing their results with other published studies that used electromagnetic transponders, ultrasound, fiducial seed marker implants, and CT on rails as image guidance methods [110, 111]. A study performed by Moseley et al. retrospectively compared the patient adjustments based on kV CBCT compared to two orthogonal MV portal images and confirmed that there was a high correlation for the measured isocenter shifts between these systems [98].

Advances in ultrasound technology have resulted in noninvasive and nonionizing methods of daily prostate localization that do not require the implantation of fiducial markers. Boda-Heggemann et al. demonstrated that transabdominal ultrasound improved the daily repositioning accuracy compared to the use of skin marks or bony anatomy [112]. The development of 3D ultrasound imaging modalities has improved IGRT via ultrasound by providing physicians with data regarding the daily changes in the target dimensions and location, while automatically calculating geometric beam coverage to ensure the accuracy of the treatment delivery.

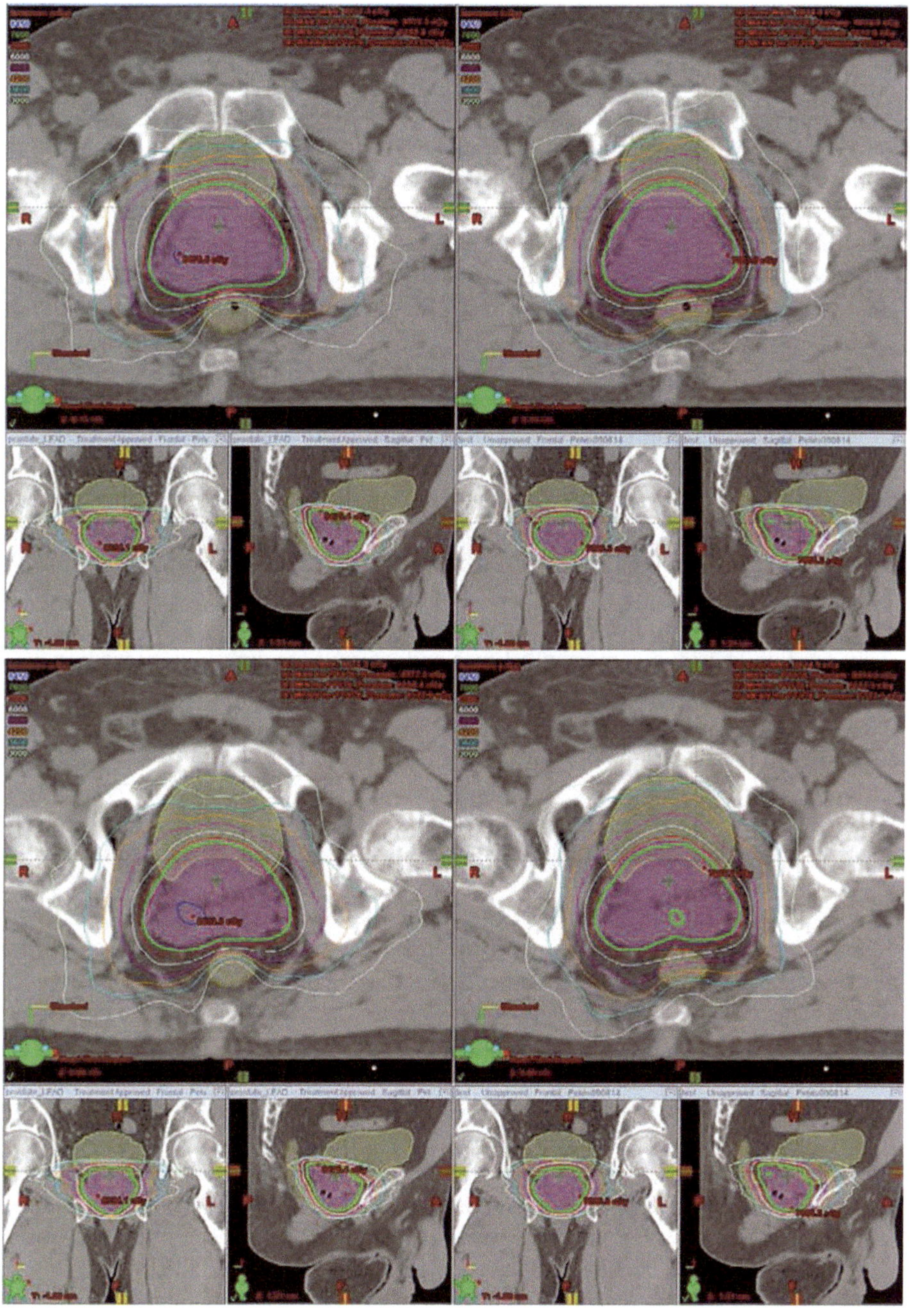

Fig. 19.4 Dose distribution and DVH comparison at multiple axial slices for a typical optimized IMRT plan (*left*) compared to an initial IMRT plan (*right*). Attention is given to the 30 Gy isodose line, which is pushed well within the rectum

19.11 Image-Guided Treatment Delivery (IGRT): Intrafraction Motion

Real-time tracking of the prostate via fiducials using fluoroscopy, electromagnetic transponders, or transperineal ultrasound [67] limits effects from intrafraction motion, which takes on considerable importance for hypofractionation. Alonso-Arrizabalaga et al. [64] found that the margins can be reduced with correction for interfraction and intrafraction prostate motion. Compared to daily online correction, there was a PTV increase of 40 % for definitive isocenter and 80 % for bony anatomy image guidance. The overlap of PTV with the rectum approximately doubles for each of these increases.

Kupelian et al. [110] reported the use of a 4D localization system for continuous, intrafraction three-dimensional isocenter tracking via electromagnetic detection of three implanted beacon transponders. In this study, 41 patients with clinical stage I–III prostate cancer treated per institutional preference had transponder setup compared to traditional laser and skin markings performed pretreatment and between beams for all patients, and during delivery for 35 of 41 patients (6 patients had large anteroposterior torso dimensions precluding tracking due to geometric constraints). Transponder stability after implantation was demonstrated with a mean standard deviation of the intertransponder distance of 0.8 mm. The study identified a seemingly random continuous motion ≥3 mm and ≥5 mm for durations exceeding 30 s in 41 % and 15 % of fractions, respectively. This intrafraction motion could have significant effects with respect to coverage, especially when treatment margins of as small as 5 mm are not uncommon.

19.12 Dose Escalation: Beyond 80 Gy

As previously discussed, IMRT has allowed the safe delivery of higher doses in the 74–80 Gy range, which has resulted in improved outcomes. Under investigation now is even further dose escalation beyond 80 Gy. Eade et al. [10] described a dose response beyond 80 Gy, the data of which were used as a rationale for the Fox Chase dose escalation hypofractionation trial described by Pollack et al. [113]. A phase II study from Memorial Sloan Kettering Cancer Center described the treatment of 1,002 patients treated to 86.4 Gy in 48 fractions from 1997 to 2008 with a median follow-up of 5.5 years; 86 % of patients were clinical stage T1–T2N0. The PTV consisted of the prostate, entire seminal vesicles, and a 10 mm circumferential margin, except posteriorly, where it was reduced to 6 mm. Weekly port films were used to verify patient positioning. Fifty-nine percent of patients received ADT. Seven-year biochemical recurrence-free survival rates based on the nadir plus 2 ng/mL definition were 98.8 %, 85.6 %, and 67.9 % ($p<0.001$) for low-, intermediate-, and high-risk patients, respectively. Seven-year actuarial distant metastasis-free survival rates were 99.4 %, 94.1 %, and 82.0 % ($p<0.001$) for low-, intermediate-, and high-risk patients, respectively. Seven-year prostate cancer-specific mortality rates using competing risk analysis were 0 %, 3.3 %, and 8.1 % ($p = 0.008$), for low-, intermediate-, and high-risk patients, respectively. Late grade 3 gastrointestinal and genitourinary toxicities at 7 years were experienced by 0.7 % and 2.2 % of patients, respectively [114]. Although this was a retrospective study with modest follow-up of a median of 5.5 years, further dose escalation beyond 80 Gy demonstrated excellent tumor control and relatively low rates of toxicity. More recently, Spratt et al. [115] described that the combination of brachytherapy plus external beam RT was superior to 86.4 Gy of external beam RT alone, illustrating the potential benefit of further dose escalation. These findings are complicated by changes in (better IGRT) methods over the period of the study and patient selection bias. Nonetheless, such data lend support for further dose escalation using IMRT to dominant tumor regions defined on MRI [54, 116], since these seem to be at the greatest risk of persistent disease [117].

19.13 Androgen Deprivation Therapy

A group of patients with early stage prostate cancer fall into the intermediate-risk prognostic group as a result of their specific PSA and/or Gleason score. In studies of primarily intermediate-risk patients, two main randomized trials support the addition of short-term ADT to RT when standard doses of RT were used (about 70 Gy). D'Amico et al. compared 6 months of ADT with RT versus RT alone in 206 men with AJCC clinical stage T1b–T2bN0M0 adenocarcinoma of the prostate with at least 1 unfavorable prognostic factor (PSA > 10 ng/mL, GS 7–10, radiographic evidence of extracapsular extension, and/or seminal vesicle invasion on MRI) who were treated between 1995 and 2001. All patients in the combined treatment arm received 3D conformal RT and ADT (consisting of a luteinizing hormone-releasing hormone (LHRH) agonist and the antiandrogen flutamide beginning 2 months prior to RT). With a median follow-up of 7.6 years, the 8-year overall survival was 74 % in the combined RT-ADT arm and 61 % in the RT alone arm ($p = 0.01$). In addition, RT alone resulted in an increased risk of prostate cancer-specific mortality (HR 4.1; 95 % CI, 1.4–12.1; $p = 0.01$), as well as all-cause mortality (HR 1.8; 95 % CI, 1.1–2.9; $p = 0.01$) compared to RT-ADT. An unplanned post-randomization subgroup analysis revealed that the increased risk in all-cause mortality in men randomized to RT alone was only found in those with no or minimal comorbidity (31 vs. 11 deaths; HR 4.2; 95 % CI, 2.1–8.5; $p < 0.001$) [118].

RTOG 94-08 compared 4 months of ADT with RT versus RT alone in 1979 men with AJCC stage T1b–T2bN0M0 adenocarcinoma of the prostate with a PSA level ≤20 ng/ml who were treated between 1994 and 2001. ADT consisted of an LHRH agonist and an antiandrogen beginning 2 months prior to RT and continued during RT. With a median follow-up of 9.1 years, the 10-year overall survival was 62 % in the combined arm and 57 % in the RT alone arm (HR 1.17; 95 % CI, 1.01–1.35; $p = 0.03$). The 10-year disease-specific mortality was 4 % in the combined arm and 8 % in the RT alone arm (HR 1.87; 95 % CI, 1.27–2.74; $p = 0.001$). The 10-year cumulative incidence of distant metastasis was 6 % in the combined arm and 8 % in the RT alone arm (HR 1.45; 95 % CI, 1.03–2.06; $p=0.04$). Of note, multivariate analysis revealed Gleason score of 7 or higher to be a negative prognostic factor for overall survival, disease-specific survival, distant metastasis, and biochemical failure. Further subgroup analysis revealed that the overall survival and disease-specific mortality benefits were greatest in the intermediate-risk category patients and were not significant in the low-risk patient group [119].

Both trials used standard RT doses of about 70 Gy and were conducted in the pre-IGRT/IMRT era. RTOG 08-15 is currently examining the role of 6 months of ADT in addition to dose-escalated RT for patients with intermediate-risk prostate cancer. Lastly, practical factors such as patient age, comorbidities including a history of congestive heart failure or prior myocardial infarction, MRI stage, and extent of tumor on diagnostic biopsies help inform whether ADT should be added to definitive RT in intermediate-risk, early stage (T1a–2bN0M0) prostate cancer patients.

References

1. Siegel R, Naishadham D, Jemal A (2012) Cancer statistics, 2012. CA Cancer J Clin 62(1):10–29
2. Eccles BK et al (2013) SABRE 1 (Surgery Against Brachytherapy – a Randomised Evaluation): feasibility randomised controlled trial (RCT) of brachytherapy vs radical prostatectomy in low-intermediate risk clinically localised prostate cancer. BJU Int 112(3):330–337
3. Kuban DA et al (2003) Long-term multi-institutional analysis of stage T1-T2 prostate cancer treated with radiotherapy in the PSA era. Int J Radiat Oncol Biol Phys 57(4):915–928
4. Pilepich MV et al (1987) Correlation of radiotherapeutic parameters and treatment related morbidity in carcinoma of the prostate–analysis of RTOG study 75–06. Int J Radiat Oncol Biol Phys 13(3):351–357
5. Hanks GE, Martz KL, Diamond JJ (1988) The effect of dose on local control of prostate cancer. Int J Radiat Oncol Biol Phys 15(6):1299–1305
6. Valicenti R et al (2000) Survival advantage from higher-dose radiation therapy for clinically localized prostate cancer treated on the Radiation Therapy Oncology Group trials. J Clin Oncol 18(14):2740–2746
7. Zelefsky MJ et al (2008) Influence of local tumor control on distant metastases and cancer related mortality after external beam radiotherapy for prostate cancer. J Urol 179(4):1368–1373; discussion 1373
8. Leibel SA et al (1994) The biological basis and clinical application of three-dimensional conformal external beam radiation therapy in carcinoma of the prostate. Semin Oncol 21(5):580–597
9. Ten Haken RK et al (1989) Boost treatment of the prostate using shaped, fixed fields. Int J Radiat Oncol Biol Phys 16(1):193–200
10. Eade TN et al (2007) What dose of external-beam radiation is high enough for prostate cancer? Int J Radiat Oncol Biol Phys 68(3):682–689
11. Kuban DA et al (2008) Long-term results of the M. D. Anderson randomized dose-escalation trial for prostate cancer. Int J Radiat Oncol Biol Phys 70(1):67–74
12. Zietman AL et al (2005) Comparison of conventional-dose vs high-dose conformal radiation therapy in clinically localized adenocarcinoma of the prostate: a randomized controlled trial. J Am Med Assoc 294(10):1233–1239
13. Peeters ST et al (2006) Dose-response in radiotherapy for localized prostate cancer: results of the Dutch multicenter randomized phase III trial comparing 68 Gy of radiotherapy with 78 Gy. J Clin Oncol 24(13):1990–1996

14. Dearnaley DP et al (2007) Escalated-dose versus standard-dose conformal radiotherapy in prostate cancer: first results from the MRC RT01 randomised controlled trial. Lancet Oncol 8(6):475–487
15. Beckendorf V et al (2011) 70 Gy versus 80 Gy in localized prostate cancer: 5-year results of GETUG 06 randomized trial. Int J Radiat Oncol Biol Phys 80(4):1056–1063
16. Viani GA, Stefano EJ, Afonso SL (2009) Higher-than-conventional radiation doses in localized prostate cancer treatment: a meta-analysis of randomized, controlled trials. Int J Radiat Oncol Biol Phys 74(5):1405–1418
17. Ling CC et al (1996) Conformal radiation treatment of prostate cancer using inversely-planned intensity-modulated photon beams produced with dynamic multileaf collimation. Int J Radiat Oncol Biol Phys 35(4):721–730
18. Marchal C et al (2004) Preliminary results of the assessment of intensity modulated radiotherapy (IMRT) for prostatic and head and neck tumors (STIC 2001). Cancer Radiother 8(Suppl 1):S121–S127
19. Nutting CM et al (2000) Reduction of small and large bowel irradiation using an optimized intensity-modulated pelvic radiotherapy technique in patients with prostate cancer. Int J Radiat Oncol Biol Phys 48(3):649–656
20. Vlachaki MT et al (2005) IMRT versus conventional 3DCRT on prostate and normal tissue dosimetry using an endorectal balloon for prostate immobilization. Med Dosim 30(2):69–75
21. Ashman JB et al (2005) Whole pelvic radiotherapy for prostate cancer using 3D conformal and intensity-modulated radiotherapy. Int J Radiat Oncol Biol Phys 63(3):765–771
22. Kao J et al (2004) Sparing of the penile bulb and proximal penile structures with intensity-modulated radiation therapy for prostate cancer. Br J Radiol 77(914):129–136
23. Ailleres N et al (2004) Pilot study of conformal intensity modulated radiation therapy for localized prostate cancer. Cancer Radiother 8(2):59–69
24. Wang-Chesebro A et al (2006) Intensity-modulated radiotherapy improves lymph node coverage and dose to critical structures compared with three-dimensional conformal radiation therapy in clinically localized prostate cancer. Int J Radiat Oncol Biol Phys 66(3):654–662
25. Perna L et al (2009) Sparing the penile bulb in the radical irradiation of clinically localised prostate carcinoma: a comparison between MRI and CT prostatic apex definition in 3DCRT, Linac-IMRT and Helical Tomotherapy. Radiother Oncol 93(1):57–63
26. Zelefsky MJ et al (2000) Clinical experience with intensity modulated radiation therapy (IMRT) in prostate cancer. Radiother Oncol 55(3):241–249
27. Zelefsky MJ et al (2001) High dose radiation delivered by intensity modulated conformal radiotherapy improves the outcome of localized prostate cancer. J Urol 166(3):876–881
28. Al-Mamgani A et al (2009) Role of intensity-modulated radiotherapy in reducing toxicity in dose escalation for localized prostate cancer. Int J Radiat Oncol Biol Phys 73(3):685–691
29. De Meerleer G et al (2004) Intensity-modulated radiotherapy as primary treatment for prostate cancer: acute toxicity in 114 patients. Int J Radiat Oncol Biol Phys 60(3):777–787
30. Kry SF et al (2005) The calculated risk of fatal secondary malignancies from intensity-modulated radiation therapy. Int J Radiat Oncol Biol Phys 62(4):1195–1203
31. Brenner DJ et al (2000) Second malignancies in prostate carcinoma patients after radiotherapy compared with surgery. Cancer 88(2):398–406
32. Brenner DJ (2006) Induced second cancers after prostate-cancer radiotherapy: no cause for concern? Int J Radiat Oncol Biol Phys 65(3):637–639
33. Nieder AM, Porter MP, Soloway MS (2008) Radiation therapy for prostate cancer increases subsequent risk of bladder and rectal cancer: a population based cohort study. J Urol 180(5):2005–2009; discussion 2009–10
34. Singh AK et al (2010) Increasing age and treatment modality are predictors for subsequent diagnosis of bladder cancer following prostate cancer diagnosis. Int J Radiat Oncol Biol Phys 78(4):1086–1094
35. Kendal WS et al (2006) Prostatic irradiation is not associated with any measurable increase in the risk of subsequent rectal cancer. Int J Radiat Oncol Biol Phys 65(3):661–668

36. Isebaert S et al (2013) Multiparametric MRI for prostate cancer localization in correlation to whole-mount histopathology. J Magn Reson Imaging 37(6):1392–1401
37. Sciarra A et al (2011) Advances in magnetic resonance imaging: how they are changing the management of prostate cancer. Eur Urol 59(6):962–977
38. Hegde JV et al (2013) Multiparametric MRI of prostate cancer: an update on state-of-the-art techniques and their performance in detecting and localizing prostate cancer. J Magn Reson Imaging 37(5):1035–1054
39. Vargas HA et al (2011) Diffusion-weighted endorectal MR imaging at 3 T for prostate cancer: tumor detection and assessment of aggressiveness. Radiology 259(3):775–784
40. Coakley FV et al (2002) Prostate cancer tumor volume: measurement with endorectal MR and MR spectroscopic imaging. Radiology 223(1):91–97
41. Mazaheri Y et al (2008) Prostate cancer: identification with combined diffusion-weighted MR imaging and 3D 1H MR spectroscopic imaging–correlation with pathologic findings. Radiology 246(2):480–488
42. Hambrock T et al (2011) Relationship between apparent diffusion coefficients at 3.0-T MR imaging and Gleason grade in peripheral zone prostate cancer. Radiology 259(2):453–461
43. Padhani AR et al (2000) Dynamic contrast enhanced MRI of prostate cancer: correlation with morphology and tumour stage, histological grade and PSA. Clin Radiol 55(2):99–109
44. Turnbull LW et al (1999) Differentiation of prostatic carcinoma and benign prostatic hyperplasia: correlation between dynamic Gd-DTPA-enhanced MR imaging and histopathology. J Magn Reson Imaging 9(2):311–316
45. Girouin N et al (2007) Prostate dynamic contrast-enhanced MRI with simple visual diagnostic criteria: is it reasonable? Eur Radiol 17(6):1498–1509
46. Schmuecking M et al (2009) Dynamic MRI and CAD vs. choline MRS: where is the detection level for a lesion characterisation in prostate cancer? Int J Radiat Biol 85(9):814–824
47. Singh AK et al (2007) Simultaneous integrated boost of biopsy proven, MRI defined dominant intra-prostatic lesions to 95 Gray with IMRT: early results of a phase I NCI study. Radiat Oncol 2:36
48. Fonteyne V et al (2008) Intensity-modulated radiotherapy as primary therapy for prostate cancer: report on acute toxicity after dose escalation with simultaneous integrated boost to intraprostatic lesion. Int J Radiat Oncol Biol Phys 72(3):799–807
49. Ippolito E et al (2012) Intensity-modulated radiotherapy with simultaneous integrated boost to dominant intraprostatic lesion: preliminary report on toxicity. Am J Clin Oncol 35(2):158–162
50. Aluwini S et al (2013) Stereotactic body radiotherapy with a focal boost to the MRI-visible tumor as monotherapy for low- and intermediate-risk prostate cancer: early results. Radiat Oncol 8:84
51. Riches SF et al (2014) Effect on therapeutic ratio of planning a boosted radiotherapy dose to the dominant intraprostatic tumour lesion within the prostate based on multifunctional MR parameters. Br J Radiol 87(1037):20130813
52. Murray LJ et al (2014) Prostate stereotactic ablative radiation therapy using volumetric modulated arc therapy to dominant intraprostatic lesions. Int J Radiat Oncol Biol Phys 89(2):406–415
53. Bauman G et al (2013) Boosting imaging defined dominant prostatic tumors: a systematic review. Radiother Oncol 107(3):274–281
54. van Lin EN et al (2006) IMRT boost dose planning on dominant intraprostatic lesions: gold marker-based three-dimensional fusion of CT with dynamic contrast-enhanced and 1H-spectroscopic MRI. Int J Radiat Oncol Biol Phys 65(1):291–303
55. Shinohara K, Roach M 3rd (2008) Technique for implantation of fiducial markers in the prostate. Urology 71(2):196–200
56. Gauthier I et al (2009) Dosimetric impact and theoretical clinical benefits of fiducial markers for dose escalated prostate cancer radiation treatment. Int J Radiat Oncol Biol Phys 74(4):1128–1133

57. Chung PW et al (2004) On-line aSi portal imaging of implanted fiducial markers for the reduction of interfraction error during conformal radiotherapy of prostate carcinoma. Int J Radiat Oncol Biol Phys 60(1):329–334
58. Schallenkamp JM et al (2005) Prostate position relative to pelvic bony anatomy based on intraprostatic gold markers and electronic portal imaging. Int J Radiat Oncol Biol Phys 63(3):800–811
59. Beltran C, Herman MG, Davis BJ (2008) Planning target margin calculations for prostate radiotherapy based on intrafraction and interfraction motion using four localization methods. Int J Radiat Oncol Biol Phys 70(1):289–295
60. Kupelian PA et al (2005) Intraprostatic fiducials for localization of the prostate gland: monitoring intermarker distances during radiation therapy to test for marker stability. Int J Radiat Oncol Biol Phys 62(5):1291–1296
61. Poggi MM et al (2003) Marker seed migration in prostate localization. Int J Radiat Oncol Biol Phys 56(5):1248–1251
62. Gill S et al (2012) Patient-reported complications from fiducial marker implantation for prostate image-guided radiotherapy. Br J Radiol 85(1015):1011–1017
63. Igdem S et al (2009) Implantation of fiducial markers for image guidance in prostate radiotherapy: patient-reported toxicity. Br J Radiol 82(983):941–945
64. Alonso-Arrizabalaga S et al (2007) Prostate planning treatment volume margin calculation based on the ExacTrac X-Ray 6D image-guided system: margins for various clinical implementations. Int J Radiat Oncol Biol Phys 69(3):936–943
65. Nederveen AJ et al (2003) Comparison of megavoltage position verification for prostate irradiation based on bony anatomy and implanted fiducials. Radiother Oncol 68(1):81–88
66. Langen KM et al (2008) Observations on real-time prostate gland motion using electromagnetic tracking. Int J Radiat Oncol Biol Phys 71(4):1084–1090
67. Abramowitz M (2012) Noninvasive real-time prostate tracking using a transperineal ultrasound approach. Int J Radiat Oncol Biol Phys 84(3):1
68. Noel C et al (2009) Prediction of intrafraction prostate motion: accuracy of pre- and post-treatment imaging and intermittent imaging. Int J Radiat Oncol Biol Phys 73(3):692–698
69. Bittner N et al (2010) Electromagnetic tracking of intrafraction prostate displacement in patients externally immobilized in the prone position. Int J Radiat Oncol Biol Phys 77(2):490–495
70. Coakley FV, Hricak H (2000) Radiologic anatomy of the prostate gland: a clinical approach. Radiol Clin N Am 38(1):15–30
71. Zackrisson B, Hugosson J, Aus G (2000) Transrectal ultrasound anatomy of the prostate and seminal vesicles in healthy men. Scand J Urol Nephrol 34(3):175–180
72. McLaughlin PW et al (2010) Radiographic and anatomic basis for prostate contouring errors and methods to improve prostate contouring accuracy. Int J Radiat Oncol Biol Phys 76(2):369–378
73. Villeirs GM, De Meerleer GO (2007) Magnetic resonance imaging (MRI) anatomy of the prostate and application of MRI in radiotherapy planning. Eur J Radiol 63(3):361–368
74. Debois M et al (1999) The contribution of magnetic resonance imaging to the three-dimensional treatment planning of localized prostate cancer. Int J Radiat Oncol Biol Phys 45(4):857–865
75. Milosevic M et al (1998) Magnetic resonance imaging (MRI) for localization of the prostatic apex: comparison to computed tomography (CT) and urethrography. Radiother Oncol 47(3):277–284
76. Smith WL et al (2007) Prostate volume contouring: a 3D analysis of segmentation using 3DTRUS, CT, and MR. Int J Radiat Oncol Biol Phys 67(4):1238–1247
77. Roach M 3rd et al (1996) Prostate volumes defined by magnetic resonance imaging and computerized tomographic scans for three-dimensional conformal radiotherapy. Int J Radiat Oncol Biol Phys 35(5):1011–1018
78. Rasch C et al (1999) Definition of the prostate in CT and MRI: a multi-observer study. Int J Radiat Oncol Biol Phys 43(1):57–66

79. van Herk M et al (1998) Automatic registration of pelvic computed tomography data and magnetic resonance scans including a full circle method for quantitative accuracy evaluation. Med Phys 25(10):2054–2067
80. Gofrit ON et al (2009) The dimensions and symmetry of the seminal vesicles. J Robot Surg 3:29–33
81. Banner MP, Hassler R (1978) The normal seminal vesiculogram. Radiology 128(2):339–344
82. Secaf E et al (1991) MR imaging of the seminal vesicles. AJR Am J Roentgenol 156(5):989–994
83. Villeirs GM et al (2005) Interobserver delineation variation using CT versus combined CT+MRI in intensity-modulated radiotherapy for prostate cancer. Strahlenther Onkol 181(7):424–430
84. Wallner KE et al (2002) Penile bulb imaging. Int J Radiat Oncol Biol Phys 53(4):928–933
85. van der Wielen GJ, Mulhall JP, Incrocci L (2007) Erectile dysfunction after radiotherapy for prostate cancer and radiation dose to the penile structures: a critical review. Radiother Oncol 84(2):107–113
86. Buyyounouski MK et al (2004) The radiation doses to erectile tissues defined with magnetic resonance imaging after intensity-modulated radiation therapy or iodine-125 brachytherapy. Int J Radiat Oncol Biol Phys 59(5):1383–1391
87. Brown MW et al (2007) An analysis of erectile function after intensity modulated radiation therapy for localized prostate carcinoma. Prostate Cancer Prostatic Dis 10(2):189–193
88. Macdonald AG et al (2005) Predictive factors for erectile dysfunction in men with prostate cancer after brachytherapy: is dose to the penile bulb important? Int J Radiat Oncol Biol Phys 63(1):155–163
89. Merrick GS et al (2005) Erectile function after prostate brachytherapy. Int J Radiat Oncol Biol Phys 62(2):437–447
90. Roach M et al (2004) Penile bulb dose and impotence after three-dimensional conformal radiotherapy for prostate cancer on RTOG 9406: findings from a prospective, multi-institutional, phase I/II dose-escalation study. Int J Radiat Oncol Biol Phys 60(5):1351–1356
91. Poon DM et al (2013) Dosimetric advantages and superior treatment delivery efficiency of RapidArc over conventional intensity-modulated radiotherapy in high-risk prostate cancer involving seminal vesicles and pelvic nodes. Clin Oncol (R Coll Radiol) 25(12):706–712
92. Kopp RW et al (2011) VMAT vs. 7-field-IMRT: assessing the dosimetric parameters of prostate cancer treatment with a 292-patient sample. Med Dosim 36(4):365–372
93. Iori M et al (2008) Dose-volume and biological-model based comparison between helical tomotherapy and (inverse-planned) IMAT for prostate tumours. Radiother Oncol 88(1):34–45
94. Korreman S et al (2010) The European Society of Therapeutic Radiology and Oncology-European Institute of Radiotherapy (ESTRO-EIR) report on 3D CT-based in-room image guidance systems: a practical and technical review and guide. Radiother Oncol 94(2):129–144
95. Landoni V et al (2006) A study of the effect of setup errors and organ motion on prostate cancer treatment with IMRT. Int J Radiat Oncol Biol Phys 65(2):587–594
96. Alasti H et al (2001) Portal imaging for evaluation of daily on-line setup errors and off-line organ motion during conformal irradiation of carcinoma of the prostate. Int J Radiat Oncol Biol Phys 49(3):869–884
97. Vigneault E et al (1997) Electronic portal imaging device detection of radioopaque markers for the evaluation of prostate position during megavoltage irradiation: a clinical study. Int J Radiat Oncol Biol Phys 37(1):205–212
98. Moseley DJ et al (2007) Comparison of localization performance with implanted fiducial markers and cone-beam computed tomography for on-line image-guided radiotherapy of the prostate. Int J Radiat Oncol Biol Phys 67(3):942–953
99. Roeske JC et al (1995) Evaluation of changes in the size and location of the prostate, seminal vesicles, bladder, and rectum during a course of external beam radiation therapy. Int J Radiat Oncol Biol Phys 33(5):1321–1329

100. Zelefsky MJ et al (1999) Quantification and predictors of prostate position variability in 50 patients evaluated with multiple CT scans during conformal radiotherapy. Radiother Oncol 50(2):225–234
101. Antolak JA et al (1998) Prostate target volume variations during a course of radiotherapy. Int J Radiat Oncol Biol Phys 42(3):661–672
102. Beard CJ et al (1996) Analysis of prostate and seminal vesicle motion: implications for treatment planning. Int J Radiat Oncol Biol Phys 34(2):451–458
103. de Crevoisier R et al (2005) Increased risk of biochemical and local failure in patients with distended rectum on the planning CT for prostate cancer radiotherapy. Int J Radiat Oncol Biol Phys 62(4):965–973
104. Heemsbergen WD et al (2007) Increased risk of biochemical and clinical failure for prostate patients with a large rectum at radiotherapy planning: results from the Dutch trial of 68 GY versus 78 Gy. Int J Radiat Oncol Biol Phys 67(5):1418–1424
105. Kupelian PA et al (2008) Impact of image guidance on outcomes after external beam radiotherapy for localized prostate cancer. Int J Radiat Oncol Biol Phys 70(4):1146–1150
106. Hanna SA et al (2012) Role of intra- or periprostatic calcifications in image-guided radiotherapy for prostate cancer. Int J Radiat Oncol Biol Phys 82(3):1208–1216
107. Ryan D et al (2009) Prostate positioning errors associated with two automatic registration based image guidance strategies. J Appl Clin Med Phys 10(4):3071
108. Bylund KC et al (2008) Analysis of interfraction prostate motion using megavoltage cone beam computed tomography. Int J Radiat Oncol Biol Phys 72(3):949–956
109. Gayou O, Miften M (2008) Comparison of mega-voltage cone-beam computed tomography prostate localization with online ultrasound and fiducial markers methods. Med Phys 35(2):531–538
110. Kupelian P et al (2007) Multi-institutional clinical experience with the Calypso System in localization and continuous, real-time monitoring of the prostate gland during external radiotherapy. Int J Radiat Oncol Biol Phys 67(4):1088–1098
111. Scarbrough TJ et al (2006) Comparison of ultrasound and implanted seed marker prostate localization methods: implications for image-guided radiotherapy. Int J Radiat Oncol Biol Phys 65(2):378–387
112. Boda-Heggemann J et al (2008) Accuracy of ultrasound-based (BAT) prostate-repositioning: a three-dimensional on-line fiducial-based assessment with cone-beam computed tomography. Int J Radiat Oncol Biol Phys 70(4):1247–1255
113. Pollack A et al (2013) Randomized trial of hypofractionated external-beam radiotherapy for prostate cancer. J Clin Oncol 31(31):3860–3868
114. Spratt DE et al (2013) Long-term survival and toxicity in patients treated with high-dose intensity modulated radiation therapy for localized prostate cancer. Int J Radiat Oncol Biol Phys 85(3):686–692
115. Spratt DE et al (2014) Comparison of high-dose (86.4 Gy) IMRT vs combined brachytherapy plus IMRT for intermediate-risk prostate cancer. BJU Int 114(3):360–367
116. Pickett B et al (1999) Static field intensity modulation to treat a dominant intra-prostatic lesion to 90 Gy compared to seven field 3-dimensional radiotherapy. Int J Radiat Oncol Biol Phys 44(4):921–929
117. Pucar D et al (2007) Clinically significant prostate cancer local recurrence after radiation therapy occurs at the site of primary tumor: magnetic resonance imaging and step-section pathology evidence. Int J Radiat Oncol Biol Phys 69(1):62–69
118. D'Amico AV et al (2008) Androgen suppression and radiation vs radiation alone for prostate cancer: a randomized trial. J Am Med Assoc 299(3):289–295
119. Jones CU et al (2011) Radiotherapy and short-term androgen deprivation for localized prostate cancer. N Engl J Med 365(2):107–118

Intensity-Modulated Radiation Therapy for Locally Advanced Prostate Cancer

20

Takashi Mizowaki

Keywords
Locally advanced prostate cancer • Intensity-modulated radiation therapy • Dose escalation • Pelvic lymph node metastasis

20.1 Introduction

Prostate cancer is the most common cancer in males in the United States and Europe [1, 2]. The incidence of prostate cancer has also increased recently in many Asian countries [3], and it is now the fourth most common site of male cancers in Japan after the stomach, colon/rectum, and lungs [4]. Furthermore, its rate of increase is the fastest of all male cancers among Japanese males. In addition, there are a significant number of patients with locally advanced prostate cancer in Japan [5–7], whereas the number of patients with advanced disease has declined significantly in the United States [6]. Therefore, the effective management of locally advanced prostate cancer is particularly important in Japan.

There are many therapeutic options for prostate cancer such as active surveillance, radical prostatectomy, external beam radiation therapy (EBRT), brachytherapy, and hormone therapy (HT). Of these, EBRT is well established as a major option for the definitive treatment of prostate cancer and is the most common therapy used for locally advanced disease [8–10]. With the development of sophisticated EBRT techniques, such as three-dimensional conformal radiotherapy (3D-CRT) and intensity-modulated radiation therapy (IMRT), significant dose escalation can be

T. Mizowaki, M.D., Ph.D. (✉)
Department of Radiation Oncology and Image-applied Therapy,
Graduate School of Medicine, Kyoto University, 54 Shogoinkawahara-cho, Sakyo-ku,
Kyoto 606-8507, Japan
e-mail: mizo@kuhp.kyoto-u.ac.jp

Y. Nishimura, R. Komaki (eds.), *Intensity-Modulated Radiation Therapy*,
DOI 10.1007/978-4-431-55486-8_20

achieved safely and has been applied to many cases [10]. IMRT can also reduce the risk of late rectal bleeding significantly compared with conventional irradiation techniques and 3D-CRT [11–13].

However, there are only a small number of reports describing the long-term clinical outcomes of patients treated with IMRT for locally advanced or TanyN1M0 cases because IMRT has been mainly applied to localized prostate cancer in the United States.

20.2 Locally Advanced Prostate Cancer

20.2.1 Dose Escalation

It is well established that local dose escalation to the prostate improves progression-free survival [14, 15]. A meta-analysis of seven randomized controlled dose escalation studies of EBRT for prostate cancer revealed significant improvements in the number of failure-free cases using high-dose radiotherapy (HDRT) compared with conventional-dose radiotherapy ($p<0.001$) [15]. Subgroup analysis also demonstrated that HDRT was beneficial in patients classified as being at low ($p=0.007$), intermediate ($p<0.0001$), and high risk ($p<0.0001$) of biochemical failure.

However, these reports included mainly cases of localized (T1-T2N0M0) prostate cancer, and few studies have targeted only locally advanced diseases (T3-T4N0M0). The currently available randomized dose escalation studies for prostate cancer are summarized in Table 20.1. Shipley et al. compared the outcome of 75.6 and 67.2 Gy in 202 patients with T3-4N0M0 cases and reported a significant improvement in the local control rate with high-dose irradiation [16]. Although this study used a conventional radiation technique, the data suggested that local dose escalation is beneficial not only for localized disease but also for locally advanced cases. Six additional, more recent, trials [17–22] also demonstrated significantly improved biochemical control using dose escalation with modern conformal radiation delivery techniques. However, the number of T3-4N0M0 cases among the patients enrolled in these studies was 0–45 %. Therefore, the true impact of dose escalation with modern, sophisticated radiation techniques in patients with locally advanced disease remains unclear and should be studied further.

Studies assessing dose escalation using IMRT in patients with locally advanced prostate cancer are even rarer. Most dose escalation studies treating prostate cancer with IMRT [23–28] are summarized in Table 20.2. The incidence of T3-4N0M0 disease was very low (3.6–12 %), except for two reports with rates of 35 and 48 %. Although a study by Takeda et al. included 48 % of patients with T3N0M0 disease, 45 of these 141 patients (36 %) received pelvic lymph node dissection to confirm metastatic disease [28]. Therefore, there is likely to be a strong selection bias in this patient cohort. In addition, they did not report the number of patients who were diagnosed with T3N0M0 disease, but then excluded from the study after re-diagnosis with N1 disease following pelvic lymph node dissection.

Table 20.1 Summary of randomized dose escalation studies in nonmetastatic prostate cancer

Study [Reference]	Patient number	Stage of cancer	Rate of T3-T4 disease	Dose (Gy)	Hormonal therapy	Median f/u (year)	bNED	DSS	OS	Late GI toxicity	Late GU toxicity
Shipley [16]	202	T3-4N0M0	100 %	75.6 67.2	No	5.1	N.S. (LC improved)	N.S.	N.S.	Increased	Increased (stricture)
Sathya [17]	138	T2-3N0M0	39–40 %	40 + implant 66	No	8.2	Improved ($p = 0.024$)	–	N.S.	Slightly increased ($p = 0.09$)	N.S.
Zietman [18]	393	T1b-T2bN0M0 PSA < 15	0 %	79.2 70.2	No	5.5	Improved ($p < 0.01$)	–	N.S.	Increased ($p = 0.05$)	N.S.
MRC RT01 [19]	843	T1b-T3aN0M0 PSA ≤ 50	43–45 %	74 64	3–6 M NA-HT	5.3	Improved ($p = 0.007$)	–	N.S.	Increased ($p < 0.05$)	N.S.
Dutch [20]	669	T1a-4N0M0 PSA < 60	37 %	78 68	NA-/A-HT (n = 143)	9.2	Improved ($p < 0.05$)	N.S. (Improved in PSA = 8–18	N.S.	Increased ($p = 0.04$)	N.S.
MDACC [21]	301	T1b-3N0M0	17 %	78 70	No	10	Improved in PSA > 10 ($p = 0.013$)	Improved in PSA > 10 ($p = 0.05$)	N.S.	Increased ($p = 0.013$)	N.S.
GETUG 06 [22]	306	T1b-T3aN0M0 PSA < 60	12–13 %	80 70	No	5.1	Improved ($p = 0.036$)	–	–	N.S. ($p = 0.22$)	Increased ($p = 0.05$)

Table 20.2 Summary of dose escalation studies by IMRT in nonmetastatic prostate cancer with the median follow-up period of>4.5 years

Study [Reference]	Patient number	Stage of cancer	Rate of high-risk disease	Rate of T3-T4 disease	Dose (Gy)	Hormonal therapy	Median f/u (year)	bNED survival rate of high risk	OS	Late GI toxicity rate (≥grade 2)	Late GU toxicity rate (≥grade 2)
Alicikus [23]	170	T1c-4N0M0	19 %	5 %	81	No A-HT	8.3	62 % at 10Y	-	3 % at 10Y	16 % at 10Y
Spratt [24]	1002	T1c-T4N0M0	34 %	12 %	86.4	Yes (59 %)	5.5	68 % at 7Y		4 % at 7Y	21 % at 7Y
Vora [25]	303	T1a-3bN0M0	23 %	4.6 %	75.6 (70.2–77.4)	Yes (35 %)	7.6	53 % at 9Y (bNED rate)		12 %	24 %
Gadia [26]	140	T1a-3bN0M0	29 %	3.6 %	74–80	-	4.8	75 % at 5Y		15.7 %	20.2 %
Sia [27]	125	T1-4N0M0	65 %	35 %	74	Yes (variated)	5.5	76 % at 5Y		1.4 % at 5Y	2.2 % at 5Y
Takeda [28]	141	T1-3N0M0[a]	74 %	48 %	76 or 80 Gy	Yes (88 %)	5	82 % at 5Y (bNED rate)		6 %	6.3 %

[a]Pelvic lymph node dissection was performed in 45 patients (32 %) to rule out metastatic disease

Zelefsky et al. reported the long-term outcome of 296 patients with T3 N0M0 prostate cancer treated with 3D-CRT or IMRT using delivered doses ranging from 66 to 86.4 Gy [29]. They reported comparable tumor control and survival outcomes with the surgical series and concluded that the potential of conformal radiotherapy as a standard management option for locally advanced prostate cancer had been confirmed. However, this study included patients treated with suboptimal doses (<70 Gy), and the number of patients treated using IMRT among the 296 cases was not reported. In addition, 36 % of patients were treated with EBRT alone, which is not consistent with the current standard approach of EBRT combined with HT for patients with locally advanced prostate cancer [30]. There are also currently no reports describing the long-term clinical outcomes of patients with locally advanced prostate cancer treated using high-dose IMRT combined with HT.

20.2.2 Prophylactic Pelvic Lymph Node Irradiation

A recent study performed extended pelvic lymph node dissection and reported that >10 % of patients with clinically localized prostate cancer harbor pelvic lymph node metastases [31]. Furthermore, pelvic lymph node metastases were found in 30–40 % of high-risk patients who received extended pelvic lymph node dissection [32]. Therefore, an increased incidence of occult pelvic lymph node metastases is expected in patients with locally advanced prostate cancer, and it is reasonable to expect an improved outcome in patients at high risk of pelvic lymph node metastases by applying prophylactic radiation therapy to the pelvic lymph nodes.

However, the three prospective randomized trials performed to date have failed to demonstrate progression-free and survival benefits of prophylactic irradiation to the pelvic lymph nodes [33–35]. In those studies, the total dose delivered to the prostate was suboptimal (65–70 Gy) in relation to the current standard of care. This is because if 45–50 Gy had already been delivered to the pelvic lymph node regions using conventional techniques, including 3D-CRT, the prostate dose has to be limited to 70 Gy to avoid a higher risk of late radiation injury to the rectum. Only a subset analysis in the RTOG 94–13 trial, which compared both prostate-only radiation (PORT) with whole pelvic radiation (WPRT) and short-term neoadjuvant hormonal therapy (NA-HT) with adjuvant hormonal therapy (A-HT), suggested a slight progression-free benefit in patients treated with WPRT combined with NAHT compared with the other three arms [33]. In contrast, the beneficial effects of dose escalation to the prostate on biochemical recurrence-free outcomes are more consistent and clear, as described in Sect. 20.2.1 [15, 14]. Therefore, based on the available data, it would be reasonable to recommend selecting local dose escalation over prophylactic pelvic node irradiation using 3D-CRT.

On the other hand, IMRT can realize local dose escalation to both the prostate and pelvic lymph node areas concurrently using the simultaneous integrated boost (SIB) IMRT technique [36]. In this approach, the effect of pelvic irradiation can be superimposed on improved local control using high-dose local irradiation;

therefore, the effect of pelvic irradiation might be significant. To date, SIB-IMRT has been applied mainly to high-risk patients and appears to be acceptable and promising in terms of both short- to intermediate-term clinical outcomes and toxicity [37–39]. However, its true clinical benefits need be validated in future prospective randomized trials.

20.2.3 Combined Hormonal Therapy

A meta-analysis of randomized controlled trials demonstrated that combined HT with EBRT improves survival outcomes significantly compared with EBRT alone in patients with intermediate- and high-risk diseases [30]. However, there are several limitations to the studies included in this meta-analysis. First, for the most part, conventional radiation techniques were used. In addition, the total dose delivered to the prostate ranged from 65 to 70 Gy, which is considerably lower than the modern standard of care using more sophisticated radiation techniques such as 3D-CRT and IMRT. In addition, the optimal duration and timing of HT remain controversial. The clear survival advantages reported in a study comparing EBRT followed by long-term HT with EBRT alone [40] were small or insignificant when compared with short-term HT plus EBRT [41, 42].

Another important unsolved issue regarding HT is the optimal timing for initiating salvage hormonal therapy (S-HT) in patients who develop recurrence after EBRT. It would be expected that if the initiation of S-HT is delayed, a poor prognosis is likely compared with patients who are treated at an earlier date. Consistent with this, it was reported that the early initiation of S-HT improved the prognoses of patients with recurrent disease after definitive EBRT for prostate cancer [43, 44]. However, the timing of the initiation of S-HT has not been defined in any previous studies assessing the impact of HT on EBRT. In addition, in the RTOG 86–10 trial, which explored the effect of short-term NA-HT in patients with bulky localized tumors, S-HT was not initiated in ~30 % of recurrent cases until distant metastases had become obvious, and S-HT was not started in >70 % of patients until their PSA values exceeded 20 ng/ml [44]. This suggests that S-HT was delayed in the radiotherapy alone or short-term HT arms in previous randomized trials, although the exact data were not reported.

Intermittent hormonal therapy (I-HT) has become a popular treatment modality to delay the development of castration resistance and reduce the side effects and cost of HT by replacing continuous HT (C-HT). Recently published meta-analyses and systematic reviews have reported that I-HT is comparable with C-HT in terms of overall survival in patients with recurrent, locally advanced, or metastatic prostate cancer [45–47]. This suggests that treating locally advanced prostate cancer with EBRT plus short-term HT, with an early salvage policy, which is similar to performing I-HT after completing short-term NA-HT plus EBRT, could achieve comparable survival outcomes to those patients who had received long-term A-HT after EBRT. This is consistent with data reported from Kyoto University using

IMRT combined with short-term NA-HT under an early salvage policy that we describe below.

Therefore, the optimal duration of HT in combination with high-dose EBRT using modern radiation techniques should be confirmed in a prospective study in which S-HT is initiated during the early phase of recurrence.

20.2.4 Hypofractional Approaches

Because the α/β value of prostate cancer has been reported to be lower than those of normal structures adjacent to the prostate, the fractionation sensitivity differential (tumor/normal tissue) theoretically favors the use of hypofractionated radiotherapy schedules. However, α/β and the application of a hypofractioned radiation regimen to prostate cancer remain controversial [48]. Recently, two studies based on large-scale data confirmed that the α/β for prostate cancer is low [49, 50]. An assembled study assessing seven datasets including ~6,000 cases confirmed that the value of α/β for the pooled data was 1.4 (95 % confidence interval [CI]=0.9–2.2) Gy [49]. There was no significant difference between the α/β value for the three risk groups (low, intermediate, and high risk), and androgen deprivation did not affect the value of α/β. Another study analyzed the PSA data of 5,093 patients from six institutions treated for localized prostate cancer using EBRT without planned androgen deprivation and estimated that the α/β ratio was 1.55 Gy (95 % confidence band, 0.46–4.52 Gy) [50].

Therefore, a hypofractional approach is theoretically more favorable for both biochemical control and late toxicity. This approach is also favorable in terms of patient convenience and cost-effectiveness. However, recent randomized studies with relatively long-term follow-up periods, in which high-dose deliveries in conventional fraction sizes (76/80 Gy at 2 Gy per fraction) were compared with a hypofractional approach (70.2 Gy at 2.7-/63 Gy at 3.1 per fraction), failed to show improved biochemical control using hypofractional regimens [51, 52]. In both studies, the biochemical outcomes and late radiation toxicities were comparable between fractionation schedules. Although more data and longer follow-up periods are necessary to draw a definitive conclusion, a hypofractional approach might be beneficial to patients in terms of a shorter treatment time and lower cost if the outcomes were at least comparable with conventional fractionations.

20.2.5 High-Dose IMRT for Locally Advanced Cases at Kyoto University

At Kyoto University, we currently perform image-guided (IG) IMRT with prostate-based positional correction based on cone-beam CT using Vero4DRT (MHI-TM2000) (Mitsubishi Heavy Industries, Ltd., Tokyo, Japan, and BrainLAB, Feldkirchen, Germany) [53, 54] for localized (T1-T2N0M0) prostate cancer. However, we have used bony structure-based IG-IMRT using Novalis (BrainLAB, Feldkirchen,

Germany) for locally advanced cases. This is because the theoretically reasonable margin reduction associated with switching from bony structure-based IGRT to prostate position-based IGRT (5 mm reduction in the anterior-posterior and cranial-caudal directions, 3 mm in the left-right directions) [55] resulted in a significant degradation in the biochemical control rate [56]. We are particularly concerned with the possible higher risk of margin reduction in locally advanced (T3-T4N0M0) cases.

We reported the IMRT outcomes of T3-4N0M0 prostate cancer combined with NA-HT previously [57]. Between January 2003 and May 2006, 103 Japanese patients with T3-4N0M0 adenocarcinoma of the prostate were definitively treated with IMRT. The median age of the patients was 71 years (range 51–80). Initial PSA (iPSA) values were 5–179 ng/ml (mean, 37; median, 26 ng/ml). Gleason score (GS) was distributed as follows: 6 in eight patients, 7 in 49, and ≥8 in 46. NA-HT (range, 4–15 months; median, 6) was performed in all cases. Seventy-eight Gy in 39 fractions was administered to the prostate and seminal vesicles in 94 patients, whereas the doses were reduced to 70 or 74 Gy in nine patients due to factors such as anticoagulant therapy. A detailed description of the IMRT treatment planning was reported previously [58, 59]. No patients received A-HT, and a uniform S-HT policy was generally applied. S-HT was initiated once the PSA value exceeded 4 ng/ml in a monotone increasing manner or when any clinical failure was detected.

The median follow-up period was 68 months (range 21–89). The estimated biochemical relapse-free survival (bRFS) rate based on the Phoenix definition using the Kaplan-Meyer method was 69 % (95 %CI = 59–78) at 5 years (Fig. 20.1). Among age, NA-HT duration, GS, iPSA, and T-factor, only iPSA and T-factor were significant independent factors for predicting bRFS outcome in both univariate and multivariate analyses. The 5-year bRFS rates for T3a and T3b-T4 cases were 78 (95 %CI = 68–88) and 50 % (95 %CI = 32–68), respectively ($p = 0.0089$; Fig. 20.2).

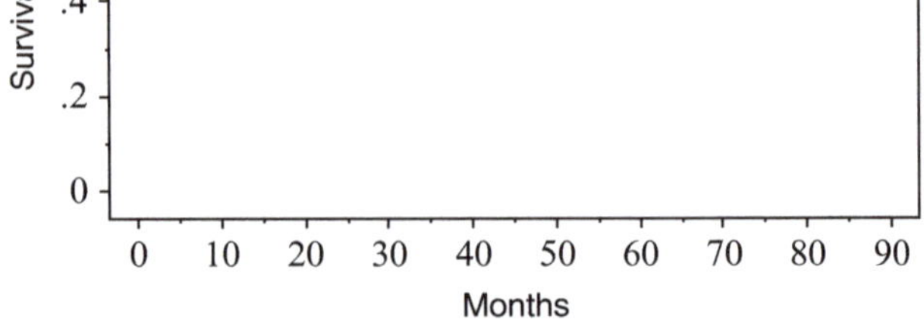

Fig. 20.1 Kaplan-Meier curve estimating PSA failure-free survival rate based on the Phoenix definition. The 5-year survival rate was 69 % (95 %CI = 59–78)

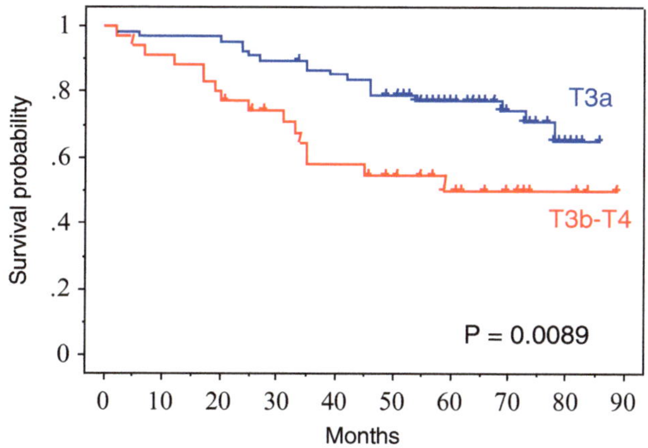

Fig. 20.2 Kaplan-Meier curve estimating PSA failure-free survival rate by T-stage based on the Phoenix definition. The 5-year survival rate was significantly better in T3a cases (78 %, 95 %CI=68–88) compared with T3b–T4 (50 %, 95 %CI=32–68; $p=0.0089$)

Fig. 20.3 Kaplan-Meier curve estimating the salvage hormonal therapy-free (S-HT-free) rate. The 5-year S-HT-free rate was 75 % (95 %CI, 68–83)

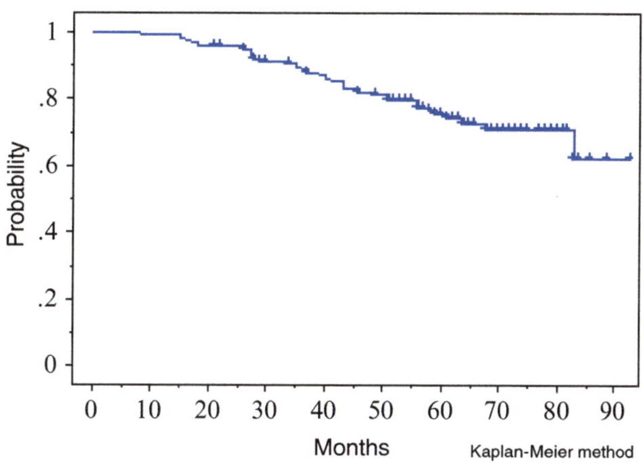

The 5-year salvage HT-free rate was 75 % (95 %CI=68–83; Fig. 20.3), and the prostate cancer-specific and overall survival rates were 98 (95 %CI=95–100) and 95 % (95 %CI=91–99), respectively (Figs. 20.4 and 20.5).

Therefore, high-dose localized IMRT in patients with locally advanced prostate cancer combined with NA-HT using an early salvage policy (PSA >4 ng/ml) resulted in both a high salvage-free rate (75 % at 5 years) and excellent survival outcomes in an intermediate follow-up period. This approach might be an alternative to uniformly administering long-term A-HT, although longer follow-up and prospective studies are needed to confirm this hypothesis. Typical PSA transitions in patients who maintained a long-term HT-free status are presented below in Sects. 20.4.1 and 20.4.2.

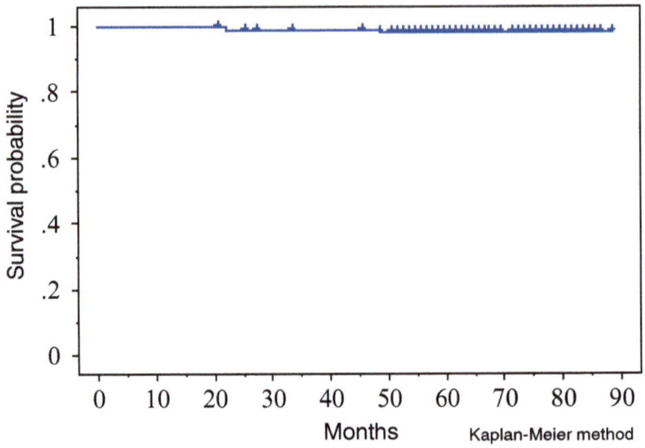

Fig. 20.4 Kaplan-Meier curve estimating the disease-specific survival (DSS) rate. The 5-year DSS rate was 98 % (95 %CI, 95–100 %)

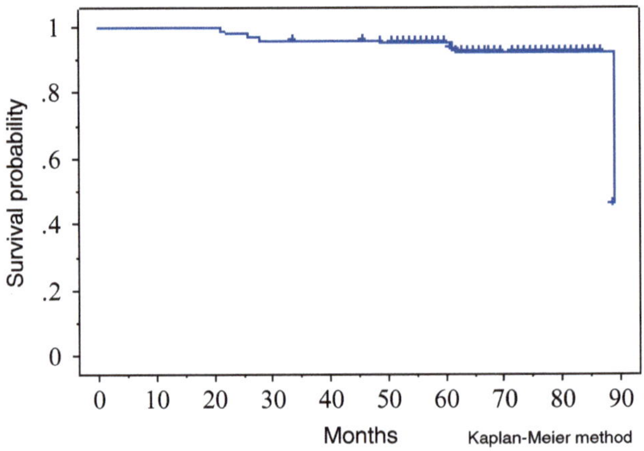

Fig. 20.5 Kaplan-Meier curve estimating the overall survival (OAS) rate. The 5-year OAS rate was 95 % (95 %CI, 91–99 %)

We also applied whole pelvic (WP) SIB-IMRT to locally advanced tumors with very high-risk features such as PSA >30 ng/ml or T3b [60]. We maintained the same total dose and fraction size as to the prostate (78 Gy in 2 Gy per fraction) so that the impact of adding WP irradiation could be assessed by comparing the data with historical controls who received the same dose of localized irradiation alone. Therefore, the pelvic lymph node area was simultaneously irradiated using the SIB technique at 1.5 Gy per fraction up to 58.5 Gy. The interim outcomes of this approach are reported below. The policies for combined NA-HT and S-HT were the same as for localized IMRT. Therefore, A-HT was not administered to this cohort of 46 patients with T3-T4N0M0 disease. With a median follow-up period of 44 months (range,

18–72), the bRFS rate according to the Phoenix definition and salvage-free rates at 42 months were 78 (95 %CI=65–91 %) and 80 % (95 %CI=68–93 %), respectively. Despite the very high-risk nature of these tumors (median iPSA, 59.4 ng/ml), these interim outcomes are encouraging. A typical case is presented below in Section 21.4.3.

20.3 Node-Positive (N1) Prostate Cancer

20.3.1 EBRT for Node-Positive (N1) Prostate Cancer

Node-positive, but distant metastasis-negative (TanyN1M0), prostate cancer is relatively rare in patients diagnosed with prostate cancer at their first visit [5, 61]. The prognosis of patients with TanyN1M0 disease is generally poor [62]. Because TanyN1M0 is relatively rare, there are few prospective studies available. In addition, the number of cases included in retrospective studies was not always sufficient to draw definitive conclusions. Nevertheless, it is known that EBRT plus long-term hormonal therapy resulted in better survival outcomes compared with EBRT alone, radical prostatectomy alone, or immediate HT alone [62]. Pollack et al. reported that the disease-free and overall survival rates at 5–10 years were ~20 and 50 %, respectively, in patients treated with EBRT, radical prostatectomy, or hormone therapy alone [62]. In contrast, the respective survival rates in patients treated with combined hormonal therapy plus EBRT were ~50 and 70 %.

A recent report clearly demonstrated that the survival outcomes of patients with N1M0 disease were improved significantly compared with those with M1 disease based on 14,000 cases registered in the Surveillance, Epidemiology, and End Results (SEER) database [63]. Based on the 14,697 stage IV cases in 17 SEER registries, the 5-year overall survival rates for T4N0M0, TxN1M0, and TxN0M1 cases were 59, 79, and 22 % ($p<0.001$), and the 5-year prostate cancer-specific survival rates were 71, 84, and 35 % ($p<0.001$), respectively. Because the survival rates were significantly better in patients with TxN1M0 compared with TxN0M1 disease, local therapy to the prostate and pelvic lymph nodes is considered to be an essential part of the optimal treatment of patients with lymph node-positive tumors. Nevertheless, this intensive treatment regimen is unnecessary in the substantial number of patients with a slow natural history or high competing death risk due to a coexisting illness [64].

20.3.2 Whole Pelvic IMRT for Node-Positive (N1) Prostate Cancer

It is not possible to safely realize both high-dose local irradiation to the prostate and pelvic lymph nodes using conventional EBRT techniques. However, IMRT overcomes this challenge via SIB-IMRT [65]. It was reported that SIB-IMRT could be applied safely to prostate cancer, although its long-term oncological efficacy should be validated with longer follow-up [37–39, 66].

At Kyoto University, WP SIB-IMRT has also been used to treat patients with cTanyN1M0 disease. Similar to WP SIB-IMRT for T3-T4N0M0 cases, 78, 66.3, and 58.5 Gy were administered simultaneously to the prostate and seminal vesicles, metastatic lymph nodes, and prophylactic pelvic lymph node area in 39 fractions, respectively (case 4, Sect. 20.4.4).

20.4 Case Presentations

20.4.1 Case 1: cT3aN0M0, iPSA 55.4 ng/ml, Gleason Score 3 + 4 = 7

A 73-year-old male was diagnosed with T3aN0M0 prostate cancer. His initial PSA (iPSA) was 55.4 ng/ml, and Gleason score was 3 + 4. The prescribed NA-HT consisted of goserelin plus bicalutamide for 6 months, followed by localized irradiation with 78 Gy that was delivered using the 5-field dynamic multileaf collimator (DMLC) IMRT technique with a 15-MV X-ray.

The clinical target volume (CTV) was defined as the prostate plus the proximal 2/3 of the seminal vesicles (CTV_PSV). A 9-mm margin was added to the CTV_PSV to create the planning target volume (PTV, defined as PTV_PSV; Fig. 20.6). Seventy-eight Gy in 39 fractions was administered to the PTV_PSV. The dose distribution and dose-volume histogram (DVH) data are shown in Figs. 20.7 and 20.8, respectively. No A-HT was given after the completion of IMRT. His changing

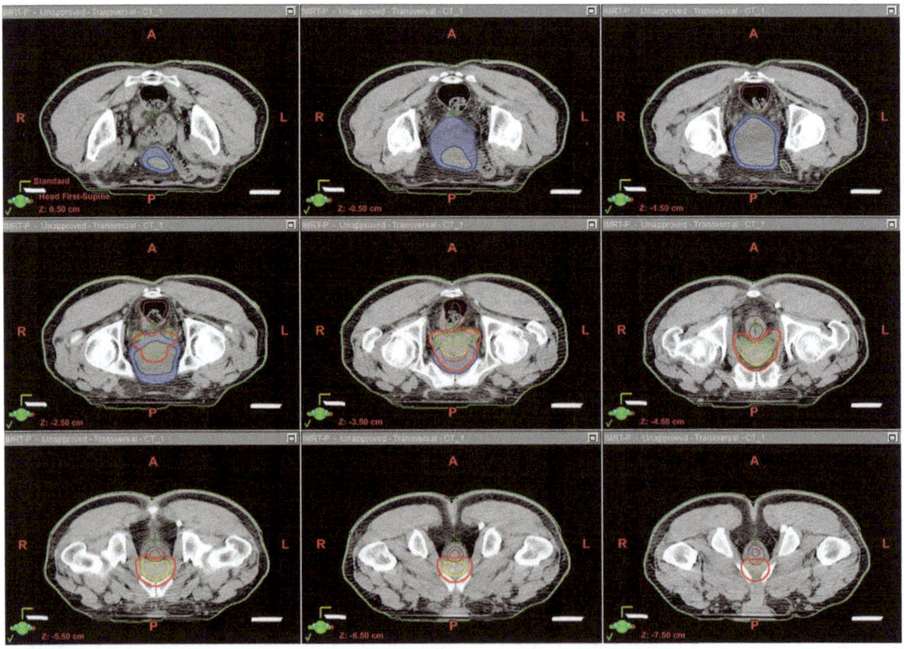

Fig. 20.6 The targets and organs at risk for case 1

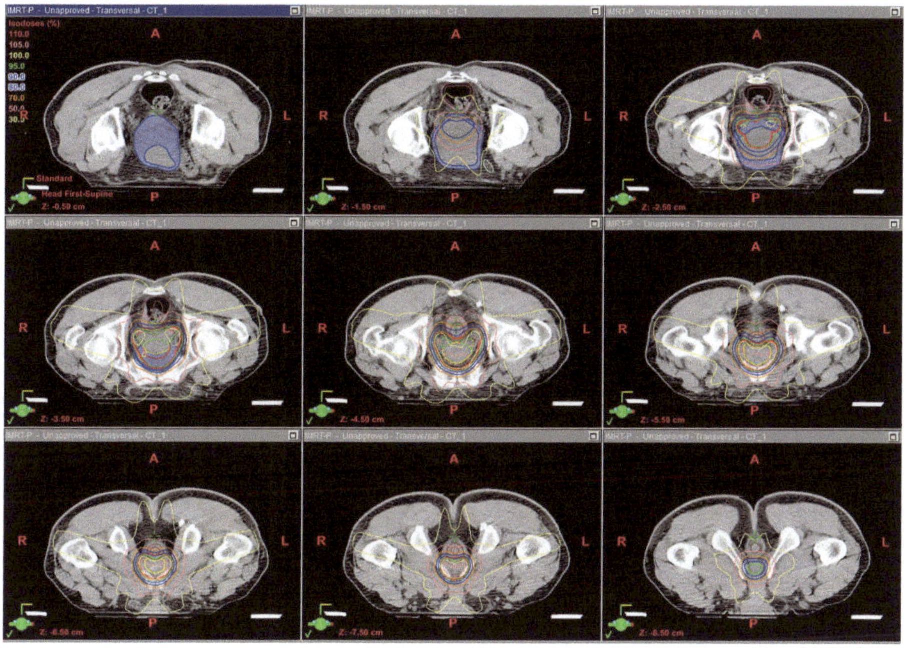

Fig. 20.7 Dose distributions of the IMRT plan for case 1

PSA values are shown in Fig. 20.9. His PSA levels remained <0.2 ng/ml during a 10-year follow-up period after IMRT, despite no additional treatment (Fig. 20.9).

20.4.2 Case 2: cT4N0M0, iPSA 104 ng/ml, Gleason Score 4 + 5 = 9

A 60-year-old male presented with T4N0M0 prostate cancer. Bladder neck invasion was confirmed based on cystoscopic examination. His iPSA was 104 ng/ml, and Gleason score was 4 + 5. The administered NA-HT consisted of luteinizing hormone-releasing hormone (LH-RH) agonists for 6 months.

The CTV_PSV was defined as the prostate, the proximal 2/3 of the seminal vesicles, and the bladder neck. The PTV_PSV was the same as those used in case 1. The IMRT plan using the DMLC technique consisted of 5 fields of 15-MV X-ray. A total of 78 Gy in 39 fractions was administered to the PTV_PSV. No A-HT was given after the completion of IMRT. The dose distribution is shown in Fig. 20.10. The transition in his PSA values is shown in Fig. 20.11. After completing IMRT, his PSA increased gradually and reached 2.24 ng/ml 6.3 years after IMRT, which was judged as a PSA failure based on the Phoenix definition. However, the PSA levels did not rise continually, but stabilized at ~3.5 ng/ml 9 years after IMRT. This patient has remained in good health with no symptoms and HT-free status.

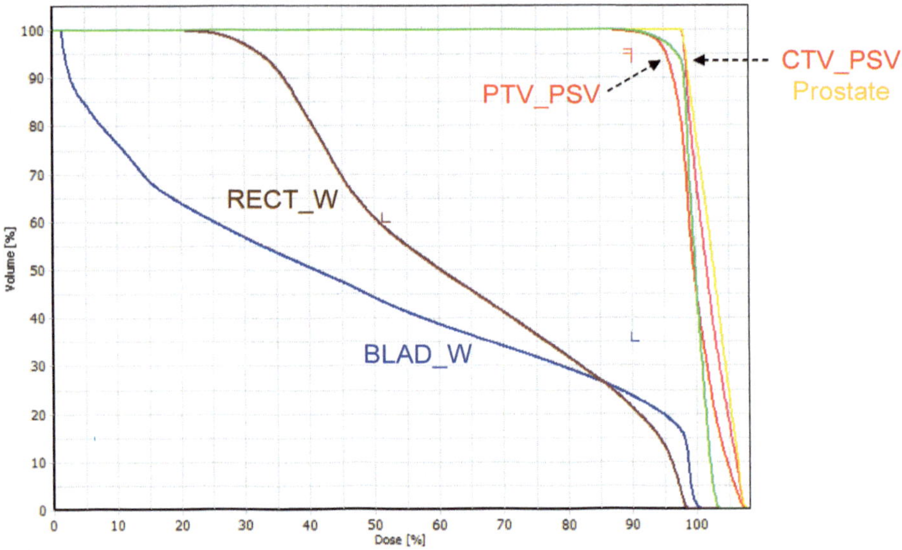

Structure	Index	Value	Structure	Index	Value
PTV_PSV	Min dose [%]	83.80	Prostate	Mean dose [%]	102.70
	Max dose [%]	108.10		D50% [%]	102.74
	D98% [%]	93.99	BLAD_W	V40Gy [%]	43.43
	D2% [%]	106.57		V70Gy [%]	23.73
	V90% [%]	99.58	RECT_W	V40Gy [%]	59.24
	D95% [%]	95.43		V60Gy [%]	34.71
	Mean dose [%]	100.10		V70Gy [%]	21.45
	D50% [%]	99.70		V78Gy [%]	0.00
CTV_PSV	D95% [%]	98.37			
	Mean dose [%]	101.90			
	D98% [%]	98.15			
	D50% [%]	101.55			

Fig. 20.8 Dose-volume histogram data for case 1. *BLAD_W* bladder wall, *RECT_W* rectal wall

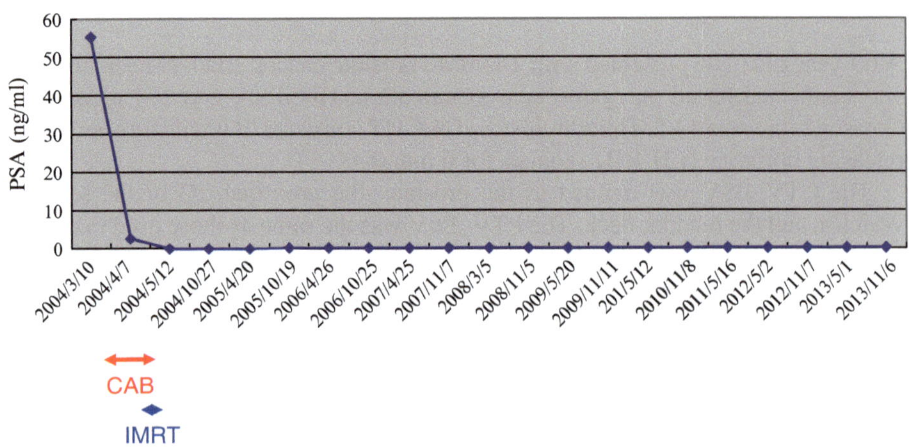

Fig. 20.9 PSA transition of case 1. *CAB* complete androgen blockade, *IMRT* intensity-modulated radiation therapy

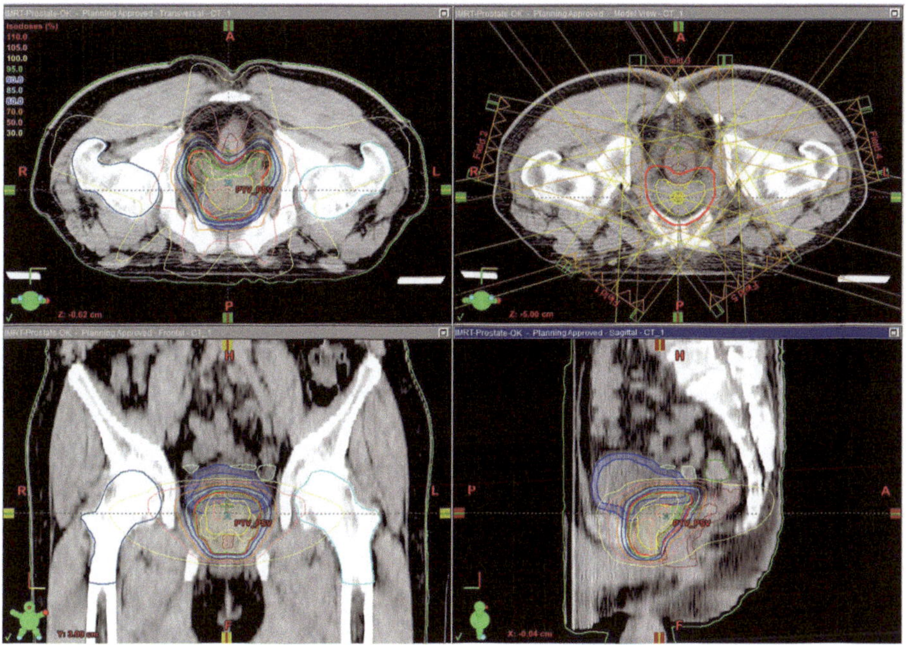

Fig. 20.10 Dose distributions of the IMRT plan for case 2

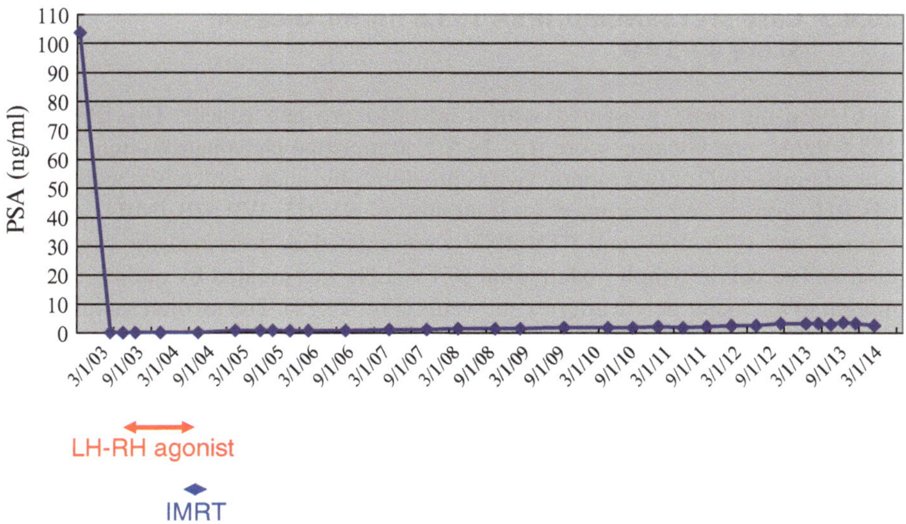

Fig. 20.11 PSA transition of case 2. *LH-RH* luteinizing hormone-releasing hormone, *IMRT* intensity-modulated radiation therapy

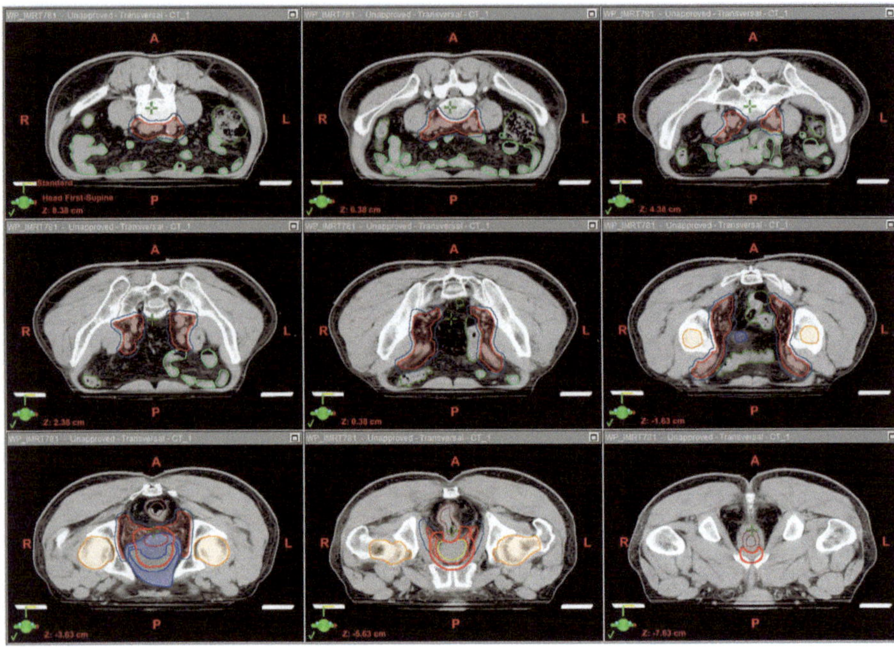

Fig. 20.12 The targets and organs at risk for case 3

20.4.3 Case 3: cT3aN0M0, iPSA 153.8 ng/ml, Gleason Score 4 + 5 = 9

A 61-year-old male presented with T3aN0M0 prostate cancer. His iPSA was 153.8 ng/ml, and Gleason score was 4 + 5. Estramustine phosphate sodium hydrate was administered for two months by his previous physician, which was switched to LH-RH agonists and continued for 9 months as NA-HT. WP SIB-IMRT was performed, and CTV_PSV and PTV_PSV were created as described for case 1 and case 2. The pelvic lymph node region (CTV_LN) was created by adding a 7-mm margin to the major pelvic arteries and veins (Fig. 20.12). The median sacral lymph nodes were not included in the CTV_LN because the recommendations of clinical and pathological studies on prostate cancer 2010 (4th edition) [67] did not include the median sacral lymph nodes in the regional lymph nodes of prostate cancer. PTV_LN was created by adding a 5-mm margin to the CTV_LN. The WP SIB-IMRT plan using the DMLC technique consisted of 7 fields of 15-MV X-ray. A total of 78 and 58.5 Gy in 39 fractions were administered to the PTV_PSV and PTV_LN, respectively (Fig. 20.13). Although no A-HT was administered after WP SIB-IMRT, his PSA remained low (<0.4 ng/ml) throughout the 6.5-year follow-up period after IMRT (Fig. 20.14).

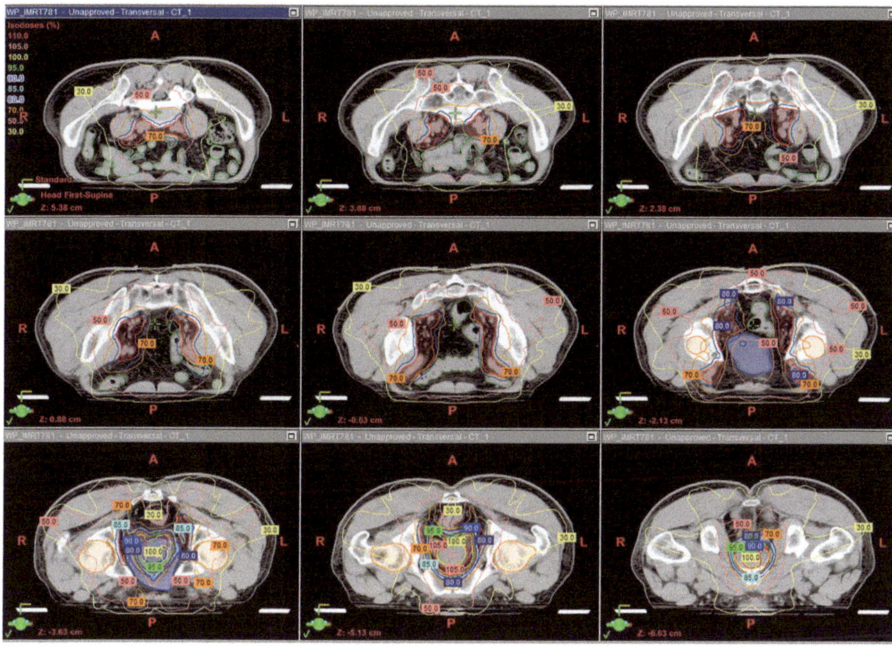

Fig. 20.13 Dose distributions of the whole pelvic simultaneous integrated boost IMRT plan for case 3

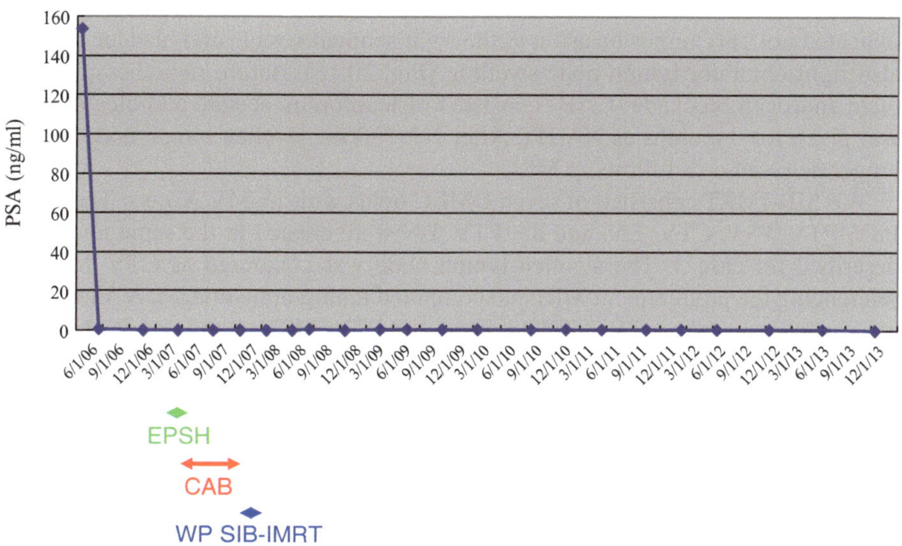

Fig. 20.14 PSA transition of case 3. *EPSH* estramustine phosphate sodium hydrate, *CAB* complete androgen blockade, *WP SIB-IMRT* whole pelvic simultaneous integrated boost intensity-modulated radiation therapy

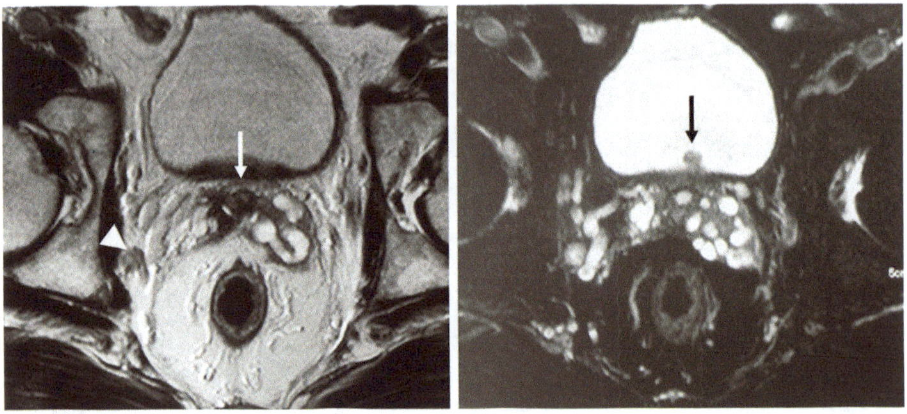

Fig. 20.15 Pretreatment MRI images for case 4. *Left*, invasion to the base of the seminal vesicles (*white arrow*) and a swollen right obturator lymph node (*white arrowhead*) were detected. *Right*, bladder invasion (*black arrow*) was apparent

20.4.4 Case 4: cT4N1M0, iPSA 21.1 ng/ml, Gleason Score 4+5=9

A 73-year-old male presented with T4N1M0 prostate cancer. His iPSA was 21.1 ng/ml, and Gleason score was 4+5. Pretreatment magnetic resonance imaging (MRI) indicated not only tumor invasion to the right seminal vesicle and bladder wall but also right obturator lymph node swelling (Fig. 20.15). Before radiotherapy, complete androgen blockade (CAB) consisted of leuprorelin acetate and bicalutamide was given for 6 months as NA-HT. After NA-HT, the swollen lymph node was no longer detected on a follow-up MRI.

WP SIB-IMRT consisted of seven DMLC fields with 15-MV X-rays. The CTV_PSV, PTV_PSV, CTV_LN, and the PTV_LN were created in the same manner as described for case 3. The swollen lymph node was contoured as CTV_MET by referencing the pretreatment MRI and computed tomography images. A 5-mm margin was added to the CTV_MET to create the PTV_MET. A total of 78, 66.3 and 58.5 Gy were administered simultaneously in 39 fractions to the PTV_PSV, PTV_MET, and PTV_LN, respectively (Fig. 20.16). CAB was continued during WP SIB-IMRT, and A-HT consisting of leuprorelin acetate alone was continued for 2 years after WP SIB-IMRT. His PSA levels were maintained ~0.3 ng/ml 5.4 years even after stopping A-HT (Fig. 20.17).

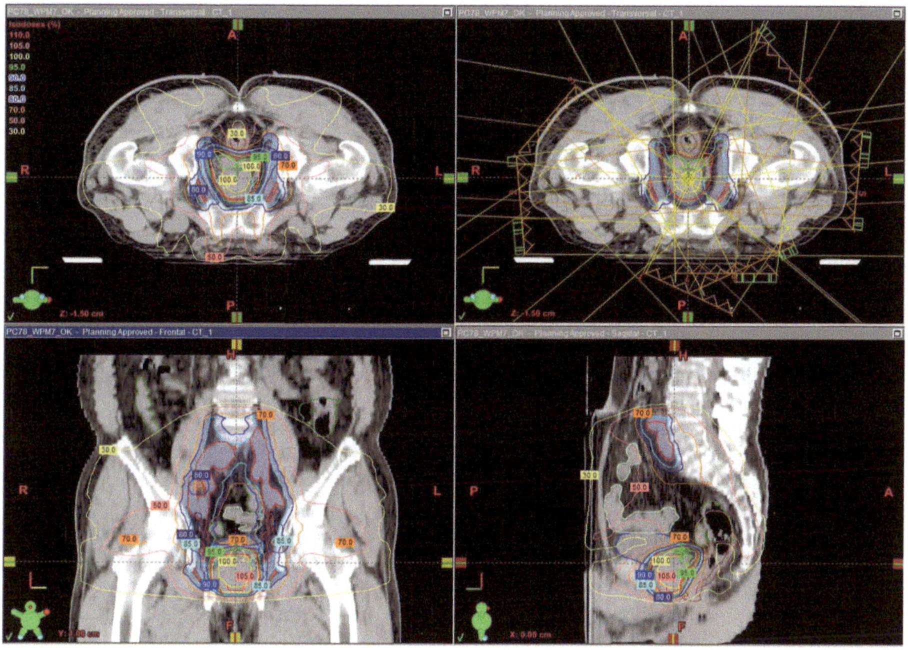

Fig. 20.16 Dose distributions of the whole pelvic simultaneous integrated boost IMRT plan for case 4

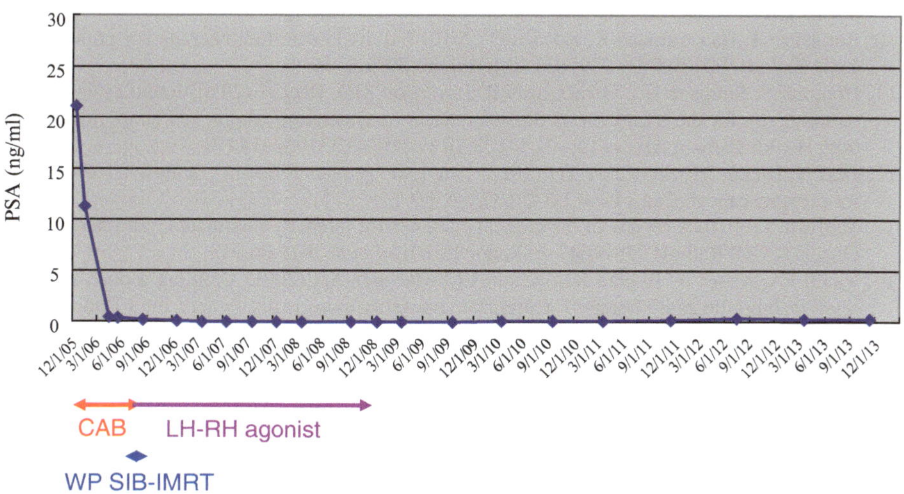

Fig. 20.17 PSA transition of case 4. *CAB* complete androgen blockade, *WP SIB-IMRT* whole pelvic simultaneous integrated boost intensity-modulated radiation therapy

References

1. Jemal A, Siegel R, Ward E, Murray T, Xu J, Thun MJ (2007) Cancer statistics, 2007. CA Cancer J Clin 57(1):43–66
2. Ferlay J, Steliarova-Foucher E, Lortet-Tieulent J, Rosso S, Coebergh JW, Comber H, Forman D, Bray F (2013) Cancer incidence and mortality patterns in Europe: estimates for 40 countries in 2012. Eur J Cancer 49(6):1374–1403. doi:10.1016/j.ejca.2012.12.027
3. Namiki M, Akaza H, Lee SE, Song JM, Umbas R, Zhou L, Lee BC, Cheng C, Chung MK, Fukagai T, Hinotsu S, Horie S (2010) Prostate Cancer Working Group report. Jpn J Clin Oncol 40(Suppl 1):i70–i75. doi:10.1093/jjco/hyq130
4. Services CfCCaI (2014) Cancer rates in Japan. http://ganjoho.jp/public/statistics/pub/statistics01.html
5. Cancer Registration Committee of the Japanese Urological A (2005) Clinicopathological statistics on registered prostate cancer patients in Japan: 2000 report from the Japanese Urological Association. Int J Urol 12(1):46–61. doi:10.1111/j.1442-2042.2004.00984.x
6. Tanaka N, Fujimoto K, Hirayama A, Yoneda T, Yoshida K, Hirao Y (2010) Trends of the primary therapy for patients with prostate cancer in Nara uro-oncological research group (NUORG): a comparison between the CaPSURE data and the NUORG data. Jpn J Clin Oncol 40(6):588–592. doi:10.1093/jjco/hyq008
7. Ogawa K, Nakamura K, Sasaki T, Onishi H, Koizumi M, Araya M, Shioyama Y, Okamoto A, Mitsumori M, Teshima T, Japanese Patterns of Care Study Working Subgroup of Prostate C (2008) Radical external beam radiotherapy for prostate cancer in Japan: differences in the patterns of care among Japan, Germany, and the United States. Radiat Med 26(2):57–62. doi:10.1007/s11604-007-0195-6
8. Network NCC (2014) NCCN clinical practice guidelines in oncology: prostate cancer version 1.2014. http://www.nccn.org/
9. Heidenreich A, Bastian PJ, Bellmunt J, Bolla M, Joniau S, van der Kwast T, Mason M, Matveev V, Wiegel T, Zattoni F, Mottet N (2014) EAU guidelines on prostate cancer part 1: screening, diagnosis, and local treatment with curative intent-update 2013. Eur Urol 65(1):124–137. doi:10.1016/j.eururo.2013.09.046
10. Budiharto T, Haustermans K, Kovacs G (2010) External beam radiotherapy for prostate cancer. J Endourol 24(5):781–789. doi:10.1089/end.2009.0436
11. Hummel S, Simpson EL, Hemingway P, Stevenson MD, Rees A (2010) Intensity-modulated radiotherapy for the treatment of prostate cancer: a systematic review and economic evaluation. Health Technol Assess 14(47):1–108, iii-iv. doi:10.3310/hta14470
12. Ohri N, Dicker AP, Showalter TN (2012) Late toxicity rates following definitive radiotherapy for prostate cancer. Can J Urol 19(4):6373–6380
13. Staffurth J (2010) A review of the clinical evidence for intensity-modulated radiotherapy. Clin Oncol (R Coll Radiol) 22(8):643–657. doi:10.1016/j.clon.2010.06.013
14. Vicini FA, Abner A, Baglan KL, Kestin LL, Martinez AA (2001) Defining a dose–response relationship with radiotherapy for prostate cancer: is more really better? Int J Radiat Oncol Biol Phys 51(5):1200–1208
15. Viani GA, Stefano EJ, Afonso SL (2009) Higher-than-conventional radiation doses in localized prostate cancer treatment: a meta-analysis of randomized, controlled trials. Int J Radiat Oncol Biol Phys 74(5):1405–1418. doi:10.1016/j.ijrobp.2008.10.091
16. Shipley WU, Verhey LJ, Munzenrider JE, Suit HD, Urie MM, McManus PL, Young RH, Shipley JW, Zietman AL, Biggs PJ et al (1995) Advanced prostate cancer: the results of a randomized comparative trial of high dose irradiation boosting with conformal protons compared with conventional dose irradiation using photons alone. Int J Radiat Oncol Biol Phys 32(1):3–12. doi:10.1016/0360-3016(95)00063-5
17. Sathya JR, Davis IR, Julian JA, Guo Q, Daya D, Dayes IS, Lukka HR, Levine M (2005) Randomized trial comparing iridium implant plus external-beam radiation therapy with external-beam radiation therapy alone in node-negative locally advanced cancer of the prostate. J Clin Oncol 23(6):1192–1199. doi:10.1200/JCO.2005.06.154

18. Zietman AL, DeSilvio ML, Slater JD, Rossi CJ Jr, Miller DW, Adams JA, Shipley WU (2005) Comparison of conventional-dose vs high-dose conformal radiation therapy in clinically localized adenocarcinoma of the prostate: a randomized controlled trial. JAMA 294(10):1233–1239. doi:10.1001/jama.294.10.1233
19. Dearnaley DP, Sydes MR, Graham JD, Aird EG, Bottomley D, Cowan RA, Huddart RA, Jose CC, Matthews JH, Millar J, Moore AR, Morgan RC, Russell JM, Scrase CD, Stephens RJ, Syndikus I, Parmar MK, collaborators RT (2007) Escalated-dose versus standard-dose conformal radiotherapy in prostate cancer: first results from the MRC RT01 randomised controlled trial. Lancet Oncol 8(6):475–487. doi:10.1016/S1470-2045(07)70143-2
20. Al-Mamgani A, van Putten WL, Heemsbergen WD, van Leenders GJ, Slot A, Dielwart MF, Incrocci L, Lebesque JV (2008) Update of Dutch multicenter dose-escalation trial of radiotherapy for localized prostate cancer. Int J Radiat Oncol Biol Phys 72(4):980–988. doi:10.1016/j.ijrobp.2008.02.073
21. Kuban DA, Levy LB, Cheung MR, Lee AK, Choi S, Frank S, Pollack A (2011) Long-term failure patterns and survival in a randomized dose-escalation trial for prostate cancer. Who dies of disease? Int J Radiat Oncol Biol Phys 79(5):1310–1317. doi:10.1016/j.ijrobp.2010.01.006
22. Beckendorf V, Guerif S, Le Prise E, Cosset JM, Bougnoux A, Chauvet B, Salem N, Chapet O, Bourdain S, Bachaud JM, Maingon P, Hannoun-Levi JM, Malissard L, Simon JM, Pommier P, Hay M, Dubray B, Lagrange JL, Luporsi E, Bey P (2011) 70 Gy versus 80 Gy in localized prostate cancer: 5-year results of GETUG 06 randomized trial. Int J Radiat Oncol Biol Phys 80(4):1056–1063. doi:10.1016/j.ijrobp.2010.03.049
23. Alicikus ZA, Yamada Y, Zhang Z, Pei X, Hunt M, Kollmeier M, Cox B, Zelefsky MJ (2011) Ten-year outcomes of high-dose, intensity-modulated radiotherapy for localized prostate cancer. Cancer 117(7):1429–1437. doi:10.1002/cncr.25467
24. Spratt DE, Pei X, Yamada J, Kollmeier MA, Cox B, Zelefsky MJ (2013) Long-term survival and toxicity in patients treated with high-dose intensity modulated radiation therapy for localized prostate cancer. Int J Radiat Oncol Biol Phys 85(3):686–692. doi:10.1016/j.ijrobp.2012.05.023
25. Vora SA, Wong WW, Schild SE, Ezzell GA, Andrews PE, Ferrigni RG, Swanson SK (2013) Outcome and toxicity for patients treated with intensity modulated radiation therapy for localized prostate cancer. J Urol 190(2):521–526. doi:10.1016/j.juro.2013.02.012
26. Gadia R, Leite ET, Gabrielli FG, Marta GN, Arruda FF, Abreu CV, Hanna SA, Haddad CK, Silva JF, Carvalho HA, Garicochea B (2013) Outcomes of high-dose intensity-modulated radiotherapy alone with 1 cm planning target volume posterior margin for localized prostate cancer. Radiat Oncol 8(1):285. doi:10.1186/1748-717X-8-285
27. Sia J, Joon DL, Viotto A, Mantle C, Quong G, Rolfo A, Wada M, Anderson N, Rolfo M, Khoo V (2011) Toxicity and long-term outcomes of dose-escalated intensity modulated radiation therapy to 74Gy for localised prostate cancer in a single Australian centre. Cancers (Basel) 3(3):3419–3431. doi:10.3390/cancers3033419
28. Takeda K, Takai Y, Narazaki K, Mitsuya M, Umezawa R, Kadoya N, Fujita Y, Sugawara T, Kubozono M, Shimizu E, Abe K, Shirata Y, Ishikawa Y, Yamamoto T, Kozumi M, Dobashi S, Matsushita H, Chida K, Ishidoya S, Arai Y, Jingu K, Yamada S (2012) Treatment outcome of high-dose image-guided intensity-modulated radiotherapy using intra-prostate fiducial markers for localized prostate cancer at a single institute in Japan. Radiat Oncol 7:105. doi:10.1186/1748-717X-7-105
29. Zelefsky MJ, Yamada Y, Kollmeier MA, Shippy AM, Nedelka MA (2008) Long-term outcome following three-dimensional conformal/intensity-modulated external-beam radiotherapy for clinical stage T3 prostate cancer. Eur Urol 53(6):1172–1179. doi:10.1016/j.eururo.2007.12.030
30. Bria E, Cuppone F, Giannarelli D, Milella M, Ruggeri EM, Sperduti I, Pinnaro P, Terzoli E, Cognetti F, Carlini P (2009) Does hormone treatment added to radiotherapy improve outcome in locally advanced prostate cancer?: meta-analysis of randomized trials. Cancer 115(15):3446–3456. doi:10.1002/cncr.24392
31. Abdollah F, Cozzarini C, Sun M, Suardi N, Gallina A, Passoni NM, Bianchi M, Tutolo M, Fossati N, Nini A, Dell'oglio P, Salonia A, Karakiewicz P, Montorsi F, Briganti A (2013)

Assessing the most accurate formula to predict the risk of lymph node metastases from prostate cancer in contemporary patients treated with radical prostatectomy and extended pelvic lymph node dissection. Radiother Oncol 109(2):211–216. doi:10.1016/j.radonc.2013.05.029
32. Heidenreich A, Ohlmann CH, Polyakov S (2007) Anatomical extent of pelvic lymphadenectomy in patients undergoing radical prostatectomy. Eur Urol 52(1):29–37. doi:10.1016/j.eururo.2007.04.020
33. Millar J, Boyd R, Sutherland J (2008) An update of the phase III trial comparing whole pelvic to prostate only radiotherapy and neoadjuvant to adjuvant total androgen suppression: updated analysis of RTOG 94–13, with emphasis on unexpected hormone/radiation interactions: in regard to Lawton et al. (Int J Radiat Oncol Biol Phys 2007;69:646–655.). Int J Radiat Oncol Biol Phys 71(1):316. doi:10.1016/j.ijrobp.2008.01.009, author reply 316
34. Pommier P, Chabaud S, Lagrange JL, Richaud P, Lesaunier F, Le Prise E, Wagner JP, Hay MH, Beckendorf V, Suchaud JP, Pabot du Chatelard PM, Bernier V, Voirin N, Perol D, Carrie C (2007) Is there a role for pelvic irradiation in localized prostate adenocarcinoma? Preliminary results of GETUG-01. J Clin Oncol 25(34):5366–5373. doi:10.1200/JCO.2006.10.5171
35. Asbell SO, Martz KL, Shin KH, Sause WT, Doggett RL, Perez CA, Pilepich MV (1998) Impact of surgical staging in evaluating the radiotherapeutic outcome in RTOG #77-06, a phase III study for T1BN0M0 (A2) and T2N0M0 (B) prostate carcinoma. Int J Radiat Oncol Biol Phys 40 (4):769–782. doi:S0360-3016(97)00926-7 [pii]
36. Cavey ML, Bayouth JE, Colman M, Endres EJ, Sanguineti G (2005) IMRT to escalate the dose to the prostate while treating the pelvic nodes. Strahlenther Onkol 181(7):431–441. doi:10.1007/s00066-005-1384-9
37. Bayley A, Rosewall T, Craig T, Bristow R, Chung P, Gospodarowicz M, Menard C, Milosevic M, Warde P, Catton C (2010) Clinical application of high-dose, image-guided intensity-modulated radiotherapy in high-risk prostate cancer. Int J Radiat Oncol Biol Phys 77(2):477–483. doi:10.1016/j.ijrobp.2009.05.006
38. McCammon R, Rusthoven KE, Kavanagh B, Newell S, Newman F, Raben D (2009) Toxicity assessment of pelvic intensity-modulated radiotherapy with hypofractionated simultaneous integrated boost to prostate for intermediate- and high-risk prostate cancer. Int J Radiat Oncol Biol Phys 75(2):413–420. doi:10.1016/j.ijrobp.2008.10.050
39. Sanguineti G, Endres EJ, Parker BC, Bicquart C, Little M, Chen G, Berilgen J (2008) Acute toxicity of whole-pelvis IMRT in 87 patients with localized prostate cancer. Acta Oncol 47(2):301–310. doi:10.1080/02841860701558849
40. Bolla M, Van Tienhoven G, Warde P, Dubois JB, Mirimanoff R-O, Storme G, Bernier J, Kuten A, Sternberg C, Billiet I, Torecilla JL, Pfeffer R, Cutajar CL, Van der Kwast T, Collette L (2010) External irradiation with or without long-term androgen suppression for prostate cancer with high metastatic risk: 10-year results of an EORTC randomised study. Lancet Oncol 11(11):1066–1073. doi:10.1016/s1470-2045(10)70223-0
41. Bolla M, de Reijke TM, Van Tienhoven G, Van den Bergh AC, Oddens J, Poortmans PM, Gez E, Kil P, Akdas A, Soete G, Kariakine O, van der Steen-Banasik EM, Musat E, Pierart M, Mauer ME, Collette L, Group ERO, Genito-Urinary Tract Cancer G (2009) Duration of androgen suppression in the treatment of prostate cancer. N Engl J Med 360(24):2516–2527. doi:10.1056/NEJMoa0810095
42. Horwitz EM, Bae K, Hanks GE, Porter A, Grignon DJ, Brereton HD, Venkatesan V, Lawton CA, Rosenthal SA, Sandler HM, Shipley WU (2008) Ten-year follow-up of radiation therapy oncology group protocol 92–02: a phase III trial of the duration of elective androgen deprivation in locally advanced prostate cancer. J Clin Oncol 26(15):2497–2504. doi:10.1200/JCO.2007.14.9021
43. Mydin AR, Dunne MT, Finn MA, Armstrong JG (2013) Early salvage hormonal therapy for biochemical failure improved survival in prostate cancer patients after neoadjuvant hormonal therapy plus radiation therapy–a secondary analysis of Irish clinical oncology research group 97–01. Int J Radiat Oncol Biol Phys 85(1):101–108. doi:10.1016/j.ijrobp.2012.03.001
44. Shipley WU, Desilvio M, Pilepich MV, Roach M 3rd, Wolkov HB, Sause WT, Rubin P, Lawton CA (2006) Early initiation of salvage hormone therapy influences survival in patients who

failed initial radiation for locally advanced prostate cancer: a secondary analysis of RTOG protocol 86–10. Int J Radiat Oncol Biol Phys 64(4):1162–1167. doi:10.1016/j.ijrobp.2005.09.039
45. Botrel TE, Clark O, Dos Reis RB, Pompeo AC, Ferreira U, Sadi MV, Bretas FF (2014) Intermittent versus continuous androgen deprivation for locally advanced, recurrent or metastatic prostate cancer: a systematic review and meta-analysis. BMC Urol 14(1):9. doi:10.1186/1471-2490-14-9
46. Niraula S, Le LW, Tannock IF (2013) Treatment of prostate cancer with intermittent versus continuous androgen deprivation: a systematic review of randomized trials. J Clin Oncol 31(16):2029–2036. doi:10.1200/JCO.2012.46.5492
47. Sciarra A, Abrahamsson PA, Brausi M, Galsky M, Mottet N, Sartor O, Tammela TL, Calais da Silva F (2013) Intermittent androgen-deprivation therapy in prostate cancer: a critical review focused on phase 3 trials. Eur Urol 64(5):722–730. doi:10.1016/j.eururo.2013.04.020
48. Miles EF, Lee WR (2008) Hypofractionation for prostate cancer: a critical review. Semin Radiat Oncol 18(1):41–47. doi:10.1016/j.semradonc.2007.09.006
49. Miralbell R, Roberts SA, Zubizarreta E, Hendry JH (2012) Dose-fractionation sensitivity of prostate cancer deduced from radiotherapy outcomes of 5,969 patients in seven international institutional datasets: alpha/beta = 1.4 (0.9-2.2) Gy. Int J Radiat Oncol Biol Phys 82(1):e17–e24. doi:10.1016/j.ijrobp.2010.10.075
50. Proust-Lima C, Taylor JM, Secher S, Sandler H, Kestin L, Pickles T, Bae K, Allison R, Williams S (2011) Confirmation of a low alpha/beta ratio for prostate cancer treated by external beam radiation therapy alone using a post-treatment repeated-measures model for PSA dynamics. Int J Radiat Oncol Biol Phys 79(1):195–201. doi:10.1016/j.ijrobp.2009.10.008
51. Pollack A, Walker G, Horwitz EM, Price R, Feigenberg S, Konski AA, Stoyanova R, Movsas B, Greenberg RE, Uzzo RG, Ma C, Buyyounouski MK (2013) Randomized trial of hypofractionated external-beam radiotherapy for prostate cancer. J Clin Oncol 31(31):3860–3868. doi:10.1200/JCO.2013.51.1972
52. Norkus D, Karklelyte A, Engels B, Versmessen H, Griskevicius R, De Ridder M, Storme G, Aleknavicius E, Janulionis E, Valuckas KP (2013) A randomized hypofractionation dose escalation trial for high risk prostate cancer patients: interim analysis of acute toxicity and quality of life in 124 patients. Radiat Oncol 8(1):206. doi:10.1186/1748-717X-8-206
53. Kamino Y, Takayama K, Kokubo M, Narita Y, Hirai E, Kawawda N, Mizowaki T, Nagata Y, Nishidai T, Hiraoka M (2006) Development of a four-dimensional image-guided radiotherapy system with a gimbaled X-ray head. Int J Radiat Oncol Biol Phys 66(1):271–278. doi:10.1016/j.ijrobp.2006.04.044
54. Takayama K, Mizowaki T, Kokubo M, Kawada N, Nakayama H, Narita Y, Nagano K, Kamino Y, Hiraoka M (2009) Initial validations for pursuing irradiation using a gimbals tracking system. Radiother Oncol 93(1):45–49. doi:10.1016/j.radonc.2009.07.011
55. Skarsgard D, Cadman P, El-Gayed A, Pearcey R, Tai P, Pervez N, Wu J (2010) Planning target volume margins for prostate radiotherapy using daily electronic portal imaging and implanted fiducial markers. Radiat Oncol 5:52. doi:10.1186/1748-717X-5-52
56. Engels B, Soete G, Verellen D, Storme G (2009) Conformal arc radiotherapy for prostate cancer: increased biochemical failure in patients with distended rectum on the planning computed tomogram despite image guidance by implanted markers. Int J Radiat Oncol Biol Phys 74(2):388–391. doi:10.1016/j.ijrobp.2008.08.007
57. Mizowaki T, Takayama K, Norihisa Y, Yano S, Kamoto T, Nakamura E, Kamba T, Inoue T, Ogawa O, Hiraoka M (2009) High Dose Local Irradiation to T3-4N0M0 Prostate Cancer with Intensity-modulated Radiotherapy Combined with Neoadjuvant Hormonal Therapy. Int J Radiat Oncol Biol Phys 75(3S):S353–S354
58. Norihisa Y, Mizowaki T, Takayama K, Miyabe Y, Matsugi K, Matsuo Y, Narabayashi M, Sakanaka K, Nakamura A, Nagata Y, Hiraoka M (2012) Detailed dosimetric evaluation of intensity-modulated radiation therapy plans created for stage C prostate cancer based on a planning protocol. Int J Clin Oncol 17(5):505–511. doi:10.1007/s10147-011-0324-1

59. Zhu S, Mizowaki T, Nagata Y, Takayama K, Norihisa Y, Yano S, Hiraoka M (2005) Comparison of three radiotherapy treatment planning protocols of definitive external-beam radiation for localized prostate cancer. Int J Clin Oncol 10(6):398–404. doi:10.1007/s10147-005-0519-4
60. Mizowaki T, Norihisa Y, Ogura M, Takayama K, Kamba T, Inoue T, Shimizu Y, Kamoto T, Ogawa O, Hiraoka M (2012) Interim outcomes of a high-dose whole pelvic IMRT for very high-risk group of patients with locally advanced prostate cancer. Int J Radiat Oncol Biol Phys 84 (3, Suppl 1):S360
61. Varenhorst E, Garmo H, Holmberg L, Adolfsson J, Damber JE, Hellstrom M, Hugosson J, Lundgren R, Stattin P, Tornblom M, Johansson JE (2005) The National Prostate Cancer Register in Sweden 1998–2002: trends in incidence, treatment and survival. Scand J Urol Nephrol 39(2):117–123. doi:10.1080/00365590510007793
62. Pollack A, Horwitz EM, Movsas B (2003) Treatment of prostate cancer with regional lymph node (N1) metastasis. Semin Radiat Oncol 13(2):121–129. doi:10.1053/srao.2003.50011
63. Hsiao W, Moses KA, Goodman M, Jani AB, Rossi PJ, Master VA (2010) Stage IV prostate cancer: survival differences in clinical T4, nodal and metastatic disease. J Urol 184(2):512–518. doi:10.1016/j.juro.2010.04.010
64. Verhagen PC, Schroder FH, Collette L, Bangma CH (2010) Does local treatment of the prostate in advanced and/or lymph node metastatic disease improve efficacy of androgen-deprivation therapy? A systematic review. Eur Urol 58(2):261–269. doi:10.1016/j.eururo.2010.05.027
65. Mohan R, Wu Q, Manning M, Schmidt-Ullrich R (2000) Radiobiological considerations in the design of fractionation strategies for intensity-modulated radiation therapy of head and neck cancers. Int J Radiat Oncol Biol Phys 46(3):619–630
66. Fonteyne V, Lumen N, Ost P, Van Praet C, Vandecasteele K, De Gersem IW, Villeirs G, De Neve W, Decaestecker K, De Meerleer G (2013) Hypofractionated intensity-modulated arc therapy for lymph node metastasized prostate cancer: early late toxicity and 3-year clinical outcome. Radiother Oncol 109(2):229–234. doi:10.1016/j.radonc.2013.08.006
67. The Japanese Urological Association, The Japan Society of Pathology, The Japan Radiological Society (2010) General rule for clinical and pathological studies on prostate cancer 2010, 4th edn. Kanehara, Tokyo

Gynecologic Malignancies

21

Daniel R. Simpson, Anthony J. Paravati,
Catheryn M. Yashar, Loren K. Mell, and Arno J. Mundt

Keywords
Intensity-modulated radiation therapy • Dose escalation • Gynecologic cancer

21.1 Introduction

Just over a decade ago, intensity-modulated radiation therapy (IMRT) was rarely used in the treatment of patients with gynecologic cancers. Since that time, however, IMRT adoption has proceeded at a tremendous pace. Today, IMRT is frequently used to treat gynecologic cancer patients at many centers [1–4]. In the United States, in particular, IMRT is becoming increasingly commonplace, notably in the treatment of cervical and endometrial cancers. In a 2002 survey of American Radiation Oncologists, 15 % of respondents reported having treated a gynecologic cancer patient with IMRT [5]. In a follow-up survey conducted 2 years later, this percentage had increased to 35 %, with gynecologic tumors representing the fastest growing disease site undergoing IMRT in the country [6]. The increasing utilization of IMRT in gynecologic cancers in the United States was also seen in a recent Surveillance, Epidemiology, and End Results (SEER)-Medicare analysis in which the percentage of endometrial cancer patients undergoing adjuvant irradiation who received IMRT was found to have increased from 3.3 % in 2002 to 23.2 % in 2007 [7].

Paralleling its rapid clinical adoption is the growing academic interest in gynecologic IMRT. Since the initial dosimetric study in 2000 suggesting a benefit to IMRT,

D.R. Simpson, M.D. (✉) • A.J. Paravati, M.D., MBA • C.M. Yashar, M.D.
L.K. Mell, M.D. • A.J. Mundt, M.D.
Department of Radiation Medicine and Applied Sciences, University of California, San Diego, 3855 Health Sciences Dr. #0865, La Jolla, CA 92093-0865, USA
e-mail: drsimpson@ucsd.edu

Y. Nishimura, R. Komaki (eds.), *Intensity-Modulated Radiation Therapy*,
DOI 10.1007/978-4-431-55486-8_21

a large number of dosimetric and, more recently, clinical outcome studies have been published in a wide variety of gynecologic tumors using a range of IMRT techniques at centers throughout the world. Gynecologic IMRT has also been the subject of multiple prospective clinical trials including a multicenter phase II trial conducted by the Radiation Therapy Oncology Group (RTOG) [8]. Today, IMRT is included in a wide number of cooperative group trials and is the focus of multiple trials [8–11] including a large multicenter, multinational clinical trial [9].

The purpose of this chapter is to provide the reader with a general overview of the role of IMRT in gynecologic malignancies as well as a summary of the published dosimetric and clinical data supporting its use, including the growing number of prospective clinical trials. Technical issues and considerations regarding its optimal use are presented. Novel and future uses of IMRT in these patients are also discussed.

21.2 Rationale for IMRT

A strong rationale exists for IMRT in gynecologic cancers. While effective, conventional radiation therapy (RT) has a number of limitations in the treatment of gynecologic cancers. First of all, traditional RT fields encompass large volumes of normal tissues, exposing patients to a wide variety of treatment-related toxicities, primarily related to the gastrointestinal (GI) tract, including diarrhea and malabsorption of vitamins, lactose, and bile acids [12, 13]. Acute and chronic genitourinary (GU) problems may also develop [12, 14]. Moreover, given that much of the total body bone marrow (BM) reserve is located within the pelvic bones [15] (and thus in the pelvic fields), hematologic toxicity may also occur, particularly in women receiving chemotherapy during or following pelvic irradiation [16, 17]. Unsurprisingly, RT-related toxicities are even more prevalent when more comprehensive treatment fields are used, owing to the inclusion of even greater volumes of normal tissues [18, 19]. Highly conformal IMRT plans offer the potential to reduce the risk of toxicity in these patients by improving the sparing of the surrounding small bowel, bladder, rectum, and other organs and may potentially translate to improved patient quality of life.

Conventional treatment techniques also limit the ability to deliver higher and potentially more efficacious doses to select patients at increased risk of recurrence, for example, those with involved lymph nodes and gross unresectable disease [20]. Highly conformal IMRT planning may allow dose escalation to these sites, potentially improving tumor control and patient outcomes. The ability to dose escalate using IMRT also provides a potential alternative to brachytherapy in select patients. Brachytherapy occupies an important role in the treatment of many gynecologic cancers, notably cervical cancer; however, it is not always possible or even feasible, particularly in the elderly or in patients with unfavorable anatomy. In such cases, only modest additional doses are possible with conventional techniques resulting in poor outcomes [21, 22]. Even when brachytherapy is feasible, coverage of bulky disease may be inadequate and patient outcomes poor. IMRT may represent a potential means of "repairing" unacceptable brachytherapy implants.

21.3 Dosimetric Studies

A large number of preclinical studies of IMRT in gynecologic cancer patients have been reported. These studies provide valuable support for the potential benefits and optimal use of IMRT in these patients as well as help guide the design of future clinical trials. Preclinical studies fall into three main groups: conventional dose analyses, dose escalation reports, and studies of IMRT as a replacement/alternative to brachytherapy.

21.3.1 Conventional Dose Delivery

Multiple investigators have compared conventional versus IMRT planning in gynecologic patients undergoing conventional dose RT. To date, the majority have focused on patients undergoing pelvic irradiation. An initial study by Roeske et al. at the University of Chicago included 10 women with cervical and endometrial cancer undergoing pelvic RT and found that IMRT reduced the volume of bladder, rectum, and small bowel receiving the prescription dose by 23, 23, and 50 %, respectively, compared to conventional planning (Fig. 21.1) [23]. Heron and coworkers noted that IMRT reduced the volume of the bladder, rectum, and small bowel receiving > 30 Gy by 36, 66, and 56 %, respectively, compared to conventional techniques [24]. Others have reported similar results [25, 26].

While initially proposed to spare the bowel, bladder, and rectum, increasing interest has been focused more recently on the ability of IMRT to spare the pelvic BM. Brixey et al. was the first to report that IMRT planning could spare the pelvic BM, noting significant reductions in the volume of pelvic BM irradiated, particularly the BM in the iliac crests [27]. Interestingly, this reduction was achieved even though the pelvic BM was *not* intentionally avoided in the optimization process. Unsurprisingly, even greater BM sparing was seen by these same investigators when the iliac crests were included as an avoidance structure [28]. Mell and colleagues subsequently demonstrated that IMRT planning could reduce the dose to other pelvic BM sites, including the lumbosacral spine (Fig. 21.2) [29].

Comparable to studies focused on pelvic RT, several series have reported significant dosimetric benefits to IMRT planning in women treated with pelvic and para-aortic fields [30–34]. Portelance and colleagues at Washington University compared IMRT plans with traditional two- and four-field extended-field radiation therapy (EFRT) plans in 10 locally advanced cervical cancer patients [30]. Although comparable target coverage was seen, IMRT planning resulted in better sparing of the surrounding normal tissues. Compared to two-field EFRT plans, IMRT planning reduced the volume of the bowel, bladder, and rectum receiving the prescription dose by 61, 96, and 71 %, respectively. Compared to four-field EFRT plans, corresponding reductions were 60, 93, and 56 %, respectively. Lian and coworkers compared conventional EFRT plans in 10 endometrial cancer

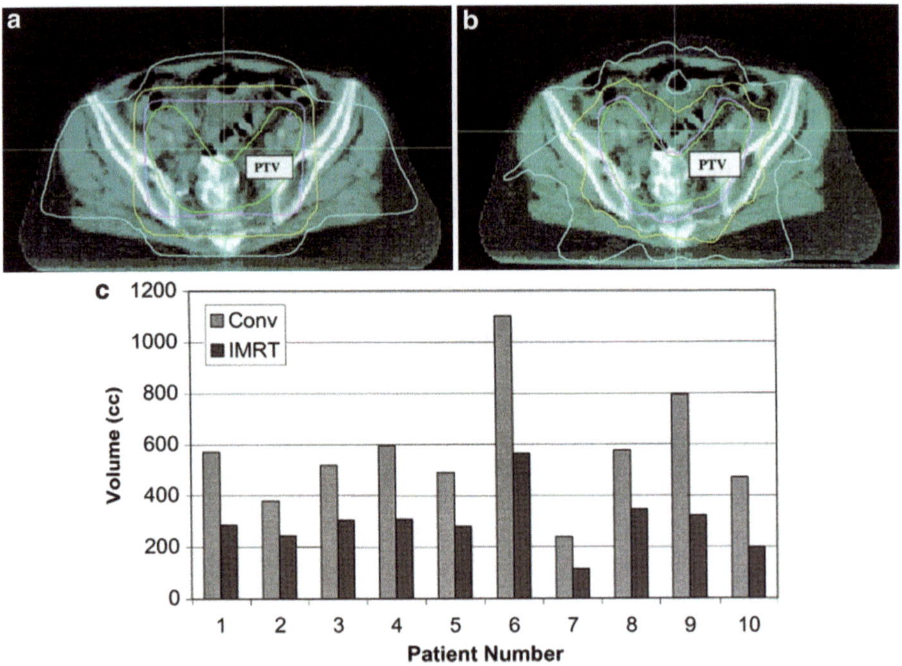

Fig. 21.1 Axial computed tomography images showing isodose distributions for conventional (*top left*) and intensity-modulated radiation therapy (*top right*) plans for a patient with a gynecologic malignancy. The graph (*bottom*) demonstrates a comparison of the absolute volume of small bowel irradiated with the two plans [23]

patients to various IMRT planning approaches [32]. Overall, IMRT planning resulted in superior target coverage and significant reductions in normal tissue doses. BM-sparing intensity-modulated EFRT (IM-EFRT) approaches have also been proposed [33, 34].

Data comparing conventional and IMRT planning in patients undergoing pelvic-inguinal irradiation and whole abdominal RT (WART) are more limited. Beriwal and coworkers at the University of Pittsburgh compared conventional and IMRT planning in 15 vulvar cancer patients undergoing pelvic-inguinal irradiation [35]. While target coverage was comparable, IMRT reduced the volume of the small bowel, bladder, and rectum receiving $\geq$30 Gy by 27 %, 26 %, and 41 %, respectively. Unfortunately, no significant difference was seen in the sparing of the femoral heads (an important organ at risk in these patients), perhaps due to the small patient numbers studied. Investigators at Memorial Sloan Kettering Cancer Center compared IMRT planning to conventional large field techniques in 10 endometrial cancer patients treated with WART [36]. IMRT planning resulted in better coverage of the peritoneal cavity, and a 60 % reduction in the volume of pelvic bones (a surrogate for pelvic BM) irradiated. Others have reported similar dosimetric benefits to IMRT planning in women undergoing WART [37].

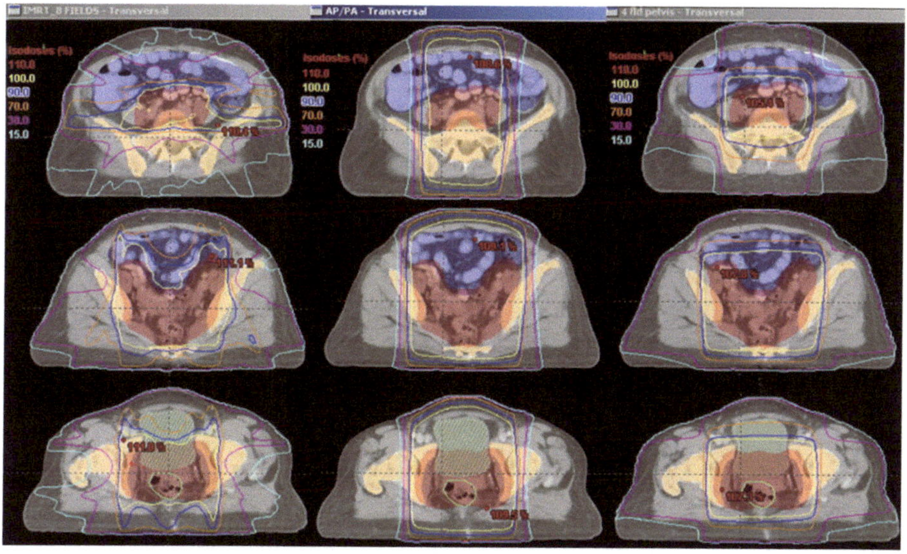

Fig. 21.2 Axial computed tomography images showing a comparison of isodose distributions for intensity-modulated (*left*), anteroposterior-posteroanterior (*middle*), and four-field conformal (*right*) radiotherapy plans in a patient with cervical cancer. The intensity-modulated plan provides superior conformity with lower dose to the bone marrow and normal tissue compared to the other two plans [29]

Yang and coworkers performed a systematic literature review and meta-analysis of 13 published studies comparing conventional and IMRT planning in cervical or endometrial cancer patients treated using standard (non-escalated) doses [38]. Ten studies reported on the irradiated volume of small bowel, nine on rectum, eight on bladder, and six on BM. These series included 222 patients treated with IMRT and 233 undergoing conventional RT. The authors calculated the pooled average percent of irradiated volumes for various normal tissues and found statistically significant sparing of the rectum at doses of $\geq$ 30 Gy and the small bowel at doses $\geq$ 40 Gy. Overall, the pooled average irradiated volumes of the bladder and pelvic BM were consistently lower using IMRT than conventional planning, with the greatest differences seen at higher doses. None of these differences reached statistical significance. However, the small number of studies which included BM as an organ at risk limited the statistical power of this analysis.

21.3.2 Dose Escalation

The use of IMRT to deliver higher than conventional doses in gynecologic cancer patients has been the subject of multiple reports. Most attention to date has focused on patients with involved para-aortic lymph nodes, given the difficulty of safely delivering

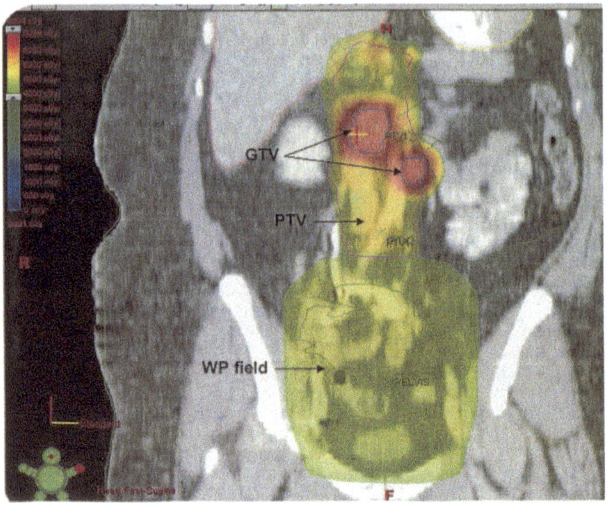

Fig. 21.3 A coronal computed tomography image demonstrating the dose distribution for an extended-field intensity-modulated radiotherapy plan with a simultaneous integrated boost technique in a patient with cervical cancer with para-aortic nodal involvement. The involved para-aortic nodes were treated to 60 Gy (in 2.4 Gy fractions), while the uninvolved nodal sites were simultaneously treated to 45 Gy (in 1.8 Gy fractions) [34]

high doses to these sites using conventional techniques. Investigators at Washington University evaluated dose escalation to enlarged para-aortic lymph nodes using a simultaneous integrated boost (SIB) approach, whereby the doses to the involved and uninvolved nodes were 59.4 Gy (in 1.8 Gy fractions) and 50.4 Gy (in 1.53 Gy fractions), respectively [39]. On average, 97.6 % of the involved nodes received 100 % of the prescription dose, while the dose to surrounding normal tissues was maintained to acceptable levels. Ahmed and colleagues presented an alternative SIB approach whereby 60 Gy (in 2.4 Gy fractions) was delivered to involved para-aortic nodes, while the uninvolved nodal sites were simultaneously treated to 45 Gy (in 1.8 Gy fractions) (Fig. 21.3) [34]. Compared to both two- and four-field conventional approaches, IMRT resulted in reduced doses to the bilateral kidneys, spinal cord, and, at dose levels $\geq$ 40 Gy, BM. While IMRT planning reduced the volume of small bowel irradiated to high doses, the difference did not reach statistical significance.

IMRT may also allow the delivery of higher than conventional pelvic doses in cervical cancer patients following hysterectomy. Patients with high-risk features (positive lymph nodes, positive margins, parametrial involvement), in particular, remain at risk of increased recurrence following surgery despite the delivery of conventional dose pelvic RT [40], suggesting a potential role for and benefit to dose escalation. D'Souza and colleagues compared conventional and IMRT treatment plans in 10 high-risk patients undergoing adjuvant pelvic RT and noted that IMRT planning maintained acceptable dose levels to the bowel, bladder, and rectum despite escalating the total pelvic dose from 45 to 54 Gy [41]. Du et al. explored the use of dose-escalated pelvic

IMRT using the so-called reduced field IMRT approach, whereby 30 Gy was initially delivered to the uterus, cervix, upper vagina, paracervical and parametrial tissues, uterosacral region, and pelvic lymph nodes followed by a 30 Gy boost to the regional nodes and paracervical and parametrial tissues [42]. The IMRT plans were compared with conventional RT approaches covering the same target volumes; however, the total dose was intentionally limited to 45–55 Gy in the conventional RT group. Overall, the mean target volume dose was significantly higher in the IMRT patients (61.5 Gy vs 50.8 Gy, $p=0.046$). Moreover, IMRT planning resulted in better dose conformity to the target and improved sparing of the rectum, bladder, and small intestine.

21.3.3 Brachytherapy Alternative/Replacement

The use of IMRT as a possible alternative to (or replacement for) brachytherapy is arguably one of the most controversial topics in all of gynecologic IMRT and has been the subject of multiple debates [43, 44]. While several authors have demonstrated that IMRT planning is capable of producing a pear-shaped dose distribution mimicking the dose distribution achieved with brachytherapy, integral normal tissue doses with brachytherapy are lower due to the steep dose falloff [45, 46]. Investigators at the Princess Margaret Hospital compared an IMRT boost with both a 3D conformal and a four-field box boost plan in 12 patients with cervical (8), endometrial (2), or vaginal (2) cancer who were not candidates for brachytherapy [47]. The authors found improved conformity with IMRT planning with a reduction of 22 and 19 % in the volume of rectum and bladder receiving the highest doses, respectively. Aydogan and colleagues compared IMRT and high-dose rate (HDR) brachytherapy planning in 10 endometrial cancer patients and noted better dose uniformity and improved bladder and rectal sparing with IMRT (Fig. 21.4) [48]. In select unfavorable anatomy cases, IMRT planning may even result in *better* target coverage [49]. In contrast, Sharma and colleagues compared IMRT and interstitial brachytherapy plans in 12 locally advanced cervical cancer patients and found that brachytherapy resulted in better target coverage and an improved conformity index [50].

As opposed to the above approaches in which an IMRT boost is delivered *after* the completion of external beam RT, Guerrero and coworkers proposed a novel SIB technique whereby the boost is incorporated into the initial treatment [51]. As envisioned, the regional pelvic nodes receive 45 Gy in 1.8 fractions, while the cervical tumor is treated to 70–77.5 Gy in 2.8–3.1 Gy fractions, which is felt to be radiobiologically equivalent to 45 Gy whole pelvic RT followed by 30 Gy HDR brachytherapy. The SIB approach achieved better bowel and bladder sparing than a sequential IMRT boost, with significant shortening of the overall treatment course. While appealing, a concern with this approach is that the decision not to use brachytherapy must be made at diagnosis. There may be some patients in whom brachytherapy would never be feasible, but in most, it is preferable to assess the clinical response to pelvic RT prior to deciding on whether or not to perform brachytherapy.

IMRT has also been proposed as a possible *adjunct* to brachytherapy. Assenholt et al. evaluated an applicator-guided IMRT boost combined with brachytherapy to

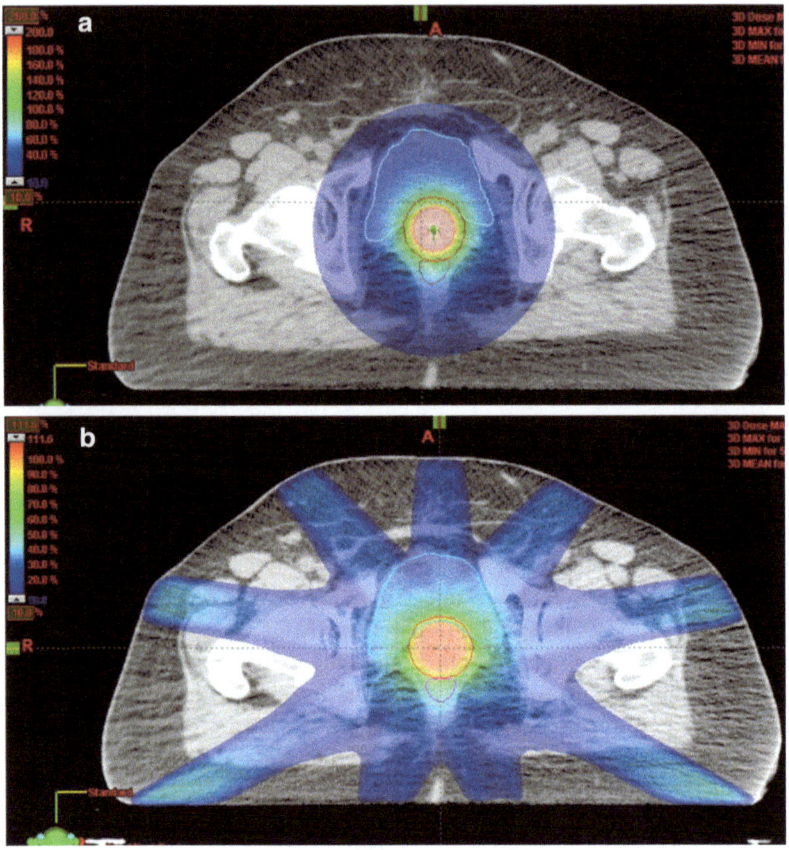

Fig. 21.4 Axial computed tomography images with dose distributions for a high-dose rate vaginal cylinder brachytherapy (*top*) and intensity-modulated radiotherapy (*bottom*) plan in a patient with endometrial cancer. The intensity-modulated plan provides superior dose conformity and improved bladder and rectal sparing at higher isodose levels [48]

improve coverage of large or topographically unfavorable tumors [52]. IMRT planning was able to improve but did not replace the dose given by intracavitary brachytherapy. Duan and coworkers compared conventional brachytherapy, optimized brachytherapy, and optimized brachytherapy with an IMRT boost and found that HDR brachytherapy plus IMRT achieved better target coverage where conventional brachytherapy dose was suboptimal [53]. Others have proposed using an SIB-IMRT approach in bulky cervical cancers prior to brachytherapy [54].

21.4 Clinical Outcome Studies

A large (and growing) number of clinical studies have been published in recent years evaluating the outcome of gynecologic cancer patients treated with IMRT. By far, the lion's share has been single institution reports focusing on cervical and/or

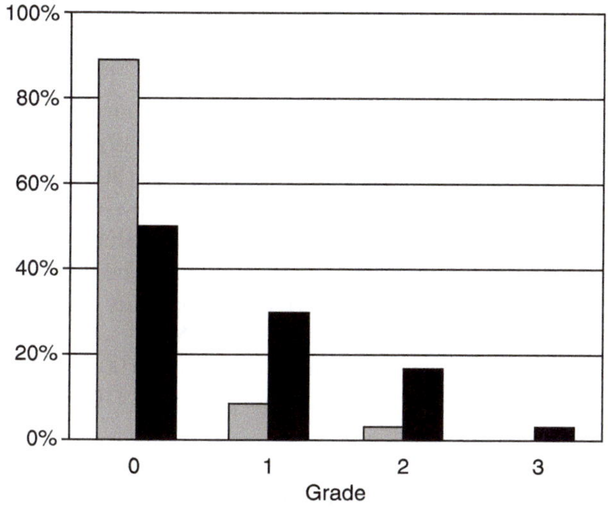

Fig. 21.5 A comparison of chronic gastrointestinal toxicity in gynecologic cancer patients treated with intensity-modulated (*gray bars*) and conventional radiotherapy (*black bars*). Overall, the patients treated with intensity-modulated radiation therapy had a lower rate of chronic gastrointestinal toxicity (11 % vs. 50 %, $p = 0.001$) [57]

endometrial cancer patients treated with conventional dose pelvic IMRT. However, an increasing number of reports have included patients treated with more comprehensive fields, with dose-escalated techniques or with IMRT in lieu of brachytherapy. Moreover, outcome studies have also begun to appear using IMRT in other tumor sites, including vaginal and ovarian cancers.

21.4.1 Pelvis

In a series of reports [55, 56], Mundt and colleagues at the University of Chicago were the first to report on the outcome of gynecologic cancer patients treated with pelvic IMRT using conventional doses. Their initial study focused on a mixed cohort of 40 cervical and endometrial cancer patients undergoing pelvic IMRT (median dose, 45 Gy) either following surgery or as definitive treatment. Compared to 40 patients treated with conventional techniques, IMRT patients experienced less grade ≥ 2 acute GI toxicity (60 % vs. 91 %, $p = 0.002$). Although less grade ≥ 2 GU toxicity was seen in the IMRT group (10 % vs. 20 %), this difference did not reach statistical significance ($p = 0.22$). In a follow-up report [57], these investigators compared rates of chronic GI toxicity in 36 IMRT and 30 conventional RT patients. The groups were well balanced in terms of age, site, stage, chemotherapy, radiation dose, and brachytherapy, except for a higher frequency of surgery in the IMRT group. Overall, IMRT patients had a lower rate of chronic GI toxicity (11 % vs. 50 %, $p = 0.001$) (Fig. 21.5). On multivariate analysis, this reduction remained statistically significant ($p = 0.01$).

Hasselle and coworkers recently presented the combined experience of the University of Chicago and the University of California San Diego (UCSD) using pelvic IMRT in cervical cancer patients [58]. Overall, 111 patients underwent pelvic IMRT between 2000 and 2007 either following surgery (22) or combined with chemotherapy and intracavitary brachytherapy (89). Overall, 95 patients (86 %) received concomitant chemotherapy and 71 (80 %) of the definitive patients underwent brachytherapy. At a median follow-up of 27 months, the three-year actuarial overall survival (OS) and disease-free survival (DFS) rates of the entire group were 78 % and 69 %, respectively. The three-year pelvic control rates for stage IB–IIA and stage IIB–IVA patients undergoing definitive treatment were 94.7 % and 70.8 %, respectively. Moreover, the actuarial three-year pelvic control in the postoperative patients was 100 %. Overall treatment was well tolerated; estimates of acute and late grade ≥ 3 GI or GU toxicities were 2 % and 7 %, respectively.

Investigators at Washington University have also reported on their experience using IMRT in cervical cancer patients, comparing the outcomes of 135 patients undergoing IMRT (primarily pelvic IMRT) with those seen in 317 treated with conventional RT [59]. Controlling for a variety of clinical and treatment factors, the IMRT patients were found to have a better cause-specific survival (CSS) ($p<0.0001$) and OS ($p<0.0001$) than the conventional RT patients. A nonsignificant trend favoring IMRT was also seen between the two groups in terms of recurrence-free survival ($p=0.07$). Patients treated with IMRT developed fewer grade ≥ 3 late GI or GU sequelae (6 % vs. 17 %, $p=0.002$). It should be noted that the differences in OS and CSS were unexpected and potentially biased by other differences between the groups. Furthermore, it is difficult to attribute these differences in survival to treatment technique alone given the similar pelvic recurrence rates. The initial Washington University experience using adjuvant pelvic IMRT in endometrial cancer was published earlier [60]. Overall, 19 women with stages IB–IVB endometrial cancer received pelvic IMRT following surgery. None developed acute grade ≥3 acute GI or GU toxicity.

More recently, investigators at the Memorial Sloan Kettering Cancer Center have published their experience using adjuvant pelvic IMRT in patients with cervical and endometrial cancer. Folkert et al. treated 34 high-risk cervical cancer patients following surgery with a median total dose of 50.4 Gy [61]. All patients received concomitant cisplatin. At a median follow-up of 44 months, the five-year actuarial OS and DFS rates of the entire group were 91.1 % and 91.2 %, respectively. Grade ≥3 acute GI and GU sequelae occurred in only 2.9 % and 0 %, respectively. No patients developed grade ≥3 chronic toxicity. In a separate report, Shih and colleagues treated 46 stages I–III endometrial cancer patients with adjuvant pelvic IMRT (median dose, 50.4 Gy) and in 30 patients (66 %) with adjuvant chemotherapy [62]. At a median follow-up of 52 months, the five-year actuarial DFS and OS rates were 88 % and 97 %, respectively. Only two patients developed grade 3 or higher GI toxicity (one acute and one chronic). No significant GU acute or chronic toxicity was seen. Others have reported favorable outcomes using adjuvant pelvic IMRT in cervical and endometrial cancer patients [63–65].

Considerable experience using pelvic IMRT in gynecologic cancer patients has been reported from centers in Asia [66–70]. Chen et al. at the Taichung Veterans General Hospital in Taiwan treated 109 stage IB–IVA cervical cancer patients with IMRT (88 % pelvic IMRT) combined with weekly chemotherapy [66]. At a median follow-up of 32.5 months, the three-year DFS and OS rates were 67.6 % and 78.2 %, respectively. Three patients (2.7 %) developed grade ≥3 acute GI toxicity. Late grade 3 or higher GI and GU sequelae developed in 4.6 % and 6.4 % of patients, respectively. In a subsequent report, the risk of late complications in 83 IMRT and 237 non-IMRT intact cervical cancer patients was compared [67]. Overall, IMRT patients had lower rates of grade ≥2 (23 % vs. 30 %) and grade ≥3 (8 % vs. 12 %) GI or GU late sequelae; however, neither difference reached statistical significance ($p=0.24$ and $p=0.33$). Other investigators in Asia have reported improved toxicity profiles using adjuvant pelvic IMRT in cervical cancer patients undergoing pelvic irradiation following surgery [69].

Chen and colleagues have also published an analysis of 101 stage IA–IIIC2 endometrial cancer patients treated with either IMRT (65) or conventional (36) RT at their institution and noted significant reductions in GI and GU toxicity using IMRT [68]. Rates of grades 2 and 3 acute GI and GU toxicity in the conventional group were 55.6 % and 11.1 % and 19.4 % and 8.3 %, respectively. Corresponding rates in the IMRT group were 27.7 % and 6.2 % and 16.9 % and 0 %, respectively. Grade ≥3 late GI and GU toxicities were seen in 2.8 % and 2.8 % of the conventional RT group. No grade ≥3 late toxicities developed in the IMRT patients. No differences were seen in terms of DFS, OS, or local control between the two groups. In contrast, a recent SEER study from the United States found a comparable rate of GI and GU sequelae between endometrial cancer patients receiving adjuvant pelvic IMRT and those treated with conventional techniques, except for a *higher* rate of bowel obstruction in the IMRT group [7]. However, details about the quality of the IMRT plans or extent of nodal surgery in the two groups were not known, both of which may have influenced the results.

While most published reports to date of pelvic IMRT have been in patients with cervical and endometrial cancers, emerging data have appeared in other disease sites [70–72]. Investigators at Stanford University treated 10 vaginal cancer patients with IMRT (predominantly pelvic IMRT) combined in most patients with concomitant chemotherapy and intracavitary brachytherapy [72]. None of these patients developed a locoregional recurrence or experienced grade 3 or higher toxicity. The experience at MD Anderson Cancer Center treating vaginal cancers unable to undergo brachytherapy with IMRT alone is discussed later in this chapter [73].

Despite the observation that patients undergoing pelvic IMRT and chemotherapy often experience less hematologic toxicity even when the BM is not included in the optimization process [27], hematologic toxicity rates remain high in most series which do not intentionally spare the BM [61, 64, 65, 68] or, at best, only modestly reduce the BM dose [60]. For example, Chen and colleagues treated 68 high-risk cervical cancer patients with chemotherapy and pelvic RT planned using either conventional (35) or IMRT (33) techniques [66]. The pelvic BM was not included in the IMRT planning. Grade 2 and 3 hematologic sequelae occurred in 31 % and 9 %, respectively, of the conventional RT patients. Corresponding rates in the IMRT patients were 27 % and 6 %, respectively ($p=0.72$). In contrast, Du and coworkers included BM as a constraint in the planning process and found less grade 2 or higher leukopenia (4 % vs. 10 %, $p=0.026$) compared to that seen in patients treated with conventional techniques [42]. Efforts to optimize BM-sparing approaches in patients undergoing pelvic IMRT or more comprehensive fields are discussed further in the subsequent section.

Limited published data are available on the use of dose-escalated pelvic IMRT approaches. The report from Du et al. described earlier is the largest experience to date [42]. They found that patients treated with dose-escalated IMRT had lower rates of proctitis ($p=0.001$), enteritis ($p=0.03$), cystitis ($p=0.001$), and dermatitis ($p=0.04$) compared to a control group of 60 conventional pelvic RT patients (treated without dose escalation). The utility of dose-escalated pelvic IMRT is the subject of an ongoing phase II clinical trial in India [74]. Others have explored the use of SIB techniques to escalate dose to either the cervix [49] or involved lymph nodes [75].

Schwarz and colleagues performed a feasibility study of IMRT (92 % pelvic IMRT) in 24 postoperative high-risk cervical cancer patients [76]. Eighteen (75 %) received concomitant chemotherapy; all underwent vaginal brachytherapy. Patients received higher than conventional pelvic doses (median dose, 51.2 Gy); two patients with gross disease underwent local boosts (10–15 Gy). Organs at risk included in the optimization process included bladder, bowel, and pelvic bones (surrogate for BM). IMRT and brachytherapy were completed as planned in all patients. Acute grade 3 GI toxicity occurred in 50 % of patients (five anorexia, four diarrhea, three nausea), predominantly in patients receiving concomitant chemotherapy. Only one patient developed a grade 3 acute GU toxicity. Grade 3 and 4 hematologic sequelae occurred in 63 % and 21 % of concurrent chemoradiotherapy patients, respectively.

Investigators from Chiang Mai Hospital in Thailand enrolled 15 locally advanced cervical cancer patients on a prospective feasibility study of pelvic IMRT and image-guided brachytherapy [77]. All patients received 45 Gy in 1.8 Gy daily fractions with concurrent cisplatin followed by image-guided brachytherapy (7 Gy$\times$4 prescribed to the high-risk CTV). All patients completed treatment as planned. Most acute toxicities were mild with no patients developing grade $\geq$3 acute sequelae. At a median follow-up of 14 months, only one patient has developed a local recurrence. No significant late sequelae were noted.

Investigators at UCSD enrolled 31 patients (19 gynecologic cancers) on a prospective trial evaluating the ability of IMRT to reduce dose to *functional* BM, as opposed to constraining dose to the total pelvic BM [78]. Functional BM was identified using both ^{18}F-fluorodeoxyglucose (FDG) positron emission tomography (PET)/computed tomography (CT) and an investigational magnetic resonance imaging (MRI) protocol, IDEAL IQ, which identifies regions of high BM cellularity. The intersection of BM subregions with SUV values above the mean and those with the fat fraction below the mean was used to identify functional BM sites (Fig. 21.6). Two types of IMRT plans were compared: standard BM-sparing plans in which the total BM (using the pelvic bones as a surrogate) was included in the optimization process and functional BM-sparing plans in which only the functional

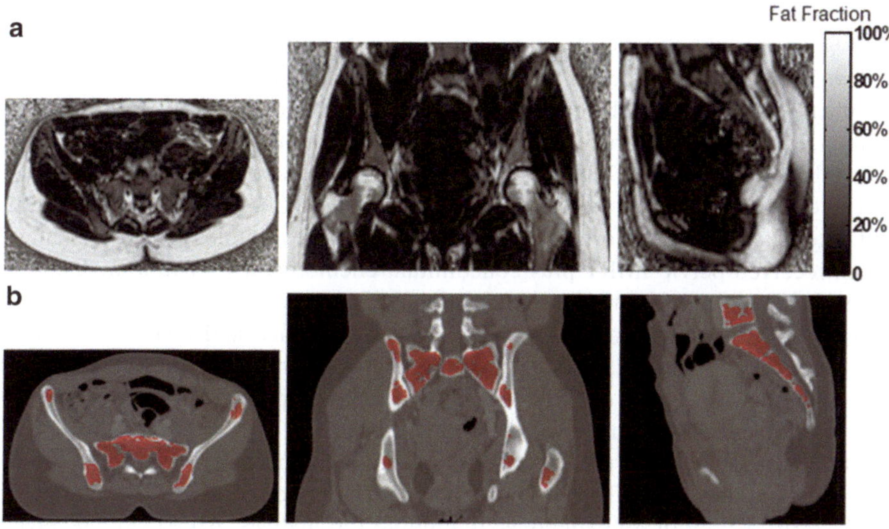

Fig. 21.6 Axial, coronal, and sagittal views of fat fraction maps (**a**) of pelvic bone marrow (BM) in a cervical cancer patient before treatment. Segmented functional BM regions are superimposed onto the simulation computed tomography images (**b**) [78]

BM subregions were included. In the 19 gynecology patients, the mean functional BM V_{10} and V_{20} were 85 % vs. 94 % ($p<0.0001$) and 70 % vs. 82 % ($p<0.0001$), respectively, for functional BM-sparing IMRT versus standard BM-sparing IMRT. Of 10 subjects treated with functional BM-sparing techniques with concomitant chemotherapy, three (30 %) experienced acute grade ≥3 hematologic toxicity.

Mabuchi and colleagues presented a phase I trial in high-risk cervical cancer patients treated with adjuvant pelvic IMRT (50.4 Gy) and concurrent weekly carboplatin and escalating doses of paclitaxel (initially 35 mg/m^2 and increasing by 5 mg/m^2 intervals) [79]. Nine women were enrolled with dose-limiting toxicity (DLTs) occurring at the second dose level (40 mg/m^2/week). No patients treated with 35 mg/m^2/week developed dose-limiting toxicity. Two of the three DLTs were hematologic.

In 2006, the RTOG launched a prospective phase II trial (RTOG 0418) evaluating adjuvant pelvic IMRT in endometrial and cervical cancer patients. The prescribed dose in all patients was 50.4 Gy in 1.8 Gy daily fractions; however, while the endometrial cancer patients received IMRT alone, the cervical cancer patients received concomitant weekly cisplatin. The results of the cervical [80] and endometrial [8, 81] cancer patients have been reported separately. Overall, 53 endometrial cancer patients were eligible for analysis. While the primary endpoint of reproducibility of the IMRT approach was achieved, considerable treatment planning deviations were noted. The proportion of cases in which doses to critical normal tissues exceeded protocol criteria were bladder (67 %), rectum (76 %), bowel (17 %), and femoral heads (33 %). Seven cases (17 %) had a dose to the small bowel that exceeded the prescribed dose constraints by >10 %. Overall, a nonsignificant absolute reduction of 12 % was seen in acute GI toxicity compared to historical controls. However, the trial was not powered to detect such a difference. It thus remains unclear whether a significant difference could have been detected in a larger cohort. In a separate abstract, these investigators reported excellent tumor control rates in these patients, with a three-year OS and DFS of 91 % and 91 %, respectively [8].

The outcome of the cervical cancer patients on RTOG 0418 has only been presented to date in abstract form [80]. Forty patients were eligible for analysis, all of whom received pelvic IMRT plus concomitant chemotherapy. With a median follow-up of 2.7 years, the two-year actuarial DFS and OS rates were 86.9 % and 94.6 %, respectively. Two-year actuarial local recurrence and distant-metastasis rates were 10.6 % and 10.3 %, respectively. In an earlier abstract [82], the authors noted that grade ≥2 bowel toxicity occurred in 22.5 % of patients which was statistically significantly lower than the hypothesized rate of 40 % ($p=0.04$).

Barillot and coworkers presented the results of a prospective phase II trial evaluating adjuvant pelvic IMRT in patients with stage IB (grade 3), IC, or II endometrial cancer [10]. A total of 46 patients were eligible for analysis. All patients received adjuvant pelvic IMRT to a dose of 45 Gy in 25 fractions. Thirty-six patients (75 %) received an additional vaginal vault HDR brachytherapy boost. The primary endpoint was the incidence of acute grade ≥2 GI toxicity. Thirteen patients (27 %) developed at least one acute grade 2 GI toxicity. There were no ≥ 3 GI toxicities observed. Nine patients (19 %) developed grade 2 GU toxicity. The authors noted that the incidence and timing of the GI toxicity were similar to those reported in RTOG 0418 [81].

Two prospective randomized trials both from India have been published evaluating pelvic IMRT in gynecologic patients (Table 21.1) [74, 83]. Gandhi and colleagues performed a phase III randomized trial in 44 stage IIB–IIIB squamous cell cervical cancer patients radiotherapy [83]. Patients received 50.4 Gy in 28 Gy fractions delivered with either conventional or IMRT with concomitant chemotherapy. All patients subsequently underwent brachytherapy (7 Gy × 3, Point A). While no difference was seen in the median treatment duration (9.1 weeks in both arms), the only patients who had treatment breaks and/or delays due to diarrhea or low blood counts were in the conventional group. IMRT patients had less grade ≥2 acute GI toxicity (32 % vs. 64 %. p 0.03) and less grade ≥2 emesis (9 % vs. 36 %, $p=0.03$). No difference was seen in the frequency of either acute GU or hematologic toxicity. At a median follow-up of 21.6 months, the incidence of grade ≥2 chronic GI toxicity was also lower in the IMRT group (5 % vs. 23 %, $p=0.011$). No differences were seen in DFS or OS rates between the two groups.

Shrivastava and colleagues presented the preliminary results of a phase IIB randomized trial comparing conventional pelvic (40 Gy in 20 fractions) with dose-escalated pelvic IMRT (50 Gy) [74]. All 86 patients received concomitant chemotherapy and planned HDR brachytherapy. No differences were seen in terms of compliance or response between the two treatment arms.

However, despite the escalation of the pelvic dose, the IMRT patients experienced fewer high grade GI and hematologic toxicities. Two IMRT patients developed grade

Table 21.1 Randomized trials comparing pelvic intensity-modulated radiation therapy to conformal radiation therapy in patients with cervix cancer

Author	Number of patients	Phase	Median follow-up (mos.)	Results
Gandhi et al. [83]	44	III	21	No difference in OS or DFS; less grade ≥2 acute and late GI toxicity in IMRT arm; no difference in GU or hematologic toxicity
Shrivastava et al. [74]	86	IIB	17	Less acute high grade GI and hematologic toxicity with IMRT despite dose escalation

IMRT intensity-modulated radiation therapy, *OS* overall survival, *DFS* disease-free survival, *CRT* conventional radiation therapy, *GI* gastrointestinal, *GU* genitourinary

3 GU toxicity. At a median follow-up of 17 months, six conventional and three IMRT patients have developed recurrent disease. Patients are continuing to be observed for late toxicities and disease recurrence.

Currently, several multicenter prospective clinical trials include and/or are assessing IMRT in gynecologic cancer patients. While earlier cooperative groups trials conducted by the GOG and RTOG previously did not allow IMRT, it is now permitted in several ongoing studies including GOG 0249, GOG 0258, GOG 0263, and RTOG 0724. Based on the favorable results of RTOG 0418, the RTOG has initiated the RTOG 1203 trial, which randomizes postoperative endometrial and cervical cancer patients to conventional or pelvic IMRT. Mell and colleagues have recently launched the International Evaluation of Radiotherapy Technology Effectiveness in Cervical Cancer (INTERTECC) trial [9], a phase II/III trial designed to test the efficacy of IMRT in the treatment of cervical cancer patients, for both definitive and adjuvant approaches.

21.4.2 Pelvis and Para-aortic Nodes

Multiple investigators have reported on the outcome of gynecologic cancer patients treated with IM-EFRT. Liang and colleagues presented their experience using prophylactic IM-EFRT in 32 stage IB2–IIIB cervical cancer patients with positive pelvic and negative para-aortic lymph nodes [84]. The prescribed dose to the para-aortic region was 40 Gy in 25 fractions. All patients received concurrent cisplatin chemotherapy and brachytherapy. Acute grade ≥ 3 GI and GU toxicities were seen in 6.2 % and 3.1 % of patients, respectively. Two grade 3 late sequelae were noted (one GI and one GU). Compared to historical controls, the IMRT patients demonstrated improved three-year actuarial OS (87 % vs. 62 %, $p=0.02$), DFS (82 % vs. 54 %, $p=0.02$), and distant-metastasis free survivals (79 % vs. 57 %, $p=0.01$). Others have reported similarly favorable outcomes in patients treated with prophylactic IM-EFRT [33].

Several investigators have explored dose-escalated IM-EFRT approaches, using either sequential [85–88] or integrated [33, 88, 89] boost techniques. Salama et al. treated 13 endometrial or cervical cancer patients with IM-EFRT [85]. All initially received 45 Gy to the pelvis/para-aortic regions with concomitant chemotherapy followed by a boost of 9 Gy to the involved para-aortic nodes. No patients experienced grade ≥3 acute GI or GU toxicity. At a median follow-up of 11 months, the one-year actuarial in-field control rate was 90 %. Jensen et al. updated this experience with a total of 21 patients and reported an 18-month rate of locoregional control of 90 % [86]. Acute grade ≥3 GI and GU toxicity occurred in four and zero patients, respectively. The two-year incidence of late grade ≥3 GU toxicity was 4.8 %. No patients experienced grade ≥3 late GI toxicity. Beriwal et al. treated 36 stage IB2–IVA cervical cancer patients with IM-EFRT (45 Gy) using an SIB boost approach to boost involved para-aortic nodes to 55–60 Gy in 2.2–2.4 Gy daily fractions [88]. At a median follow-up of 18 months, the two-year actuarial locoregional control, DFS, OS, and grade ≥3 toxicity rates for the entire group were 80 %, 51 %, 65 %, and 10 %, respectively. Acute grade ≥3 GI and GU toxicities occurred in 2.8 % and 2.8 % of patients, respectively.

Investigators at the Brigham and Women's Hospital reported on the outcomes of 32 gynecologic cancer patients (22 endometrial, 10 cervical cancer) who underwent sequential IMRT boosts to involved lymph node regions and concurrent chemotherapy [89]. Twelve patients had pelvic nodal boosts, 13 para-aortic nodal boosts, and 7 both. The median nodal size was 2.5 cm (range, 1.4–4.2 cm), and the median total dose was 63 Gy (range, 54–68.4 Gy). At a median follow-up of 21.8 months, the two-year nodal control was 85 %. Treatment was well tolerated, with a two-year actuarial late grade ≥3 toxicity rate of 14 %.

Marnitz and colleagues evaluated the feasibility of a SIB-IMRT approach in cervical cancer patients with positive pelvic and/or para-aortic lymph nodes undergoing RT and chemotherapy [75]. Of 40 patients, 29 (72.5 %) had documented enlarged lymph nodes and underwent pre-RT lymph node dissections. IMRT plans were generated to treat the pelvis and/or para-aortic region to 50.4 Gy (in 1.8 Gy fractions) and involved lymph nodes to 59.36 (in 2.12 Gy fractions). All patients received concurrent chemotherapy, predominantly weekly cisplatin. Overall, treatment was well tolerated with only two grade 3 acute GI toxicities (one diarrhea, one nausea) and no grade 3 GU toxicity. Tumor control rates were not reported. Others have reported favorable results using dose-escalated IMRT as salvage therapy in women with recurrent disease in the para-aortic region [90].

Prospective trials have also been performed in para-aortic node-positive cervical cancer patients [91] and in endometrial cancer patients undergoing IMRT in lieu of brachytherapy [92]. Investigators at the Shandong Tumor Hospital and Institute completed a novel trial in 60 para-aortic positive cervical cancer patients who were serially assigned to either conventional dose (45–50 Gy) conformal para-aortic irradiation or high-dose (58–68 Gy) IMRT. Overall, IMRT patients experienced less acute grade ≥3 myelosuppression (4 % vs. 19 %, $p=0.005$), dermatitis (0 % vs. 6 %, $p=0.04$), and GI toxicity (4 % vs. 19 %, $p=0.005$) and a higher complete

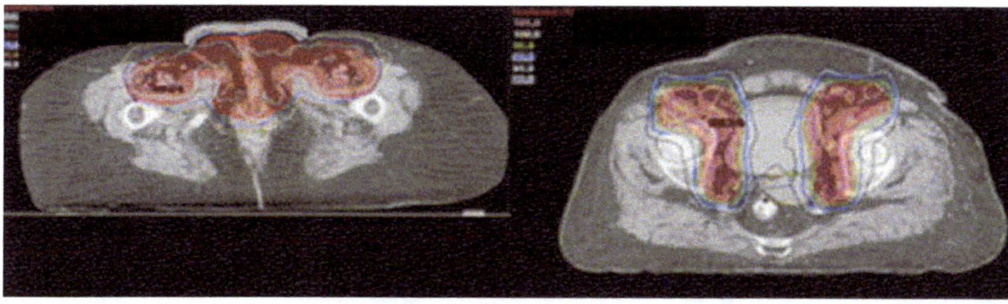

Fig. 21.7 Axial computed tomography images showing isodose distributions for an intensity-modulated radiation therapy plan covering the inguinal (*left*) and pelvic (*right*) lymph nodes in a patient with locally advanced vulvar cancer undergoing preoperative radiotherapy [35]

response rate (57 % vs. 28 %, $p=0.02$) than the conventional RT patients. At a median follow-up of 28 months, the three-year OS rate was superior in the IMRT group (36 % vs. 16 %, $p=0.016$). Chronic enterocolitis was lower in the IMRT group (0 % vs. 19 %, $p=0.001$).

21.4.3 Pelvis and Inguinal Nodes

The sole published experience using pelvic-inguinal IMRT in gynecologic cancer patients is from the University of Pittsburgh. In a series of reports [35, 93], Beriwal et al. presented the outcomes of locally advanced vulvar cancer patients undergoing preoperative pelvic-inguinal IMRT combined with chemotherapy followed by planned surgery (Fig. 21.7). In their most recent report [93], 42 stage I–IVA patients, all of whom required preoperative treatment, were treated with a modified Gynecologic Oncology Group (GOG) regimen of 5-fluorouracil and cisplatin with twice-daily IMRT (36) or weekly cisplatin with daily IMRT (6). The twice-daily IMRT regimen consisted of 1.6 Gy twice daily for 10 fractions, followed by 1.8 Gy daily for 7 or 8 days, followed by a planned break of 10 days, and then resumption of radiation with 1.6 Gy twice daily for 10 more fractions. The patients who underwent daily IMRT received 50.4 Gy in 1.8 Gy fractions. Surgery was performed 6–10 weeks following treatment. Overall, treatment was well tolerated with all patients completing IMRT and chemotherapy as planned. No acute grade 3 or higher GI or GU toxicities occurred. One patient developed grade 3 cutaneous toxicity. Of 41 evaluable patients, a complete clinical response was noted in 21 (51.2 %). Thirty-three patients underwent surgery of which 16 (48.5 %) had a pathologic complete response in the vulva. Of these, 15 (93.8 %) remained without disease recurrence. Of the 17 who had a pathologic partial response, 8 (47.1 %) developed a local recurrence. No patient in the series developed a grade 3 or higher late GI or GU complication.

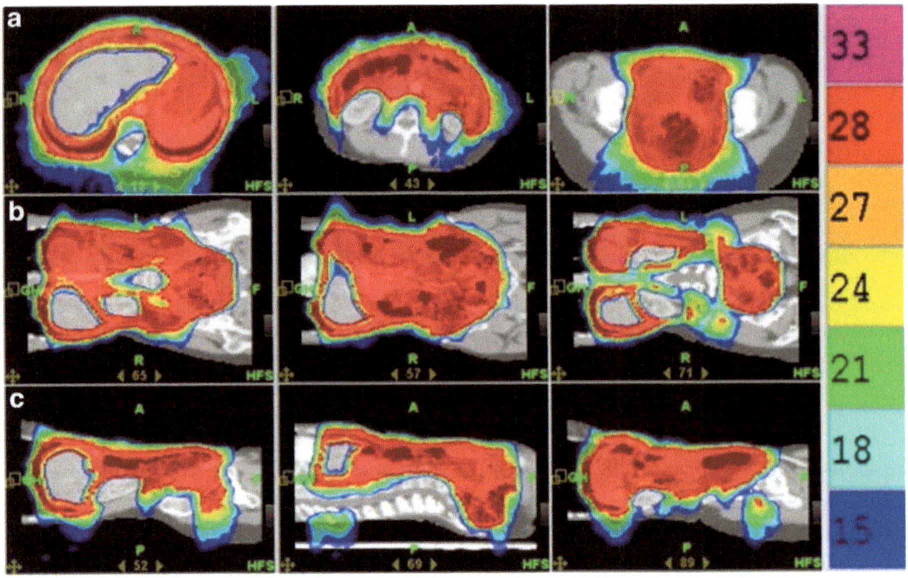

Fig. 21.8 Axial (**a**), coronal (**b**), and sagittal (**c**) computed tomography images showing dose distributions from a helical tomotherapy plan used for a patient with relapsed ovarian cancer [94]

21.4.4 Whole Abdomen

Given the decreasing use of WART in the United States and other countries, it is not surprising that clinical studies in gynecologic cancer patients undergoing intensity-modulated WART (IM-WART) are limited [91–93]. Mahantshetty et al. at Tata Memorial Center presented the outcomes of 8 relapsed ovarian cancer patients with disease confined to the abdomen and/or pelvis treated with salvage intensity-modulated WART [92]. Using an SIB approach, 25 Gy in 1 Gy fractions was delivered to the whole abdomen, while the pelvis received 45 Gy in 1.8 Gy fractions (Fig. 21.8). Treatment was well tolerated with no patient requiring significant unplanned treatment breaks. Overall, three patients developed grade 2 GI and two developed grade 2 transient liver toxicities. While all patients had been heavily pre-treated with chemotherapy, only three (37.5 %) developed grade ≥3 hematologic toxicity. No grade ≥3 GI or renal toxicity was noted. At a median follow-up of 15 months, three patients progressed in the abdomen and/or pelvis. The other 5 remained free of disease.

Rochet and coworkers presented a feasibility trial evaluating chemotherapy followed by IM-WART in newly diagnosed ovarian cancer patients [94]. Ten optimally debulked stage IIIC ovarian cancer patients were enrolled and received six cycles of carboplatin/paclitaxel chemotherapy followed by IM-WART (30 Gy in 1.5 Gy daily fractions). Overall, treatment was well tolerated with only one patient developing an acute grade 3 GI toxicity. Three patients experienced grade 3 leukopenia. While no patient developed chronic enteritis, three patients required surgery due to small-bowel obstruction (one of which was noted to have an abdominal recurrence). At a median follow-up of 23 months, the two-year actuarial DFS, OS, and local progression free survival rates were 63, 68, and 78 %. These results compared favorably to earlier studies of carboplatin/paclitaxel followed by conventional WART [95].

21.4.5 Brachytherapy Alternative/Replacement

Several investigators have reported on the outcome of gynecologic cancer patients treated with IMRT in lieu of brachytherapy [96–101]. The largest series to date was presented by Huang and coworkers at the Princess Margaret Hospital and consisted of 70 locally advanced/recurrent gynecologic cancer patients (77 % cervical cancer) who were deemed ineligible for brachytherapy, primarily due to unfavorable anatomy or tumor bulk [96]. Treatment consisted of 12–30 Gy delivered in 1.4–2 Gy/fraction. Acute grade ≥3 non-hematologic toxicities were infrequent. Late grade ≥3 sequelae included cystitis (2.8 %), enteritis (1.4 %), and fistula (1.4 %). At a median follow-up of 1.33 years, the three-year actuarial OS and DFS rates were 84.4 % and 49 %, respectively. Overall, pelvic, retroperitoneal, and distant recurrences occurred in 39, 27, and 27 % of patients. Olson and coworkers reported on 32 endometrial cancer patients treated with pelvic RT followed by either brachytherapy (24) or an IMRT/conformal (8) boost [99]. At a median follow-up of 18.6 months, no difference was seen between the OS and CSS of the brachytherapy and non-brachytherapy groups. Moreover, no difference was seen in terms of acute toxicity.

Hypofractionated IMRT boost techniques in brachytherapy ineligible patients have also been reported [97, 98, 100, 101]. Hsieh and colleagues treated nine stage IIB–IVA cervical cancer patients in whom brachytherapy was felt not to be feasible with an IMRT boost of 16–27 Gy in 5–9 fractions [101]. Treatment was well tolerated with no grade ≥3 acute toxicities. Only one patient developed a late grade ≥2 toxicity. The three-year actuarial OS, DFS, and local recurrence-free survivals were 46.9, 25.9, and 77.8 %. Molla et al. reported on 16 patients treated with a hypofractionated boost (94 % IMRT) for definitive (4 Gy × 5 fractions) or adjuvant therapy (7 Gy × 2 fractions) (Fig. 21.9) [97]. At a median follow-up of 12.6 months, only one patient (6.2 %) failed locally. Treatment was well tolerated with no grade ≥3 acute sequelae and only one late GI toxicity. In a follow-up report, these investigators updated their experience in 26 postoperative patients [98]. At a median follow-up of 47 months, the three-year local control and OS rates were 96 % and 95 %, respectively. No grade ≥3 acute or chronic sequelae were seen. Kemmerer et al. treated 11 endometrial cancer patients with a 6 Gy × 5 twice weekly boost (82 % IMRT). At 18 months, the OS and DFS rates were 57 % and 68 %, respectively [100].

The sole series using IMRT in place of brachytherapy in patients with vaginal cancer was presented by investigators from the MD Anderson Cancer Center [73]. Twenty-three patients (eight stages I–II, 15 stages III–IVB) ineligible for brachytherapy received treatment with IMRT alone. The total dose including pelvic RT

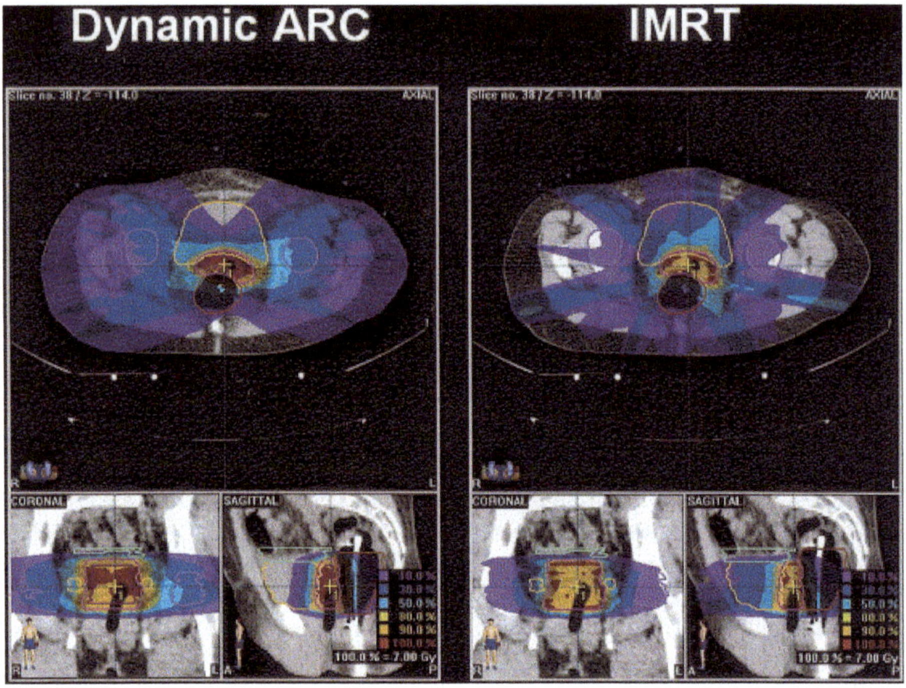

Fig. 21.9 Computed tomography images showing dose distributions for dynamic-arc (*left*) and fixed-field intensity-modulated radiation therapy (*right*) hypofractionated boost plans used for gynecologic cancer patients unable to undergo brachytherapy [97]

(when given) was 65.1 Gy (range, 61–70 Gy). At a median follow-up of 35 months, the five-year OS rates for stage I–II, III, IVA, and IVB patients were 100 %, 82 %, 28 %, and 0 %, respectively. Five-year pelvic control rates for stages I, II, III, IVA, and IVB patients were 100 %, 66 %, 86 %, 50 %, and 0 %, respectively.

Macchia et al. performed a prospective phase I trial in 12 stage IB–IC endometrial cancer patients undergoing IMRT in place of vaginal brachytherapy [92]. Two dose levels were evaluated: 5 Gy × 5 and 6 Gy × 5. No patients at either dose level experienced a DLT. No grade 2 or higher late toxicity was noted. The authors are now evaluating the efficacy of the higher dose level in a phase II trial.

21.5 Technical Issues and Considerations

21.5.1 Patient Selection

When IMRT was first introduced, only a subset of gynecologic cancer patients was considered eligible at many centers. By far, the most commonly treated patients were cervical and endometrial cancer patients undergoing adjuvant pelvic

RT. However, even among this group, exceptions existed, notably markedly obese patients due to the inability to obtain full external contours and concerns over setup accuracy [56]. Over time, however, indications for gynecologic IMRT have grown considerably, and nearly all patients are now considered candidates, even those treated with comprehensive treatment volumes as well as women with gross residual disease requiring higher than conventional doses.

Considerable controversy has long existed regarding the use of IMRT in *intact* cervical cancer patients [4, 102]. In fact, at many prestigious centers, IMRT is commonly used in postoperative patients but not in women with intact disease. Multiple concerns are often cited; however, the major issue is clearly internal organ motion. Nevertheless, such concerns can be addressed with the use of proper treatment margins and daily in-room imaging. Moreover, the favorable published outcome results from experienced investigators clearly support the use of IMRT in these patients [58, 59, 66].

Another controversial issue is the use of IMRT as a replacement/alternative for brachytherapy. Despite promising dosimetric [47, 49] and clinical [96, 98, 101] results, particularly in early-stage patients treated with hypofractionated techniques, IMRT should not be considered equivalent to nor used in lieu of brachytherapy in women eligible for brachytherapy. Concerns have been raised that technologies such as IMRT have contributed to an overall decline in brachytherapy use in patients with cervical cancer treated in the United States [103, 104]. Care should be taken to strictly limit its use to women in whom brachytherapy truly cannot be performed.

21.5.2 Simulation

Gynecologic cancer patients undergoing IMRT are typically immobilized in the supine position and undergo CT simulation with thin (3–5 mm) slices. Custom immobilization devices are recommended and can reduce setup error to <5 mm [105]. Several reports have suggested that prone positioning alone [106, 107] or combined with a small-bowel displacement system [108] may improve small-bowel sparing in patients undergoing IMRT. However, Beriwal and colleagues compared supine and prone positioning in patients undergoing pelvic IMRT and did not note any significant differences in either bowel dose or in rates of GI toxicity [109].

While most investigators perform CT simulation, more sophisticated simulation approaches, notably PET/CT simulation, are used at select centers and are particularly useful in patients with involved lymph nodes. Patient legs should be placed together in a neutral position except in women undergoing pelvic-inguinal IMRT where a frog-leg position may be used; arms are placed overhead in women treated with IM-EFRT and IM-WART. Contrast may be administered to aid in the delineation of the target and normal tissues, including, at some centers, vaginal and rectal contrast [61]. Intravenous contrast is particularly useful since the vasculature is used as a surrogate for the lymph nodes in the planning process. If oral contrast is used, it should be corrected for in the treatment planning process [110]. Some investigators place fiducial markers within the cervix and/or vaginal cuff [88]; others utilize removable vaginal markers [63, 76]. Wires can be used to outline enlarged inguinal nodes and/or gross disease, and bolus is placed over the vulva in vulvar cancer patients undergoing preoperative IMRT [35, 111]. Most centers simulate patients with a comfortably full bladder; others perform two scans (full and empty bladder) in order to generate an integrated target volume (ITV) with planning performed using the full bladder scan [8, 81, 82].

21.5.3 Target and Normal Tissue Delineation

Following simulation, a gross tumor volume (GTV) and clinical target volume (CTV) are contoured on the planning scan. The GTV should include all demonstrable disease, including involved enlarged lymph nodes. A variety of imaging modalities can be used to aid in target delineation, with growing attention on PET [59, 91] and MRI [99]. Investigators at Washington University base their target volume definition on PET and contour a metabolically active tumor volume (MTV), specified at the 40 % threshold level (Fig. 21.10) [59].

In general, the CTV in gynecologic cancer patients undergoing pelvic IMRT consists of the uterus/cervix (if present), upper vagina, paracervical and parametrial tissues, and pelvic lymph nodes. The most superior extent of the CTV is typically placed 1–1.5 cm inferior to the L4–L5 interspace to account for planning target volume (PTV) expansions. Detailed descriptions of CTVs in various gynecologic malignancies are summarized in Table 21.2. At some centers, a single CTV is drawn, whereas at others several CTVs are delineated. As noted above, at some centers, an integrated target volume (ITV) is generated by fusing empty and full bladder planning CT scans, encompassing contours of the cervix (or vaginal cuff in postoperative patients) on both scans, with patients treated with a full bladder or, at other centers, an empty bladder as maintaining a full bladder has not been shown to be reproducible.

Unsurprisingly, considerable variability has been seen in CTV delineation even among experienced radiation oncologists [121, 122]. Fortunately, consensus guidelines and atlases for CTV delineation in cervical and endometrial cancer patients undergoing postoperative pelvic IMRT have been published by the RTOG [117] (Fig. 21.11) and utilized in the RTOG 0418 trial. More recently, the RTOG has developed guidelines for CTV delineation in women with intact cervical cancer (Fig. 21.12) [118]. The Japan Clinical Oncology Group (JCOG) has similarly published contouring guidelines in cervical patients undergoing adjuvant [123] and definitive [119] IMRT. Instructional contouring videos are available online for the multinational INTERTECC clinical trial [124]. Careful analyses of patterns of failure in patients treated using these guidelines will help optimize target design in these patients.

No guidelines for CTV delineation have been published in other gynecologic sites or in patients treated with more comprehensive fields. However, multiple investigators have described their approaches in patients undergoing pelvic-inguinal IMRT [35, 111], IM-EFRT [85, 88], and IM-WART [94, 125]. Several

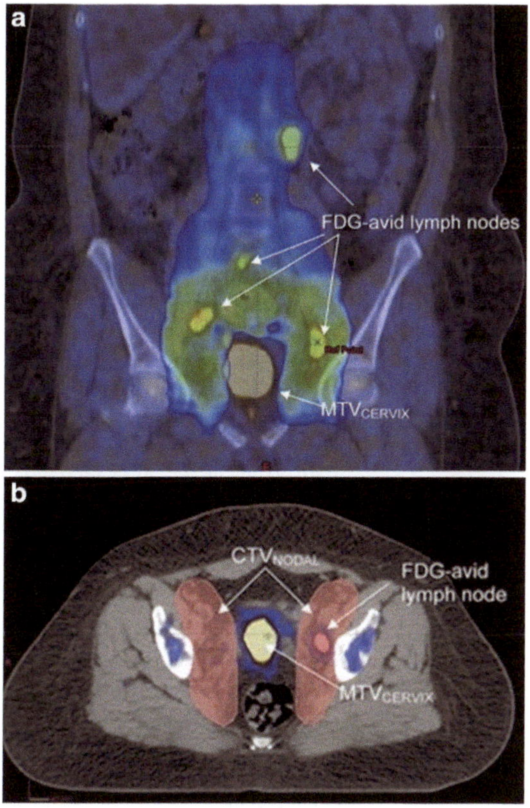

Fig. 21.10 Fused computed tomography and F-18 fluorodeoxyglucose (FDG) positron emission tomography (PET) images with dose distributions (*top*) and clinical target volumes (CTV) (*bottom*) from an intensity-modulated radiotherapy plan for a patient with cervical cancer. The metabolically active tumor volume (MTV) is contoured based on the fused PET image [59]

investigators utilizing IMRT in lieu of brachytherapy also include descriptions of the target design [100, 101]. An international survey has also been performed in vulvar cancer patients treated with pelvic-inguinal IMRT describing target delineation approaches [126].

No consensus exists regarding the optimal PTV margin in gynecologic cancer patients undergoing IMRT, accounting for setup uncertainty and organ motion. In general, most investigators utilize tight margins (0.5–1 cm) around regional lymph nodes and larger margins (1–2 cm or greater) around structures (uterus, cervix) subject to internal organ motion. Multiple studies have been performed evaluating internal organ motion in gynecologic cancer patients and have noted a high degree of interfraction organ motion with recommended planning margins up to 4 cm to fully encompass the CTV for all fractions [120, 127–131]. Collen et al. evaluated ten patients with three times weekly megavoltage CT imaging and found the largest motion to occur in the anteroposterior and superior-inferior directions [129]. They recommended anterior, posterior, right, left, superior, and inferior margins around the uterus and cervix of 19, 19, 13, 13, 29, and 19 mm and 17, 12, 8,9,15, and 9 mm, respectively, to achieve 95 % coverage. Jhingran and colleagues observed motion of seeds placed in the vaginal apex of 24 postoperative patients and found median

Table 21.2 Suggested target volumes for gynecologic malignancies

Site	Target volumes	Definition	Planning volumes	Definition[a]
Cervix	CTV1	GTV, cervix, and uterus or vaginal cuff	PTV1	CTV1+15 mm
	CTV2	Parametrial/paravaginal tissues. Include parauterine fat, ovaries, and proximal vagina for definitive cases. For extensive vaginal involvement, the entire vagina should be included	PTV2	CTV2+10 mm
	CTV3	Common iliac and external and internal iliac nodal regions and presacral regions	PTV3	CTV3+7 mm
Uterine Fundus	CTV1	Vaginal cuff	PTV1	CTV1+15 mm
	CTV2	Paravaginal/parametrial tissues, proximal vagina	PTV2	CTV2+10 mm
	CTV3	Common iliac and external and internal iliac nodal regions. If there is cervical stromal involvement, the presacral region is also included	PTV3	CTV3+7 mm
Vulva	CTV1	GTV plus remaining uninvolved vulva and adjacent soft tissues	PTV1	CTV1+10 mm
	CTV2	Common iliac and external and internal iliac nodal regions and bilateral inguinofemoral lymph nodes. If there is vaginal involvement, the presacral lymph nodes should be included. If there is anal/rectal involvement, the perirectal lymph nodes should be included	PTV2	CTV2+7 mm
Vagina	CTV1	GTV plus a minimum of 3 cm of vagina superiorly and inferiorly	PTV1	CTV1+15 mm
	CTV2	Paravaginal/parametrial tissues adjacent to CTV 1 as well as the common iliac, external, and internal iliac nodal regions and presacral regions. For lower one-third vaginal involvement the bilateral inguinofemoral lymph nodes should also be included	PTV2	CTV2+7 mm
Ovary	CTV	Entire peritoneal cavity from the top of the diaphragms to the bottom of the obturator foramina, as well as the outer 1 cm of liver surface and the pelvic and para-aortic lymph nodes	PTV	CTV+10 mm

Adapted from Lee et al. [112–116] and guidelines from Radiation Therapy Oncology Group, the Gyn IMRT consortium, and the Japan Clinical Oncology Group [117–119]
CTV clinical target volume, *GTV* gross tumor volume, *PTV* planning target volume
[a]Based on work from Khan et al. [120]

maximal displacements of 0.59, 1.46, and 1.2 cm in the left-right, anteroposterior, and superior-inferior directions [131]. In contrast, intra-fraction motion has been shown to be more limited [127].

Khan and coworkers modeled interfraction CTV variations in 50 intact cervical cancer patients undergoing pelvic IMRT with daily CBCT imaging utilizing over

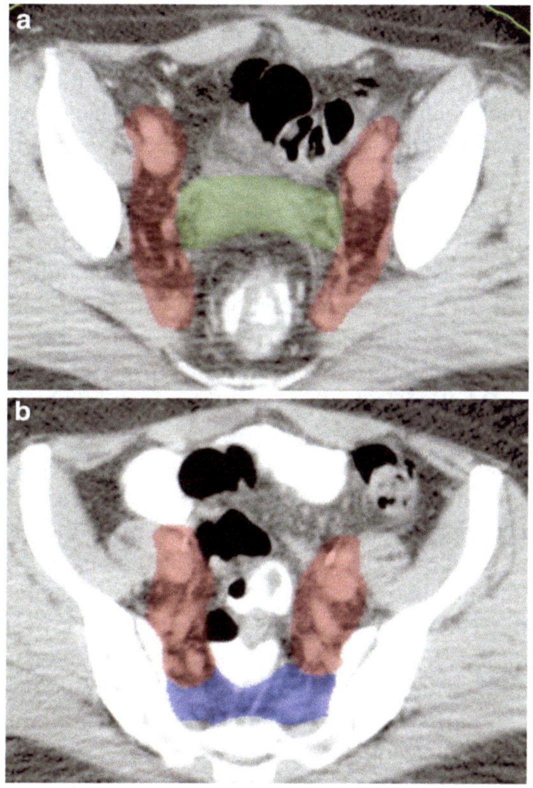

Fig. 21.11 Computed tomography-based clinical target volumes for patients with endometrial and cervical cancer undergoing postoperative intensity-modulated radiation therapy based on the Radiation Therapy Oncology Group guidelines [117]. Representative contours of the external and internal iliac lymph nodes (*red*), parametrial and vaginal tissues (*green*), and presacral region (*blue*) are shown

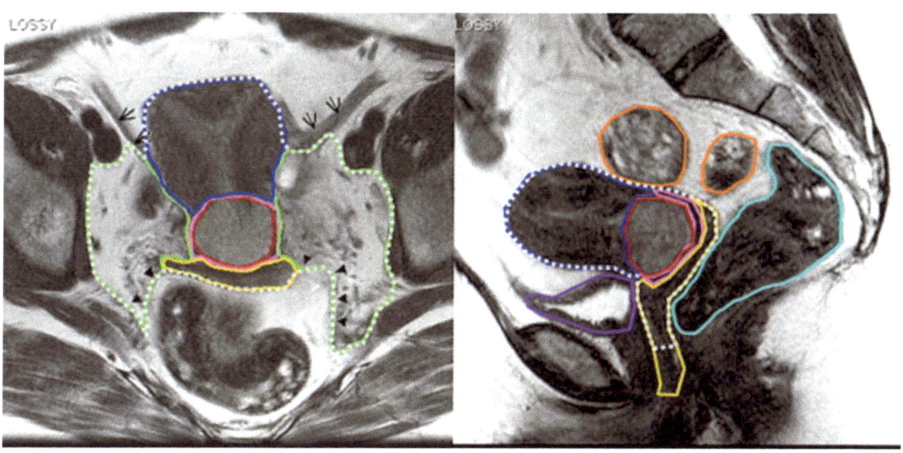

Fig. 21.12 Magnetic resonance imaging-based clinical target volumes for patients with cervical cancer undergoing definitive intensity-modulated radiation therapy based on the Radiation therapy Oncology Group guidelines [118]. Representative contours of the gross tumor volume (*red*), cervix (*pink*), uterus (*blue*), vagina (*yellow*), parametrium (*green*), bladder (*purple*), rectum (*light blue*), and sigmoid (*orange*) are shown on T_2-weighted axial (*left*) and sagittal (*right*) images

700 surface landmarks [120]. Their recommended PTV margins around the cervix/uterus, parametrial/vaginal tissues, and lymph nodes were 1.5 cm, 1.0 cm, and 0.7 cm respectively. Less is known about optimal PTV margins in patients undergoing more comprehensive IMRT volumes. In patients undergoing IM-WART, PTV margins up to 2.5 cm have been used along the diaphragm to account for respiratory-induced motion [37].

Depending on the volume treated, multiple normal tissues are contoured and included as organs at risk (OARs) in the IMRT planning process (Fig. 21.13). In pelvic IMRT patients, most investigators delineate the bowel, bladder, and rectum [58, 61, 133]. Some contour the sigmoid colon as a separate structure [64, 77]. Pelvic BM has been increasingly included in recent years, particularly in women receiving concomitant chemotherapy [58, 59, 122]. In patients undergoing IM-EFRT or IM-WART, the liver and kidneys should be included. At some centers, the stomach, duodenum, spinal cord, and spine are routinely contoured as well. The femoral heads are included in women undergoing pelvic-inguinal IMRT [35, 93].

While different investigators may include the same normal tissues in the planning process, they often disagree on *how* they should be delineated. To address this issue, the RTOG recently convened an expert panel to develop normal tissue contouring guidelines (Table 21.3) [132]. While these guidelines address contouring of the rectum, sigmoid, bowel, bladder, uterus, ovaries, and femoral heads, recommendations for pelvic BM contouring were not included. Most centers that include BM as an avoidance structure use the pelvic bones as a surrogate. However, others favor contouring the intramedullary cavity within the bones [132, 133]. Krishnatry and coworkers compared the two contouring methods in 47 cervical cancer patients treated with IMRT and found that the intramedullary cavity was a better surrogate for active BM and correlated more closely with hematologic toxicity [134]. In the future, functional imaging may help further to optimize delineation of active BM sites [78, 135]. Little consensus exists on the contouring of organs outside the pelvis. Most investigators treating ovarian cancer patients with IM-WART consistently *exclude* a 1 cm outer rim of liver from the normal liver contours [94, 125]. Increasing attention is also being focused on including the duodenum as an OAR in patients undergoing dose-escalated IM-EFRT [87, 136].

21.5.4 Treatment Planning

A large number of IMRT approaches have been used successfully in gynecologic cancer patients, including fixed-field techniques (typically 6–9 fields) [33, 58, 87, 92], arc therapy [65, 125], and helical tomotherapy [76, 101]. In the survey of INTERTECC participants, the fixed-field approach was used at nearly 90 % of centers [122]. The percentage of respondents using arc therapy, helical tomotherapy, and compensator-based techniques were 39 %, 17 %, and 6 %, respectively. Bouchard and coworkers described a novel aperture-based IMRT technique which achieved comparable target coverage and normal tissue sparing as conventional fixed-field approaches but utilized nearly 60 % fewer monitor units (MUs) [63,

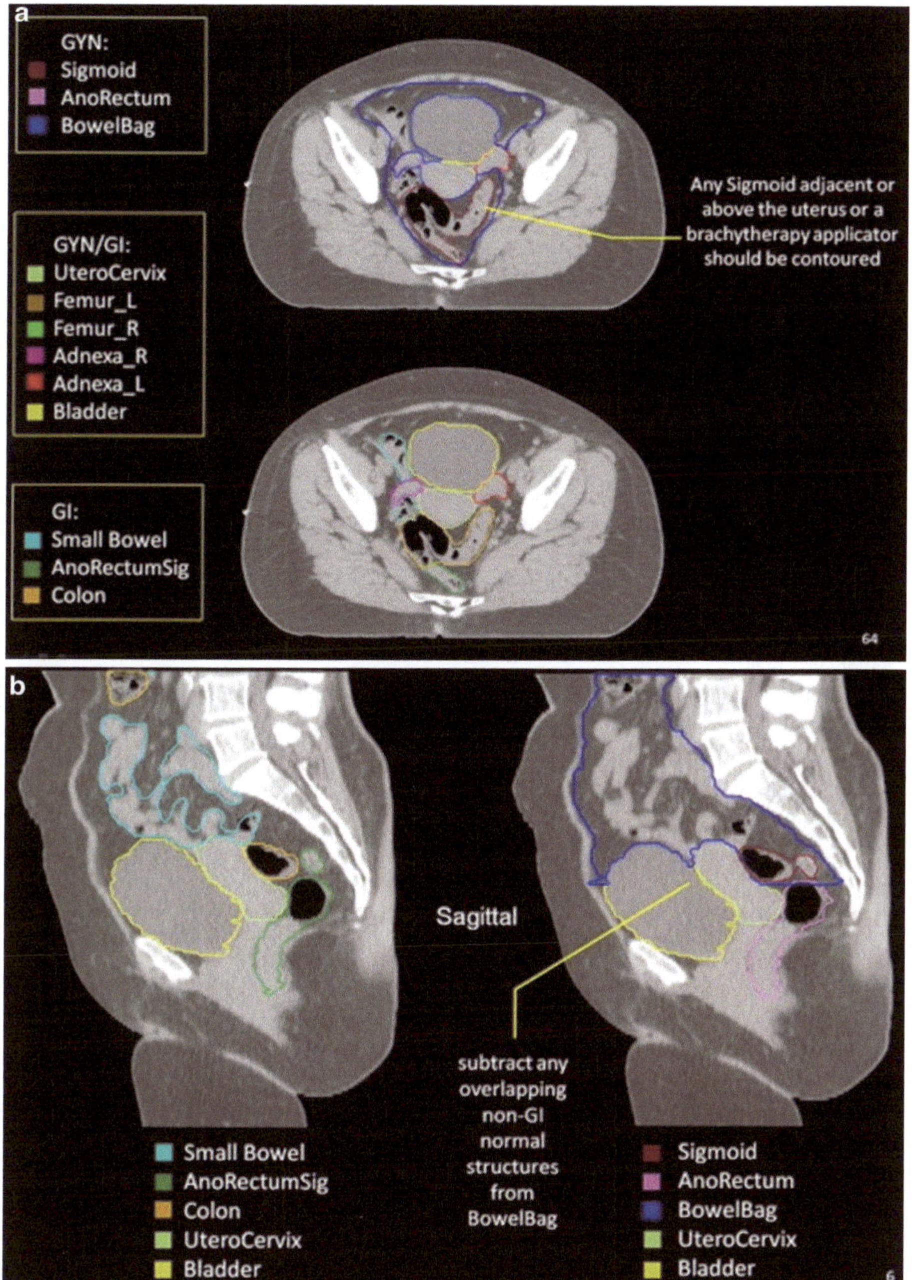

Fig. 21.13 Normal tissue contours on axial (*top*) and sagittal (*bottom*) computed tomography images from Gay et al. [132]

Table 21.3 Contouring guidelines for organs at risk

Organ	Definition and description
Bowel	Space encompassing the outermost loops of bowel from the level of the L4–L5 interspace to the sigmoid flexure. Includes the sigmoid colon and ascending/descending colon present in the pelvis. In women with intact cervical cancer, bowel loops posterior to the uterus in the lower pelvis within the PTV are not included
Sigmoid	Bowel continuing where the rectum contour ended. Stops before connecting to the ascending colon laterally. Contoured when a brachytherapy applicator rests in the uterus. Any sigmoid adjacent or above the uterus, as well as the brachytherapy applicator, should be contoured
Rectum	Inferiorly from the lowest level of the ischial tuberosities (right or left). Contouring ends superiorly before the rectum loses its round shape in the axial plane and connects anteriorly with the sigmoid
Bladder	Outer bladder wall inferiorly from its base and superiorly to the dome.
Bone Marrow	The pelvic bones serve as a surrogate for the pelvic bone marrow. Regions included are the os coxae, L5 vertebral body, entire sacrum, acetabulae, and proximal femora superiorly from the superior border of L5 or the iliac crest (whichever is more superior) and inferiorly to the ischial tuberosities
Liver	Entire liver excluding a 1 cm outer border
Kidneys	Entire kidney parenchyma
Uterus	The uterus and cervix as one structure
Ovaries and Fallopian Tubes	Right and left ovaries and fallopian tubes
Femoral Heads	The proximal femur inferiorly from the lowest level of the ischial tuberosities (right or left) and superiorly to the top of the ball of the femur, including the trochanters

Adapted in part from Gay et al. [132]
PTV planning target volume

133]. A wide variety of photon energies are used for gynecologic IMRT, ranging from 6 to 23 MV. However, most centers treat patients with either 6 or 10 MV beams. Lower energy beams are appealing due to better target conformality [23] and a potentially lower risk of second malignancies [133].

In most patients, conventional total doses (45–50.4 Gy) are delivered in standard fraction sizes (1.8–2 Gy/day). However, as noted earlier, IMRT may be used to deliver higher doses to high-risk sites such as involved lymph nodes and/or residual/recurrent disease. If delivered using a SIB approach, hypofractionated approaches can be used delivering 55–60 Gy in 2.12–2.4 Gy fractions [35, 75, 88]. In patients treated with IMRT in lieu of brachytherapy, fraction sizes of 4–7 Gy have been used [97, 100]. Fraction sizes below 1.8 Gy are rarely prescribed. A notable exception is in ovarian cancer patients at select centers who receive 25 Gy in 1 Gy fractions to the upper abdomen while the pelvis receives 45 Gy in 1.8 Gy fractions [137].

The optimal dose-volume constraints for the PTV and normal tissues in gynecologic cancer patients undergoing IMRT remain unclear. Investigators at Washington University reported the use of the following constraints in patients undergoing pelvic IMRT: PTV (100 % to receive 95 % of the prescription dose), small bowel (<40 %

Table 21.4 Normal dose constraints used in multicenter trials involving intensity-modulated radiation therapy for gynecologic cancer

	Constraints	
Structures	RTOG 0724	INTERTECC
Bowel	30 % receiving ≤40 Gy	$V_{45} \leq 250$ cc; maximum dose < 115 %
Rectum	60 % receiving ≤40 Gy	Maximum dose < 115 %
Bone marrow	NS	$V_{10} < 90$ %; $V_{20} < 75$ %
Bladder	35 % receiving ≤45 Gy	Maximum dose < 115 %
Femoral head	NS	Maximum dose < 115 %
Spinal cord	Max dose ≤45 Gy	NS
Kidneys	2/3 of each ≤18 Gy	NS

Table adapted from Jensen et al. [150]
RTOG Radiation Therapy oncology group, *INTERTECC* International Evaluation of Radiotherapy Technology Effectiveness in Cervical Cancer, *Vx* volume receiving "X" Gy, *NS* not specified

to receive ≥30 Gy), rectum (<40 % to receive ≥40 Gy), and femoral heads (<40 % to receive ≥30 Gy) [59]. In the original series from the University of Chicago, <40 % of the small bowel, rectum, and bladder were constrained to receive ≥36 Gy, ≥ 40 Gy, and ≥40 Gy, respectively. Moreover, >95 % of the PTV needed to receive >95 % of the prescribed dose [56]. Dose-volume constraints used in ongoing trials of patients treated with pelvic IMRT are summarized in Table 21.4.

Detailed normal tissue complication probability (NTCP) studies have been performed in gynecologic cancer patients undergoing pelvic IMRT, which shed light on the optimal dose-volume constraints for various normal tissues. Most attention to date has been focused on the small bowel [138, 139] and pelvic BM. Simpson and coworkers [138] recently validated the association between the small bowl V_{45} and the risk of acute GI toxicity in a cohort of 50 cervical cancer patients undergoing pelvic IMRT which was initially noted by Roeske et al. at the University of Chicago [139]. The recent Quantitative Analyses of Normal Tissue Effects in the Clinic (QUANTEC) effort currently recommends constraining the V_{45} to <195 cc to reduce the incidence of acute small-bowel toxicity below 10 % in patients treated with pelvic irradiation [140]. In an analysis of 37 cervical cancer patients treated with pelvic IMRT and concomitant cisplatin, Mell and coworkers initially found a strong correlation between the pelvic BM V_{10} and V_{20} with acute hematologic toxicity [141, 142] and recently validated these findings in an independent cohort of patients [142]. Currently, these investigators recommended constraining the pelvic BM V_{10} and V_{20} to ≤90 % and ≤75 %, respectively. A post hoc analysis of patients on the RTOG 0418 trial noted a correlation between hematologic toxicity and pelvic BM V_{40} [143]. Cervical cancer patients with a BM V_{40} above 37 % had a higher rate of grade ≥2 hematologic toxicity (75 % vs. 40 %, $p=0.025$) than those with a V_{40} below 37 %. However, none of these patients underwent BM-sparing IMRT limiting the number of patients with low BM doses.

Limited NTCP analyses have been performed in patients treated with more comprehensive volumes [87, 136]. Verma and colleagues evaluated dosimetric predictors of duodenal toxicity in 105 gynecologic cancer patient with gross para-aortic lymph nodes who received total doses between 60 and 66 Gy [136]. Overall, nine patients (8.6 %) developed grade ≥2 duodenal toxicity confirmed on endoscopy. Duodenal V_{55} was the most significant factor correlated with duodenal sequelae. Patients with a V_{55} above 15 cm^3 had a higher rate of severe duodenal toxicity (49 % vs. 7 %, $p<0.01$) than those with a V_{55} below 15 cm^3. Duodenal V_{55} remained significant on multivariate analysis.

21.5.5 Treatment Delivery

Gynecologic IMRT can be delivered on all commercially available treatment machines, including linear accelerators and helical tomotherapy machines. On a linear accelerator, IMRT plans can be delivered using multi-leaf collimators (MLC) or customized compensators [122]. Regardless of the machine used, it is important to verify treatment with in-room image-guided RT (IGRT) given the highly conformal treatment plans and rapid dose falloff. Many types of in-room IGRT approaches have been used for both patient setup and target localization including electronic portal imaging devices (EPID) [88], kilovoltage planar imaging [92], and megavoltage [101] or kilovoltage [100] cone-beam CT (CBCT) imaging. Volumetric imaging is particularly useful to verify inclusion of the cervix and uterus in cervical cancer patients undergoing definitive IMRT. In-room volumetric imaging is particularly important in patients receiving higher than conventional doses to involved sites or in patients treated with IMRT in lieu of brachytherapy. In the Princess Margaret series of patients unfit for brachytherapy who received high-dose central boosts using IMRT, Huang and colleagues noted a higher pelvic control in patients treated with daily IGRT (72 % vs. 42 %, $p=0.04$) compared to those treated without daily image guidance [96].

Prior to (and throughout) the delivery of IMRT, rigorous quality assurance (QA) is essential to ensure proper delivery of the treatment plans. At our institution, the accuracy of treatment is verified daily using both kilovoltage planar and, in patients with intact cervical cancer, CBCT imaging. A variety of QA procedures are also performed including an independent MU verification calculation. A full discussion of QA techniques and procedures used in gynecologic IMRT is beyond the scope of this chapter, and interested readers are referred elsewhere [144].

21.6 Future Directions

From its humble beginnings, gynecologic IMRT has developed considerably over the last decade and is rapidly becoming increasingly commonplace in patients treated at many centers throughout the world. In the coming years, additional clinical trials will be performed helping to define the role of this novel technology. Given that cervical cancer is exceedingly common throughout the world, it is hoped that many of these trials will be multinational efforts like the ongoing INTERTECC trial. It is also likely that advanced imaging techniques will play an increasingly important role in these patients, including novel PET and MRI approaches. Given the growing interest in particle therapy, it is hoped that studies will be performed evaluating the role of protons in these patients [145]. Finally, given the significant changes that are known to occur in patients with bulky cervical cancer during a course of chemoradiotherapy [146], it is hoped that new tools and techniques will be developed to allow the rapid generation of adaptive IMRT plans providing the means to treat the anatomy of the day not simply the anatomy on the initial planning scan. Several novel approaches have been proposed and hopefully will soon make their way into the clinic, potentially improving the quality and delivery of IMRT in these patients [147–149].

References

1. Mell LK, Mundt AJ (2008) Intensity-modulated radiation therapy in gynecologic cancers: growing support, growing acceptance. Cancer J 14:198–199
2. Mazeron R, Dumas I, El Khouri C et al (2014) IMRT in cervix cancer: towards a new standard? Cancer Radiother 18:154–160
3. Fernandez-Ots A, Crook J (2013) The role of intensity modulated radiotherapy in gynecological radiotherapy: present and future. Rep Pract Oncol Radiother 18:363–370
4. Wagner A, Jhingran A, Gaffney D (2013) Intensity modulated radiotherapy in gynecologic cancers: hope, hype or hyperbole? Gynecol Oncol 130:229–236
5. Mell LK, Roeske JC, Mundt AJ (2003) A survey of intensity-modulated radiation therapy use in the United States. Cancer 98:204–211
6. Mell LK, Mehrotra AK, Mundt AJ (2005) Intensity-modulated radiation therapy use in the U.S., 2004. Cancer 104:1296–1303
7. Wright JD, Deutsch I, Wilde ET et al (2013) Uptake and outcomes of intensity-modulated radiation therapy for uterine cancer. Gynecol Oncol 130:43–48
8. Jhingran A, Winter K, Portelance L et al (2011) Efficacy and safety of IMRT after surgery in patients with endometrial cancer: RTOG 0418 Phase II study. Int J Radiat Oncol Biol Phys 81:S45
9. International Evaluation of Radiotherapy Technology Effectiveness in Cervical Cancer (INTERTECC). Available from: http://clinicaltrials.gov/show/NCT01554397. Accessed 25 Apr 2014
10. Barillot I, Tavernier E, Peignaux K, et al. (2014) Impact of post operative intensity modulated radiotherapy on acute gastro-intestinal toxicity for patients with endometrial cancer: results of the phase II RTCMIENDOMETRE French multicentre trial. Radiother Oncol 111:138–143
11. RTOG 1203: Standard versus intensity-modulated pelvic radiation therapy in treating patients with endometrial or cervical cancer. Available from: http://clinicaltrials.gov/show/NCT01672892. Accessed 25 Apr 2014
12. Perez CA, Breaux S, Bedwinek JM et al (1984) Radiation therapy alone in the treatment of carcinoma of the uterine cervix. II. Analysis of complications. Cancer 54:235–246
13. Snijders-Keilholz A, Griffioen G, Davelaar J et al (1993) Vitamin B12 malabsorption after irradiation for gynaecological tumours. Anticancer Res 13:1877–1881
14. Creutzberg CL, van Putten WL, Koper PC et al (2000) Surgery and postoperative radiotherapy versus surgery alone for patients with stage-1 endometrial carcinoma: multicentre randomised trial. PORTEC Study Group. Post operative radiation therapy in endometrial carcinoma. Lancet 355:1404–1411
15. Ellis RE (1961) The distribution of active bone marrow in the adult. Phys Med Biol 5:255–258
16. Keys HM, Bundy BN, Stehman FB et al (1999) Cisplatin, radiation, and adjuvant hysterectomy compared with radiation and adjuvant hysterectomy for bulky stage IB cervical carcinoma. N Engl J Med 340:1154–1161

17. Lhomme C, Fumoleau P, Fargeot P et al (1999) Results of a European organization for research and treatment of cancer/Early Clinical Studies Group phase II trial of first-line irinotecan in patients with advanced or recurrent squamous cell carcinoma of the cervix. J Clin Oncol 17:3136–3142
18. Rose PG, Cha SD, Tak WK et al (1992) Radiation therapy for surgically proven para-aortic node metastasis in endometrial carcinoma. Int J Radiat Oncol Biol Phys 24:229–233
19. Gibbons S, Martinez A, Schray M et al (1991) Adjuvant whole abdominopelvic irradiation for high risk endometrial carcinoma. Int J Radiat Oncol Biol Phys 21:1019–1025
20. Stock RG, Chen AS, Flickinger JC et al (1995) Node-positive cervical cancer: impact of pelvic irradiation and patterns of failure. Int J Radiat Oncol Biol Phys 31:31–36
21. Ferreira PR, Braga-Filho A, Barletta A et al (1999) Radiation therapy alone in stage III-B cancer of the uterine cervix–a 17-year old experience in southern Brazil. Int J Radiat Oncol Biol Phys 45:441–446
22. Barraclough LH, Swindell R, Livsey JE et al (2008) External beam boost for cancer of the cervix uteri when intracavitary therapy cannot be performed. Int J Radiat Oncol Biol Phys 71:772–778
23. Roeske JC, Lujan A, Rotmensch J et al (2000) Intensity-modulated whole pelvic radiation therapy in patients with gynecologic malignancies. Int J Radiat Oncol Biol Phys 48:1613–1621
24. Heron DE, Gerszten K, Selvaraj RN et al (2003) Conventional 3D conformal versus intensity-modulated radiotherapy for the adjuvant treatment of gynecologic malignancies: a comparative dosimetric study of dose-volume histograms small star, filled. Gynecol Oncol 91:39–45
25. Ahamad A, D'Souza W, Salehpour M et al (2005) Intensity-modulated radiation therapy after hysterectomy: comparison with conventional treatment and sensitivity of the normal-tissue-sparing effect to margin size. Int J Radiat Oncol Biol Phys 62:1117–1124
26. Chen Q, Izadifar N, King S et al (2001) Comparison of IMRT with 3-D conformal radiotherapy for gynecologic malignancies. Int J Radiat Oncol Biol Phys 51:S332
27. Brixey CJ, Roeske JC, Lujan AE et al (2002) Impact of intensity-modulated radiotherapy on acute hematologic toxicity in women with gynecologic malignancies. Int J Radiat Oncol Biol Phys 54:1388–1396
28. Lujan AE, Mundt AJ, Yamada SD et al (2003) Intensity-modulated radiotherapy as a means of reducing dose to bone marrow in gynecologic patients receiving whole pelvic radiotherapy. Int J Radiat Oncol Biol Phys 57:516–521
29. Mell LK, Tiryaki H, Ahn KH et al (2008) Dosimetric comparison of bone marrow-sparing intensity-modulated radiotherapy versus conventional techniques for treatment of cervical cancer. Int J Radiat Oncol Biol Phys 71:1504–1510
30. Portelance L, Chao KS, Grigsby PW et al (2001) Intensity-modulated radiation therapy (IMRT) reduces small bowel, rectum, and bladder doses in patients with cervical cancer receiving pelvic and para-aortic irradiation. Int J Radiat Oncol Biol Phys 51:261–266
31. Hermesse J, Devillers M, Deneufbourg JM et al (2005) Can intensity-modulated radiation therapy of the paraaortic region overcome the problems of critical organ tolerance? Strahlenther Onkol 181:185–190
32. Lian J, Mackenzie M, Joseph K et al (2008) Assessment of extended-field radiotherapy for stage IIIC endometrial cancer using three-dimensional conformal radiotherapy, intensity-modulated radiotherapy, and helical tomotherapy. Int J Radiat Oncol Biol Phys 70:935–943
33. Zhang G, He F, Fu C et al (2014) Definitive extended field intensity-modulated radiotherapy and concurrent cisplatin chemosensitization in the treatment of IB2-IIIB cervical cancer. J Gynecol Oncol 25:14–21
34. Ahmed RS, Kim RY, Duan J et al (2004) IMRT dose escalation for positive para-aortic lymph nodes in patients with locally advanced cervical cancer while reducing dose to bone marrow and other organs at risk. Int J Radiat Oncol Biol Phys 60:505–512
35. Beriwal S, Heron DE, Kim H et al (2006) Intensity-modulated radiotherapy for the treatment of vulvar carcinoma: a comparative dosimetric study with early clinical outcome. Int J Radiat Oncol Biol Phys 64:1395–1400

36. Hong L, Alektiar K, Chui C et al (2002) IMRT of large fields: whole-abdomen irradiation. Int J Radiat Oncol Biol Phys 54:278–289
37. Rochet N, Sterzing F, Jensen A et al (2008) Helical tomotherapy as a new treatment technique for whole abdominal irradiation. Strahlenther Onkol 184:145–149
38. Yang B, Zhu L, Cheng H et al (2012) Dosimetric comparison of intensity modulated radiotherapy and three-dimensional conformal radiotherapy in patients with gynecologic malignancies: a systematic review and meta-analysis. Radiat Oncol 7:197
39. Mutic S, Malyapa RS, Grigsby PW et al (2003) PET-guided IMRT for cervical carcinoma with positive para-aortic lymph nodes-a dose-escalation treatment planning study. Int J Radiat Oncol Biol Phys 55:28–35
40. Peters WA 3rd, Liu PY, Barrett RJ 2nd et al (2000) Concurrent chemotherapy and pelvic radiation therapy compared with pelvic radiation therapy alone as adjuvant therapy after radical surgery in high-risk early-stage cancer of the cervix. J Clin Oncol 18:1606–1613
41. D'Souza WD, Ahamad AA, Iyer RB et al (2005) Feasibility of dose escalation using intensity-modulated radiotherapy in posthysterectomy cervical carcinoma. Int J Radiat Oncol Biol Phys 61:1062–1070
42. Du XL, Tao J, Sheng XG et al (2012) Intensity-modulated radiation therapy for advanced cervical cancer: a comparison of dosimetric and clinical outcomes with conventional radiotherapy. Gynecol Oncol 125:151–157
43. Alektiar KM (2002) Can intensity-modulated radiation therapy replace brachytherapy in the management of cervical cancer? Point. Brachytherapy 1:191–192
44. Mundt AJ, Roeske JC (2002) Can intensity-modulated radiation therapy replace brachytherapy in the management of cervical cancer? Counterpoint. Brachytherapy 1:192–194
45. Malhotra HK, Avadhani JS, deBoer SF et al (2007) Duplicating a tandem and ovoids distribution with intensity-modulated radiotherapy: a feasibility study. J Appl Clin Med Phys 8:2450
46. Gielda BT, Shah AP, Marsh JC et al (2011) Helical tomotherapy delivery of an IMRT boost in lieu of interstitial brachytherapy in the setting of gynecologic malignancy: feasibility and dosimetric comparison. Med Dosim 36:206–212
47. Chan P, Yeo I, Perkins G et al (2006) Dosimetric comparison of intensity-modulated, conformal, and four-field pelvic radiotherapy boost plans for gynecologic cancer: a retrospective planning study. Radiat Oncol 1:13
48. Aydogan B, Mundt AJ, Smith BD et al (2006) A dosimetric analysis of intensity-modulated radiation therapy (IMRT) as an alternative to adjuvant high-dose-rate (HDR) brachytherapy in early endometrial cancer patients. Int J Radiat Oncol Biol Phys 65:266–273
49. Low DA, Grigsby PW, Dempsey JF et al (2002) Applicator-guided intensity-modulated radiation therapy. Int J Radiat Oncol Biol Phys 52:1400–1406
50. Sharma DN, Gandhi AK, Sharma S et al (2013) Interstitial brachytherapy vs. intensity-modulated radiation therapy for patients with cervical carcinoma not suitable for intracavitary radiation therapy. Brachytherapy 12:311–316
51. Guerrero M, Li XA, Ma L et al (2005) Simultaneous integrated intensity-modulated radiotherapy boost for locally advanced gynecological cancer: radiobiological and dosimetric considerations. Int J Radiat Oncol Biol Phys 62:933–939
52. Assenholt MS, Petersen JB, Nielsen SK et al (2008) A dose planning study on applicator guided stereotactic IMRT boost in combination with 3D MRI based brachytherapy in locally advanced cervical cancer. Acta Oncol 47:1337–1343
53. Duan J, Kim RY, Elassal S et al (2008) Conventional high-dose-rate brachytherapy with concomitant complementary IMRT boost: a novel approach for improving cervical tumor dose coverage. Int J Radiat Oncol Biol Phys 71:765–771
54. Schefter TE, Kavanagh BD, Wu Q et al (2002) Technical considerations in the application of intensity-modulated radiotherapy as a concomitant integrated boost for locally-advanced cervix cancer. Med Dosim 27:177–184
55. Mundt AJ, Roeske JC, Lujan AE et al (2001) Initial clinical experience with intensity-modulated whole-pelvis radiation therapy in women with gynecologic malignancies. Gynecol Oncol 82:456–463

56. Mundt AJ, Lujan AE, Rotmensch J et al (2002) Intensity-modulated whole pelvic radiotherapy in women with gynecologic malignancies. Int J Radiat Oncol Biol Phys 52:1330–1337
57. Mundt AJ, Mell LK, Roeske JC (2003) Preliminary analysis of chronic gastrointestinal toxicity in gynecology patients treated with intensity-modulated whole pelvic radiation therapy. Int J Radiat Oncol Biol Phys 56:1354–1360
58. Hasselle MD, Rose BS, Kochanski JD et al (2011) Clinical outcomes of intensity-modulated pelvic radiation therapy for carcinoma of the cervix. Int J Radiat Oncol Biol Phys 80:1436–1445
59. Kidd EA, Siegel BA, Dehdashti F et al (2010) Clinical outcomes of definitive intensity-modulated radiation therapy with fluorodeoxyglucose-positron emission tomography simulation in patients with locally advanced cervical cancer. Int J Radiat Oncol Biol Phys 77:1085–1091
60. Tierney RM, Powell MA, Mutch DG et al (2007) Acute toxicity of postoperative IMRT and chemotherapy for endometrial cancer. Radiat Med 25:439–445
61. Folkert MR, Shih KK, Abu-Rustum NR et al (2013) Postoperative pelvic intensity-modulated radiotherapy and concurrent chemotherapy in intermediate- and high-risk cervical cancer. Gynecol Oncol 128:288–293
62. Shih KK, Milgrom SA, Abu-Rustum NR et al (2013) Postoperative pelvic intensity-modulated radiotherapy in high risk endometrial cancer. Gynecol Oncol 128:535–539
63. Nadeau S, Bouchard M, Germain I et al (2007) Postoperative irradiation of gynecologic malignancies: improving treatment delivery using aperture-based intensity-modulated radiotherapy. Int J Radiat Oncol Biol Phys 68:601–611
64. Beriwal S, Jain SK, Heron DE et al (2006) Clinical outcome with adjuvant treatment of endometrial carcinoma using intensity-modulated radiation therapy. Gynecol Oncol 102:195–199
65. Vandecasteele K, Tummers P, Makar A et al (2012) Postoperative intensity-modulated arc therapy for cervical and endometrial cancer: a prospective report on toxicity. Int J Radiat Oncol Biol Phys 84:408–414
66. Chen CC, Lin JC, Jan JS et al (2011) Definitive intensity-modulated radiation therapy with concurrent chemotherapy for patients with locally advanced cervical cancer. Gynecol Oncol 122:9–13
67. Chen SW, Liang JA, Hung YC et al (2013) Does initial 45Gy of pelvic intensity-modulated radiotherapy reduce late complications in patients with locally advanced cervical cancer? A cohort control study using definitive chemoradiotherapy with high-dose rate brachytherapy. Radiol Oncol 47:176–184
68. Chen CC, Wang L, Lu CH, et al. (2013) Comparison of clinical outcomes and toxicity in endometrial cancer patients treated with adjuvant intensity-modulated radiation therapy or conventional radiotherapy. J Formos Med Assoc 113:949–955
69. Chen MF, Tseng CJ, Tseng CC et al (2007) Clinical outcome in posthysterectomy cervical cancer patients treated with concurrent Cisplatin and intensity-modulated pelvic radiotherapy: comparison with conventional radiotherapy. Int J Radiat Oncol Biol Phys 67:1438–1444
70. An JS, Huang MN, Wu LY et al (2013) Study of the radiotherapy modality for patients with stage IIb-IIIb cervical stump cancer. Zhonghua Fu Chan Ke Za Zhi 48:654–658
71. Brown AP, Jhingran A, Klopp AH et al (2013) Involved-field radiation therapy for locoregionally recurrent ovarian cancer. Gynecol Oncol 130:300–305
72. Hiniker SM, Roux A, Murphy JD et al (2013) Primary squamous cell carcinoma of the vagina: prognostic factors, treatment patterns, and outcomes. Gynecol Oncol 131:380–385
73. Fisher CM, Klopp AH, Jhingran A (2011) Vaginal cancer outcomes using intensity modulated radiation therapy for definitive treatment at UT MD Anderson Cancer Center. Int J Radiat Oncol Biol Phys 81:S480–S481
74. Shrivastava SJ, Mahantshetty U, Jamema S et al (2009) A phase II randomized trial comparing intensity modulated radiation therapy (IMRT) with conventional radiation therapy in stage IIB cervical carcinoma: an audit. Int J Radiat Oncol Biol Phys 75:S84–S85

75. Marnitz S, Kohler C, Burova E et al (2012) Helical tomotherapy with simultaneous integrated boost after laparoscopic staging in patients with cervical cancer: analysis of feasibility and early toxicity. Int J Radiat Oncol Biol Phys 82:e137–e143
76. Schwarz JK, Wahab S, Grigsby PW (2011) Prospective phase I-II trial of helical tomotherapy with or without chemotherapy for postoperative cervical cancer patients. Int J Radiat Oncol Biol Phys 81:1258–1263
77. Tharavichitkul E, Wanwilairat S, Chakrabandhu S et al (2013) Image-guided brachytherapy (IGBT) combined with whole pelvic intensity-modulated radiotherapy (WP-IMRT) for locally advanced cervical cancer: a prospective study from Chiang Mai University Hospital, Thailand. J Contemp Brachyther 5:10–16
78. Liang Y, Bydder M, Yashar CM et al (2013) Prospective study of functional bone marrow-sparing intensity modulated radiation therapy with concurrent chemotherapy for pelvic malignancies. Int J Radiat Oncol Biol Phys 85:406–414
79. Mabuchi S, Takahashi R, Isohashi F et al (2013) A phase I study of concurrent weekly carboplatin and paclitaxel combined with intensity-modulated pelvic radiotherapy as an adjuvant treatment for early-stage cervical cancer patients with positive pelvic lymph nodes. Int J Gynecol Cancer 23:1279–1286
80. Portelance L, Moughan J, Jhingran A et al (2011) A phase ii multi-institutional study of post-operative pelvic intensity modulated radiation therapy (IMRT) with weekly cisplatin in patients with cervical carcinoma: two year efficacy results of the RTOG 0418. Int J Radiat Oncol Biol Phys 81:S3
81. Jhingran A, Winter K, Portelance L et al (2012) A phase II study of intensity modulated radiation therapy to the pelvis for postoperative patients with endometrial carcinoma: radiation therapy oncology group trial 0418. Int J Radiat Oncol Biol Phys 84:e23–e28
82. Portelance L, Winter K, Jhingran A et al (2009) Post-operative pelvic intensity modulated radiation therapy (IMRT) with chemotherapy for patients with cervical carcinoma/RTOG 0418 Phase II study. Int J Radiat Oncol Biol Phys 75:S640–S641
83. Gandhi AK, Sharma DN, Rath GK et al (2013) Early clinical outcomes and toxicity of intensity modulated versus conventional pelvic radiation therapy for locally advanced cervix carcinoma: a prospective randomized study. Int J Radiat Oncol Biol Phys 87:542–548
84. Liang JA, Chen SW, Hung YC, et al. (2013) Low-dose, prophylactic, extended-field, intensity-modulated radiotherapy plus concurrent weekly cisplatin for patients with stage IB2-IIIB cervical cancer, positive pelvic lymph nodes, and negative para-aortic lymph nodes. Int J Gynecol Cancer 24:901–907
85. Salama JK, Mundt AJ, Roeske J et al (2006) Preliminary outcome and toxicity report of extended-field, intensity-modulated radiation therapy for gynecologic malignancies. Int J Radiat Oncol Biol Phys 65:1170–1176
86. Jensen LG, Hasselle MD, Rose BS et al (2013) Outcomes for patients with cervical cancer treated with extended-field intensity-modulated radiation therapy and concurrent cisplatin. Int J Gynecol Cancer 23:119–125
87. Poorvu PD, Sadow CA, Townamchai K et al (2013) Duodenal and other gastrointestinal toxicity in cervical and endometrial cancer treated with extended-field intensity modulated radiation therapy to paraaortic lymph nodes. Int J Radiat Oncol Biol Phys 85:1262–1268
88. Beriwal S, Gan GN, Heron DE et al (2007) Early clinical outcome with concurrent chemotherapy and extended-field, intensity-modulated radiotherapy for cervical cancer. Int J Radiat Oncol Biol Phys 68:166–171
89. Townamchai K, Lee LJ, Poorvu PD et al (2012) Clinical outcomes with dose escalation using intensity modulated radiation therapy for node-positive endometrial and cervical cancer. Int J Radiat Oncol Biol Phys 84:S454
90. Shirvani SM, Klopp AH, Likhacheva A et al (2013) Intensity modulated radiation therapy for definitive treatment of paraaortic relapse in patients with endometrial cancer. Pract Radiat Oncol 3:e21–e28

91. Du XL, Sheng XG, Jiang T et al (2010) Intensity-modulated radiation therapy versus para-aortic field radiotherapy to treat para-aortic lymph node metastasis in cervical cancer: prospective study. Croat Med J 51:229–236
92. Macchia G, Cilla S, Ferrandina G et al (2010) Postoperative intensity-modulated radiotherapy in low-risk endometrial cancers: final results of a Phase I study. Int J Radiat Oncol Biol Phys 76:1390–1395
93. Beriwal S, Coon D, Heron DE et al (2008) Preoperative intensity-modulated radiotherapy and chemotherapy for locally advanced vulvar carcinoma. Gynecol Oncol 109:291–295
94. Rochet N, Sterzing F, Jensen AD et al (2010) Intensity-modulated whole abdominal radiotherapy after surgery and carboplatin/taxane chemotherapy for advanced ovarian cancer: phase I study. Int J Radiat Oncol Biol Phys 76:1382–1389
95. Dinniwell R, Lock M, Pintilie M et al (2005) Consolidative abdominopelvic radiotherapy after surgery and carboplatin/paclitaxel chemotherapy for epithelial ovarian cancer. Int J Radiat Oncol Biol Phys 62:104–110
96. Huang F, Chan P, Yan J et al (2010) The intensity modulated radiotherapy alternative: a boost for gynecological cancers not suited to brachytherapy. Int J Radiat Oncol Biol Phys 78:S405
97. Molla M, Escude L, Nouet P et al (2005) Fractionated stereotactic radiotherapy boost for gynecologic tumors: an alternative to brachytherapy? Int J Radiat Oncol Biol Phys 62:118–124
98. Jorcano S, Molla M, Escude L et al (2010) Hypofractionated extracranial stereotactic radiotherapy boost for gynecologic tumors: a promising alternative to high-dose rate brachytherapy. Technol Cancer Res Treat 9:509–514
99. Olson AK, Bhatia S, Duncan C et al (2010) Definitive radiation therapy for the treatment of inoperable endometrial cancer: pelvic radiation followed by HDR brachytherapy or IMRT/conformal boost. Int J Radiat Oncol Biol Phys 78:S421–S422
100. Kemmerer E, Hernandez E, Ferriss JS et al (2013) Use of image-guided stereotactic body radiation therapy in lieu of intracavitary brachytherapy for the treatment of inoperable endometrial neoplasia. Int J Radiat Oncol Biol Phys 85:129–135
101. Hsieh CH, Tien HJ, Hsiao SM et al (2013) Stereotactic body radiation therapy via helical tomotherapy to replace brachytherapy for brachytherapy-unsuitable cervical cancer patients – a preliminary result. Onco Targets Ther 6:59–66
102. Randall ME, Ibbott GS (2006) Intensity-modulated radiation therapy for gynecologic cancers: pitfalls, hazards, and cautions to be considered. Semin Radiat Oncol 16:138–143
103. Han K, Milosevic M, Fyles A et al (2013) Trends in the utilization of brachytherapy in cervical cancer in the United States. Int J Radiat Oncol Biol Phys 87:111–119
104. Tanderup K, Eifel PJ, Yashar CM et al (2014) Curative radiation therapy for locally advanced cervical cancer: brachytherapy is NOT optional. Int J Radiat Oncol Biol Phys 88:537–539
105. Haslam JJ, Lujan AE, Mundt AJ et al (2005) Setup errors in patients treated with intensity-modulated whole pelvic radiation therapy for gynecological malignancies. Med Dosim 30:36–42
106. Adli M, Mayr NA, Kaiser HS et al (2003) Does prone positioning reduce small bowel dose in pelvic radiation with intensity-modulated radiotherapy for gynecologic cancer? Int J Radiat Oncol Biol Phys 57:230–238
107. Stromberger C, Kom Y, Kawgan-Kagan M et al (2010) Intensity-modulated radiotherapy in patients with cervical cancer. An intra-individual comparison of prone and supine positioning. Radiat Oncol 5:63
108. Huh SJ, Kang MK, Han Y (2004) Small bowel displacement system-assisted intensity-modulated radiotherapy for cervical cancer. Gynecol Oncol 93:400–406
109. Beriwal S, Jain SK, Heron DE et al (2007) Dosimetric and toxicity comparison between prone and supine position IMRT for endometrial cancer. Int J Radiat Oncol Biol Phys 67:485–489
110. Hashem R, Winter K, Ruo P et al (2012) Defining the impact of the use of oral contrast in pelvic intensity modulated radiation therapy (IMRT) – An RTOG 0418 secondary analysis. Int J Radiat Oncol Biol Phys 84:S18–S19

111. Beriwal S, Shukla G, Shinde A et al (2013) Preoperative intensity modulated radiation therapy and chemotherapy for locally advanced vulvar carcinoma: analysis of pattern of relapse. Int J Radiat Oncol Biol Phys 85:1269–1274
112. Mundt AJ, Yashar C, Mell LK (2013) Endometrial cancer. In: Lee NY, Lu JJ (eds) Target volume delineation and field setup: a practical guide for conformal and intensity-modulated radiation therapy. Springer, Berlin/New York, p 188
113. Mundt AJ, Yashar C, Mell LK (2013) Cervical cancer. In: Lee NY, Lu JJ (eds) Target volume delineation and field setup: a practical guide for conformal and intensity-modulated radiation therapy. Springer, Berlin/New York, pp 178–179
114. Mundt AJ, Yashar C, Mell LK (2013) Ovarian cancer. In: Lee NY, Lu JJ (eds) Target volume delineation and field setup: a practical guide for conformal and intensity-modulated radiation therapy. Springer, Berlin/New York, p 197
115. Mundt AJ, Yashar C, Mell LK (2013) Vaginal cancer. In: Lee NY, Lu JJ (eds) Target volume delineation and field setup: a practical guide for conformal and intensity-modulated radiation therapy. Springer, Berlin/New York, p 203
116. Mundt AJ, Yashar C, Mell LK (2013) Vulvar cancer. In: Lee NY, Lu JJ (eds) Target volume delineation and field setup: a practical guide for conformal and intensity-modulated radiation therapy. Springer, Berlin/New York, p 209
117. Small W Jr, Mell LK, Anderson P et al (2008) Consensus guidelines for delineation of clinical target volume for intensity-modulated pelvic radiotherapy in postoperative treatment of endometrial and cervical cancer. Int J Radiat Oncol Biol Phys 71:428–434
118. Lim K, Small W Jr, Portelance L et al (2011) Consensus guidelines for delineation of clinical target volume for intensity-modulated pelvic radiotherapy for the definitive treatment of cervix cancer. Int J Radiat Oncol Biol Phys 79:348–355
119. Toita T, Ohno T, Kaneyasu Y et al (2011) A consensus-based guideline defining clinical target volume for primary disease in external beam radiotherapy for intact uterine cervical cancer. Jpn J Clin Oncol 41:1119–1126
120. Khan A, Jensen LG, Sun S et al (2012) Optimized planning target volume for intact cervical cancer. Int J Radiat Oncol Biol Phys 83:1500–1505
121. Fyles A, Lim K, Small W Jr et al (2009) Variability in delineation of clinical target volumes for cervix cancer intensity-modulated pelvic radiotherapy. Int J Radiat Oncol Biol Phys 75:S83–S84
122. Jensen LG, Mahantshetty U, Shi K et al (2011) Survey of IMRT practices in centers participating in the international evaluation of radiotherapy technology effectiveness in cervical cancer (IntERTECC) trial. Int J Radiat Oncol Biol Phys 81:S457–S458
123. Japan Clinical Oncology Group, Toita T, Ohno T et al (2010) A consensus-based guideline defining the clinical target volume for pelvic lymph nodes in external beam radiotherapy for uterine cervical cancer. Jpn J Clin Oncol 40:456–463
124. Contouring guidelines for the INTERTECC trial. Available from: http://radonc.ucsd.edu/RESEARCH/IRTOC/Pages/videos.aspx. Accessed 4 Apr 2014
125. Duthoy W, De Gersem W, Vergote K et al (2003) Whole abdominopelvic radiotherapy (WAPRT) using intensity-modulated arc therapy (IMAT): first clinical experience. Int J Radiat Oncol Biol Phys 57:1019–1032
126. King B, Barkati A, Fyles A et al (2011) Current practice of IMRT to treat carcinoma of the vulva: results of an international survey. Int J Radiat Oncol Biol Phys 81:S45–S46
127. Chan P, Dinniwell R, Haider MA et al (2008) Inter- and intrafractional tumor and organ movement in patients with cervical cancer undergoing radiotherapy: a cinematic-MRI point-of-interest study. Int J Radiat Oncol Biol Phys 70:1507–1515
128. Kaatee RS, Olofsen MJ, Verstraate MB et al (2002) Detection of organ movement in cervix cancer patients using a fluoroscopic electronic portal imaging device and radiopaque markers. Int J Radiat Oncol Biol Phys 54:576–583
129. Collen C, Engels B, Duchateau M et al (2010) Volumetric imaging by megavoltage computed tomography for assessment of internal organ motion during radiotherapy for cervical cancer. Int J Radiat Oncol Biol Phys 77:1590–1595

130. van de Bunt L, Jurgenliemk-Schulz IM, De Kort GA et al (2008) Motion and deformation of the target volumes during IMRT for cervical cancer: what margins do we need? Radiother Oncol 88:233–240
131. Jhingran A, Salehpour M, Sam M et al (2012) Vaginal motion and bladder and rectal volumes during pelvic intensity-modulated radiation therapy after hysterectomy. Int J Radiat Oncol Biol Phys 82:256–262
132. Gay HA, Barthold HJ, O'Meara E et al (2012) Pelvic normal tissue contouring guidelines for radiation therapy: a Radiation Therapy Oncology Group consensus panel atlas. Int J Radiat Oncol Biol Phys 83:e353–e362
133. Bouchard M, Nadeau S, Gingras L et al (2008) Clinical outcome of adjuvant treatment of endometrial cancer using aperture-based intensity-modulated radiotherapy. Int J Radiat Oncol Biol Phys 71:1343–1350
134. Mahantshetty U, Krishnatry R, Chaudhari S et al (2012) Comparison of 2 contouring methods of bone marrow on CT and correlation with hematological toxicities in non-bone marrow-sparing pelvic intensity-modulated radiotherapy with concurrent cisplatin for cervical cancer. Int J Gynecol Cancer 22:1427–1434
135. Rose BS, Liang Y, Lau SK et al (2012) Correlation between radiation dose to (1)(8) F-FDG-PET defined active bone marrow subregions and acute hematologic toxicity in cervical cancer patients treated with chemoradiotherapy. Int J Radiat Oncol Biol Phys 83:1185–1191
136. Verma J, Sulman EP, Jhingran A et al (2014) Dosimetric predictors of duodenal toxicity after intensity modulated radiation therapy for treatment of the para-aortic nodes in gynecologic cancer. Int J Radiat Oncol Biol Phys 88:357–362
137. Shetty UM, Shankar S, Engineer R et al (2013) Image-guided intensity-modulated whole abdominal radiation therapy in relapsed epithelial ovarian cancers: a feasibility study. J Cancer Res Ther 9:17–21
138. Simpson DR, Song WY, Moiseenko V et al (2012) Normal tissue complication probability analysis of acute gastrointestinal toxicity in cervical cancer patients undergoing intensity modulated radiation therapy and concurrent cisplatin. Int J Radiat Oncol Biol Phys 83:e81–e86
139. Roeske JC, Bonta D, Mell LK et al (2003) A dosimetric analysis of acute gastrointestinal toxicity in women receiving intensity-modulated whole-pelvic radiation therapy. Radiother Oncol 69:201–207
140. Kavanagh BD, Pan CC, Dawson LA et al (2010) Radiation dose-volume effects in the stomach and small bowel. Int J Radiat Oncol Biol Phys 76:S101–S107
141. Mell LK, Kochanski JD, Roeske JC et al (2006) Dosimetric predictors of acute hematologic toxicity in cervical cancer patients treated with concurrent cisplatin and intensity-modulated pelvic radiotherapy. Int J Radiat Oncol Biol Phys 66:1356–1365
142. Rose BS, Aydogan B, Liang Y et al (2011) Normal tissue complication probability modeling of acute hematologic toxicity in cervical cancer patients treated with chemoradiotherapy. Int J Radiat Oncol Biol Phys 79:800–807
143. Klopp AH, Moughan J, Portelance L et al (2013) Hematologic toxicity in RTOG 0418: a phase 2 study of postoperative IMRT for gynecologic cancer. Int J Radiat Oncol Biol Phys 86:83–90
144. Ting J (2005) Commissioning and dosimetric quality assurance. In: Mundt AJ, Roeske J (eds) Intensity modulated radiation therapy: a clinical perspective. BC Decker Inc., Hamilton, pp 186–197
145. Song WY, Huh SN, Liang Y et al (2010) Dosimetric comparison study between intensity modulated radiation therapy and three-dimensional conformal proton therapy for pelvic bone marrow sparing in the treatment of cervical cancer. J Appl Clin Med Phys 11:3255
146. van de Bunt L, van der Heide UA, Ketelaars M et al (2006) Conventional, conformal, and intensity-modulated radiation therapy treatment planning of external beam radiotherapy for cervical cancer: the impact of tumor regression. Int J Radiat Oncol Biol Phys 64:189–196

147. Oh S, Stewart J, Moseley J et al (2014) Hybrid adaptive radiotherapy with on-line MRI in cervix cancer IMRT. Radiother Oncol 110:323–328
148. Ahmad R, Bondar L, Voet P et al (2013) A margin-of-the-day online adaptive intensity-modulated radiotherapy strategy for cervical cancer provides superior treatment accuracy compared to clinically recommended margins: a dosimetric evaluation. Acta Oncol 52:1430–1436
149. Stewart J, Lim K, Kelly V et al (2010) Automated weekly replanning for intensity-modulated radiotherapy of cervix cancer. Int J Radiat Oncol Biol Phys 78:350–358
150. Jensen LG, Mundt A, Mell LK (2011) Intensity-modulated radiotherapy for gynecologic malignancies. In: Mundt A, Yashar C, Mell LK (eds) Gynecologic cancer (radiation medicine rounds), Vol 2. Demos Medical, New York, p 344

Pediatric Cancers

22

Lynn Million and Marian Axente

Keywords
Pediatric • Cancer • IMRT

22.1 Feasibility of Intensity Modulated Radiation Therapy in Pediatric Cancers

The role of intensity modulated radiation therapy (IMRT) is now firmly established in pediatric radiation oncology and has been incorporated as an essential treatment modality in contemporary pediatric oncology clinical trials. Specific target volume and dose guidelines rely on the use of IMRT and IGRT to achieve normal tissue dose constraints [1–3]. However, in its implementation phase, practical concerns were raised including the use of multiple fields, longer corresponding treatment times and increased risk of patient motion, and uncertainty in spatial delivery of highly complex dose distributions with sharp dose falloff. The size difference between pediatric and adult patients is an important consideration as the differences between the normal tissues and target volumes are proportionally more narrow (Fig. 22.1). The small target volumes and required steep dose gradients to avoid normal tissues using IMRT introduce highly complex fluence optimization problems to produce acceptable dose distributions. Lee et al. reported that "when the critical structures are relatively small compared with the PTV, dose reduction to critical structures is accompanied by inferior scores in conformal coverage and homogeneity" [4]. However, improved collimation systems with thinner MLCs [5], optimization of leaf motion in concordance with dose rate, and gantry speed for arc delivery, such as in volumetric modulated arc

L. Million, M.D. (✉) • M. Axente, Ph.D.
Department of Radiation Oncology, Stanford University Medical Center, 875 Blake Wilbur Drive, Stanford, CA 94305, USA
e-mail: lmillion@stanford.edu

Y. Nishimura, R. Komaki (eds.), *Intensity-Modulated Radiation Therapy*,
DOI 10.1007/978-4-431-55486-8_22

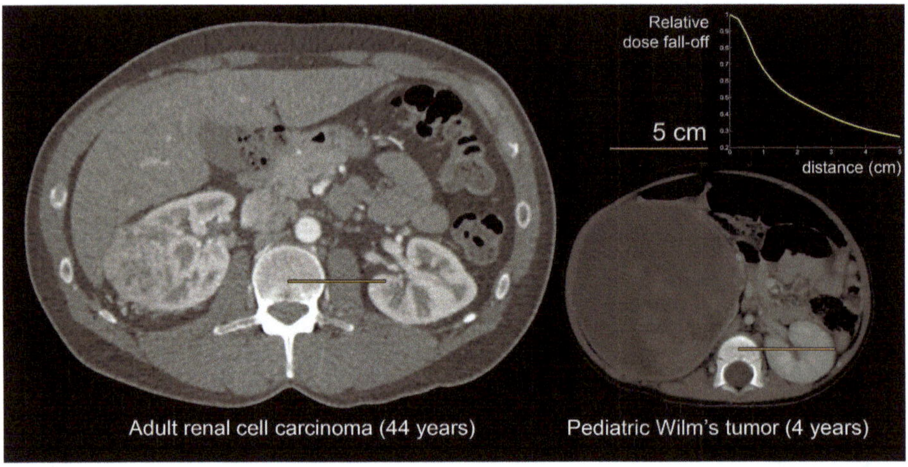

Fig. 22.1 Axial CT images of adult and pediatric patient treated with IMRT. Images are at the same scale (*orange bar*). Since the physical characteristics of the delivery hardware are independent of the patient anatomy, adults benefit from larger distances between targets and tissues at risk. In the upper right corner, there is a representative 5 cm dose falloff for an IMRT plan (6 MV flattening filter free, arc delivery, 2.5 mm MLC)

therapy (VMAT) [6], can significantly increase target coverage for even in the smallest of targets, improve dose falloff, and generally reduce the delivery time, while decreasing the dose to adjacent normal tissue [7].

For treatment delivery, IMRT requires rigid immobilization devices which may include anesthesia and daily confirmation of setup using image-guided radiation therapy (IGRT) to ensure the spatial accuracy of the treatment (Fig. 22.2). The additional procedures added to the already prolonged treatment times for static IMRT led investigators to query whether IMRT was practical in pediatric and young adult patients. Furthermore, motion compensation techniques such as breath hold and respiratory gating, allowing for beam-on timing to be controlled and correlated to the phase or amplitude of the patients breathing, were not deemed feasible for pediatric cases. However, literature has shown that motion compensation is applicable for pediatric cases, at the expense of longer treatment times [8, 9]. Optimized simulation workflows, improved anesthesia solutions, increased experience with IGRT and gating practices, and optimized IMRT plans with arc delivery technology [10], aided by the introduction of high-dose-rate clinical beams (≥10 Gy/min) [11], have the potential to reduce the end-to-end time in the treatment room for children, while reducing radiation exposure to normal tissues.

The therapeutic advantages of IMRT to conform therapeutic doses to the target volume while avoiding critical organs and normal tissues were rationale enough to study its feasibility in pediatric cases. In 2004, Penagaricano et al. reported their clinical experience treating five pediatric malignancies using IMRT and indicated that the technique was a viable alternative to conventional 3D treatment [12]. In 2006, Bhatnager et al. reported on their IMRT experience while treating 22 pediatric

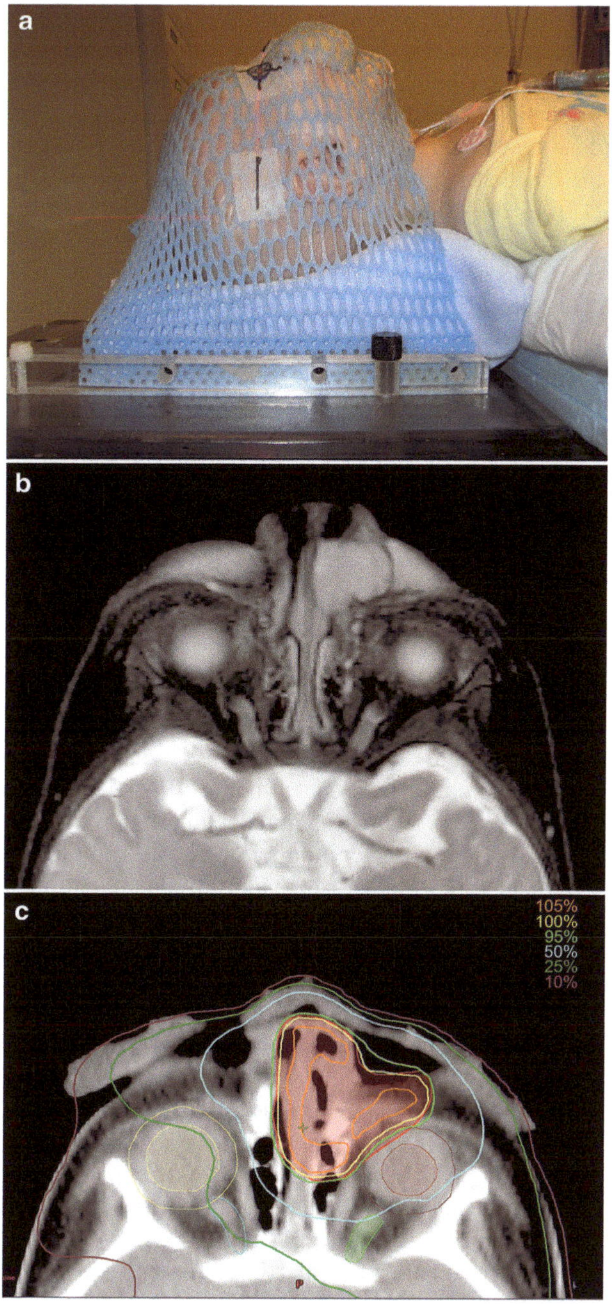

Fig. 22.2 IMRT process for pediatric patient with rhabdomyosarcoma of the nostril. (**a**) Treatment setup under full anesthesia; thermoplastic mask is used to immobilize the patient and the custom bolus, offering a high degree of positioning reproducibility. (**b**) Patient pretreatment MRI images are available for fusion to treatment planning CT to accurately delineate the span of the tumor and adjacent normal tissues. (**c**) Planning CT with dose distribution for VMAT delivery with 6 MV photons (target is delineated in red, with transparent red filling; also present right/left globes and corresponding 3 mm expansions and optic nerves)

tumors involving different sites. Immediate benefits were observed in significant dose reduction for OARs (6–42 % of the planned target dose) which was demonstrated in the pituitary, brainstem, cochlea, optic nerves, and lens during intracranial irradiation. Spinal cord and parotid dose reductions of up to 50 % of the planned dose were observed in head and neck cases, while for pelvis treatments, the dose to the bladder, rectum, and small bowel was reduced by 22–63 % of the target dose [13]. In 2009, Sterzing et al. reported a single institution IMRT experience in children, with long-term outcomes spanning 9 years, validating that local control was not compromised using IMRT in the pediatric population, and there was a significant reduction in normal tissue exposure in the high-dose radiation volume [14]. Other observed clinical benefits of IMRT include fewer cutaneous and subcutaneous late effects such as pigment changes, hair loss, telangiectasia, and subcutaneous fibrosis. The significant reduction in superficial dose to the cutaneous and subcutaneous regions has been demonstrated in dosimetric studies [15] and attributed to an increased number of incident beams, spreading dose over more skin area which inherently lowers the dose buildup in the superficial tissues.

22.1.1 The Main Challenge: Extraneous Radiation Exposure of Normal Tissues

Extraneous dose is the unavoidable low dose radiation which the patient receives outside the planned target volume. In all radiation treatment modalities, including IMRT, normal tissues receive low dose radiation from head leakage, patient internal scatter, and collimator scatter sources (Fig. 22.3). The main disadvantage of IMRT is that a larger volume of unintended normal tissues are exposed to low dose radiation as a result of higher monitor units with complex multi-field plans and increased treatment times and greater neutron contamination for higher energy treatments [18]. Some authors argue that the extraneous low dose radiation may contribute to a higher rate of secondary malignancy(s). While the estimation of radiation-induced secondary malignancy(s) is an area of active research, radiation biologists have modeled risks for second malignancy(s) in the low dose region to be approximately "8 % per Gy probability for a fatal cancer to develop due to radiation above spontaneous incidence. An estimated 0.75–1 % of surviving patients would be expected to develop a second malignancy as a consequence of IMRT, approximately double the incidence of secondary malignancies due to conventional therapy" [19–22].

In addition to second malignancy(s), radiation-induced sequelae to radiosensitive organs outside the radiation field need to be considered. Organs/tissues which may be highly susceptible to extraneous low dose radiation include gonads, breast tissue, thyroid, and lens (infertility or sterility, cessation or abnormal growth, hypothyroidism, cataracts) [23, 24]. Irreversible radiation-induced injury to the gonads can significantly alter quality of life for childhood cancer survivors who desire an offspring later in life. Since the gonads are one of the most radiosensitive organs, even exposure to low dose radiation may inadvertently contribute to infertility or sterility [25]. Thyroid and lens are also very radiosensitive structures which can be affected by very low dose radiation. Breast tissue is highly susceptible to the

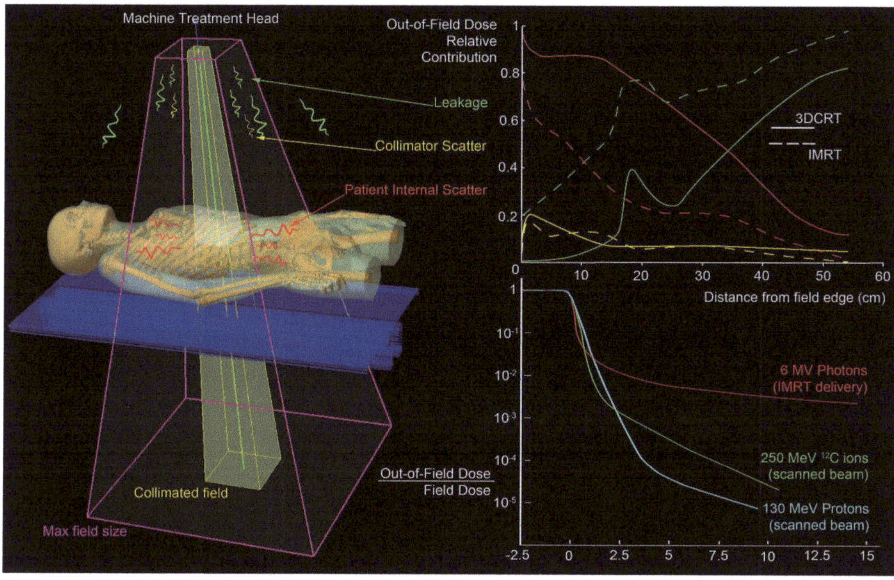

Fig. 22.3 Extraneous or out-of-field dose has three sources (leakage, patient internal scattering, collimator scattering) and is defined as the dose that the patient receives outside the treatment field. Leakage increases for IMRT delivery versus 3DCRT, the opposite is true for patient internal scatter, while collimator scatter contribution is relatively constant as seen in the graph in the upper right corner (Data adapted from Ref. [16]). The out-of-field dose is significantly reduced using charged particles versus photons as indicated in the graph in the lower right corner (Data adapted from Ref. [17])

mutagenic effects of ionizing radiation in young females. The risk for breast cancer in Hodgkin's lymphoma survivors who receive chest irradiation remains high even for patients treated with conformal techniques such as IMRT. Recent reports show that the majority of female childhood cancer survivors who receive low dose (median 14 Gy) radiation therapy to large volumes (e.g., whole-lung irradiation) have a significantly elevated risk for developing breast cancer [26].

While studies show that differences between IMRT and 3DCRT regarding peripheral extraneous doses received by gonads, thyroid, breast tissue, and lens in children and young adults are minimal [27], they should always be considered, calculated, and/or directly measured for IMRT plans. Dose distributions therefore may influence clinical decisions where pediatric radiation oncologists may consider 3DCRT versus IMRT to protect sensitive structures from the low dose irradiation. However, it is important to know that the overall absolute extra-target peripheral dose for the two modalities is influenced by different "competing" dominant effects (Fig. 22.3). For 3DCRT patient, internal scatter is larger resulting in higher out-of-field doses closer to the target region, while with IMRT, there is an increased head leakage corresponding to higher peripheral doses far from the target region [28]. Therefore, the increased monitor units delivered with IMRT should not be treated as a single, accurate indicator of increased out-of-field peripheral irradiation and subsequent low dose exposure of normal tissues [28].

22.1.2 Strategies to Minimize the Out-of-Field Dose

Given the observed benefits of organ-/tissue-sparing capabilities of IMRT in the high-dose region, several strategies have been developed and demonstrated to minimize the unintended out-of-field dose from IMRT delivery. Using coplanar IMRT beams may be one technique to minimize the internal scatter dose. Kan et al. reported on five cases with different pediatric malignancies in the head and neck planned with both coplanar and noncoplanar IMRT techniques. It was observed that peripheral doses were 1.8–2.5 times higher while using the noncoplanar beams [29]. Another technique to reduce out-of-field dose is the removal of the flattening filter, which removes a neutron contamination source associated with high-energy photons interacting in the filter, reduces head leakage, and reduces the delivery time. Cashmore et al. tested this hypothesis with a linear accelerator outfitted with conventional (flattened) and flattening-filter-free modes. IMRT treatment plans for pediatric intracranial treatments were delivered to phantoms using both approaches. Measurements indicated a 23–70 % reduction in peripheral doses (from thyroid to gonads) using the flattening-filter-free modes [30].

The reduction in beam-on time (number of monitor units) is also a method to minimize out-of-field peripheral doses. Besides significantly increasing target coverage and reducing the delivery time, VMAT delivery of IMRT plans has been shown to generally match or even reduce the treatment monitor units compared to conventional static field IMRT plans [31]. In complex pediatric pelvic cases, VMAT reduced the treatment time by 78 % and monitor units by 25 % compared with standard IMRT as reported by Matuszak et al. [32]. The reduction in out-of-field peripheral dose associated with the delivery and reduction in monitor units was as high as threefold for thyroid during pelvic irradiation [31]. There is a variety of commercially available treatment planning systems, each offering optimization solutions that can be employed in reducing extraneous dose in pediatric patients. Attalla et al. reviewed current planning systems offering IMRT optimization. In this study, IMRT plans using step-and-shoot were designed for pediatric head and neck and CNS cases necessitating simultaneous integrated boosts. The authors observed that while three different commercial treatment planning systems were used to obtain optimized IMRT plans achieving the same clinical objectives, using the same energy, number, and direction of beams, the resulting plan quality was not comparable. They found one system was superior producing more efficient plans, fewer segments and MUs, shorter treatment delivery times, and better conformality [33]. Dose painting (DP), simultaneous integrated boost (SIB), and simultaneous modulated accelerated radiotherapy (SMART) are planning techniques which allow highly customized, highly conformal dose distributions while treating multiple targets to different dose levels and sparing more normal tissue [34, 35].

Due to its complex resulting spatial distributions of dose deposition and high gradients between targets and adjacent normal tissues, IGRT is necessary when using IMRT. Even with anesthesia minimizing the risk for movement during treatment, the precision in executing the treatment plan relies on accurate alignment with daily in-room verification using kV imaging. The use of three-dimensional alignment using cone beam computed tomography (CBCT) adds additional accuracy of setup alignment, particularly in the head and neck region. Additionally, it aids in the decision as to whether replanning is necessary for tumors regressing during the course of treatment or for anatomical changes during the course of radiation such as rapid weight loss. The benefit from frequent CBCT portal imaging must be carefully weighed against the risk for exposure to nontherapeutic ionizing radiation. Daily IGRT imaging doses should be recorded for the cumulative exposure above the prescribed therapeutic dose [36]. The dose contribution from imaging is generally less than 2–3 % of the prescribed dose and is usually neglected from total dose summations [37]. However, for pediatric patients, sensitive structures such as the testes or ovaries may be affected if daily kV planar imaging or CBCT was to be used. For example, testicular doses from kV imaging can be 3–4 times higher than the actual incidental dose from pelvic irradiation treatments. Reducing the imaging field of view to exclude the testes may significantly decrease dose to this region [38, 39].

22.1.3 Future Directions

While refinements in delivery methods, increasing positional accuracy, and decreasing delivery time are critical in minimizing extraneous radiation dose exposure in pediatric patients, the physics of photons' interaction with tissue does not allow for complete avoidance of normal tissue exposure to low dose radiation. The pediatric oncology community has been investigating intensity modulated proton therapy (IMPT) as the step beyond IMRT in radiotherapy [40]. Many proponents of proton therapy to treat childhood cancer refer to the desirable beam characteristics which minimizes the out-of-field dose (see Fig. 22.3). Protons, unlike photons, deposit most of their energy at the distal end of their range (Bragg peak). This implies practically nonexistent exit dose, while the entrance dose is lower compared to photon attenuation in tissue [41].

There are two main delivery methods in proton therapy: passive scattering and spot scanning. Passive scattering employs beam scatterers to spread out the input beam laterally, metal apertures to collimate the beam, and range modifying devices to spread out the Bragg peak. In principle, passive scattering employs both beam energy and intensity modulation. However, IMPT refers to the second delivery method. Spot scanning, also referred to as pencil beam scanning or modulated scanning, delivers the treatment layer by layer in raster format scanning a pencil beam using powerful magnets. Depth is changed by switching energy, hence changing the range of the protons. Modern delivery systems are increasingly efficient, allowing for very quick delivery of dose even for large target volumes [41].

The primary advantage that high-energy proton therapy has over photon therapy is reducing normal tissue dose [41]. Several studies have modeled the relative risk associated with normal tissue irradiation for both proton treatments and current state-of-the-art IMRT. Athar et al. reported that the potential organ-specific second cancer lifetime attributable risks would be from unintended internal scatter or leakage out-of-field low dose irradiation from 6 MV IMRT and passive scattering proton therapy. The modeling study included data from patients ranging in age from 9 months to 14 years old and one adult and two treatment sites (brain and mid-spine). The lifetime attributable risk for developing a thyroid cancer after treatment to a brain lesion in a 4-year-old patient was estimated at 1.1 % for IMRT versus 0.3 % for proton therapy. The lifetime attributable risk for developing a bladder cancer after treatment to a mid-spine field was estimated to be 0.2 % with IMRT and 0.02 % with proton therapy, suggesting a distinct advantage to proton beam especially with regard to organs at risk further away from the field [42]. Lifetime attributable risks were also modeled to assess the risk for developing radiation-induced tumors within the path of the beam. Using whole-body phantoms for a 4- and 14-year-old child, plans were generated for optic glioma and vertebral body Ewing's sarcoma. The lifetime attributable risks were modeled for these cases, and the results showed that the risks associated with proton therapy were lower by a factor of 2–10 in the case of protons [43]. In both cases, long-term follow-up is needed to confirm the premise that proton beam therapy is associated with fewer radiation-induced secondary malignancy(s) [44].

Greco et al. reviewed the current trends in proton therapy for children and concluded that IMPT has the potential to "yield superior dose distributions to photon IMRT, with the added advantage of a significant reduction in the volume of healthy normal tissues exposed to low-to-medium doses" [45]. Lomax et al. emphasized the versatility of scanning beam IMPT, where the individual Bragg peaks can be delivered from any field and can be distributed in 3D throughout the target volume, providing an increasing amount of degrees of freedom for designing dose distributions when compared to IMRT or conventional proton therapy [46].

There are problems to be solved with IMPT. The protons relative biological effectiveness (RBE) is spatially variable across the energy deposition curve and maximized at the Bragg peak [47]. Furthermore, the uncertainty in the range of protons in tissue is affected by imaging, patient setup, beam delivery, and dose calculation techniques [48]. High-energy charged particles such as protons yield a significant neutron spectrum in their interaction with the beam line components and the patients' body, increasing the out-of-field dose [49, 50]. A major hurdle to overcome is that proton facilities are very costly to build and operate, limiting the feasibility of using this modality in many places around the world.

22.2 IMRT Clinical Applications by Tumor Type

22.2.1 Central Nervous System Tumors

22.2.1.1 Medulloblastoma

Medulloblastoma is the most common malignant childhood brain tumor, representing 15–20 % of all central nervous system (CNS) tumors, with the highest incidence in children between 4 and 7 years of age. The cerebellum is the most common site of origin, and there is a distinct propensity for cerebrospinal fluid (CSF) dissemination. Metastatic disease at diagnosis occurs in approximately

30 % of patients; however, with aggressive therapy, two-thirds of the patients are long-term survivors [51].

Craniospinal irradiation (CSI) with a boost to the primary site has long been part of the multimodality management, which includes surgery and chemotherapy. Conventional CSI therapy for standard risk medulloblastoma patients consists of 3DCRT with two opposed lateral whole-brain fields matched to a posterior spine field to comprehensively cover the CSF with doses as high as 36 Gy, followed by a selective boost to treat the entire posterior fossa (if primary site) using doses as high as 54 Gy [52]. This treatment is associated with severe significant late effects including trunk shortening (spine), severe ototoxicity (cochlea), xerostomia (parotid), permanent hair loss (scalp), endocrine (hypothalamic-pituitary axis), and neurocognitive effects (temporal lobes/hippocampus) [53]. Children's Oncology Group (COG) protocols are testing the safety of reducing CSI doses and limiting boost target volumes to the tumor bed plus a circumscribed margin. Further refinement of the boost volumes using natural barriers to tumor spread such as bony calvarium or tentorium can further reduce boost target volumes. The optimal PTV volumes and doses for CSI and primary site boost for medulloblastoma have yet to be validated in prospective clinical trials.

Incorporating IMRT in CSI treatments results in higher conformality, dose homogeneity, and normal tissue sparing – primarily in the cochleae, temporal lobe and hypothalamic-pituitary tract but also organs in the beam pathway targeting the spinal CSF (the thyroid, heart, lung, esophagus, vertebral bodies, esophagus, stomach, kidneys, liver, small bowel, and gonads). The feasibility of CSI-IMRT has been confirmed by several investigators [54–58]. CSI-IMRT allows for a dosimetric match at the junction of cranial and spinal fields and increased dose homogeneity across the matched fields when compared to 3DCRT. Furthermore, there is significant reduction in exposure to normal tissues while providing superior coverage of the craniospinal axis (Fig. 22.4). Further reduction in normal tissue exposure is reported with removing flattening filters [59]. Enhanced reliability of setup with kV image guidance and using quicker delivery techniques such as VMAT delivery with high-dose rate beams reduce beam-on time while allowing for greater reliability of delivering the prescribed radiation dose [54, 60].

Long-term follow-up of patients is important to prove that highly conformal plans which limit dose to critical structures translate to objective clinical reduction in late effects. Severe ototoxicity resulting in hearing loss has been a significant problem for children with medulloblastoma receiving radiotherapy, particularly when combined with cisplatinum-based chemotherapy. The boost dose to the posterior fossa using 3DCRT results in doses to the cochlea surpassing the prescribed dose of 54 Gy. Studies using audiometric tests as indicators for reduction in grade 3–4 hearing loss have been conducted in pediatric patients with brain tumors, including children with medulloblastoma treated with IMRT. A mean cumulative cochlear dose of less than 35 Gy is recommended after statistics indicated low incidence of ototoxicity with doses of 30 Gy or less and increased at greater than 40–45 Gy [61–63].

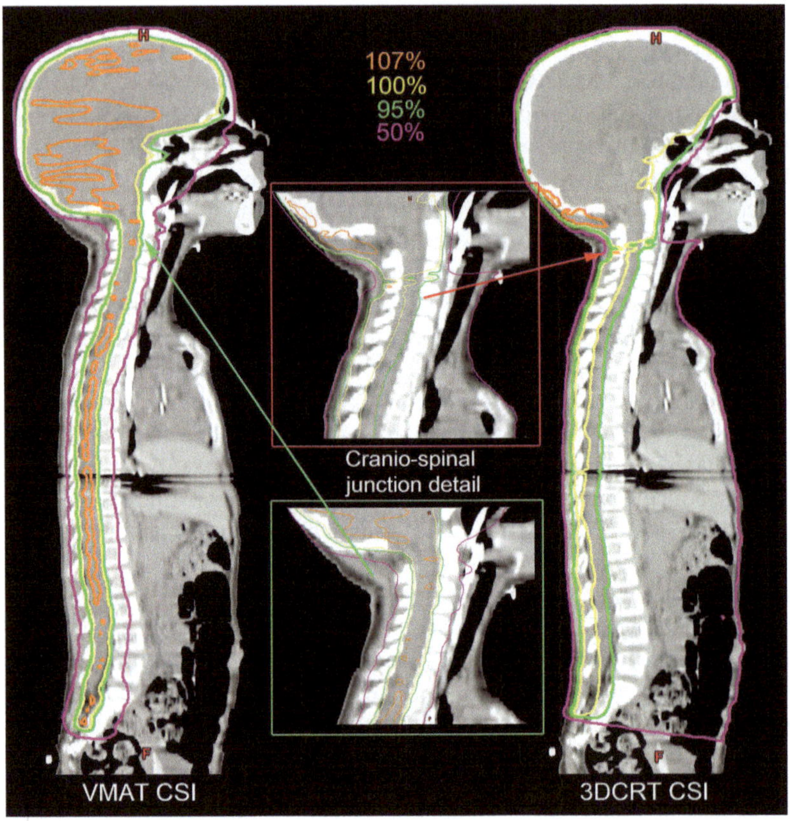

Fig. 22.4 Craniospinal irradiation dose distribution comparison between IMRT delivery (VMAT) and 3DCRT for a prescription of 23.4 Gy, with 6 MV photons, for a supine patient. IMRT plan is much more conformal, while the 3DCRT is more uniform. Furthermore, the junction between the cranial and spinal field is done dosimetrically with IMRT versus the classic dose feathering using match-moves in 3DCRT cases

One of the most devastating effects of radiation therapy for medulloblastoma is impaired neurocognitive function, particularly for younger patients [64, 65]. Cumulative doses from standard 3DCRT to the entire posterior fossa result in substantial dose to the parietal, occipital, and temporal lobes, as well as the hypothalamic-pituitary axis. IMRT boost to the primary tumor site after standard CSI significantly reduces ototoxicities associated with 3DCRT techniques [66]. However, some have questioned whether this strategy may inadvertently increase the cumulative whole-brain doses due to the increased volume of exposure in the low dose range from the IMRT boost resulting in inferior neurocognitive outcomes. This hypothesis was tested by studying cognitive impairment in 25 patients with medulloblastoma treated with 3DCRT only versus IMRT used for the boost after standard CSI. There was no statistical difference between cohorts when assessing for long-term neurocognitive decline thus supporting the use of IMRT, which reduces normal tissues toxicities [21].

Studies have shown that the hippocampus, part of the temporal lobe, is involved with memory and is highly sensitive to radiation [67]. The degree of memory impairment associated with radiation therapy has been shown to be correlated to the mean hippocampal dose (relative to prescription dose). Investigators have systematically studied this using comprehensive standardized assessments of motor speed/dexterity, verbal memory, visual perception, vocabulary, and visual-spatial working memory and found a direct positive correlation between neurocognitive dysfunction and the amount of dose received by the subventricular zone, hippocampus, temporal lobes, and cerebrum [68]. Hippocampal-sparing IMRT techniques are being introduced into radiation therapy as a mechanism to minimize late neurocognitive effects. However, an unknown factor is whether this places patients at higher risk for relapse due to underdosing the adjacent ventricular system which may harbor tumor cells [69–72]. A specific detailed planning strategy is exemplified in the studies by Gondi et al. for hippocampal-sparing whole-brain radiotherapy using IMRT [73, 74]. Comparisons between 3DCRT and IMRT (standard and VMAT) have shown 12–23 % reduction in dose to the hippocampus with IMRT techniques with a corresponding reduction in the calculated risk for memory impairment. Prospective pediatric clinical trials would be important to validate this technique in the treatment of medulloblastoma.

Of note, in long-term follow-up of medulloblastoma patients, it is important to remember that transient imaging changes resembling leptomeningeal disease in the posterior fossa often occur on MRI surveillance specifically after IMRT boosts. To distinguish radiation-induced changes from recurrent disease, the timing of radiation therapy and location of boost are important features for the radiologist to be aware of, as these radiographic findings usually resolve within 6 months [75].

22.2.1.2 Glioma

Pediatric gliomas are histologically indistinguishable from their adult counterparts. However, clinical behavior and locations differ in the pediatric age group. Low-grade gliomas in young children are usually pilocytic astrocytoma tumors and found along the optic nerves, brainstem, and cerebellum. In older children, diffuse astrocytoma tumors more commonly arise in the cerebellum followed by cerebrum, deep midline structures, optic pathway, and brainstem. Low-grade gliomas are frequently cured with surgery. Radiotherapy is selectively administered for unresectable or progressive low-grade gliomas with doses of 50–54 Gy. Long-term cures are possible; therefore, highly conformal margins with IMRT are recommended to minimize late effects. A tight anatomically constrained expansion on the GTV (i.e., 5 mm margin to CTV) has been recommended as failure patterns are typically in the high-dose volume [76, 77].

Pediatric high-grade gliomas frequently occur in the brainstem where surgery is not feasible and prognosis is very poor. Radiation therapy is transiently effective but requires high doses of radiation therapy (59.4 Gy). The optimal margin for expansion on the GTV when using IMRT has not been established, but 1–1.5 cm margin to CTV has been used with the option for boost after 45 Gy if resection is not feasible.

22.2.1.3 Craniopharyngioma

Craniopharyngioma is a tumor arising in the sellar or suprasellar region from remnants of Rathke's pouch with the highest incidence in the pediatric population occurring in those 5–14 years of age [78–80]. Radiation therapy is indicated for unresectable tumors, after partial resection, or recurrent disease. Doses of 54 Gy are prescribed using conformal margins for expansion on the solid and cystic component of the tumor of 0.5–1 cm margin to the CTV [81]. Determining target volumes for craniopharyngiomas is challenging as these tumors are prone to change during radiation therapy because of cystic components which are frequently associated with these tumors. Since IMRT is reliant on strict margins around GTV, any change in the tumor volume may under-/overdose the tumor or adjacent normal tissues. Close surveillance during treatment with MRI, CBCT, and repeat planning is recommended to account for observed tumor volume fluctuations, sometimes greater than 25 % [82].

22.2.1.4 Ependymoma

Ependymoma arises from the lining of the ventricular system with the highest incidence between 0 and 4 years of age, the fourth ventricle being the most common location. Postoperative radiation therapy is delivered after resection in selected patients with residual disease and/or high-grade tumors. Coverage to high doses (59.4 Gy) is frequently recommended for disease control as local recurrence is the predominant pattern of failure. Recommended margins for expansion include at most 1.5 cm margin for CTV from the operative bed and residual disease [83–85]. IMRT has been proven to increase the tumor control probability relative to 3DCRT and reduces dose to adjacent normal tissues. Given the possibility for daily image guidance and IMRT treatments, safe margin reductions to less than 1 cm have been recommended for future clinical trials to further minimization of normal tissue exposure [86].

Radiation therapy for ependymoma tumors, as well as for other tumors in close proximity to the hypothalamic-pituitary axis, increases the risk for associated growth hormone deficiency. Merchant et al. investigated the variation of growth hormone levels after radiotherapy in a cohort of children with ependymoma and demonstrated positive correlations between the time to develop growth hormone deficiency and cumulative dose to the hypothalamus. The authors determined that cumulative mean hypothalamic doses of 16 Gy are associated with a 50 % risk of developing growth hormone deficiency at 5 years [87]. Maximizing hypothalamic avoidance with IMRT delivery techniques is possible [88], inherently minimizing radiation therapy-associated growth deficiencies.

22.2.1.5 Germ Cell Tumors

Intracranial germ cell tumors (germinoma and non-germinomatous germ cell tumors) are rare, commonly occurring in the pineal and suprasellar region with a male preponderance. The role of IMRT for localized germinoma tumors is currently evolving toward limiting the extent of irradiation to the whole ventricular system, using doses in the range of 18–30.6 Gy followed by a conformal boost (23.4 Gy) to the primary site [89]. Compared to 3DCRT, whole ventricular system IMRT reduces the volume of normal brain irradiated in the higher isodose range (25–100 % of the prescription dose) by 0.7–16 %, without increasing the volumes irradiated in the lower dose component (5–10 %) [90]. VMAT delivery of IMRT for whole ventricle treatments can be utilized for more complex treatment geometries, delivering dose faster, with fewer monitor units, at the expense of increased low dose exposure of the normal brain [91]. Non-germinomatous germ cell tumors are less sensitive to radiotherapy. Traditional CSI treatments and prescription doses of 36 Gy with a boost to the primary site (up to 54 Gy) continue to play an important role in the disease control [92].

22.2.2 Sarcoma

Pediatric sarcomas comprise a diverse set of histologic subtypes occurring in children from infancy to adolescence and occur in all anatomic locations. Rhabdomyosarcoma (RMS) accounts for approximately 50 % of pediatric sarcomas. Other sarcomas are broadly categorized under the umbrella term: non-rhabdomyosarcoma soft tissue sarcomas (NRSTS). Synovial sarcoma and malignant peripheral nerve sheath tumor (MPNST) are among the most frequent histologic subtypes of NRSTS seen in children and young adults. Unlike RMS where a combination of surgery, chemotherapy, and radiation therapy remains the preferred treatment strategy, the surgery is the primary therapy for all NRSTS. Radiation therapy is selectively administered either postoperatively for high-grade tumors and positive margins or preoperatively for unresectable sarcomas or cases where radiotherapy is indicated based on size, grade, or location of the tumor [93]. For RMS, radiation doses range from 36–50.4 Gy depending on the extent of surgical resection, whereas NRSTS require higher doses. IMRT is the treatment of choice in pediatric sarcomas in most locations to achieve dose conformality while limiting normal tissues exposure.

Lin et al. reviewed the COG experience on dosimetric differences between IMRT and 3DCRT and whether there is an effect on local control for intermediate-risk pediatric patients with RMS. While dose coverage was improved with IMRT, there were no statistical differences between the rates of control between the two modalities [94]. Current pediatric oncology trials rely on multimodality imaging (MRI, PET-CT, or CT scans) obtained at the time of initial diagnosis to identify initial gross tumor volumes. Narrow margins are used for CTV expansion (1 cm) with a daily setup margin for uncertainty of 0.3–0.5 cm. Further refinements of the target volume can include field reduction to the post-induction chemotherapy volume for non-infiltrative tumors, using natural barriers, such as bone and skin, to define the limits of the CTV and shaping target volumes around critical structures that will surpass normal tissue dose constraints. Macdonald et al. reported on their experience with IMRT in the treatment of unresected pediatric head and neck RMS using limited margins and median doses of up to 50.4 Gy, with an option to reduce the treatment margins after an initial 36 Gy based on post-induction chemotherapy imaging studies. This strategy proved to be successful, providing excellent local control (100 % at 3 years follow-up) with significant normal tissue reductions particularly within the difficult-to-protect critical structures of the head and neck [95].

The optimal volumes and prescription dose levels have yet to be defined for NRSTS. A recent COG phase III NRSTS trial suggests that postoperative doses of 55.8 Gy for high-grade large tumors (>5 cm) yield very good local control for negative or microscopic positive surgical margins. In the preoperative setting, when combined with intensive chemotherapy, doses as low as 45 Gy have achieved excellent local control [96]. Krasin et al. prospectively studied the use of smaller margins when using IMRT and IGRT for pediatric NRSTS. Results indicated that using 2 cm anatomically constrained (bone, fascia) margins for CTV expansion on the GTV and an additional 0.4–1 cm for PTV expansion resulted in excellent local control at 3 years [97].

Sequential cone-down approaches have been shown to be beneficial for pediatric patients with sarcoma. IMRT planning for sequential boosts is a complicated process, since adequate target coverage with respect to normal tissue protection needs to be achieved for different plans, while the composite doses still need to be acceptable. This is a laborious process requiring a lengthy trial-and-error process or the use of complex optimization platforms [98]. DP and SIB are IMRT techniques which facilitate treatment of multiple target volumes to different prescription levels using the same number of fractions. This approach allows for better control of dose distribution by minimizing heterogeneity as compared to sequential cone-down techniques. The use of nonconventional dose per fraction has not been studied in the pediatric population. As fraction size has been demonstrated to be a critical radiation parameter in contributing to undesirable late effects, altered fractionation schemes would need to be tested in controlled clinical trial. Memorial Sloan Kettering Cancer Center reported a novel IMRT treatment approach for pediatric patients with sarcomas involving the thoracic region. While concomitantly delivering whole-lung or hemithorax lung irradiation for lung metastasis, DP IMRT was employed to treat the primary site. The recorded dose per fraction to the lungs was low (0.55–0.88 Gy), while the primary site received 1.8 Gy per fraction. A general decrease in the mean dose to the esophagus (15 %), heart (31 %), spinal cord (15 %), and liver (19 %) was observed using the proposed DP technique compared to standard techniques, with no local failures [35].

IMRT techniques are useful in the treatment of retroperitoneal sarcomas, which typically are aggressive, infiltrative, and difficult to resect. Organs in the vicinity of the surgical resection site (the kidneys, liver, small bowel, rectum, bladder, bone, and gonads) have dose tolerances that are commonly difficult to meet with 3DCRT. IMRT techniques allow for minimization of dose to these organs by maximizing conformity of the dose distribution and integrating motion compensation and SIB planning strategies [99]. Desmoplastic small round cell tumor is another aggressive pediatric sarcoma characterized by diffuse spread throughout the peritoneal region in the abdomen and pelvis. Multimodality therapy frequently includes whole abdomen-pelvis irradiation (WAPI). MD Anderson reported on their experience using WAPI IMRT as a technique to limit radiation dose to the bone marrow. A dose of 30 Gy was delivered postoperatively with some patients receiving a simultaneous boost (6–10 Gy) to sites of gross residual disease and showed that this technique lowered the dose to bone marrow containing regions by 25 % compared with 3DCRT [100].

The value of IMRT in the treatment of extremity NRSTS has not been carefully studied. In postoperative radiotherapy, extremities are difficult to immobilize, and day-to-day setup geometry is complicated as extremities are prone to fluctuations in size due to muscle atrophy from limb disuse and soft tissue swelling from postoperative seroma or lymphedema. Daily IGRT including CBCT allows for increased setup accuracy. In turn sharper dose distributions can be delivered without the risk of under-/overdosing: better joint protection, better control of dose to the femoral heads and neurovascular bundles, and dose sculpting around natural barriers such as bone. Radiation therapy dose to skin and subcutaneous tissue can be more carefully controlled as well. Reducing acute side effects such as radiation dermatitis or surgical wound-related complications [101, 102]. Long-term benefits of IMRT potentially include preventing joint fibrosis, reducing the risk of avascular necrosis of the femoral head, protecting skin to minimize subcutaneous fibrosis and lower extremity edema, and reducing the risk for bone fracture [102].

22.2.3 Other Select Pediatric Tumors

22.2.3.1 Nasopharynx Carcinoma

Nasopharynx carcinoma is a rare occurrence in children and adolescents. High doses of radiation (>60 Gy) are used to treat the primary site. As cervical lymph nodes are often involved with carcinoma, the planning target volumes for these treatments are large, frequently encompassing both sides of the neck, which contributes to serious radiation-induced late effects [103]. SIB and DP, as used in adult head neck cancers, are IMRT strategies used to limit normal tissue exposure while ensuring local control; however, they are not incorporated in clinical trials due to the altered dose per fraction in the high-dose target volume. The prescription dose levels need to be carefully considered for acute and late effects. For example, in a retrospective study of 34 pediatric and young adults patients (ages 8–20 years) treated with SIB-IMRT, it was observed that for prescription levels of 64–68 Gy, high-grade late effects only occurred in two patients for grade 3 ototoxicity (<10 %), with no observed grade 4 toxicities [104]. A different pattern was observed in the similarly treated cohort (ages 10–17 years) with all patients experiencing grade 3–4 acute and long-term toxicities for prescription dose levels of 61–66 Gy [105]. While the latter study involved only five patients, implying the possibility of a sampling bias in the chosen patient population, the compromise between achieving therapeutic doses proven in the adult population and the possibility of radiation-associated complications and late effects needs to be further studied in the pediatric population.

22.2.3.2 Neuroblastoma

Neuroblastoma is a childhood tumor arising from the neural crest cells. It occurs at locations related to the distribution of the sympathetic nervous system (the neck, chest, abdomen, and pelvis). Multimodality treatment strategies involving chemotherapy, surgery, and, in advanced cases, stem cell transplant have been proven effective [106]. The role of radiation therapy is limited for this disease. However, for high-risk neuroblastoma, a common indication is low dose radiation therapy (20–30 Gy) to the primary tumor site and sometimes metastatic locations, specifically after autologous myeloablative stem cell transplant. The adrenal gland is the most frequent primary site, and IMRT is efficiently used to protect the adjacent kidney(s) and liver. Some patients will have to undergo a nephrectomy procedure; therefore, limiting dose to contralateral kidney is critical. Narrow expansion margins (1.5–2 cm) are used to limit doses to bone marrow containing regions and small bowel.

Due to the proximity to the diaphragm, respiratory motion should be accounted for using techniques such as 4D-CT for planning and respiratory gating. St. Jude Children's Research Hospital has reported on their experience using IMRT with motion compensation and daily IGRT for pediatric patients with neuroblastoma. The addition of CBCT guidance for treatment setup reproducibility was demonstrated to reduce setup uncertainty from >5 mm to less than 2 mm. Daily alignment and highly conformal IMRT dose distributions allowed for margin reduction in these cases, inherently reducing the volume of irradiated normal tissue which resulted in significant reduction in acute toxicity, while for short follow-up, there were no skeletal asymmetry and no abnormal liver presentations nor any kidney dysfunctions [107–110].

22.2.3.3 Wilms' Tumor

Wilms' tumor is a highly curable pediatric kidney tumor commonly occurring at a very young age. Approximately 15 % of patients will present with metastatic disease, lung involvement being the most common site. COG treatment protocols selectively incorporate whole-lung irradiation for patients presenting with metastatic disease to the lungs. As Wilms' tumor is highly curable even in advanced stages, many of these patients have been followed over 3-4 decades. Cardiovascular injury [111] has been increasingly seen as a late effect after a whole-lung irradiation despite the low doses delivered (12 Gy). Cardiac-sparing whole-lung irradiation techniques have been reported using IMRT in patients with Wilms' tumor showing superior dose distributions with significant reduction of dose to the heart [112].

References

1. Paulino AC, Skwarchuk M (2002) Intensity-modulated radiation therapy in the treatment of children. Med Dosim 27(2):115–120. doi:10.1016/S0958-3947(02)00093-6. [pii]
2. Teh BS, Mai WY, Grant WH 3rd, Chiu JK, Lu HH, Carpenter LS, Woo SY, Butler EB (2002) Intensity modulated radiotherapy (IMRT) decreases treatment-related morbidity and potentially enhances tumor control. Cancer Investig 20(4):437–451

3. Schwartz CL (2004) Health status of adult long-term survivors of childhood cancer: a report from the Childhood Cancer Survivor Study. J Pediatr 144(3):407–408
4. Lee EK, Fox T, Crocker I (2006) Simultaneous beam geometry and intensity map optimization in intensity-modulated radiation therapy. Int J Radiat Oncol Biol Phys 64(1):301–320. doi:10.1016/j.ijrobp.2005.08.023
5. Dhabaan A, Elder E, Schreibmann E, Crocker I, Curran WJ, Oyesiku NM, Shu HK, Fox T (2010) Dosimetric performance of the new high-definition multileaf collimator for intracranial stereotactic radiosurgery. J Appl Clin Med Phys 11(3):3040
6. Otto K (2008) Volumetric modulated arc therapy: IMRT in a single gantry arc. Med Phys 35(1):310–317
7. Tanyi JA, Summers PA, McCracken CL, Chen Y, Ku LC, Fuss M (2009) Implications of a high-definition multileaf collimator (HD-MLC) on treatment planning techniques for stereotactic body radiation therapy (SBRT): a planning study. Radiat Oncol 4:22. doi:10.1186/1748-717X-4-22. [pii]
8. Sontag M, Merchant T, Burnham B, Shah A, Kun L (1998) Clinical experience with a system for pediatric respiratory gated radiotherapy. Int J Radiat Oncol Biol Phys 42(1 Suppl), p 40
9. Ahmed RS, Shen S, Ove R, Duan J, Fiveash JB, Russo SM (2007) Intensity modulation with respiratory gating for radiotherapy of the pleural space. Med Dosim 32(1):16–22. doi:10.1016/j.meddos.2006.10.002. S0958-3947(06)00151-8 [pii]
10. Holt A, Van Gestel D, Arends MP, Korevaar EW, Schuring D, Kunze-Busch MC, Louwe RJ, van Vliet-Vroegindeweij C (2013) Multi-institutional comparison of volumetric modulated arc therapy vs. intensity-modulated radiation therapy for head-and-neck cancer: a planning study. Radiat Oncol 8:26. doi:10.1186/1748-717X-8-26. 1748-717X-8-26 [pii]
11. Thomas EM, Popple RA, Prendergast BM, Clark GM, Dobelbower MC, Fiveash JB (2013) Effects of flattening filter-free and volumetric-modulated arc therapy delivery on treatment efficiency. J Appl Clin Med Phys 14(6):4328. doi:10.1120/jacmp.v14i6.4328
12. Penagaricano JA, Papanikolaou N, Yan Y, Ratanatharathorn V (2004) Application of intensity-modulated radiation therapy for pediatric malignancies. Medical Dosim 29(4):247–253. doi:10.1016/j.meddos.2004.04.007
13. Bhatnagar A, Deutsch M (2006) The Role for intensity modulated radiation therapy (IMRT) in pediatric population. Technol Cancer Res Treat 5(6):591–595
14. Sterzing F, Stoiber EM, Nill S, Bauer H, Huber P, Debus J, Munter MW (2009) Intensity modulated radiotherapy (IMRT) in the treatment of children and adolescents–a single institution's experience and a review of the literature. Radiat Oncol 4:37. doi:10.1186/1748-717x-4-37
15. Abou-Elenein HS, Attalla EM, Ammar H, Eldesoky I, Farouk M, Shoer S (2013) The impact of intensity modulated radiotherapy on the skin dose for deep seated tumors. Chin-Ger J Clin Oncol 12(4):194–198. doi:10.1007/s10330-012-1127-1
16. Kry SF (2010) Non-target dose from radiotherapy: magnitude, evaluation, and impact. American Association of Medical Physicists, College Park, MD
17. Kaderka R, Schardt D, Durante M, Berger T, Ramm U, Licher J, La Tessa C (2012) Out-of-field dose measurements in a water phantom using different radiotherapy modalities. Phys Med Biol 57(16):5059–5074. doi:10.1088/0031-9155/57/16/5059
18. Klein EE, Maserang B, Wood R, Mansur D (2006) Peripheral doses from pediatric IMRT. Med Phys 33(7):2525. doi:10.1118/1.2207252
19. Hall EJ, Wuu CS (2003) Radiation-induced second cancers: the impact of 3D-CRT and IMRT. Int J Radiat Oncol Biol Phys 56(1):83–88. doi:10.1016/S0360301603000737 [pii]
20. Followill D, Geis P, Boyer A (1997) Estimates of whole-body dose equivalent produced by beam intensity modulated conformal therapy. Int J Radiat Oncol Biol Phys 38(3):667–672. doi:10.1016/S0360301697000126 [pii]
21. Jain N, Krull KR, Brouwers P, Chintagumpala MM, Woo SY (2008) Neuropsychological outcome following intensity-modulated radiation therapy for pediatric medulloblastoma. Pediatr Blood Cancer 51(2):275–279. doi:10.1002/pbc.21580

22. Lin SH, Wang L, Myles B, Thall PF, Hofstetter WL, Swisher SG, Ajani JA, Cox JD, Komaki R, Liao Z (2012) Propensity score-based comparison of long-term outcomes with 3-dimensional conformal radiotherapy vs intensity-modulated radiotherapy for esophageal cancer. Int J Radiat Oncol Biol Phys 84(5):1078–1085. doi:10.1016/j.ijrobp.2012.02.015. S0360-3016(12)00227-1 [pii]
23. Schwartz CL (1999) Long-term survivors of childhood cancer: the late effects of therapy. Oncologist 4(1):45–54
24. Hudson MM, Mertens AC, Yasui Y, Hobbie W, Chen H, Gurney JG, Yeazel M, Recklitis CJ, Marina N, Robison LR, Oeffinger KC (2003) Health status of adult long-term survivors of childhood cancer: a report from the Childhood Cancer Survivor Study. JAMA 290(12):1583–1592. doi:10.1001/jama.290.12.1583. 290/12/1583 [pii]
25. Wallace WH, Anderson RA, Irvine DS (2005) Fertility preservation for young patients with cancer: who is at risk and what can be offered? Lancet Oncol 6(4):209–218. doi:10.1016/S1470-2045(05)70092-9 [pii]
26. Moskowitz CS, Chou JF, Wolden SL, Bernstein JL, Malhotra J, Friedman DN, Mubdi NZ, Leisenring WM, Stovall M, Hammond S, Smith SA, Henderson TO, Boice JD, Hudson MM, Diller LR, Bhatia S, Kenney LB, Neglia JP, Begg CB, Robison LL, Oeffinger KC (2014) Breast cancer after chest radiation therapy for childhood cancer. J Clin Oncol. doi:10.1200/JCO.2013.54.4601 [pii]
27. Koshy M, Paulino AC, Marcus RB Jr, Ting JY, Whitaker D, Davis LW (2004) Extra-target doses in children receiving multileaf collimator (MLC) based intensity modulated radiation therapy (IMRT). Pediatr Blood Cancer 42(7):626–630. doi:10.1002/pbc.20030
28. Mansur DB, Klein EE, Maserang BP (2007) Measured peripheral dose in pediatric radiation therapy: a comparison of intensity-modulated and conformal techniques. Radiother Oncol 82(2):179–184. doi:10.1016/j.radonc.2007.01.002
29. Kan MW, Leung LH, Kwong DL, Wong W, Lam N (2010) Peripheral doses from noncoplanar IMRT for pediatric radiation therapy. Medical Dosim 35(4):255–263. doi:10.1016/j.meddos.2009.07.003
30. Cashmore J, Ramtohul M, Ford D (2011) Lowering whole-body radiation doses in pediatric intensity-modulated radiotherapy through the use of unflattened photon beams. Int J Radiat Oncol Biol Phys 80(4):1220–1227. doi:10.1016/j.ijrobp.2010.10.002
31. Jia MX, Zhang X, Yin C, Feng G, Li N, Gao S, Liu da W (2014) Peripheral dose measurements in cervical cancer radiotherapy: a comparison of volumetric modulated arc therapy and step-and-shoot IMRT techniques. Radiat Oncol 9:61. doi:10.1186/1748-717X-9-61. 1748-717X-9-61 [pii]
32. Matuszak MM, Yan D, Grills I, Martinez A (2010) Clinical applications of volumetric modulated arc therapy. Int J Radiat Oncol Biol Phys 77(2):608–616. doi:10.1016/j.ijrobp.2009.08.032
33. Eldesoky I, Attalla EM, Elshemey WM, Zaghloul MS (2012) A comparison of three commercial IMRT treatment planning systems for selected paediatric cases. J Appl Clin Med Phys 13(2):3742. doi:10.1120/jacmp.v13i2.3742
34. Dogan N, King S, Emami B, Mohideen N, Mirkovic N, Leybovich LB, Sethi A (2003) Assessment of different IMRT boost delivery methods on target coverage and normal-tissue sparing. Int J Radiat Oncol Biol Phys 57(5):1480–1491. doi:10.1016/S0360301603015694 [pii]
35. Yang JC, Wexler LH, Meyers PA, Happersett L, La Quaglia MP, Wolden SL (2013) Intensity-modulated radiation therapy with dose-painting for pediatric sarcomas with pulmonary metastases. Pediatr Blood Cancer 60(10):1616–1620. doi:10.1002/pbc.24502
36. Walter C, Boda-Heggemann J, Wertz H, Loeb I, Rahn A, Lohr F, Wenz F (2007) Phantom and in-vivo measurements of dose exposure by image-guided radiotherapy (IGRT): MV portal images vs. kV portal images vs. cone-beam CT. Radiother Oncol 85(3):418–423. doi:10.1016/j.radonc.2007.10.014. S0167-8140(07)00527-0 [pii]

37. Deng J, Chen Z, Roberts KB, Nath R (2012) Kilovoltage imaging doses in the radiotherapy of pediatric cancer patients. Int J Radiat Oncol Biol Phys 82(5):1680–1688. doi:10.1016/j.ijrobp.2011.01.062. S0360-3016(11)00237-9 [pii]
38. King CR, Maxim PG, Hsu A, Kapp DS (2010) Incidental testicular irradiation from prostate IMRT: it all adds up. Int J Radiat Oncol Biol Phys 77(2):484–489. doi:10.1016/j.ijrobp.2009.04.083. S0360-3016(09)00723-8 [pii]
39. Deng J, Chen Z, Yu JB, Roberts KB, Peschel RE, Nath R (2012) Testicular doses in image-guided radiotherapy of prostate cancer. Int J Radiat Oncol Biol Phys 82 (1):e39–e47. doi:10.1016/j.ijrobp.2011.01.071. S0360-3016(11)00262-8 [pii]
40. Merchant TE, Farr JB (2014) Proton beam therapy: a fad or a new standard of care. Curr Opin Pediatr 26(1):3–8. doi:10.1097/MOP.0000000000000048
41. Smith AR (2009) Vision 20/20: proton therapy. Med Phys 36(2):556–568
42. Athar BS, Paganetti H (2011) Comparison of second cancer risk due to out-of-field doses from 6-MV IMRT and proton therapy based on 6 pediatric patient treatment plans. Radiother Oncol 98(1):87–92. doi:10.1016/j.radonc.2010.11.003
43. Paganetti H, Athar BS, Moteabbed M, Adams JA, Schneider U, Yock TI (2012) Assessment of radiation-induced second cancer risks in proton therapy and IMRT for organs inside the primary radiation field. Phys Med Biol 57(19):6047–6061. doi:10.1088/0031-9155/57/19/6047
44. Daw NC, Mahajan A (2013) Photons or protons for non-central nervous system solid malignancies in children. In: American Society of Clinical Oncology educational book/ASCO American Society of Clinical Oncology meeting, pp 354–359. doi:10.1200/EdBook_AM.2013.33.e354
45. Greco C, Wolden S (2007) Current status of radiotherapy with proton and light ion beams. Cancer 109(7):1227–1238. doi:10.1002/cncr.22542
46. Lomax AJ, Bohringer T, Bolsi A, Coray D, Emert F, Goitein G, Jermann M, Lin S, Pedroni E, Rutz H, Stadelmann O, Timmermann B, Verwey J, Weber DC (2004) Treatment planning and verification of proton therapy using spot scanning: initial experiences. Med Phys 31(11):3150–3157
47. Paganetti H (1998) Calculation of the spatial variation of relative biological effectiveness in a therapeutic proton field for eye treatment. Phys Med Biol 43(8):2147–2157
48. Paganetti H (2012) Range uncertainties in proton therapy and the role of Monte Carlo simulations. Phys Med Biol 57(11):R99–R117. doi:10.1088/0031-9155/57/11/R99
49. Chen Y, Ahmad S (2009) Evaluation of inelastic hadronic processes for 250 MeV proton interactions in tissue and iron using GEANT4. Radiat Prot Dosim 136(1):11–16. doi:10.1093/rpd/ncp149. ncp149 [pii]
50. Yan X, Titt U, Koehler AM, Newhauser WD (2002) Measurement of neutron dose equivalent to proton therapy patients outside of the proton radiation field. Nucl Instrum Methods Phys Res Sect A 476(1–2):429–434. http://dx.doi.org/10.1016/S0168-9002(01)01483-8
51. Gerber NU, Mynarek M, von Hoff K, Friedrich C, Resch A, Rutkowski S (2014) Recent developments and current concepts in medulloblastoma. Cancer Treat Rev 40(3):356–365. doi:10.1016/j.ctrv.2013.11.010. S0305-7372(13)00263-6 [pii]
52. Packer RJ, Gajjar A, Vezina G, Rorke-Adams L, Burger PC, Robertson PL, Bayer L, LaFond D, Donahue BR, Marymont MH, Muraszko K, Langston J, Sposto R (2006) Phase III study of craniospinal radiation therapy followed by adjuvant chemotherapy for newly diagnosed average-risk medulloblastoma. J Clin Oncol 24(25):4202–4208. doi:10.1200/JCO.2006.06.4980. 24/25/4202 [pii]
53. Mulhern RK, Merchant TE, Gajjar A, Reddick WE, Kun LE (2004) Late neurocognitive sequelae in survivors of brain tumours in childhood. Lancet Oncol 5(7):399–408. doi:10.1016/S1470-2045(04)01507-4. S1470204504015074 [pii]
54. Pichandi A, Ganesh KM, Jerrin A, Balaji K, Sridhar PS, Surega A (2014) Cranio spinal irradiation of medulloblastoma using high precision techniques – a dosimetric comparison. Technol Cancer Res Treat. doi:10.7785/tcrt.2012.500421

55. Lee YK, Brooks CJ, Bedford JL, Warrington AP, Saran FH (2012) Development and evaluation of multiple isocentric volumetric modulated arc therapy technique for craniospinal axis radiotherapy planning. Int J Radiat Oncol Biol Phys 82(2):1006–1012. doi:10.1016/j.ijrobp.2010.12.033. S0360-3016(11)00021-6 [pii]
56. Fogliata A, Bergstrom S, Cafaro I, Clivio A, Cozzi L, Dipasquale G, Hallstrom P, Mancosu P, Navarria P, Nicolini G, Parietti E, Pesce GA, Richetti A, Scorsetti M, Vanetti E, Weber DC (2011) Cranio-spinal irradiation with volumetric modulated arc therapy: a multi-institutional treatment experience. Radiother Oncol 99(1):79–85. doi:10.1016/j.radonc.2011.01.023. S0167-8140(11)00071-5 [pii]
57. Brodin NP, Vogelius IR, Bjork-Eriksson T, Munck Af Rosenschold P, Maraldo MV, Aznar MC, Specht L, Bentzen SM (2014) Optimizing the radiation therapy dose prescription for pediatric medulloblastoma: minimizing the life years lost attributable to failure to control the disease and late complication risk. Acta Oncol 53(4):462–470. doi:10.3109/0284186X.2013.858824
58. Beltran C, Gray J, Merchant TE (2012) Intensity-modulated arc therapy for pediatric posterior fossa tumors. Int J Radiat Oncol Biol Phys 82(2):e299–e304. doi:10.1016/j.ijrobp.2010.11.024. S0360-3016(10)03573-X [pii]
59. Anchineyan P, Mani GK, Amalraj J, Karthik B, Anbumani S (2014) Use of flattening filter-free photon beams in treating medulloblastoma: a dosimetric evaluation. ISRN Oncol 2014:769698. doi:10.1155/2014/769698
60. Beltran C, Gray J, Merchant TE (2012) Intensity-modulated arc therapy for pediatric posterior fossa tumors. Int J Radiat Oncol Biol Phys 82(2):e299–e304. doi:10.1016/j.ijrobp.2010.11.024
61. Huang E, Teh BS, Strother DR, Davis QG, Chiu JK, Lu HH, Carpenter LS, Mai WY, Chintagumpala MM, South M, Grant WH 3rd, Butler EB, Woo SY (2002) Intensity-modulated radiation therapy for pediatric medulloblastoma: early report on the reduction of ototoxicity. Int J Radiat Oncol Biol Phys 52(3):599–605
62. Paulino AC, Lobo M, Teh BS, Okcu MF, South M, Butler EB, Su J, Chintagumpala M (2010) Ototoxicity after intensity-modulated radiation therapy and cisplatin-based chemotherapy in children with medulloblastoma. Int J Radiat Oncol Biol Phys 78(5):1445–1450. doi:10.1016/j.ijrobp.2009.09.031
63. Hua C, Bass JK, Khan R, Kun LE, Merchant TE (2008) Hearing loss after radiotherapy for pediatric brain tumors: effect of cochlear dose. Int J Radiat Oncol Biol Phys 72(3):892–899. doi:10.1016/j.ijrobp.2008.01.050. S0360-3016(08)00229-0 [pii]
64. Merchant TE, Pollack IF, Loeffler JS (2010) Brain tumors across the age spectrum: biology, therapy, and late effects. Semin Radiat Oncol 20(1):58–66. doi:10.1016/j.semradonc.2009.09.005. S1053-4296(09)00066-6 [pii]
65. Palmer SL, Armstrong C, Onar-Thomas A, Wu S, Wallace D, Bonner MJ, Schreiber J, Swain M, Chapieski L, Mabbott D, Knight S, Boyle R, Gajjar A (2013) Processing speed, attention, and working memory after treatment for medulloblastoma: an international, prospective, and longitudinal study. J Clin Oncol 31 (28):3494–3500. doi:10.1200/JCO.2012.47.4775. JCO.2012.47.4775 [pii]
66. Polkinghorn WR, Dunkel IJ, Souweidane MM, Khakoo Y, Lyden DC, Gilheeney SW, Becher OJ, Budnick AS, Wolden SL (2011) Disease control and ototoxicity using intensity-modulated radiation therapy tumor-bed boost for medulloblastoma. Int J Radiat Oncol Biol Phys 81(3):e15–e20. doi:10.1016/j.ijrobp.2010.11.081. S0360-3016(11)00133-7 [pii]
67. Rola R, Raber J, Rizk A, Otsuka S, VandenBerg SR, Morhardt DR, Fike JR (2004) Radiation-induced impairment of hippocampal neurogenesis is associated with cognitive deficits in young mice. Exp Neurol 188(2):316–330. doi:10.1016/j.expneurol.2004.05.005. S0014488604001748 [pii]
68. Redmond KJ, Mahone EM, Terezakis S, Ishaq O, Ford E, McNutt T, Kleinberg L, Cohen KJ, Wharam M, Horska A (2013) Association between radiation dose to neuronal progenitor cell

niches and temporal lobes and performance on neuropsychological testing in children: a prospective study. Neuro-Oncol 15(3):360–369. doi:10.1093/neuonc/nos303. nos303 [pii]
69. Awad R, Fogarty G, Hong A, Kelly P, Ng D, Santos D, Haydu L (2013) Hippocampal avoidance with volumetric modulated arc therapy in melanoma brain metastases – the first Australian experience. Radiat Oncol 8:62. doi:10.1186/1748-717X-8-62. 1748-717X-8-62 [pii]
70. Brodin NP, Munck Af Rosenschold P, Blomstrand M, Kiil-Berthlesen A, Hollensen C, Vogelius IR, Lannering B, Bentzen SM, Bjork-Eriksson T (2014) Hippocampal sparing radiotherapy for pediatric medulloblastoma: impact of treatment margins and treatment technique. Neuro-Oncol 16 (4):594–602. doi:10.1093/neuonc/not225. not225 [pii]
71. Han G, Liu D, Gan H, Denniston KA, Li S, Tan W, Hu D, Zhen W, Wang Z (2014) Evaluation of the dosimetric feasibility of hippocampal sparing intensity-modulated radiotherapy in patients with locally advanced nasopharyngeal carcinoma. PLoS One 9(2):e90007. doi:10.1371/journal.pone.0090007. PONE-D-13-52010 [pii]
72. Oskan F, Ganswindt U, Schwarz SB, Manapov F, Belka C, Niyazi M (2014) Hippocampus sparing in whole-brain radiotherapy. A review. Strahlenther Onkol 190(4):337–341. doi:10.1007/s00066-013-0518-8
73. Gondi V, Tome WA, Mehta MP (2010) Why avoid the hippocampus? A comprehensive review. Radiother Oncol 97(3):370–376. doi:10.1016/j.radonc.2010.09.013
74. Gondi V, Tolakanahalli R, Mehta MP, Tewatia D, Rowley H, Kuo JS, Khuntia D, Tome WA (2010) Hippocampal-sparing whole-brain radiotherapy: a "how-to" technique using helical tomotherapy and linear accelerator-based intensity-modulated radiotherapy. Int J Radiat Oncol Biol Phys 78(4):1244–1252. doi:10.1016/j.ijrobp.2010.01.039
75. Muscal JA, Jones JY, Paulino AC, Bertuch AA, Su J, Woo SY, Mahoney DH Jr, Chintagumpala M (2009) Changes mimicking new leptomeningeal disease after intensity-modulated radiotherapy for medulloblastoma. Int J Radiat Oncol Biol Phys 73(1):214–221. doi:10.1016/j.ijrobp.2008.03.056
76. Merchant TE, Kun LE, Wu S, Xiong X, Sanford RA, Boop FA (2009) Phase II trial of conformal radiation therapy for pediatric low-grade glioma. J Clin Oncol 27 (22):3598–3604. doi:10.1200/JCO.2008.20.9494. JCO.2008.20.9494 [pii]
77. Paulino AC, Mazloom A, Terashima K, Su J, Adesina AM, Okcu MF, Teh BS, Chintagumpala M (2013) Intensity-modulated radiotherapy (IMRT) in pediatric low-grade glioma. Cancer 119(14):2654–2659. doi:10.1002/cncr.28118
78. Bunin GR, Surawicz TS, Witman PA, Preston-Martin S, Davis F, Bruner JM (1998) The descriptive epidemiology of craniopharyngioma. J Neurosurg 89 (4):547–551. doi:10.3171/jns.1998.89.4.0547
79. Fernandez-Miranda JC, Gardner PA, Snyderman CH, Devaney KO, Strojan P, Suarez C, Genden EM, Rinaldo A, Ferlito A (2012) Craniopharyngioma: a pathologic, clinical, and surgical review. Head Neck 34(7):1036–1044. doi:10.1002/hed.21771
80. Minniti G, Esposito V, Amichetti M, Enrici RM (2009) The role of fractionated radiotherapy and radiosurgery in the management of patients with craniopharyngioma. Neurosurg Rev 32(2):125–132. doi:10.1007/s10143-009-0186-4. discussion 132
81. Merchant TE, Kun LE, Hua CH, Wu S, Xiong X, Sanford RA, Boop FA (2013) Disease control after reduced volume conformal and intensity modulated radiation therapy for childhood craniopharyngioma. Int J Radiat Oncol Biol Phys 85(4):e187–e192. doi:10.1016/j.ijrobp.2012.10.030. S0360-3016(12)03729-7 [pii]
82. Beltran C, Naik M, Merchant TE (2010) Dosimetric effect of target expansion and setup uncertainty during radiation therapy in pediatric craniopharyngioma. Radiother Oncol 97(3):399–403. doi:10.1016/j.radonc.2010.10.017
83. Merchant TE, Chitti RM, Li C, Xiong X, Sanford RA, Khan RB (2010) Factors associated with neurological recovery of brainstem function following postoperative conformal radiation therapy for infratentorial ependymoma. Int J Radiat Oncol Biol Phys 76(2):496–503. doi:10.1016/j.ijrobp.2009.01.079. S0360-3016(09)00342-3 [pii]

84. Merchant TE, Fouladi M (2005) Ependymoma: new therapeutic approaches including radiation and chemotherapy. J Neuro-Oncol 75(3):287–299. doi:10.1007/s11060-005-6753-9
85. Landau E, Boop FA, Conklin HM, Wu S, Xiong X, Merchant TE (2013) Supratentorial ependymoma: disease control, complications, and functional outcomes after irradiation. Int J Radiat Oncol Biol Phys 85 (4):e193–e199. doi:10.1016/j.ijrobp.2012.10.033. S0360-3016(12)03733-9 [pii]
86. Beltran C, Naik M, Merchant TE (2010) Dosimetric effect of setup motion and target volume margin reduction in pediatric ependymoma. Radiother Oncol 96(2):216–222. doi:10.1016/j.radonc.2010.02.031. S0167-8140(10)00167-2 [pii]
87. Merchant TE, Rose SR, Bosley C, Wu S, Xiong X, Lustig RH (2011) Growth hormone secretion after conformal radiation therapy in pediatric patients with localized brain tumors. J Clin Oncol 29 (36):4776–4780. doi:10.1200/JCO.2011.37.9453. JCO.2011.37.9453 [pii]
88. Elson A, Bovi J, Kaur K, Maas D, Sinson G, Schultz C (2014) Effect of treatment modality on the hypothalamic-pituitary function of patients treated with radiation therapy for pituitary adenomas: hypothalamic dose and endocrine outcomes. Front Oncol 4:73. doi:10.3389/fonc.2014.00073
89. Calaminus G, Kortmann R, Worch J, Nicholson JC, Alapetite C, Garre ML, Patte C, Ricardi U, Saran F, Frappaz D (2013) SIOP CNS GCT 96: final report of outcome of a prospective, multinational nonrandomized trial for children and adults with intracranial germinoma, comparing craniospinal irradiation alone with chemotherapy followed by focal primary site irradiation for patients with localized disease. Neuro-Oncol 15 (6):788–796. doi:10.1093/neuonc/not019. not019 [pii]
90. Sakanaka K, Mizowaki T, Hiraoka M (2012) Dosimetric advantage of intensity-modulated radiotherapy for whole ventricles in the treatment of localized intracranial germinoma. Int J Radiat Oncol Biol Phys 82(2):e273–e280. doi:10.1016/j.ijrobp.2011.04.007. S0360-3016(11)00521-9 [pii]
91. Sakanaka K, Mizowaki T, Sato S, Ogura K, Hiraoka M (2013) Volumetric-modulated arc therapy vs conventional fixed-field intensity-modulated radiotherapy in a whole-ventricular irradiation: a planning comparison study. Med Dosim 38(2):204–208. doi:10.1016/j.meddos.2013.01.004. S0958-3947(13)00006-X [pii]
92. Calaminus G, Bamberg M, Jurgens H, Kortmann RD, Sorensen N, Wiestler OD, Gobel U (2004) Impact of surgery, chemotherapy and irradiation on long term outcome of intracranial malignant non-germinomatous germ cell tumors: results of the German Cooperative Trial MAKEI 89. Klin Padiatr 216(3):141–149. doi:10.1055/s-2004-822626
93. Million L, Donaldson SS (2012) Resectable pediatric nonrhabdomyosarcoma soft tissue sarcoma: which patients benefit from adjuvant radiation therapy and how much? ISRN Oncol 2012:341408. doi:10.5402/2012/341408
94. Lin C, Donaldson SS, Meza JL, Anderson JR, Lyden ER, Brown CK, Morano K, Laurie F, Arndt CA, Enke CA, Breneman JC (2012) Effect of radiotherapy techniques (IMRT vs. 3D-CRT) on outcome in patients with intermediate-risk rhabdomyosarcoma enrolled in COG D980-- report from the Children's Oncology Group. Int J Radiat Oncol Biol Phys 82(5):1764–1770. doi:10.1016/j.ijrobp.2011.01.036. S0360-3016(11)00203-3 [pii]
95. McDonald MW, Esiashvili N, George BA, Katzenstein HM, Olson TA, Rapkin LB, Marcus RB, Jr. (2008) Intensity-modulated radiotherapy with use of cone-down boost for pediatric head-and-neck rhabdomyosarcoma. Int J Radiat Oncol Biol Phys 72(3):884–891. doi:10.1016/j.ijrobp.2008.01.058. S0360-3016(08)00319-2 [pii]
96. Spunt SL, Million L, Anderson JR et al Risk-based treatment for nonrhabdomyosarcoma soft tissue sarcomas (NRSTS) in patients under 30 years of age: Children's Oncology Group study ARST0332. In: Abstracts of the 2014 American Society of Clinical Oncology annual meeting, Chicago, 30 May–3 June, 2014
97. Krasin MJ, Davidoff AM, Xiong X, Wu S, Hua CH, Navid F, Rodriguez-Galindo C, Rao BN, Hoth KA, Neel MD, Merchant TE, Kun LE, Spunt SL (2010) Preliminary results from a prospective study using limited margin radiotherapy in pediatric and young adult patients with high-grade nonrhabdomyosarcoma soft-tissue sarcoma. Int J Radiat Oncol Biol Phys 76(3):874–878. doi:10.1016/j.ijrobp.2009.02.074. S0360-3016(09)00503-3 [pii]

98. Popple RA, Prellop PB, Spencer SA, De Los Santos JF, Duan J, Fiveash JB, Brezovich IA (2005) Simultaneous optimization of sequential IMRT plans. Med Phys 32(11):3257–3266
99. Roeder F, Schulz-Ertner D, Nikoghosyan AV, Huber PE, Edler L, Habl G, Krempien R, Oertel S, Saleh-Ebrahimi L, Hensley FW, Buechler MW, Debus J, Koch M, Weitz J, Bischof M (2012) A clinical phase I/II trial to investigate preoperative dose-escalated intensity-modulated radiation therapy (IMRT) and intraoperative radiation therapy (IORT) in patients with retroperitoneal soft tissue sarcoma. BMC Cancer 12:287. doi:10.1186/1471-2407-12-287. 1471-2407-12-287 [pii]
100. Pinnix CC, Fontanilla HP, Hayes-Jordan A, Subbiah V, Bilton SD, Chang EL, Grosshans DR, McAleer MF, Sulman EP, Woo SY, Anderson P, Green HL, Mahajan A (2012) Whole abdominopelvic intensity-modulated radiation therapy for desmoplastic small round cell tumor after surgery. Int J Radiat Oncol Biol Phys 83(1):317–326. doi:10.1016/j.ijrobp.2011.06.1985. S0360-3016(11)02915-4 [pii]
101. O'Sullivan B, Griffin AM, Dickie CI, Sharpe MB, Chung PW, Catton CN, Ferguson PC, Wunder JS, Deheshi BM, White LM, Kandel RA, Jaffray DA, Bell RS (2013) Phase 2 study of preoperative image-guided intensity-modulated radiation therapy to reduce wound and combined modality morbidities in lower extremity soft tissue sarcoma. Cancer 119(10):1878–1884. doi:10.1002/cncr.27951
102. Stewart AJ, Lee YK, Saran FH (2009) Comparison of conventional radiotherapy and intensity-modulated radiotherapy for post-operative radiotherapy for primary extremity soft tissue sarcoma. Radiother Oncol 93 (1):125–130. doi:10.1016/j.radonc.2009.06.010. S0167-8140(09)00322-3 [pii]
103. Hu S, Xu X, Xu J, Xu Q, Liu S (2013) Prognostic factors and long-term outcomes of nasopharyngeal carcinoma in children and adolescents. Pediatr Blood Cancer 60(7):1122–1127. doi:10.1002/pbc.24458
104. Tao CJ, Liu X, Tang LL, Mao YP, Chen L, Li WF, Yu XL, Liu LZ, Zhang R, Lin AH, Ma J, Sun Y (2013) Long-term outcome and late toxicities of simultaneous integrated boost-intensity modulated radiotherapy in pediatric and adolescent nasopharyngeal carcinoma. Chin J Cancer 32(10):525–532. doi:10.5732/cjc.013.10124. cjc.013.10124 [pii]
105. Louis CU, Paulino AC, Gottschalk S, Bertuch AA, Chintagumpala M, Heslop HE, Russell HV (2007) A single institution experience with pediatric nasopharyngeal carcinoma: high incidence of toxicity associated with platinum-based chemotherapy plus IMRT. J Pediatr Hematol Oncol 29(7):500–505. doi:10.1097/MPH.0b013e3180959af4. 00043426-200707000-00013 [pii]
106. Maris JM (2010) Recent advances in neuroblastoma. N Engl J Med 362(23):2202–2211. doi:10.1056/NEJMra0804577. 362/23/2202 [pii]
107. Pai Panandiker AS, Beltran C, Billups CA, McGregor LM, Furman WL, Davidoff AM (2013) Intensity modulated radiation therapy provides excellent local control in high-risk abdominal neuroblastoma. Pediatr Blood Cancer 60(5):761–765. doi:10.1002/pbc.24350
108. Nazmy MS, Khafaga Y (2012) Clinical experience in pediatric neuroblastoma intensity modulated radiotherapy. J Egypt Natl Cancer Inst 24(4):185–189. doi:10.1016/j.jnci.2012.10.001
109. Pai Panandiker AS, Beltran C, Gray J, Hua C (2013) Methods for image guided and intensity modulated radiation therapy in high-risk abdominal neuroblastoma. Pract Radiat Oncol 3(2):107–114
110. Beltran C, Pai Panandiker AS, Krasin MJ, Merchant TE (2010) Daily image-guided localization for neuroblastoma. J Appl Clin Med Phys 11(4):3388
111. Cotton CA, Peterson S, Norkool PA, Takashima J, Grigoriev Y, Green DM, Breslow NE (2009) Early and late mortality after diagnosis of Wilms' tumor. J Clin Oncol 27(8):1304–1309. doi:10.1200/JCO.2008.18.6981. JCO.2008.18.6981 [pii]
112. Kalapurakal JA, Zhang Y, Kepka A, Zawislak B, Sathiaseelan V, Rigsby C, Gopalakrishnan M (2013) Cardiac-sparing whole lung IMRT in children with lung metastasis. Int J Radiat Oncol Biol Phys 85(3):761–767. doi:10.1016/j.ijrobp.2012.05.036. S0360-3016(12)00752-3 [pii]

Index

Y. Nishimura, R. Komaki (eds.), *Intensity-Modulated Radiation Therapy*,
DOI 10.1007/978-4-431-55486-8

P

Q

R

FSC
www.fsc.org
MIX
Papier aus verantwortungsvollen Quellen
Paper from responsible sources
FSC® C105338

If you have any concerns about our products,
you can contact us on
ProductSafety@springernature.com

In case Publisher is established outside the EU,
the EU authorized representative is:
Springer Nature Customer Service Center GmbH
Europaplatz 3, 69115 Heidelberg, Germany

Printed by Libri Plureos GmbH
in Hamburg, Germany